The "RADIO" Handbook

By

Frank C. Jones

1936 EDITION

Reprinted by Lindsay Publications Inc, Bradley IL 60915
 ISBN 1-55918-236-9 • 2000

1 2 3 4 5 6 7 8 9 0

WARNING

Remember that the materials and methods described here are from another era. Workers were less safety conscious then, and some methods may be downright dangerous. Be careful! Use good solid judgement in your work, and think ahead.

Lindsay Publications Inc. has not tested these methods and materials and does not endorse them. Our job is merely to pass along to you information from another era. Safety is *your* responsibility.

Pacific Building, (P. O. Box 3278), San Francisco, California

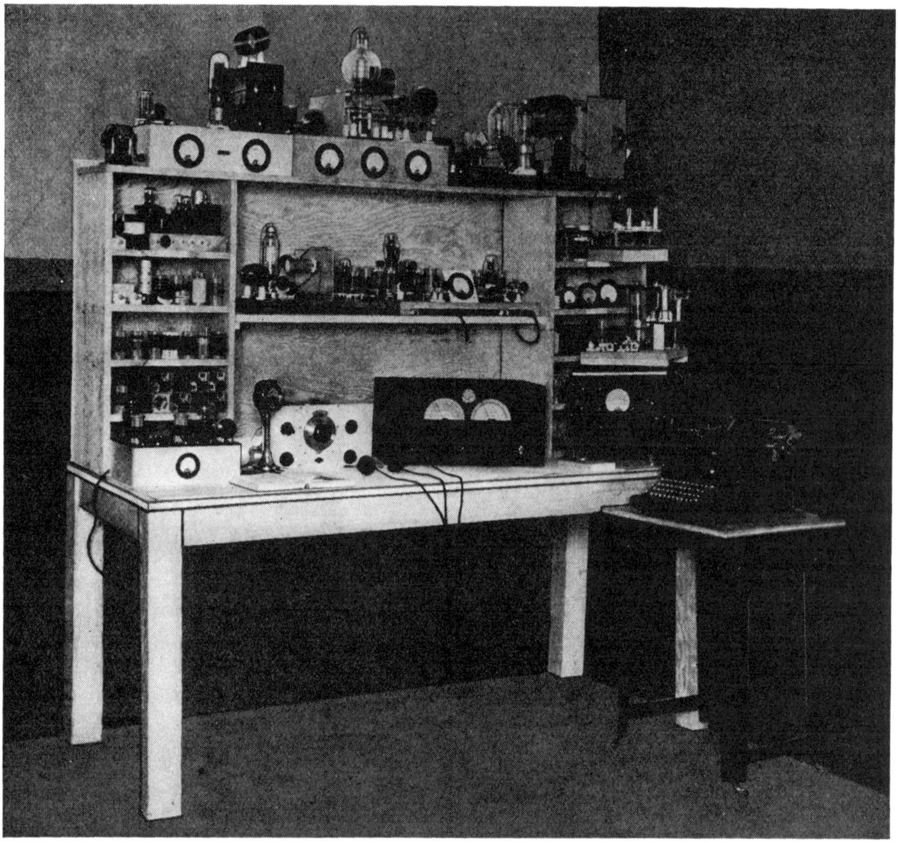

Equipment Ready for Test

Here is a corner of the Test Laboratory of THE "RADIO" HANDBOOK, where the equipment described in these pages is given a thorough test on the air.

The "WHY" of a Second Edition of The "RADIO" Handbook

IN June, 1935, the publishers of the monthly magazine, "RADIO," the international technical authority for amateur radio, released the first edition of "The 'RADIO' Handbook." The work was printed in a comparatively small edition in order to determine the requirements of the amateur and attention was given primarily to the practical side of high-frequency communication. Too often, however, theory and practice do not go hand in hand in a field in which such rapid progress is being made as that of amateur short-wave voice and telegraph communication. The art is progressing so rapidly and so many innovations appear each month that it is a herculean task to give the reader, in such compact form, the exciting panorama of the romantic field of communication.

But, much of the "new" which is presented to an unsuspecting public as perfection itself has proved by exhaustive testing to be, unfortunately, much less than perfection. Frank C. Jones, the engineer who developed the greater part of the new equipment described in this, the second edition of "The 'RADIO' Handbook," brought to light many new theories which were rigidly tested "on the air." A theory was not enough. Paper plans sometimes have a way of surprising the "paper planner" unpleasantly, so the standard procedure adopted was first to conceive, then engineer, then design, then construct, if necessary then reconstruct again and again and, finally, when exhaustive tests under actual operating conditions both with local stations and with others in the far corners of the world have shown the worth of the conceptions they found their way into "The 'RADIO' Handbook." There is nothing impractical, unsafe or insane in the pages of this new edition. Theory gives way to fact. The laboratory is the proving-grounds. All during the period of revision, the writers were in constant touch with leaders in specialized fields and we take this opportunity of thanking these men who, content to be anonymous, so unselfishly gave their time and their experience that amateur radio might be benefited. In the preparation of this Handbook the author wishes to acknowledge gratefully the assistance he received from Bernard Ephraim, Prof. F. E. Terman, John N. Hawkins, Clayton F. Bane and D. B. McGown.

The task of the writers of this book has been a pleasant one, full of romances, thrills, life! Many new fields are in these pages, information which has not previously been in print, or, was hidden away in ponderous scientific papers and weighted with the complexities of language and of vocabulary so beloved to the engineer and so baffling to the amateur.

The market is flooded with textbooks and textbooks are flooded with theory, with a maze of conflicting information floating on the flood, so that the lay reader striving to swim rather than sink is quite

One of the World's Finest Amateur Radio Stations.
W9DXX, Owned and Operated by Alice R. Bourke, Chicago, Illinois.

likely to sink rather than swim. Bewildered he is at loss to know just what to do. The very fundamentals of practical radio receiver and transmitter engineering, design and construction have, in many of the current works not only failed to keep up with the trend, but have taken a step backward. Thus, in these pages the reader will find an exhilarating and refreshing amount of information on the most useful pieces of equipment which he may desire to build or operate.

Beginning with theory, only such matter is treated as will be of practical use. No space is wasted on matters of communication, traffic handling or to the preaching of gospels of loyalty, to whomsoever the loyalty is to be pledged, for the publishers of this book are of another belief. They hold that it is the solemn duty of a publisher to be loyal to his reader, not for the reader to be loyal to the publisher.

The men who were assigned to the task of presenting this newer edition of "The 'RADIO' Handbook" are conscious of their trust which must at all times be preserved. This book has become an accepted authority the world over. The sales of the first edition were so great that a one-year printing was completely exhausted in less than three months.

The delay in getting this new edition before the public was occasioned by the rather unexpected opening of the 10-meter band. This band is suitable for extremely long distance communication, both for voice and for code, only at certain intervals, that is, during certain "cycles" of sun-spot activity which occur usually seven years apart. Because any licensed amateur can use voice communication without benefit of special-privilege governmental license, and because the fascination on the 10-meter band is so great, it was deemed wise to delay the publication date until such time the engineers had opportunity to correctly design and construct the equipment needed for this now-popular band. In these pages the reader will find, in their respective sections, a group of transmitters and receivers specially designed for 10-meter operation. There is an instrument to fit every need, and every pocketbook. Careful design resulted in the presentment of the most accurate, reliable and dependable 10-meter information anywhere available.

New tubes came into the market. New tables had to be prepared. New apparatus of all types was developed. New ideas in exciter units by Frank C. Jones, which make the all-band exciter for 1936 many times more useful and valuable than its predecessor, new receivers with metal tubes, pre-amplifiers with the new tubes, and a host of other equipment fresh from the laboratories is presented in this new edition.

The reader is asked to accept as fact the authoritative statements herein with regards to certain forms of oscillators, buffers, doublers, amplifiers and antenna systems. The wide confusion which has resulted from a conflicting opinion of the radio press is herein subject for authoritative compromise, and the answers to all the important problems are given. This handbook can be your daily guide to all that is new and all that is best in the field of high-frequency communication. Nothing of value has been omitted from its pages.

Exhaustive tests have proved that the Jones Exciter is supreme in its field. Further tests have shown that the small and inexpensive receivers for high and ultra-high frequency operation are the best that engineering skill can conceive. Coil tables and charts are unusually complete and the section devoted to Radiotelephony is unquestionably the most complete treatise of its kind ever made available for the radio amateur and experimenter. It should be stated here, that the reader will do well to scrutinize closely many of the radiotelephony circuits, because these, too, are suitable, in the main, for excellent CW telegraph operation.

Controlled-carrier systems of modulation are covered only from the standpoint of those systems which have withstood the test of time, short as that time has been. There is still much to be learned about this year-old system, but what is herein described is authoritative. There is no ballast, and the systems shown can be depended upon for reliability and performance.

The various calculating charts, prepared by Mr. Ephraim, will be of great use to those who design their own equipment. These simplified charts effect a genuine saving in cost over other equipment which gives no better result, and a study of these charts is of paramount importance to the engineering and designing fraternity.

As in the first edition, there is an extended treatment of antennas and antenna systems. Wide acclaim has come to the publisher for the simple manner in which this heretofore complex subject has been handled.

The publishers express grateful thanks to more than 19,000 of our friends who placed pre-publication orders for the book, and for their patience and sportsmanship they have shown by waiting so long for this newer edition to appear.

Should the sale, as it did in the first edition, exceed the anticipated requirements, a third printing will be released in the fall of the current year.

Those who desire to have more complete data on any particular subject or chapter are advised to communicate with the publishers, for it is the duty of a publisher to give the reader what he needs, not what the publisher desires to give him.

For the promptest attention, orders for additional copies should be addressed to:

Pacific Radio Publishing Co.,
San Francisco, California, U. S. A.

INTRODUCTION TO AMATEUR RADIO

Every person who desires to operate an amateur transmitting station must obtain a "station" and "operating" license from the district offices of the Federal Communication Commission. The locale of these district offices are given in the Appendix. A license is not required to operate a receiver. To secure the necessary licenses, the applicant must pass an examination in order to prove his technical knowledge of the theory and practice of amateur radio communication, as well as being able to copy the Continental Code at a speed of ten words per minute.

Those who desire to learn the code without the aid from others may do so by means of a CODE PRACTICE SET. Several kinds of these sets can be constructed, of which the simplest is not always the best. Any amateur will gladly assist the beginner to properly build a practice instrument. Of importance, the instrument must be so designed that a sharp, clear tone will be produced in the receivers, simulating that of a continuous-wave signal (c.w.) as commonly heard on short-wave receivers. Only a few moments are required to assemble a code practice set, of which the parts may be purchased at a modest cost from most any radio store. Should difficulty be encountered in making the instrument properly function, it should be taken to an amateur or radio service man who will make the necessary adjustments.

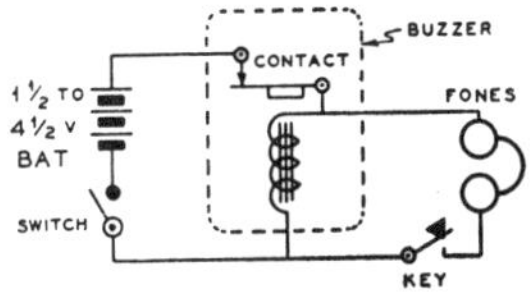

FIG. 1

The simplest code practice set is shown in Figure 1. Here, an oscillator consisting of a high-frequency buzzer, one or two dry-cells, a pair of headphones, and a telegraph key complete the instrument ensemble. The advantages of this oscillator is that the buzzer operates continuously to produce a stable audio-frequency tone. The headphones and telegraph key are connected in series across the buzzer coil. The buzzer contacts should be adjusted for the least change in the note when the key is depressed. Although the buzzer operates continuously as long as current is being supplied from the batteries, the tone is only heard in the headphones when the key is manipulated.

Another code practice set is shown in Figure 2. Here a cathode-heater type vacuum tube functions as an oscillator in what is known as a "Hartley" oscillating circuit. The 2.5 volt or 6.3 volt tubes may be substituted in the circuit with equal success. The type 76 tube is the 6.3 volt equivalent of the type 56 which draws only 0.3 ampere heating current; hence, on account of such low current consumption the tube

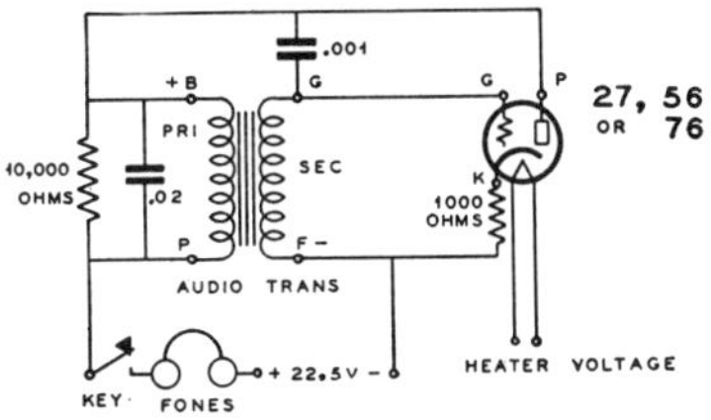

FIG. 2

Code Practice Oscillator Using Heater Tube.

can be operated from a series of four No. 6 dry-cells. The coupling transformer is of the audio-frequency type, any turns ratio. The telegraph key is in series with the headphones and the plate ("p") terminal of the audio transformer. The pitch of the note may be varied by merely increasing or decreasing the "B" voltage. It will be found that the most pleasing note will be obtained at about 22½ volts.

The capacity of the fixed condensers will also affect the pitch of the note, as well as the volume. The vacuum tube code practice oscillator is far more satisfactory than the buzzer type shown in Fig. 1. It gives a more stable tone, which can be varied in intensity and pitch to suit the individual requirements of the operator.

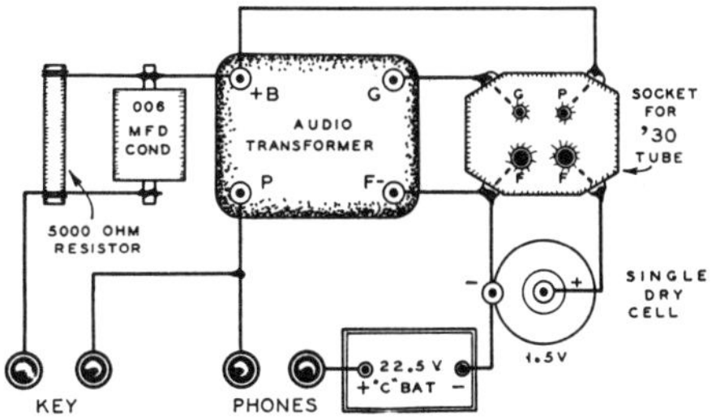

Pictorial and schematic diagrams showing how to wire a simple code practice oscillator for use with dry batteries.

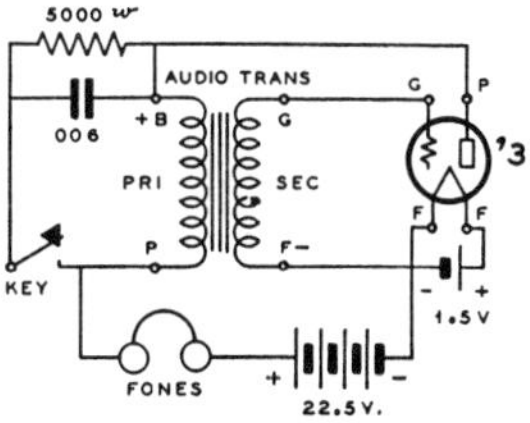

FIG. 3

A type '30 tube, a single dry-cell and a 22½ volt B battery gives good results in the code practice set shown in this diagram. The tone or note can be changed by using larger or smaller condensers and resistors than those shown in the diagram, or by varying the voltage of the "B" battery.

Code Practice Oscillator as shown in Fig. 2.

How to Master the Radio Code

There is nothing complicated about learning the code. Many young boys and girls have succeeded in attaining a code speed of ten words per minute with but few months of practice. When learning the code it is important to distinguish dots and dashes; that is, a dot or a dash must always be properly characterized, no matter how fast or slow one transmits. A dash, however, should be three times as long as a dot. Beginners commonly make the mistake of "holding the dots"—not making dots at all but a series of long and short dashes. A dot is made by one quick, sharp touch of the key, irrespective of the sending speed. A dash should never be made longer than the time required to make three quick dots. The difference between slow and fast sending is the time interval between letters which comprise words, and not between the characters which make a letter. **The spacing between letters and words may be lengthened when slow speed sending is used.** All dots and dashes should have the same "time form" no matter whether an attempt is being made to send 5 or 50 words per minute. The secret of success of good operating is in the **spacing** between letters and words, but there should be no spacing between dots and dashes which make up an individual letter. For example: take the letter A; it consists of a dot and a dash, but in code it should not be considered cryptically except as a dot and a dash. Phonetically, pronounce it as "did-daw," "did" for the dot, "daw" for the dash. Thus, the letter A becomes "did-daw," **not** dot-dash. By repeating the phonetic sounds, the letter soon becomes firmly

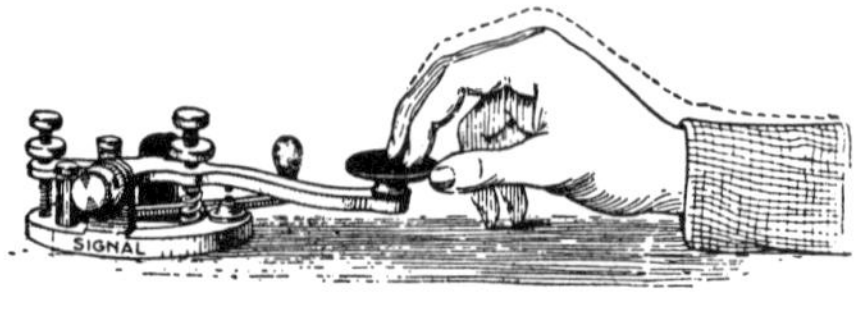

Proper grip for manipulating the key.

THE RADIO TELEGRAPH CODE

A	. -
B	- . . .
C	- . - .
D	- . .
E	.
F	. . - .
G	- - .
H	. . . .
I	. .
J	. - - -
K	- . -
L	. - . .
M	- -
N	- .
O	- - -
P	. - - .
Q	- - . -
R	. - .
S	. . .
T	-
U	. . -
V	. . . -
W	. - -
X	- . . -
Y	- . - -
Z	- - . .

Numerals, Punctuation Marks, Etc.

. - - - -	1
. . - - -	2
. . . - -	3
. . . . -	4
. . . . .	5
-	6
- - . . .	7
- - - . .	8
- - - - .	9
- - - - -	Ø

. . . - - - . . .	INTERNATIONAL DISTRESS SIGNAL
. . . . . .	PERIOD
. - . - . -	COMMA
. . - - . .	INTERROGATION
. - . . - .	QUOTATION MARKS
- - . . - -	EXCLAMATION
- - - . . .	COLON
- . - . - .	SEMICOLON
- . - - . -	PARENTHESIS
- . . - .	FRACTION BAR
. - . . .	WAIT SIGN
- . . . -	DOUBLE DASH (BREAK)
. . . .	ERROR (ERASE) SIGN
. - . - .	END OF MESSAGE
. . . - . -	END OF TRANSMISSION

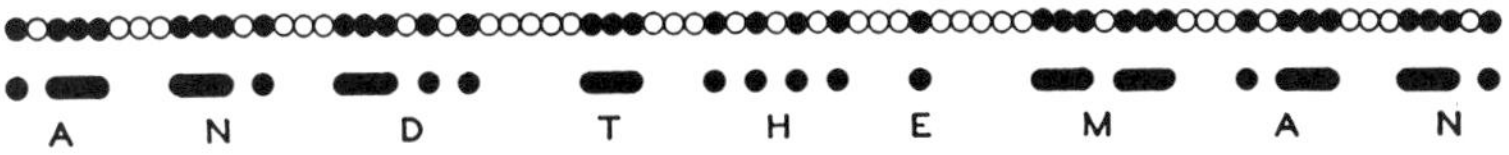

DASH EQUAL TO THREE DOTS — SPACING BETWEEN CHARACTERS EQUAL TO ONE DOT
SPACING BETWEEN LETTERS EQUAL TO ONE DASH
SPACING BETWEEN WORDS EQUAL TO FIVE DOTS

fixed in the mind. During mental repetition, no pauses should be made between the "did" and the "daw"; the two must roll smoothly into each other; thus, "diddaw." One of the greatest mistakes made by operators is in permitting a pause to come between the "did" and the "daw." To further illustrate code learning examples, take the letter B, which consists of a dash and three dots. Again, there must be no spacing between the dash and the three dots. B is "dawdiddiddid." Now, if a space is permitted to come between the daw and the the three dids, the code character will have the form of the letters T S, and not B.

Send cautiously, slowly and surely! Haste makes waste. One often hears of the operator "who falls all over himself." He becomes confused, sends faster than he can receive. Nothing is more painful than to listen to a fast, erratic operator who cannot read his own sending. How, then, can he expect others to copy his signals?

The SOUND system of learning the code has been universally proven as the best method for beginners to use. This system teaches the operator to think in terms of SOUND, instead of the more common letter formations. By thinking in such terms a letter is recognized by its characteristic sound and cadence. When the sound "diddaw" is heard, it is immediately recognized as the letter A, and not in any other form.

Correct "grip" and position of wrist for operating automatic key ("bug").

With the code practice set connected, grasp the key **gently** with the thumb, fore and index fingers being placed on the knob of the key. The illustration shows how to properly manipulate the key. Avoid cramping the hand—relax—forget entirely that the thumb and forefingers are on the key. Be interested in correctly making the telegraph signals "dids" and "daws." Send slowly, until becoming adept to the knack of sending. Do not open the key too wide, else the sending will become "choppy." On the other hand, if the key does not have sufficient play, the sending will sound "sloppy." Do not make the key too "stiff" by exerting too great a pressure on the adjusting spring. If in doubt about the correct tension, ask a more experienced operator to assist you in making the proper adjustment.

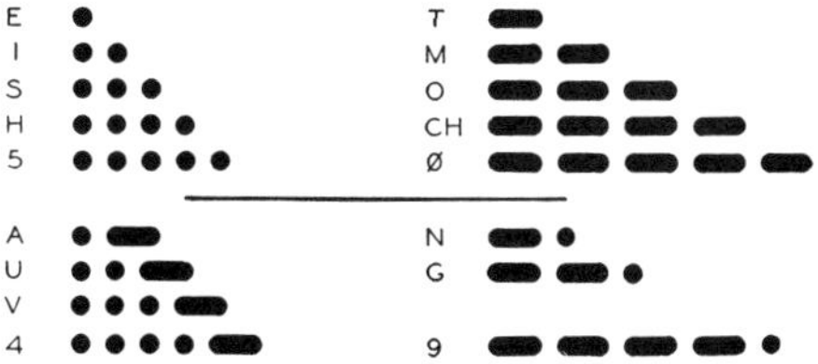

Easily Memorized Practice Letters.

Begin learning the code without assistance from others. First memorize a few letters of the alphabet, starting with the letter A. Proceed by the methods outlined in the above text until finally practicing with the telegraph key. To augment the mind's retentive powers, use a short-wave receiving set and pick out as many diddaws as can be recognized from a slow sending station. At each recognition the letter should be written down on a small pad. After becoming thoroughly acquainted with the letter A, proceed until the complete alphabet can be **instantly** recognized by sound code-formations. Later, connective words like AND, TO, OF and others should be learned. Words composed entirely of dots and dashes are excellent for practice. For example: (all dots) is, his, sis, and she; (all dashes) to, tom, otto and etc. In exercising, it is important that the letters be properly spaced lest the word structure be ruined.

The next step is to make short sentences consisting of words, some of which are all comprised of dots, others all dashes, such as "she sees Otto.." The student operator will find that by learning all code characters comprised of dots or dashes before others, that his telegraphic technique will be developed at much faster rate than it otherwise would be. Every effort should be made to copy signals heard on short-wave sets until proficiency is attained.

Throughout the day, it would be advisable to make silent repetitions of the diddaw sounds, including words, short sentences, figures, calls and other miscellany. Always think in terms of SOUND . . . in dids and daws. Soon it will be amazing to learn how simple it is to gain speed and accuracy in a relatively short period of time.

FUNDAMENTALS OF ELECTRICITY

A study of electrical or radio phenomena requires a knowledge of the electron theoretical conception of matter and energy. This theory assumes a scheme whereby very small particles of matter carrying electrical charges form the basic mechanism in electric conduction. These elemental units or electronic charges form the basis of the electron theory which has been universally accepted as the best means for coordinating present knowledge of electric phenomena.

In general, the smallest particle of matter which can exist alone is the atom. It consists of a heavy nucleus of one or more protons surrounded at planetary distances by an equal number of electrons. The outermost electrons revolve in elliptical paths around this inner nucleus. Every atom of matter has as many electrons as protons, and therefore the total number of positive and negative charges neutralize each other. This atomic system has been given the name of the **nuclear atom.** The charge retained in the nucleus of an atom is what designates its **weight,** while the attendant electrons revolving around the nucleus is that which determines the **atomic number.** Atomic numbers run from one to ninety-two, which are the ranges given to all the chemical elements.

To the electrons, or more properly to the negative electronic charges with corresponding positive charges on the protons, or positive nuclei, are ascribed properties of electric fields (the space surrounding magnets, electric charges and electric currents), considered as innate characteristics of each elemental unit. Electrons at rest produce electro-static phenomena, while electro-dynamic effects result from electrons in motion.

In all substances which are non-conductors of electricity, the electrons in the atoms are held permanently in place in fixed orbits about the nucleus, but in the atoms of all electrical conductors one or more of the electrons farthest out from the nucleus is attached rather loosely and may, by various means, be drawn away from the atom altogether. These are termed **free electrons.**

In all insulating substances the atomic structure is such that all the outlying electrons cannot be freed by external forces. This statement is relative, because, given enough external force, even the atoms of the very best insulators can be made to give up an electron.

Electromotive Force

Electricity consists of a movement of electrons through a conductor or conducting medium. To initiate the flow, a difference in electrical pressure (analogous to a hydrostatic head of water) or electromotive force must exist between the two ends of the conductor. To clarify these statements in an electronic exposition is without the scope of this text, but briefly the explanation is: The looseness with which the outer electrons are held in any atom is related to the electrical conductivity of the substance composed of this particular atom. The more loosely the free electrons are attached to their nuclei, the better the electrical conductivity. Thus, the flow of current in a conductor consists of a stream of electrons moving along the conductor, from atom to atom, in a definite direction under the influence of an outside applied force or pressure. In electrical circuits this outside force consists of an equalizing tendency on the part of the electrons which, like water, seek their level. Hence, there will be a flow of current in any conductor which possesses an excess of electrons at one point and a deficiency at another. This flow will continue until the number of electrons at all points along the conductor are equal. This equalizing force is called the **electromotive force,** abbreviated EMF, and is usually expressed in **volts.** This force is due to the non-uniform distribution of electrons in a circuit. For illustration, if a battery is placed in a closed circuit, a current of electricity will flow around the conducting medium because the battery pulls electrons into one terminal and pushes electrons out of the other. The source from which the electrons flow is called the **negative** terminal, and the point which the electrons travel to is called the **positive** terminal. The words POSITIVE and NEGATIVE have no meaning, but serve only to distinguish or differentiate between the two electrical charges. The terms were chosen many years before the electron-movement theory was established, and for a long time it was assumed, for reasons of conventionality, that current flowed from a positive terminal to a negative terminal. It is now known that the co-ordinated motion of electrons actually move in the opposite direction; that is, from the negative to positive terminals.

Electric Potential

The value of an electromotive force existing between any two points is known as the potential difference, and is measured in terms called a **volt.**

The Electric Circuit

The simplest electrical circuit consists of a source of electromotive force and a continuous path from the negative to the positive terminals through a resistance. The source voltage may be either a unidirectional (DC), or alternating (AC) current. If direct current, the voltage source maintains a constant positive and negative polarity. On the other hand, if the current be of an alternating nature the polarity of the two terminals is periodically reversed. In an alternating current circuit the direction of the electron movement reverses once each cycle. In the ordinary 60 cycle AC power line, the polarity of the AC generator reverses 60 times per second, which indicates the line FREQUENCY. Alternating and direct currents have quite different characteristics. Accordingly the study of electricity is divided into two parts; direct currents and alternating currents.

Electric Resistance

Electrons moving through a conductor continually collide with atoms of the conducting material. This impedes or slows the electron flow to such an extent that the amount of current is limited which can flow through a circuit when a given voltage is applied. This limiting effect is termed the **resistance** of the conductor; it is expressed in **ohms.** Hence, a circuit has a resistance of 1 ohm when an EMF of 1 volt will force a current of 1 ampere through it. And, in an inverted sense, a source of EMF is said to have 1 volt electrical pressure when it will establish a current of 1 ampere in a resistance of 1 ohm.

The collisions between the free electrons and the atoms move the atoms around slightly, which takes a certain amount of energy away from the electron stream. This energy heats up the conductor and explains why resistors carrying current increase their temperature.

Electric Current

Electric current describes the rate of flow of electricity through a circuit, and the unit of current flow is the **ampere.** Electric currents are measured either by their heating effects on a conductor (thermoammeters, etc.) or by their magnetic effects (moving coil and moving iron instruments).

Sources of Electricity

An electromotive force (and therefore a flow of current) can be produced either by chemical or mechanical means. All batteries produce electricity by converting energy from one form to another by means of a chemical reaction. All the common types of electrical generators transform mechanical energy into electrical energy, either by magnetic or electrostatic action.

Series and Parallel Circuits

A simple circuit can contain any number of resistances. For example, Figure 1 shows a circuit having two resistances connected in series, while that in Figure 2 has resistances connected in parallel. The current in a parallel circuit will divide between the various resistance branches, and will not be equal in each branch unless the resistance in every branch is equal. In a series circuit the current flow is equal at every point in the circuit.

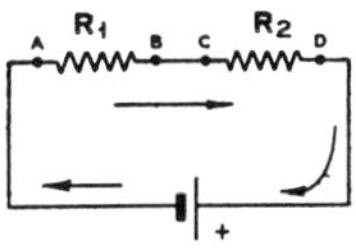

FIG. 1

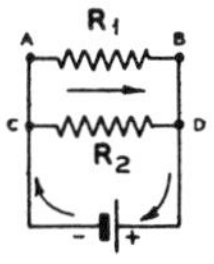

FIG. 2

Ohm's Law

The resistance of any conductor depends on the structure of the material of which it is made, together with its cross-section and length. The relationship between the electromotive force (volts), the flow of current (amperes), and the resistance impeding the flow of current (ohms) is expressed in Ohm's Law, which states: **"For any circuit or part of any circuit the current in amperes is equal to the electromotive force in volts divided by the resistance in ohms."** This relationship is usually expressed by the following three formulas:

Where I is the current in amperes,
E is the electromotive force in volts,
R is the circuit resistance in ohms.

$$E = IR \qquad I = \frac{E}{R} \qquad R = \frac{E}{I}$$

Thus, resistance equals voltage divided by current,
current equals voltage divided by resistance,
voltage equals current times resistance.

In many commonly used circuits it is found that there are resistances connected in series, in parallel or in series-parallel, as shown in Figure 3. In order to calculate the total resistance of any network composed of two or more resistances connected in any of the above three ways, the formula shown in Figure 3 is used. Note that the total resistance of resistors connected in series is larger than that of the

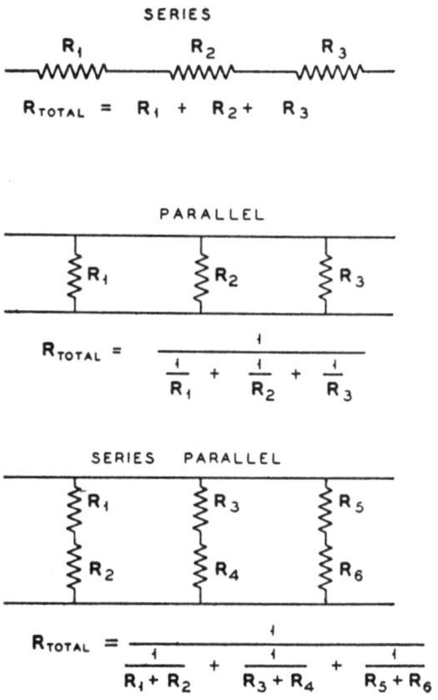

FIG. 3

highest resistance in the circuit. Also, the total resistance of resistors connected in parallel is **less** than that of the **lowest** resistance in the circuit.

Electric Power and Heating Effects

The heat generated in a conductor by the flow of current varies directly with the resistance of the conductor and as the square of the amperes of current flow. The unit of power is the watt, and equals the product of the voltage across a resistor, times the current through it. This equals the amount of electrical power transformed into heat in the resistor. Using the symbols described above, plus **W = Watts of Power,** it is found that the following relationships hold true:

$$W = EI \qquad W = I^2R \qquad W = \frac{E^2}{R}$$

Electrical power can do other forms of work besides generating heat, such as driving a motor, radiating waves from an antenna or driving a loudspeaker. Electrical power takes many different forms and can be transduced from one form to another by means of a motor-generator, or vacuum-tube.

R M A STANDARD
RESISTOR COLOR CODE

A BODY COLOR-1st. figure of resistance value.
B COLORED END-2nd figure.
C CENTER DOT-number of ciphers following first two figures.

figure	color	figure	color
0	BLACK	5	GREEN
1	BROWN	6	BLUE
2	RED	7	VIOLET
3	ORANGE	8	GRAY
4	YELLOW	9	WHITE

Electromagnetic Phenomena

A magnetic field envelopes or surrounds a conductor when an electric current is flowing. How this field is developed is explained as follows: Electrons of like charge will repel each other due to the electro-static field of force which surrounds each electron; this force is inversely proportional to the square of the distance. Thus, if the repulsion at any distance is a certain value, the repulsion at twice this distance is one-half squared, or ¼ as much. The electro-static field around any electron which is at rest, or moving with a constant velocity, can be visualized by a group of concentric equipotential circles surrounding the electron. See Figure below:

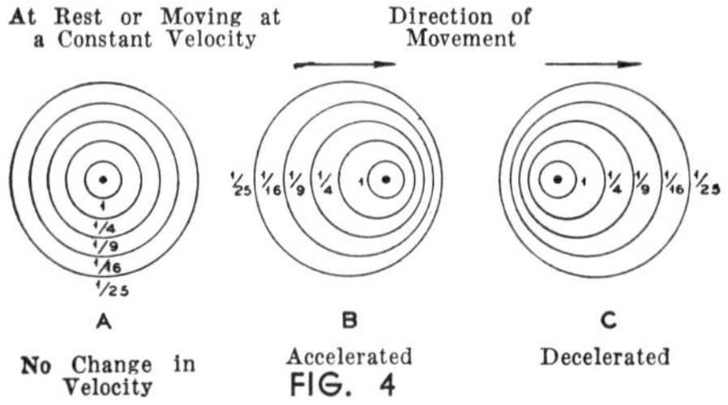

FIG. 4

When an electron moves, it must carry its field of force along with it. Hence, due to the relatively enormous volume of this field, each electron has considerable inertia. Thus, when a switch in a circuit is closed, the current does not jump instantly to the final value determined by the voltage divided by the resistance.

This gradual build-up of current in any circuit depends upon the circuit characteristics. It takes a greater length of time for current to build-up in a circuit containing a coiled wire than in one which consists of one long, straight wire. This is because the electro-static fields overlap surrounding electrons in adjacent turns of the coil. The energy stored at any point in space is proportional to the square of the electro-static intensity (or force) at that point. Thus, by coiling the wire, the energy concentration stored in the space around the coil has been materially increased, due to the increased overlap in the fields of the electrons. If the electro-static intensity at any point has been increased a hundred times over that of a point near a straight wire, the energy storage is 100 squared, or 10,000 times that of the energy stored in the space surrounding the long, straight wire. This stored energy comes from the source of power supplying the circuit, and any given current in a coil represents much more stored energy than the same current in a straight wire. Hence, for a given impressed voltage, it takes more time to start or stop the current flow in a coil than in a straight wire. Likewise, to start or stop the current flow in a coil **in a given time** requires the application of a larger voltage than would be necessary to start or stop the same current flow in a straight wire.

The inertia offered by a circuit to either an increase or a decrease in current is termed the **inductance** of the circuit. This inertia can be visualized in the following manner. When an electron is accelerated, or speeded up, its electro-static field does not instantly respond to the motion of the electron because the electro-static disturbances caused by the sudden acceleration of the electron travel outward from the electron with the speed of light. Hence, different parts of these fields are moving at different speeds, as shown in Figure 4 (B), and the concentration of energy ahead of the electron is greater than the concentration behind it. As soon as the electron attains constant velocity, its field again becomes systematically arranged. When the electron is decellerated the concentration of energy behind it becomes greater than that ahead of it, as shown in Figure 4 (C). These non-uniform concentrations of energy tend to oppose any change in the velocity of the electron, and it should be evident that the overlapping of the electron fields which occurs in the coil increases the non-uniform energy concentration which accompanies any change in the velocity of an electron, thus increasing the opposition to change, or inertia of the electron. This inertia, therefore, exerts a force opposing any change in the current through an inductance, and this opposing force is called the **back electro-motive force.**

Induction and Induced Voltages

When an alternating current is passed through a coil of wire, energy is alternately stored in the field and returned to the wire. The greater the number of turns of wire on the coil, the greater is the **magneto-motive force.** This force varies with the number of turns, the diameter of the coil and the current. MMF corresponds to magnetic pressure.

Magnetic Flux

Magnetic flux consists of the lines of magnetic force which surround any conductor. Magnetic flux might be termed magnetic current, just as magneto-motive force corresponds to magnetic voltage. The reluctance of a magnetic circuit could be described as the resistance of the magnetic path and the relationship between magnetic flux; magneto-motive force and reluctance is exactly similar to that between current, voltage and resistance, (Ohm's Law).

Magnetic flux depends on the material, cross-section and length of the magnetic circuit and varies directly as the current flowing in the circuit. Reluctance depends upon the length, cross-section, permeability and air-gap, if any, in the magnetic circuit.

Permeability

Permeability describes the difference of the magnetic properties of any magnetic substance compared with the magnetic properties of air. Iron, for example, has a permeability of approximately 3100 times that of air, which means that a given amount of magnetizing effect produced in an iron core by current flowing through a coil of wire will produce 3100 times the flux density that the same magnetizing effect would produce in air. The permeability of different iron alloys varies quite widely and permeabilities up to 10,000 can be obtained, if required. Permeability is similar to electric conductivity. However, there is one important difference—the permeability of iron is not independent of the magnetic current (flux) flowing through it, although electrical conductivity is usually independent of electric current in a wire. After a certain point is reached in the flux density of a magnetic conductor, an increase in the magnetizing field will not produce any material increase in the flux density. This point is known as the **point of saturation.** The inductance of a choke coil whose core becomes saturated declines to a very low value. This characteristic is extremely valuable in the **swinging choke** and in the **saturable reactor** used in some controlled carrier modulation systems.

The magnetizing effect of a coil is often described in **ampere-turns.** Two amperes of current flowing through one turn equals two ampere-turns, or one ampere of current flowing through two turns also equals two ampere-turns.

Mutual Inductance

When two parallel wires are placed in proximity to each other and a varying current flows through one of them, the non-uniform energy concentrations around the accelerating and decellerating electrons in the conductor carrying the varying current cause an induced electro-motive force to be applied to the free electrons in the neighboring conductor. The electro-motive force (voltage) produced in the adjacent wire is always in the same direction as the back-electro-motive force set up in the wire which is carrying the exciting current. This point helps to explain why the inductance of a circuit containing many turns of wire is greater than that of a circuit composed only of a straight wire. In a coil, each turn has a back-electro-motive force induced by the changing current within itself. In addition, it has an induced electro-motive force in the same direction, due to the changing current in the adjacent turns on each side of the portion of the coil under consideration. The self-inductance of a coil in henrys equals the induced voltage in volts across that coil when the current is varying at the rate of one ampere per second.

If a second coil is wound directly over the first coil, any change in current in the first coil will induce a voltage in the second coil, and the mutual inductance in henrys between the two coils equals the voltage induced in either coil when the current in the other is varying at the rate of one ampere per second. The unit of inductance is the **henry.**

Inductive Reactance

The principal property of an inductance is to resist any change in current through it, and therefore any inductance in a circuit will impede the flow of alternating current. The higher the frequency of the alternating voltage impressed across the inductance the lower will be the current through the coil. The current flowing through the inductance is related to the inductance in henrys and to the frequency in cycles per second.

Formula:

Where X_l is the inductive reactance in ohms,
f, the frequency in cycles per second,
L, the inductance in henrys,

$$X_l = 2 \pi fL$$

Thus, if the inductance of a coil and the frequency of the impressed alternating voltage is known, the current in any AC circuit in which there is an inductance can be determined by dividing the voltage by the inductive reactance.

Inductances can be connected in series or in parallel. The electrical effect of making such connections are quite similar to those obtained when connecting resistors in series or parallel. Inductances in series:

$$L\ total = L1 + L2 + L3,\ etc.,$$

Inductances in parallel:

$$\frac{1}{L\ total} = \frac{1}{L_1} + \frac{1}{L_2} + \frac{1}{L_3} +,\ etc.,$$

Transformers

From the foregoing, it was seen that a variation of current flowing through an inductive winding will induce a similar voltage in an adjacent winding if both are coupled within the proximity of the common magnetic circuit. This explains the operation of a **transformer.** The winding of a transformer carrying the exciting current is known as the **primary,** and the coupled coil in which is induced a voltage is known as the **secondary.** If both prmary and secondary windings have an equal number of turns which are closely coupled, and if neither of the windings are tuned by means of a capacity to resonance at the frequency of operation, the voltage across the secondary will be equal to the voltage across the primary. If the secondary has twice as many turns as the primary, the induced voltage in the secondary will be twice the exciting voltage across the primary. For any other turns ratio between the primary and secondary windings, the ratio of the secondary voltage to the primary voltage will be equal to the ratio between the number of secondary turns to the number of primary turns. These relationships hold as long as no current flows in the secondary winding, which is the case in all low-level audio-frequency circuits. When a load is connected across the secondary, as in a power transformer, or audio-output transformer, the DC resistance and the leakage reactance of the transformer windings slightly modify the voltage relationships.

Useful transformer formula:

$$\frac{Z_p}{Z_s} = \left(\frac{N_p}{N_s}\right)^2$$

Where Z_p = primary impedance
Z_s = secondary impedance
N_p = number of primary turns
N_s = number of secondary turns

Condensers and Capacitive Reactance

A condenser is a device for storing electrical energy, and in its simplest form consists of two parallel metallic plates separated by an insulator, such as air. If the two plates are connected to a DC source, one will be positively and the other negatively charged. As soon as the potential difference between the two plates becomes equal to the voltage of the DC source, the current in the circuit will cease. If the condenser is connected to an AC voltage, the current will surge back and forth every cycle, because first one plate takes on a positive charge, then the other. During that part of the cycle when one plate becomes negative, the excess of electrons driven on to this plate repels an equal number of electrons off the other plate. These electrons then travel back toward the positive terminal of the voltage source. On the next half cycle this process is reversed. No electrons actually pass through the condenser from one plate to the other, because the electrons arriving at one plate drive an equal number away from the other plate. The effect on the circuit is the same as if the electrons actually passed right through the condenser—except for the phase relation between the impressed voltage and the resulting current.

The quantity of electricity stored in a condenser is proportional to the square of the impressed voltage. The quantity stored is measured in **coulombs** or ampere-seconds. One coulomb is the quantity of electricity carried by one ampere of current flowing for one second. Hence, if the voltage changes at the rate of one volt per second and the current produced (or absorbed) is one ampere, the capacity of the circuit has one **farad;** that is, the condenser has a capacity of one farad. The farad is too large a unit for practical use, so in radio work a very small fraction of this capacity is used, the more common unit being the **micro-farad,** which is one-millionth of a farad.

The capacity of a condenser depends on the area of the plates, their spacing, and the dielectric properties of the insulator which separates the plates. For mechanical reasons, it is desirable to construct condensers with two or more plates; hence, most radio condensers consist of two paralleled sets of plates, each connected together conductively. The dielectric property varies with the insulating material which affects the ability of a condenser to store electricity. The **dielectric constant,** therefore, describes the ability of a condenser to increase its capacity over that of an air condenser.

The capacity of a condenser can be computed from the following formula:

$$C \text{ (microfarads)} = 0.8842 \frac{kA}{d} (n-1)\ 10^{-7}$$

Where k = the dielectric constant (air 1.00, mica 4.5 to 7.5)
A = area in cm^2 (one side of one plate)
d = separation in cm
n = number of plates

Condensers in Parallel and Series

Condensers can be connected in series or in parallel, but the effect is just the opposite to that of connecting inductances or resistances in series or parallel. A simple rule covering parallel or series connections is: Capacities in parallel should be added to find the total capacity, and for capacities in series the reciprocal of the sum of reciprocals must be taken. Illustrating by formula:

$$C \text{ (parallel)} = C_1 + C_2 + C_3 \text{ etc.,}$$

$$C \text{ (series)} : \frac{1}{C} = \frac{1}{C_1} + \frac{1}{C_2} + \frac{1}{C_3} \text{ etc.,}$$

Capacitative Reactance

Alternating current does not flow through a capacity without some impeding effect taking place, which is termed **capacitive reactance.** This retarding factor is inversely proportional to the frequency and the capacity of the condenser. To find the inductive reactance, the following formula is given:

$$X_c = \frac{1{,}000{,}000}{2 \pi f C}$$

Where X_c = the capacitive reactance in ohms
f = the frequency in cycles per second
C = the capacity in microfarads

Thus, if the capacity of a condenser and the frequency of the impressed alternating voltage is known, the current through any condenser can be determined by dividing the voltage by the capacitive reactance.

Impedance

When an inductance, capacity and a resistance are connected in series, the combined effect is called the impedance of the circuit.

The capacitive reactance and inductive reactance are of opposite sign, because the current through a conductor leads the impressed voltage by 90 electrical degrees, while the current through an inductance lags the voltage by 90 degrees. Thus, the current is 180 degrees out of phase with that through the inductance. The reactance of the circuit becomes $X_l - X_c$. Since the current through an inductance or capacity lags or leads that through a resistance by 90 degrees, it is necessary to take the square root of the sum of the squares to solve for the total impedance of the circuit to the flow of current

$$Z \text{ (impedance)} = \sqrt{R^2 + (X_l - X_c)^2}$$

With any two quantities known, the third can be solved from the following formulas:

$$E = IZ \qquad Z = \frac{E}{I} \qquad I = \frac{E}{Z}$$

From the equation of the impedance of a series circuit it can be seen that the impedance is equal to the resistance when the inductive reactance is equal to the capacitive reactance. This is known as **resonance.**

Alternating Current Considerations

Alternating current produces a heating effect in a resistor in spite of the fact that the current flow periodically reverses at a uniform rate of speed. To explain the theory, principle, and applications of alternating current in its many ramifications, would be taking too much of the more valuable space in this book. The student, then, is referred to texts wherein this and other information on AC phenomena can be found. Briefly, a generator produces alternating current which starts at zero, reaches maximum, returns to zero, reverses direction. and repeats the performance. This variation follows a mathematical law called a **sine wave.** The actual heating effect of this alternating current depends on the effective value of each half-sine wave. This is called the R. M. S. value and is equal to the peak value divided by 1.41, in case it is a pure sine wave. The RMS value of either voltage or current is the value read on most AC voltmeters or ammeters.

In considering alternating current the actual power is not the product of I^2Z, since the effect of either the inductance or capacity is to make the current lag or lead that through the resistance of the circuit. The lag or lead is known as the **phase angle,** and the power can be computed from the expression $P = E \times I \cos\theta$. The $\cos\theta$ represents the power factor which has a zero (unity) value in a pure (100%) resistive circuit. A perfect condenser having no resistance would have a zero power factor, which would provide a means for making comparative tests with other condensers.

One of the many interesting applications of "power factor" is in determining the effective shunt and series resistance of a condenser when the frequency of operation is known. Solutions for the determinants are:

$$\text{Series Resistance} = \frac{\text{power factor}}{2\pi f C}$$

$$R_s = \text{Shunt Resistance} = \frac{1}{2\pi f C \times \text{power factor}}$$

Eliminating the power factor term gives·

$$\text{Series Resistance} = \frac{1}{R_s(2\pi f C)^2}$$

Fundamentals of Radio

In power, telephone and telegraph lines, electricity energy is carried from the sending point to the receiving point through individual and isolated lines. All radio signals, however, utilize a common conducting medium, **the ether.** The mixing of thousands of radio signals in one conducting medium necessitates some method of selecting the desired signal and rejecting all others. This is accomplished by means of **resonant circuits** involving inductances and capacitances in series or parallel. Vacuum tubes are used to amplify the signals, while tuned circuits are used for **selecting** the desired signals.

Radiation

Radio waves are transmitted from an antenna through space in two general types of waves. One is called the **ground wave,** which follows along the surface of the ground, and for very short waves is rapidly attenuated. The ground wave is useful in long wave radio communication and also for very short distance work on ultra-short wave lengths. The other form of wave is known as the **sky wave,** since it is reflected back to the earth by ionized layers in the upper atmosphere known as the Kennelly-Heaviside Layers. The sky waves are propagated from the antenna at angles above the horizon.

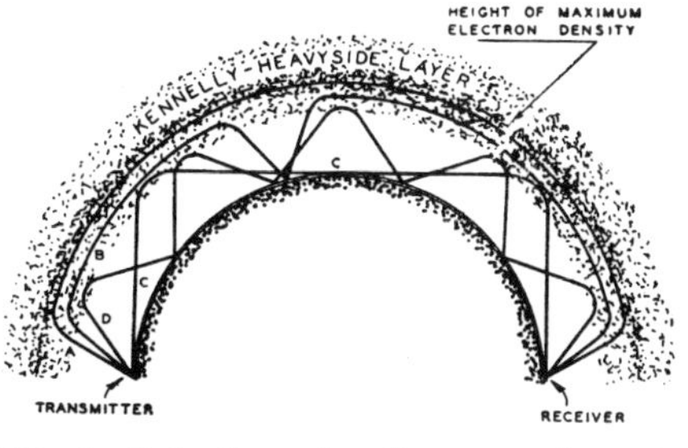

FIG. 5—Reflection of radio waves from the Heaviside Layer around the earth.

At very low angles of radiation, the waves go out practically tangent to the earth's surface, penetrate into the ionized layers and are bent back to the earth at a very distant point. Higher angles of radiation are bent back to earth at shorter distances until a certain high angle is reached for any particular frequency which will not be bent back to earth. This angle varies with the season of the year, frequency and time of day. At angles slightly less than this value at which the layer is penetrated, the radio waves can be carried around one of the upper layers to extremely great distances before being bent back to earth, no matter what the angle of propagation.

The Kennelley-Heaviside Layer is a strata of ionized air molecules, of which the ionization is due to the ultra-violet radiations from the sun. This stratospheric layer lies above the earth at distances of less than one hundred to several hundred miles elevation. The relative density of the layer lies closest to the earth, especially in the daytime and in the summer. However, it is not constant, but varies from year to year and seems to depend upon sunspot activity.

Inductance Considerations

Inductances are used in radio, audio-frequency and power circuits. An inductance used for the latter purpose can be designed from a rather simple formula:

$$L = 1.257\, N^2 P \times 10^{-8}$$

Where N = the number of turns of wire
L = the inductance in henrys
P = the permeance of the complete magnetic circuit

In most inductances, the magnetic circuit is confined by means of an iron magnetic core to the close proximity of the coil itself. For radio-frequencies some form of air-core coil is most often used. Lately, pulverized iron has been successfully employed for low and medium frequency coils, such as in intermediate-frequency transformer assemblies.

The inductance of an air-core solenoid can be calculated from the formula:

$$L_1 = N^2 d K$$

Where L_1 = the inductance in microhenrys
N = the number of turns
d = the average diameter of the coil
K = a constant depending on the ratio of the length to the coil diameter

This formula shows that the inductance of radio-frequency coils varies as the square of the number of turns and directly as the diameter of the coil.

An inductance has a certain amount of resistance due to the metallic conductor used in winding the coil. At radio-frequencies this resistance is a great many times more than the resistance would be for direct current. At radio-frequencies the current tends to concentrate at the surface of the conductor, which in effect gives an increase in the resistance. This crowding of the current density toward the surface of a conductor is known as the **"skin effect."**

The ratio between the inductive reactance of the coil and its effective resistance gives a measure of its efficiency, and is known as the "Q" of the coil. "Q," therefore, is the factor of merit of a reactance element; this factor can be determined by the following formula:

$$Q = \frac{2 \pi f L}{R}$$

Series Resonance

When an inductance, resistance and capacitance are connected in series, there will be a certain resonant frequency at which the inductive reactance is equal and opposite in effect to the capacitive reactance, and the flow of current will only be limited by effective resistance of the circuit. At higher frequencies than resonance, the capacitive reactance is less than the inductive reactance, with the result that the impedance is higher than at resonance. The same holds true at lower frequencies, except that the larger reactive term is capacitive. The reactive voltage drop across either the coil or condenser is very high at resonance, because the current is only limited by the resistance of the circuit. This reactive voltage may be several hundred times the value of the impressed voltage, as given by the expression:

$$El = \frac{E \times 2 \pi f L}{R} = \frac{E}{2 \pi f C R} = E \times Q$$

For example, if the impressed voltage is 10 volts, and if the "Q" of the coil is 100, the reactive voltage across the condenser or coil would be 1,000 volts. The sharpness of a resonance curve depends upon the "Q" of the coil, for example:

$\frac{1}{2Q}$ difference of frequency from resonance will only give 70% of the resonant current.

$\frac{1}{Q}$ difference of frequency from resonance will only give 45% of current at resonance.

Series resonance is applied to antennas, antenna feeders, and occasionally in audio-frequency and filter circuits.

Parallel Resonance

Parallel resonant circuits are used in both transmitters and receivers for purposes of selectivity or coupling between vacuum tubes. At frequencies below resonance, the inductive branch draws high current while the capacitive branch draws low current, resulting in a lagging current known as **inductive reactance.** The opposite holds true for frequencies higher than resonance. At resonance the inductive reactance is equal to the capacitive reactance, and the parallel impedance is an effectively high resistance. The parallel impedance at resonance is equal to:

$$\frac{(2\pi f L)^2}{R} = 2\pi f L Q$$

This shows that at resonance there is a resonant rise in impedance of "Q" times the reactance of either branch; meaning, for example, that a tuned radio-frequency amplifier would have more gain and also better selectivity with a high "Q" coil in the tuned coupling circuits. Since the plate impedance of an RF amplifier tube is often much greater than 100,000 ohms, it is important that inter-stage tuned circuits have a very high resonant impedance so that a good impedance match and maximum voltage step-up will be obtained.

When parallel circuits are placed across the grid or plate circuits of a transmitting amplifier tube, the impedance of the tank is greatly reduced, because of the low shunt resistance across the parallel tuned circuit. The effect of a shunt resistance is to increase the effective series resistance of the same circuit; the amount can be determined by the following formula:

$$r = \frac{1}{r_s(2\pi f C)^2}$$

Where r_s = the shunt resistance.

For example, a shunt resistance of 2,000 ohms would increase the effective series resistance of a representative tank circuit from 5 ohms to 100 ohms at a frequency of 7 megacycles. Assuming the circuit had a "Q" of 100 without any shunt load, the "Q" would be reduced to 5, due to the loading effect. The parallel impedance (from the above formulas) would be approximately 2500 ohms under load conditions, and 50,000 ohms with no load. The example brings out the effect of a resistance shunted across a parallel tuned circuit.

The resonant frequency of a parallel tuned high-Q circuit is given by the expression:

$$f = \frac{1}{2\pi\sqrt{LC}}$$

This expression is slightly in error for low-Q circuits, because the resonant frequency is affected by the effective series resistance. The sharpness of resonance is similar to that of a series resonant circuit and the same "Q" formulas can be used for determining currents at frequencies off resonance.

In many applications of a parallel tuned circuit, it is desirable to obtain a step-down ratio of impedance. A typical example is in matching a 500 ohm single wire antenna feedline to the tuned output circuit of a transmitter, as shown in Figure 6. In this case, the load is only connected across part of the parallel tuned circuit impedance in order that optimum power transfer will be obtained.

Another case of parallel resonance occurs in radio-frequency choke coils which are used to prevent radio-frequency currents from flowing into undesired circuits. The self-capacitance of the coil resonates it with its inductance to a frequency usually much lower than the operating frequency. The RF choke functions as a very small condenser of not more than two or three microfarads which presents a high impedance to RF currents. At frequencies below resonance the choke performs like an inductance having an apparent value equal to:

$$\frac{L}{1 - m^2}$$

Where m is the ratio of applied frequency to the natural resonant frequency of the coil; and L, the theoretical inductance. This apparent inductance can be very great near resonance.

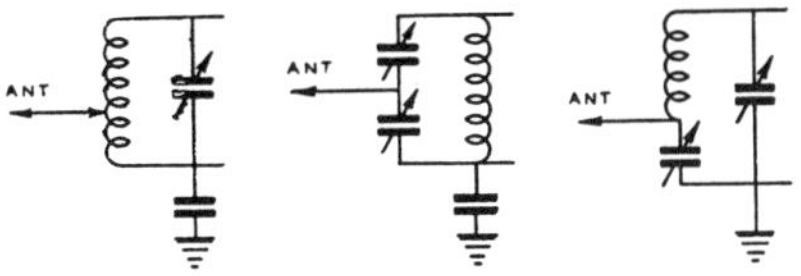

FIG. 6

Coupled Circuits

As single reactive circuits are not always employed in radio transmitting and receiving circuits, it is therefore, more common to use various combinations of coupled circuits; four simple electrical configurations are shown in Figure 7. In all of the diagrams the presence of a secondary circuit changes the impedance of the primary circuit by an amount equal to the expression:

$$\frac{(2\pi f M)^2}{Z_2}$$

The equivalent primary impedance becomes:

$$Z_1 + \frac{(2\pi f M)^2}{Z_2}$$

Where Z_1 = the series impedance of the primary alone
Z_2 = the series impedance of the secondary alone
M = the mutual inductance of the coils L_1 and L_2

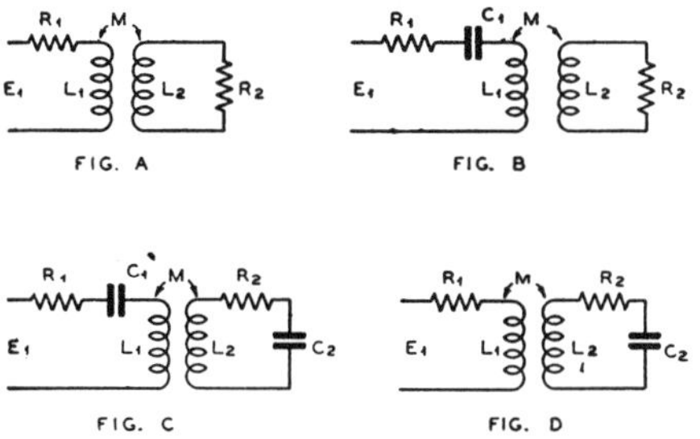

FIG. 7

Note: when Z_2 is low, such as at resonance, and M is not small, the effect on the primary is large. The effect of the secondary upon the primary circuit may be determined from the above expression when applied to the schematic diagrams shown on page 17. From these expressions it is possible to roughly analyze most any transmitter or receiving circuit.

Power transformers are a form of coupled circuits of the type shown in Figure 7-a. The difference between a power transformer and a similar RF coupled circuit is that the leakage reactance may only be about two per cent in the former case, and as high as 90 per cent in the latter case. The leakage reactance is much higher at radio-frequencies because most high-frequency coupled circuits are resonant and require very loose coupling with a very small value of M' to attain the desired result. In many cases the coupling between two or more circuits is obtained by other methods using some form of inductive or capacitive reactance, or even resistance coupling.

Band-pass circuits are special forms of parallel resonant coupled circuits. The coupling is increased until the secondary causes an extreme broadening of the resonance curve or it may even form a double resonant peak in the primary circuit. True band-pass circuits are seldom used in short-wave radio receivers or transmitters because selectivity and gain are more important to the amateur than a level frequency response over a range of frequencies.

Radio Symbols Used in Schematic Circuit Diagrams.

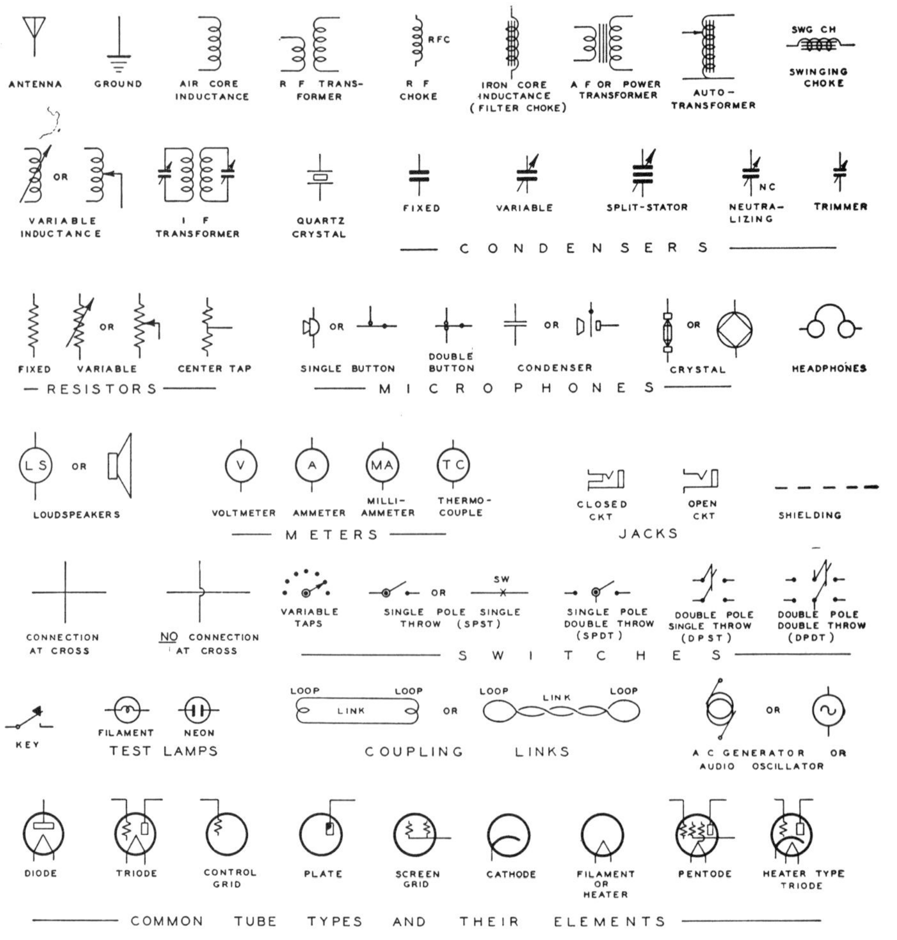

VACUUM TUBES

In radio transmission and reception, vacuum tubes are employed for the generation, detection, and amplification of radio and audio-frequency currents; in addition, electron tubes serve as power rectifiers which convert alternating current into direct current, and in special cases for controlling and inverting electric power.

The functions performed by a thermionic tube depend on the emission of electrons from a metallic surface and the flow of these electrons to other surfaces; the transition constituting an electric current Exercising a suitable control over the flow of electrons increases the adaptability of the radio tube over a wide range. In this chapter some of these applications will be cited.

An electron tube consists essentially of an evacuated glass or metal envelope in which are enclosed an electron emitting surface, called a **cathode,** and one or more additional electrodes. The connections from the various elements are carried through the tube envelope to special connectors.

Electron Emission—Cathodes

The rate of electronic motion in every atom increases if the molecular constituents of any material are subjected to thermal agitation. Hence, by heating certain metallic conductors the motion of electrons become so rapid that some of them break away from their parent atoms and are set free in space. In the absence of any external attraction, the electrons escaping from the emissive surface repel each other because they are all negatively charged. Therefore, the number of electrons leaving the emittor are limited on account of the free negatively charged electrons counteracting the escape function of new electrons. The point of electronic saturation is called the "space charge effect." When this condition is reached no further electrons will leave the emittor regardless of how much higher the temperature of the emitting surface is increased. The element from which electrons are detached in a radio vacuum tube is energized electrically by the passage of current through either a directly-heated filamentary cathode, or metallic sleeve indirectly-heated by an internal resistive element. In all modern vacuum tubes the surface of the cathode material is chemically treated to increase electronic emission. The two principal types of surface treatment include "thoriated tungsten filaments," as used in medium and high-powered transmitting tubes, and "oxide coated filaments," or cathode sleeves, such as used in most receiving tubes. Pure tungsten filaments are practically obsolete, and are only being manufactured for some types of high-power transmitting tubes where sufficient vacuum cannot be maintained for properly operating a thoriated tungsten type of filament.

Cathode Current

When a heated cathode and separate metallic plate are placed in an evacuated envelope, it is found that a few of the electrons thrown off by the cathode leave with sufficient velocity so that they reach the plate. If the plate is electrically connected back to the cathode, the electrons will flow back to the cathode, due to the difference in electrical charges caused by the electrons leaving the cathode and reaching the plate. This small current that flows is the **plate current.** If a battery, or other source of DC voltage is placed in the external circuit between the plate and cathode, so that the battery voltage places a positive potential on the plate, the flow of current from the cathode to plate will be increased. This is due to the attraction offered by the positively charged plate for any negatively charged electrons. If the positive potential on the plate is increased, the flow of electrons between the cathode and plate will also increase up to the point of **saturation.** Saturation current flows when all of the electrons leaving the cathode are attracted over to the plate, and no increase in plate voltage can increase the number being attracted to the plate. Raising the temperature of the cathode will increase the plate current on account of the electronic increment from the emitter. Operating a cathode at a temperature materially above its normal rating will shorten the life of the emitting surface. In the case of thoriated tungsten emittors, which are rather sensitive to changes in filament temperature, it is advisable to provide a close control over the filament voltage. If there is any doubt about the filament voltage, it is better to operate the filament slightly higher than normal, rather than below normal, especially if the tube is operating with high plate current.

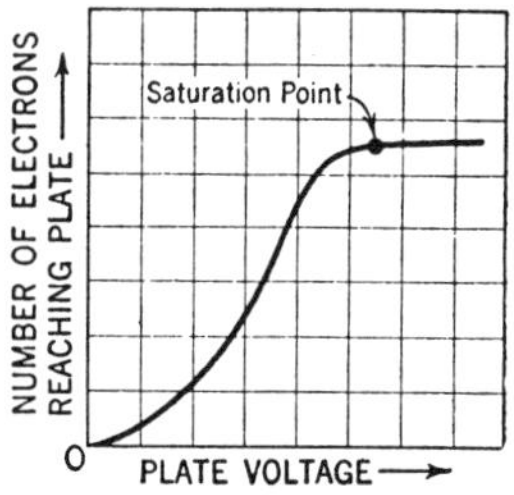

FIG. 9—Curve showing emission from a cathode.

Diode Rectification

If a negative charge is applied to the plate, the electrons in the space charge are repelled back to the cathode and no current flows in the circuit between the cathode and plate. Thus, in a vacuum tube, current can flow from the cathode to plate, but not from plate to cathode. If an alternating current is applied to the plate, current will flow only when the plate is positive with respect to the cathode. This current will be pulsating, but uni-directional. If a suitable smoothing filter is placed in the circuit, the pulsations will be smoothed out and will simulate that of a direct current. This process is known as **rectification,** it is widely applied in all radio circuits. All amplifiers employing radio tubes usually require the application of rather high positive DC potential to the plate, which of course, necessitates the stepping-up of the AC current supplied by the power mains before it is rectified and filtered. Other applications of the principle of rectification occur in radio receivers and transmitters.

Vacuum Tubes as Amplifiers

The addition of a mesh-like structure, called a **grid,** interposed between the cathode and plate in a vacuum tube allows a wide control over the electron flow from the cathode to plate. This control is made possible by applying small control voltages to the grid which either increase or decrease the plate current according to the direction of potential command. Vacuum tubes in which there are three electrodes are called triodes. Hence, when the grid is given a negative charge with respect to the cathode, it repels the electronic flow to the plate, resulting in a decreased plate current. On the other hand, if the voltage is made high enough, the plate current will be cut off. The point at which the flow ceases is called the "cut-off bias," and it depends on the grid-to-plate spacing, as well as the closeness of mesh of the grid structure. When the potential on the grid is made positive with respect to the cathode, electrons are attracted away from the space charge area surrounding the cathode and are speeded on through and past the grid structure on to the plate with increased velocity. This increases the plate current.

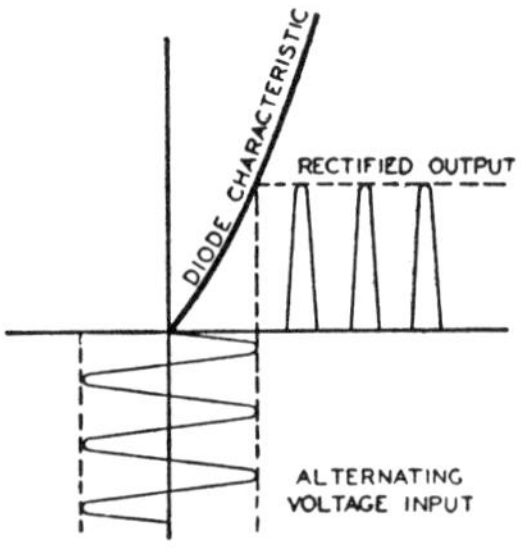

FIG. 10—Showing how a rectifier changes AC into DC.

Some of the electrons are intercepted by the grid and flow back to the cathode through the external grid circuit, but this grid current is usually quite small in comparison to the plate current. The ideal grid structure would be one that would give high acceleration to the electron flow when positive, yet would not intercept any grid current. The interception of grid current requires that the source of controlling voltage applied to the grid will supply enough power to swing the grid voltage to the required positive point, in spite of the resisting effect of the grid current.

A vacuum tube amplifies the voltage excursions of the grid by reason of the fact that the effected change in the plate current causes a similar amplified voltage drop to take place across an impedance in series with the plate circuit.

Tetrodes and Pentodes

The term "tetrode" and "pentode" indicate the presence of four and five element tubes, respectively.

A tetrode consists of a triode to which has been added a second grid between the control grid and the plate. The grid is usually maintained at a positive potential, with respect to the cathode. The purpose of this grid is two-fold: first, it accelerates the electron flow from cathode to plate, thereby improving the tube's ability to amplify voltage. Second, it provides a grounded electro-static screen between the plate and control grid, so that energy will not be fed back to the control grid through the plate-to-grid capacitance of the tube. If the amplification through the tube is high enough, this feedback, or regeneration of energy, might set the tube into self-oscillation, which would destroy its usefulness as an amplifier. This regeneration is put to work in certain detectors and in all oscillators, but its presence is undesirable in most amplifier applications. The tetrode has several disadvantages, the principal one being that the instantaneous AC plate voltage caused by the changing plate current cannot be allowed to swing to a value below the fixed positive potential on the outer, or screen grid. When the potential on the plate becomes less than the potential on the screen grid, the secondary electrons constantly being driven out of the plate by the impact of those arriving from the cathode fall into the more positive screen, instead of falling back into the plate, as they normally do. This increases the screen current, and under certain conditions, gives the tube negative resistance. This effect causes tremendous distortion in a voltage amplifier and limits the output of a power amplifier.

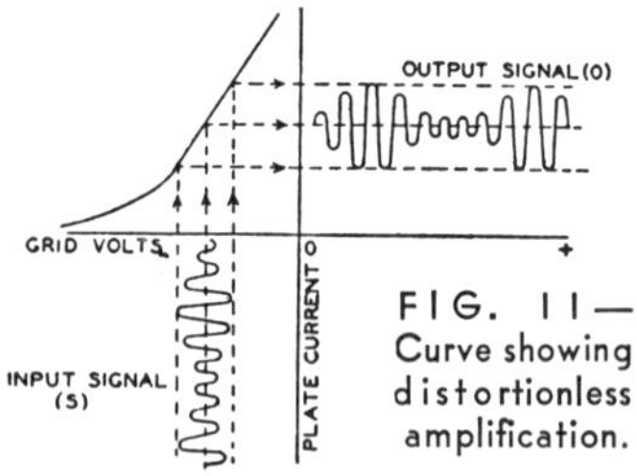

FIG. 11—Curve showing distortionless amplification.

The pentode was developed to avoid this disadvantage of the tetrode. In this development, a third grid is added between the grid and the plate for the purpose of shielding the plate from the screen grid, so that the secondary electrons emitted from the plate will be forced to fall back into the plate and are prevented from going over to the screen. This outer grid in a pentode is called the **suppressing grid** because it suppresses the secondary electrons driven out of the plate.

Pentodes are highly useful for all class A voltage and power amplifiers, although they are not as desirable as triodes in high efficiency power amplifiers (above 40% overall efficiency). The main drawback to the use of tetrodes in high efficiency power amplifiers is the fact that the presence of the additional grids raises the plate resistance somewhat more than the amplification factor. Thus, the control-grid to plate transconductance cannot be as high in a similar triode. Transconductance, as will be explained in the next few paragraphs, is the best yardstick of a vacuum tubes' ability to amplify power, particularly at high plate efficiencies required by economy considerations in the construction of radio transmitters.

Gaseous Conduction

If a diode vacuum tube is evacuated and then filled with a gas, such as mercury vapor, its characteristics and performance will differ materially from an ordinary high-vacuum type diode tube.

The principle on which depends the operation of a gas-filled rectifier is known as the **phenomenon of ionization.** Investigations have shown that electrons emitted by the hot-cathode in a mercury vapor tube are accelerated toward the anode (plate) with great velocity. These accelerated electrons move in the (electrical) force-free space between the hot-cathode and anode, in which space they collide with mercury vapor molecules. If the moving electrons attain a velocity equivalent to falling through a potential difference of 10.4 volts (for mercury), they are able to knock electrons out of the atoms with which they collide. When an electron is separated from its normal orbit it leaves not as an electron but as a **positive ion.** This freed positive ion will consequently be neutralized by the optional acquisition of a free electron. Finally the free electrons will be attracted to the anode or plate as will the positive ions that have been separated from the mercury atoms, which collectively constitute the flow of current in the tube. The passage of ions and electrons cause more of the atoms to be broken up by collision so that the vapor becomes heavily ionized and transmits a considerable amount of current. When the anode is positive, the ions are repelled and attracted to the cathode, tending to neutralize the negative space charge as long as saturation current is not being drawn. The mechanics of this electronic effect neutralize the negative space charge to such a degree that the voltage drop across the tube is reduced to a very low value and, in addition, reduces the heating of the diode plate as well as improving the voltage regulation of the circuit in which the tube is used. This greatly increases the efficiency of rectification because the voltage drop across any vacuum tube represents a waste of power.

Grid Controlled Rectifier

A controlled rectifier is a gaseous type of diode employing either mercury, neon or argon gases, and a control grid. As the grid does not perform in the same manner as in the triode, it is necessary to give a description of the controlling action.

If a grid is placed between the cathode and the plate in a gaseous tube, the starting of the plate current can be controlled. A negative bias (or an absence of the required positive bias, in positive controlled tubes) prevents the flow of electrons from starting. However, once the flow begins, and the gas has become ionized, the grid loses all control over the electron stream. After starting, the grid neither modulates, limits, nor extinguishes the discharge. Herein lies the fundamental difference between high vacuum tubes and grid controlled rectifiers. The grid can regain control and prevent the passage of current if the potential on the plate is lowered to below the ionization voltage of the conducting vapor or gas. The time for de-ionization is very short; hence, interrupting the plate current for a few micro-seconds allows the grid to regain control.

If an AC voltage is applied to the plate, the grid is permitted to regain control after every positive half-cycle when the plate goes negative. In addition, the grid can delay the start of ionization for as long a period during the positive half cycle, as long as the grid bias voltage is sufficiently negative. In this manner, the grid can control the average current flowing through the tube if both plate and grid are supplied with AC, and the phase relation between the grid and plate voltage are adjusted to either increase or decrease the frequency or time of ionization.

Grid controlled rectifiers are more commonly known by their trade names—"Thyraton" (General Electric Co.) and "Grid Glow Tubes" (Westinghouse Company). For the amateur, the tubes are quite useful in varying the output of DC power supplies. Grid controlled rectifiers are also used in keying CW transmitters, or in applying carrier control to the plate power supply of a modulated amplifier.

The use of gaseous conduction tubes are limited to very low frequencies, such as 500 cycles and lower. The tubes are unstable at high-frequencies due to the finite time required for the internal gas to deionize after each cycle of conduction.

Vacuum Tube Characteristics

The characteristics of a vacuum tube are the electrical properties which describe its ability to perform various functions. These characteristics are obtained by operating a vacuum tube under certain known electrode voltages, and then measuring the electrode currents. By plotting the change in any electrode current as any one of the electrodes voltages are likewise varied, a **characteristic curve** is obtained. When a negligible amount of impedance is inserted in the plate circuit of a tube and different DC potentials are applied to the tube electrodes, and should the variations in electrode current be graphically plotted on cross-section paper, the results are known as the tubes' **static characteristic curve.** On the other hand, if there is an impedance in the plate circuit, the plate voltage will vary with the plate current; hence, if a pure resistive impedance is placed in the plate circuit, and an AC voltage is impressed on the control-grid under various conditions of DC potentials on the electrodes, and if the variations in current are plotted on graph paper, the result will be the **dynamic characteristic curve.** This characteristic indicates the performance of a vacuum tube under actual operating conditions.

From three sets of static curves, it is possible to calculate in advance the actual performance of practically any type of vacuum tube amplifier or detector. Investigators have done a great deal of work in developing means by which the optimum operating conditions for the operation of class B and C power amplifiers can be accurately determined in advance. This information, in the form of curves or tables, will probably be made available soon by the tube manufacturers, so that proper values of bias, plate voltage, grid current and plate current can be chosen in order to obtain optimum power output and plate efficiency from any power amplifier.

Dynamic Characteristics
Amplification Factor

The amplification factor, cryptically written as either u, mu, or k, is the ratio of the change in plate voltage, plate current constant, to a change in grid voltage in the opposite direction. For example, if the plate voltage is changed 20 volts, and if it requires a change of 2 volts (opposite polarity) in the control grid voltage to hold the plate current constant, the amplification factor is 20/2 or 10. Expressed as an equation, it is:

$$u = \frac{dE_p}{dE_g}$$

Where d = any small change increment
E_p = variable component of plate voltage
Where E_g = variable component of grid voltage

Plate Resistance

The plate resistance of a vacuum tube is defined as the ratio of a small change in plate voltage to the resulting change in plate current, when the grid voltage is assumed to remain constant. For example, if a change in plate voltage of 20 volts causes a change in plate current of 10 milliamperes (ma.), the plate current resistance equals 20 divided by .01 ampere (10 ma.), or 2000 ohms. Expressed as an equation:

$$R_p = \frac{dE_p}{dI_p}$$

It is desirable to make the plate resistance of a tube as low as possible, especially in power amplifiers where the load circuit is coupled to the plate in order to make a more effective impedance match. This allows the use of a lower plate voltage than would otherwise be obtained.

Transconductance

The control grid-plate transconductance (S_m), formerly called mutual conductance, combines in one term the u and the plate resistance of a vacuum tube, and is the ratio of the first to the second. By introducing the equations given above for u and R_p in the ratio defined for transconductance, it can be seen that the S_m can also be expressed as the ratio of the change in plate current to the small change in grid voltage producing it (plate voltage constant, load resistance zero). Combining the above expressions, the formula for transconductance can be written:

$$S_m = \frac{u}{R_p} = \frac{\frac{dE_p}{dE_g}}{\frac{dE_p}{dI_p}} = \frac{dI_p}{dE_g}$$

S_m is expressed in MHOs, the unit of conductance.

Note that it is ohm spelled backwards; this is logical, since conductance is the reciprocal of resistance.

To illustrate an example of transconductance, take the case where ratio of the dI_p to dE_g equals S_m; hence, if a grid voltage change of 5 volts causes a plate current change of 10 ma., the transconductance is .04 divided by 5, or 0.002 mho.

A convenient means of determining transconductance without any calculations is to read the plate current change caused by a change of exactly one volt on the control grid. By multiplying the resulting I_p change in ma. by 1000, the S_m obtained is directly in micromhos.

Vacuum Tube Amplification

A tube amplifies by reason of the fact that a small change in grid voltage produces a larger change in plate current than would be produced by the same change in plate voltage. See Figure 11. This function can be applied in many ways, depending upon the result desired.

Vacuum tubes can be classified into two general categories, according to **application** and **operating characteristics.**

In general, vacuum tubes may be classified into four groups, according to their principal application. These are:

- Voltage amplifiers
- Power amplifiers
- Current amplifiers
- General purpose amplifiers

A voltage amplifier tube usually has a very high mu and finds its greatest use where tremendous voltage amplification is desired. This type of tube, like the type 57, must feed into a high impedance device like the grid of another vacuum tube for maximum voltage amplification. High mu tubes are used mostly as radio and intermediate-frequency amplifiers.

A power amplifier tube has a relatively low amplification factor and is used where the primary consideration requires a maximum amount of undistorted output. For maximum power transfer the load impedances must be properly matched to the plate resistance, which is generally of a low value. In power tubes, the output increases with great rapidity as the plate voltage is increased; hence, for maximum transfer, power tubes are operated with high plate voltages.

A current amplifier tube is one that gives large changes in plate current for very small changes in grid voltage; in other words, a tube having a high Sm will pass high plate currents; hence, the term "current amplifier." The use of these tubes is mostly confined to electronic industrial applications and therefore will not be discussed here.

General purpose amplifier tubes have characteristics between voltage and power amplifier tubes. The usefulness of this type of amplifier tube, in radio, is practically without end; for instance, in voltage amplification where a smaller power output is desired, and where the connecting link is a voltage step-up transformer, a general purpose triode will supply the circuit requirements. These tubes are now used extensively in class B or C power amplifier.

From the foregoing it can be seen that vacuum tubes may be employed in a wide variety of ways, depending on the result desired. In addition to the above classification there are three principal types of tube amplifiers and two secondary types. These types differ largely in the choice of bias axis, angle of plate current flow and whether the average DC plate input is constant or variable.

Class A Amplifier

The class A amplifier is biased usually in the middle of the linear portion of the dynamic characteristic curve. This is the usual condition of operating vacuum tubes, since the input impedance is then very high and very little energy is required to control the tube. In this type of amplifier, plate current flows the whole AC cycle, or 360 degrees. The average plate current waveform is independent of the signal or exciting voltage.

Class A amplifiers are used in all RF, IF and low level audio amplifiers in receivers. It is characterized by low plate efficiency and power output, but almost infinite power gain, because the control grid never goes positive and thus requires no grid driving power.

Glass B Amplifier

The class B amplifier is always biased to the point known as the "theoretical cut-off." The plate current is not zero at this point, but is quite low (no signal present on the grid). Theoretical cut-off bias equals the plate voltage divided by the u, or amplification factor (not applicable to pentodes). It can also be determined by extending the linear portion of the dynamic characteristic down to the zero plate current line and reading the negative bias intercepted at that point. In class B amplifiers, the useful plate-current flow should last for exactly 180 electrical degrees, or one-half cycle.

The class B amplifier is used as an audio power amplifier where it is too expensive to provide the required audio power output from a class A amplifier. It will also give distortionless amplification of a radio-frequency wave that has been modulated in some preceding stage of a transmitter. Class B is characterized by maximum plate efficiencies from 40 to 70 per cent, depending upon application. This type amplifier is practically a compromise between power gain and power output, when functioning as an amplifier of unmodulated radio-frequency power. For audio-frequencies, it is necessary to use two tubes in push-pull in order to eliminate high distortion. As an audio-amplifier, the plate input varies widely with the signal, but the input remains constant when amplifying a modulated radio-frequency wave. At audio-frequencies the power output is proportional to the square of the grid excitation voltage.

Class C Amplifier

The class C amplifier is biased considerably beyond the cut-off, and requires the application of a high amplitude signal voltage to carry the grid positive. Plate current flows for less than 180 degrees and the pulsating power pulses are usually quite peaked, which renders this type of amplifier unfit for distortionless amplification. However, for radio-frequency amplifiers and vacuum tube oscillators it is customary to use some type of class C amplifier. The characteristics of the amplifier render it capable of very high plate efficiency and power output, although the power gain drops as the plate efficiency and power output go up. In general, the output varies as the square of the plate voltage within limits. A common use for a class C amplifier is that of functioning as a plate modulated RF power amplifier, in which case the grid must be biased to at least twice the cut-off.

Class AB Amplifier

The class AB amplifier is biased somewhere between the class A and the class B points. Plate current flows for more than 180 degrees, but less than 360 degrees. The plate efficiency and power output are intermediate between class A and B, and tubes with low u are often adaptable to this class of service. Amplifiers of this class are almost exclusively used for audio-frequencies which are generally operated in push-pull to avoid distortion. The class AB amplifier was formerly called the class A Prime Amplifier.

Class BC Amplifier

The class BC amplifier is biased somewhat beyond the cut-off, and thus plate current flows for less than 180 degrees. The only applications of the class BC amplifier at the present time are the modulation gaining RF linear amplifier and the grid bias modulated RF amplifier. In both these amplifiers, fixed low resistance bias equal to "theoretical cut-off" is supplemented by approximately an equal amount of cathode resistor bias. This arrangement permits the angle of plate current flow to be constant and independent of the audio modulation signal, even though the actual plate current flow is less than 180 degrees. The power output, plate efficiency and power gain are intermediate between class B and C.

Detection

Detection is a process by which the audio modulation is separated from the radio-frequency signal carrier at the receiver. Detection always involves rectification or non-linear amplification of an AC current. All other types of detectors or demodulators provide exactly the same rectification except the triode, tetrode and pentode detectors which, in addition, combine the function of amplification to such an advantage that more over-all amplification can be obtained with fewer tubes.

There are two types of detectors used in radio; these are the **plate** and **grid detectors.** How each of these function are briefly described below.

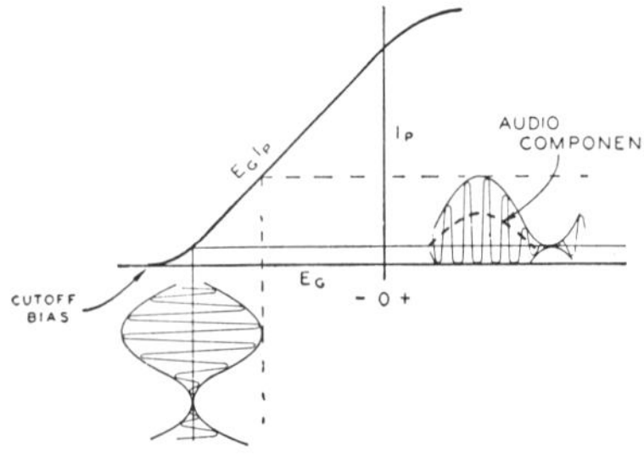

B—PLATE DETECTION

Showing how the average plate current increases.

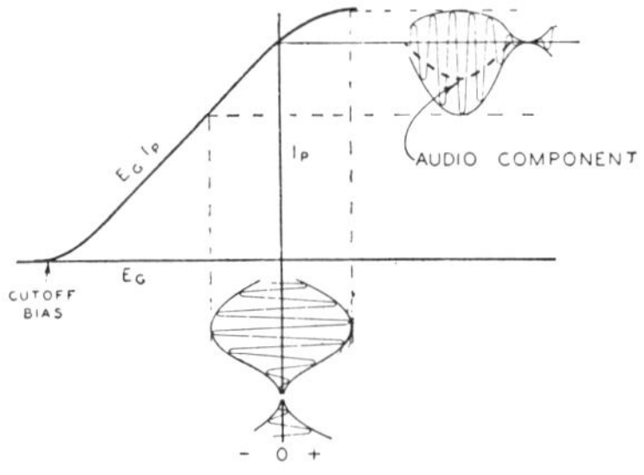

A — GRID DETECTION

Showing how the average plate current decreases.

The plate detector (or bias detector), sometimes improperly called power detector, amplifies the radio-frequency wave and then rectifies it and passes the audio-signal component on to the succeeding audio amplifier. The detector works on the lower bend in the plate current characteristic, as it is biased out close to the cut-off point. It might be called a class B amplifier. Plate detectors can be either of the weak signal or power type. The plate current is quite low in the absence of a signal and the audio component is evidenced by an increase in the average unmodulated plate current. The grid detector differs from the plate detector, as will be evidenced in the subsequent explanation.

The grid detector rectifies in the grid circuit and then amplifies the resulting audio signal. The only source of grid bias is the grid leak, so that the plate current is maximum when no signal is present. This detector works on the upper, or saturation, bend of its curve at a high plate voltage, and the demodulated signal appears as an audio-frequency **decrease** in the average plate current. However, at low plate voltage most of the rectification usually takes place as a result of the curvature in the grid characteristic. As with plate detectors, grid detectors can be either of the weak signal or power type. By proper choice of grid leak and plate voltage, distortion can be held to a small value. The grid detector absorbs some power from the preceding stage because of drawing grid current. The higher gain through the grid detector does **not** indicate that it is more sensitive. Detector sensitivity is a matter of rectification efficiency, not amplification alone.

The grid detector has an advantage when used as a regenerative detector because the grid leak usually allows a somewhat smoother control of regeneration than is possible with any form of plate or bias detector.

Oscillation

The ability of an amplifier tube to control power enables it to function as an oscillator in a suitable circuit. By coupling part of the amplified output back into the input circuit, sustained oscillations will be generated; that is, if the input voltage to the grid is of the proper magnitude and phase with respect to the plate. In general, the voltage fed back and applied to the grid must be approximately 180 degrees out of phase with the voltage across the load impedance in the plate circuit and, in addition, have sufficient magnitude to develop the necessary grid voltage. The voltage swings are limited by the circuit losses and are of a frequency depending upon circuit conditions.

When a parallel resonant circuit consisting of an inductance and a capacitance (LC) is inserted in series with the plate circuit of an amplifier tube and connected so that the potential drop across its terminals is impressed on the grid in the same tube 180 degrees out of phase, amplification of the potential across the LC circuit will result. The potential would increase to an unrestricted value were it not for the limited range of linearity of the tube characteristic and the limited voltage available on the plate. Therefore, a value will eventually be reached limiting the amplitude of oscillation. When this value is attained, the process of amplification reverses, reducing the voltage across the LC circuit as quickly as it had been raised a moment before. When the voltage across the resonant circuit reaches zero, it reverses, and is developed to another value having the same amplitude but of opposite polarity; at the point of the greatest voltage swing, amplification again reverses, and the process continues indefinitely.

The frequency range of an oscillator can be made very great; thus, by varying the circuit constants, oscillations from a few cycles per second up to many millions can be generated. One of the unique properties of an oscillator is that it can oscillate at more than one frequency at the same time; these frequencies are called **harmonics.**

One of the most common types of oscillator circuits known is called the "Hartley Oscillator," a diagram of which is shown.

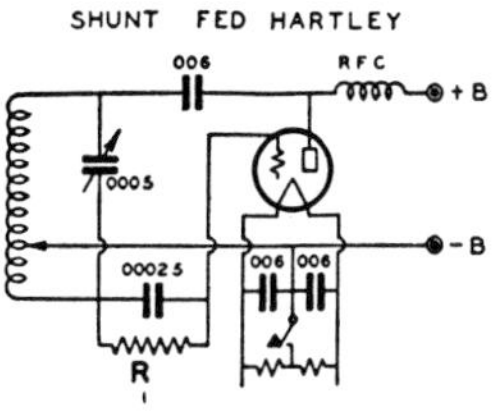

In this circuit the **plate** and **grid inductances** together with the **tank condenser** form an oscillatory circuit known as the **tank circuit.** If the condenser in this circuit be charged, then allowed to discharge through the plate and grid inductances as shown, the current flow would be alternating and of decreasing magnitude. The frequency is determined by the size of the condenser and inductances and is equal to:

$$f = \frac{1}{2\pi\sqrt{LC}}$$

Where f, the cycles per second; L, in henrys; and C, in farads.

The decrease in amplitude is due to losses in the tank circuit and to the energy delivered to the output. If sufficient energy be supplied to this circuit, during each cycle, to supply the losses and power output per cycle, the amplitude of the current would remain unchanged. The function of the vacuum tube is to deliver the required energy to the tank circuit.

As the energy stored in the tank circuit alternates, there will be a time when the grid will be charged positively with respect to the filament. A large direct-current plate current will flow under the influence of this positive grid potential, building up the field and storing energy in the plate choke (inductance). At a time of one-half cycle later the grid will be negative, thereby greatly decreasing the flow of plate current and causing the plate choke to discharge its energy into the tank circuit. This discharge occurs once each cycle and thereby the necessary energy is delivered to the tank circuit to maintain oscillations of constant magnitude.

When the grid is positive with respect to the filament, electrons leak back to the cathode via the grid leak and condenser. If the value of this RC path is high, a high effective grid bias results, making the tube function as a class C amplifier. This results in maximum output and high efficiency.

The best way to classify regenerative vacuum tube oscillators is by the feedback coupling method. All such oscillators use either capacitive or inductive coupling from the plate circuit back into the grid circuit. Usually very low-frequency oscillators (below 100 KC) use some form of inductive coupling, while high-frequency oscillators (100 to 100,000 KC) are capacitively or inductively coupled; however, for frequencies higher than 100,000 KC, only capacitive feedback is required.

At frequencies above 100,000 KC (3 meters) the effectiveness of the regenerative oscillator drops off rapidly because the time of flight from the electrons between the grid and plate becomes a large fraction of one cycle of oscillation. The losses in regenerative oscillators also become so large at these frequencies that the plate circuit is incapable of supplying the grid losses, let alone supplying power for driving an amplifier or antenna .

Thus, at frequencies above 100 MC (100,000 KC), the newer **electron orbit** oscillator is becoming more widely used. This type of oscillator can be of several forms, the more important being the Magnetron and the Barkhausen-Kurz oscillators.

RECEIVERS

Elements of Tuning Inductances

Resonant circuits are the major electrical tuning units in all amateur, communication, and broadcast receivers. The importance attached to the tuning circuit and other associated elements requires a detailed analysis; however, the following considerations are all that are necessary.

Electro-magnetic and Electro-static Coupling

When an electro-magnetic wave is intercepted by an antenna, a small radio-frequency voltage is induced in the conductor, which surges to-and-fro in an oscillatory manner. Tapping the antenna at a suitable point by a **lead-in** or **feeder** and causing the voltage to pass through an inductance will produce a current in the coil in proportion to its reactance.

Assuming that the inductance in the antenna circuit is untuned; that is, an inductance without any shunt capacity, the voltage induced across the coil will be equal to the current times the inductive resistance. Hence, anything done to increase the voltage developed across the coil will also increase its magnetic flux; and furthermore, when a secondary winding is coupled to the antenna coil a greater voltage will be induced on account of the increased flux density cutting the secondary inductors. Anything to cause the antenna voltage to increase, before being applied to the grid of the detector tube, will augment the over-all amplification of the signal strength.

Now, by changing the untuned antenna coil to a tuned parallel resonant circuit by the simple expedient of adding a variable capacity across the inductance, the voltage will no longer equal the current times the inductive reactance; but, instead, will equal the current times the ratio of the reactance and resistance. The impedance of such a circuit drops off rather rapidly at either side of resonance; the voltage, and consequently the signal diminishes proportionately. In other words, a circuit that is tuned exactly to the signal frequency will give considerably more gain than one that is untuned or that may differ in some respects from the resonant frequency by an appreciable amount.

The energy from the antenna can be connected directly across a coil and induced into another coil without being physically connected to the former; this is known as **inductive coupling**. On the other hand, if the energy is connected across the plates of a condenser, then fed to the grid side of the coupling coil, the connection is known as **electro-static coupling**. From the foregoing explanations it will be apparent that this type of coupling has no voltage gain in itself, and is therefore inferior, though possibly more convenient to use than inductive coupling.

Whenever an antenna circuit is coupled closely to the grid circuit, some electro-static coupling is bound to exist, due to the capacity between the metals in the respective coils. A combination of coupling is undesirable in most cases, since electrostatic coupling permits steep wave-front voltages, such as static and noise, to have greater paralyzing effect on the grid. Pure inductive coupling is only practicable if the separation between the two coils is made large, or through the use of an electrostatic shield, commonly known as a "Faraday screen."

In inductively coupled circuits, the amplitude of the induced voltage will depend upon the strength of the magnetic field set up, the proximity of the two "coils" and the impedance of the grid circuit to the particular frequency.

The impedance of the grid will follow the same rules set forth for the antenna circuit, since they are both parallel resonant circuits and are both maintained at resonance with the incoming frequency. At this point it is necessary to take into consideration another property of resonant circuits known as the "Q."

"Q" of Resonant Circuits

"Q" may be defined as the inductive reactance divided by the resistance. The Q of a coil is the factor of merit; the higher the Q, the better the coil. Authorities differ quite widely on the ideal shape for a coil, but, in general, agree that very long, or very short coils are to be avoided. A coil whose length is approximately equal to its diameter is often considered best.

The diameter of the wire used to form the coil also has a definite influence on the Q. Hence the wire size should be as large as possible to get into a given winding space. NOTE: Practically all the resistance in a parallel resonant circuit is contributed by the inductance; the condenser, if well designed, has negligible resistance. But nearly all the resistance in the inductance is contributed by the "skin effect." This effect increases almost directly with frequency and is introduced at high-frequencies because the current is not equally distributed throughout the conductor, but travels only on the outermost surface. Thus, in order to provide ample surface for the current to pass along, it is necessary to use a much larger size conductor than would be the case if the current was equally distributed throughout the conductor.

Round conductors are always better than flat strips because, even if the flat strip has more surface area, the fact remains that the current does not distribute evenly over the entire surface but has a maximum density at the edges, with low density on the sides.

Distributed capacity, or the capacity existing between successive turns and also between these turns and the ends, is to be avoided in any receiver coil, since this capacity has the effect of lowering the Q. Space winding is one means of lessening this effect. Where the conductor is large in diameter, "space winding" reduces the skin-effect, due to currents set up in adjacent turns. Dielectric loss due to poor insulating material in coil forms also has the bad effect of lowering the Q.

Summarizing: The ideal inductance would be one having the following properties:

1—A shape such as to make the length approximate the diameter.

2—Entirely air-supported. Since this condition is practically impossible, a compromise must be adopted taking the form of a coil support of a low-loss dielectric, such as Isolantite.

3—A wire size of ample proportions. This must also be a compromise, since with excessive wire diameters the skin-effect and distributed capacity more than offset the gain due to increased surface. For all practical purposes a wire size larger than No. 16 need not be used in receiver coil design.

4—A space type of winding. The spacing will be more or less governed by the length-to-diameter rule. In general, the spacing ought not to exceed twice the diameter of the coil.

Considering the coil and condenser as a unit (a parallel resonant circuit), it is required in good design to adhere to the following:

1—In order for the circuit Q to be as high as possible, the inductance-to-capacity ratio should be very high.

2—The tuning condenser should have excellent mechanical and electrical properties and be preferably insulated with Isolanite, or similar material. Some type of pig-tail connection or positive wiping contact must be included in the assembly for contacting the rotor; this reduces high-resistance during rotation.

Selecting a Receiver

The selection of the proper type of receiver best suited to one's needs is a problem that confronts every beginner. Incidentally, there are practically as many types of receivers as there are kinds of amateurs. No perfect receiver exists for all-around operation under all operating conditions; hence, it is largely the personal choice of the operator that governs the receiver type. All receivers represent a compromise between such factors as cost, size, accessibility, convenience, dependability, versatility, output desired and the purpose for which it is to be used.

If a receiver is to be built, instead of being purchased, and if the constructor has had no experience in receiver construction, it is advisable to first build the more simple types of receivers, using from one to three tubes, instead of the more complicated multi-tube superheterodyne receivers, which may have from six to twelve or more tubes.

The constructor who chooses the regenerative autodyne receiver must weigh the compromises involved in its design. If the receiver is located in a metropolitan area, where power lines, street cars, oil furnaces and other sources of man-made static interference are prevalent, the receiver must be particularly well shielded. If the set is battery-operated, the noise pick-up will be minimized, as no interference will be introduced through AC power lines feeding a mains-operated plate or filament supply. If the receiver is used in the country, remote from man-made static, shielding is a matter of lesser importance, and thus a somewhat simpler receiver will give entirely satisfactory results.

If a receiver is located in the neighborhood of a powerful radio transmitter, the strong radiations may block or paralyze the RF or detector circuits, making it necessary to provide a tuned stage of radio-frequency amplification or some other form of volume control to obtain satisfactory selectivity. At the same time it may also be necessary to choose a somewhat less sensitive detector circuit in order to make the detector less susceptible to overload.

One of the salient points of receiver construction is that of cost. The actual design of a receiver is a simple problem. Of course, the design may become complex if all late engineering refinements are incorporated into the construction. In general, the most elaborately designed receiver is actually more modest in cost than might otherwise be expected. Although every set builder will desire the most expensive coil forms, tuning condensers and vernier dials, it is essential to strive for a happy medium when selecting a receiver circuit which makes the best use of the parts available.

A receiver which is to operate on one band is much easier to build than one which must operate satisfactorily in the entire range of from 160 meters to 10 meters. A band-spread arrangement of condenser combinations which give excellent results on 20 meters will not be satisfactory when used to cover the 160-meter band. Thus, if the constructor desires to operate on two such widely different frequencies, a sacrifice must be made of both convenience and efficiency on one or both of these bands.

Methods of Band-Spreading

Band-spreading is an electrical means of obtaining tremendous gear reduction on the tuning condenser dial of a receiver. High-frequency receivers must cover a very wide range of frequencies and therefore it is difficult to design a dial and drive mechanism which will cover the desired ranges, yet still provide sufficient "vernier" (geared down) drive so that weak signals will not be passed over without hearing them. In newer all-wave broadcast receivers this problem is solved by the use of a two-speed dial arrangement, the low reduction being provided for rough tuning and the high reduction for fine tuning. This is usually accomplished mechanically by means of planetary gear. The system is quite satisfactory, but rather difficult to manufacture by the average amateur or experimenter. Practically the same effect can be obtained by means of electrical band-spread. Almost all receiver circuits use a variation in the capacity of the tuned circuit for tuning purposes. In order to obtain a small variation in tuning it is essential that the capacity be increased or decreased by a small amount. However, difficulty is encountered in varying the capacity of a large condenser by small increments or decrements, but in an electrical band-spreading system utilizing two tuning condensers—one large condenser to give rough tuning, the other, a very small condenser (two or three plates) may be connected in a wide variety of combinations to give the electrical effect of "fine" or "vernier" tuning. The first system is shown in Figure 1a. It is the most common system and consists of a small condenser C_2, connected directly in parallel with the large condenser C_1. In most high-frequency receivers the capacity of C_1 will be chosen so that the coil and the condenser combination will cover a frequency range of between 2-and-3-to-1. The condenser C_2 is much smaller than C_1 and will often be chosen so as to cover a band of approximately 1000KC.

Figure 1b shows a band-spread condenser in series with the main tuning condenser. Because the capacity of two condensers in series is always smaller than the capacity of the smaller of the two condensers, it will be seen that both condensers in Figure 1b must be considerably larger in capacity than the corresponding condensers in Figure 1a in order to cover the same frequency ranges. Both of the systems shown in Figures 1a and 1b have the disadvantage in that the degree of band-spread varies with the tuning of C_1, and thus if a given coil covered both 40 and 20 meters, the system may provide too much band-spread for 40 meters and not enough band-spread for 20 meters. In Figure 1c the band-spread effect can be kept constant over a wide range of frequencies by tapping the band-spread condenser across part of a coil, instead of being tapped across the entire coil, as in Figure 1a. The position of the tap varies with frequency. On the larger low-frequency coils, the tap will be placed near the top of the coil. On small high-frequency coils, the tap will be placed proportionately farther down on the coil in order to maintain an approximately constant degree of band-spread. This system has the disadvantage in that some selectivity is lost in the tuned circuit. Figure 1d shows another means of equalizing the degree of band-spread over a wide range of frequencies. C_1 is the conventional large tuning condenser of between 140 and 350 mmfd. C_2 and C_3 are both band-spread condensers. C_2 has approximately 50 mmfds. for band-spreading the 80 and 160 meter bands; C_3, from 15 to 20 mmfd., is best for use on the 40 and 20 meter bands. The proper condenser is chosen by means of switches, as shown in the accompanying figure. A disadvantage of switching is that rather long leads are required, as well as a possibility of losses in the switch contact.

METHODS OF BAND SPREADING

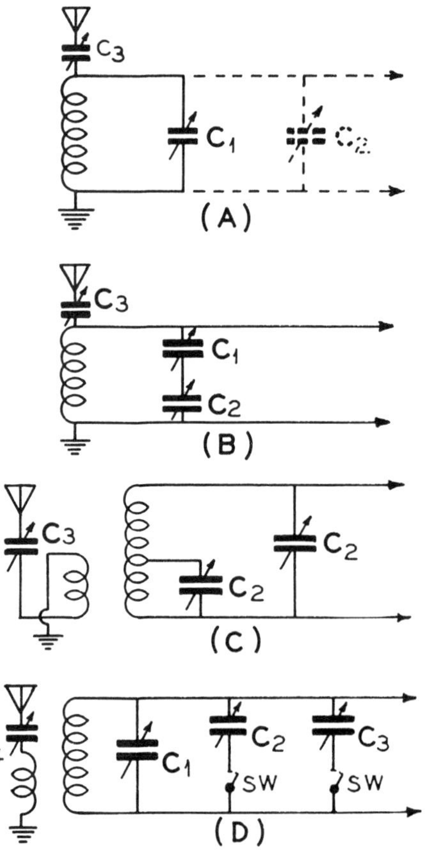

FIG. 1

Plug-in Coils

Practically all regenerative receivers use plug-in coils. This is also true of some of the highest-priced amateur receivers and commercial superheterodynes. The advantages of plug-in coils are only obtained when low-loss materials and low-loss design are featured as a complement. The very best low-loss coil form is "dry-air," or self-supported coil winding. Next best are the ceramic forms which use Isolantite, Mycalex, or their equivalents. Then follow the special mica compounds, such as the XP-53 and R-39 compounds. Whereas celluloid is a more inferior dielectric than the aforementioned materials, its advantage is that a very thin form will serve as an excellent coil support. In addition, because losses are a function of the volume of dielectric material in an electric field, the thin celluloid makes possible the construction of an extremely low-loss coil form.

Wire for Coil Winding

Bare wire, having as large a diameter as possible, is better than insulated wire in winding coils, because the larger the wire diameter, the lower will be the radio-frequency resistance. In coil winding, the space-wound method is superior to others, while grooved coil forms are undesirable on account of increasing distributed capacity. It is essential that all coils be placed as far away as possible from metallic shields or other metal bodies, such as the chassis.

Coil Winding Data for Simple Receivers

Coil winding tables vary with the size of the coil form used. The standard form is 1½ inches outside diameter. A table is given below for the number of turns required on a coil form to cover the four popular amateur bands. If forms larger than 1½ inches in diameter are on hand, obviously fewer turns will be required. Conversely, a smaller form will require a greater number of turns per coil. It is a simple matter to use the "cut and try" method when winding coils; however, the accompanying table will greatly simplify matters. It is assumed that the coils are to be wound on standard forms and tuned with a 100 uufd. midget variable condenser.

Wave-length	L1, Secondary Winding	L2, Tickler Winding
20 M	7 turns, No. 18 DCC, spaced two diameters.	4 turns, No. 22 DSC, close wound.
40 M	18 turns, No. 22 DSC wire, spaced one diameter.	Ditto.
80 M	36 turns, No. 22 DSC wire, close wound.	6 turns, No. 22 DSC, close wound.
160 M	72 turns, No. 32 DSC or SCC wire, close wound.	11 turns, No. 22 DSC or SCC, close wound.

Spacing between secondary and tickler coils to be ⅛-inch. The wire should be tightly wound on the coil forms. Insulating varnishes should be used sparingly, if at all. The most common form of coil "dope" is known as Collodion, made by diluting small pieces of celluloid in a vessel containing about an ounce of Acetone.

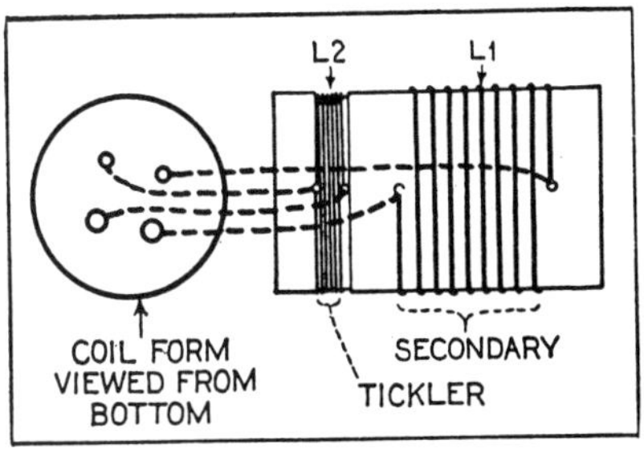

Reading from Right to Left, the coil connections are as follows: Antenna (and grid condenser), Ground, Plate, B Plus.

Tickler Winding

If the detector does not regenerate, reverse the tickler connections or add one or two turns of wire to the tickler coil, until smoothest regeneration is obtained.

The Detector in a Regenerative Autodyne

The detector is the heart of the regenerative autodyne receiver, and a wide variety of tubes may be used for this purpose, each having certain advantages and disadvantages. The four most commonly used detector tubes are the 76 and 6C6, for operation from house lighting current, and the 30 and 32 types for battery-operated sets. The 76 and 30 are triodes, while the 6C6 and 32 are screen-grid types. Screen-grid detectors are somewhat more sensitive than triodes, although are more susceptible to overload and more difficult to get going. In place of the 6C6 or 32, it is often desirable to utilize a tube with a variable mu, such as the 6D6 or 34. This type of tube is slightly less susceptible to overload than the sharp cut-off detectors, such as the 6C6 and 32. Variable mu tubes afford a smoother control of regeneration but necessitate a sacrifice in sensitivity.

The 24, 36 and 57 tubes are very similar to the 6C6. By the same token, the 39 and 58 are similar to the 6D6. Likewise the 27, 37 and 56 will act exactly like the 76 in most circuits. In the battery-operated field there is less choice, although the 99, 201A and 12A are quite similar in characteristics to the 30, and type 22 can be used in a circuit designed for a 32.

Audio Coupling

The detector can be coupled to an audio amplifier in three different ways, which are known as resistance coupling, impedance coupling, and transformer coupling.

In general, resistance coupling is the least desirable of the three methods when working out of a regenerative detector, because the question of fidelity is relatively unimportant and fidelity is the principal advantage of a resistance coupled amplifier. Resistance coupling can be used out of either triode or screen-grid detectors.

Impedance coupling (or choke coupling) is particularly recommended when working out of a screen-grid detector because it enables the full plate voltage to be applied to the detector and also has enough distributed capacity so that any radio-frequency present is easily by-passed to ground. The only disadvantage of impedance coupling is that it affords no voltage step-up, as does transformer coupling. An impedance to work out of a triode detector should be approximately 30 henrys at 15 to 20 milliamperes. An impedance designed to give best results out of a screen-grid or pentode detector should be rated at more than 250 henrys at 5 milliamperes.

Transformer coupling is unsuited when using a screen-grid or pentode detector, although it is recommended when working out of a triode detector. A step-up ratio of approximately three-to-one gives the best all-around results.

Impedance or transformer coupling sometimes gives trouble, due to fringe audio howl in a regenerative receiver. A 50,000 to 250,000 ohm resistor shunted across the impedance coil or transformer secondary will usually cure this trouble.

Audio Tubes

The choice of the audio output tube is largely dictated by the amount of audio power required. If loudspeaker operation is desired, two stages of audio amplification will ordinarily suffice; for example, a triode type 76, in the first stage, and a pentode, such as a 41, in the second stage.

If headphone operation is desired, the second stage may be eliminated and the phones connected in the plate circuit of the first amplifier stage. For loudspeaker use, pentodes are recommended, such as types 38, 41, 42, 47, 59, 89, 33, or 43. Triodes may also be used, but will require somewhat more amplification; they are the 12A, 71A, 45, 46, 2A3, 31, 120, and others.

Any of the following tubes are entirely satisfactory for headphone reception in the audio stage: 99, 30, 201A, 112A, 27, 37, 56, 76 and either of the following pentodes when connected as triodes (screen and suppressor grids tied to plate): 57 and 6C6.

Notes for Set Builders

SOCKETS: The socket material is as important as the material from which the coil forms are made, because the socket is in the direct field of the coil. In receiver construction it is essential that only the very best material is used in socket assemblies; thus, ceramic, Isolantite and other good insulators will suffice.

LEADS AND CONNECTIONS: Leads to the tube socket and tuning condenser must be short and direct, sharp bends being avoided whenever possible. All joints must be carefully soldered with rosin-core solder, and a clean, hot iron should be used for all soldering operations. Make all connecting wires mechanically secure to all connecting points and keep all wiring well remote from metal shielding and chassis.

CALCULATING FILAMENT DROPPING RESISTOR VALUES: It is important that the filaments of all tubes, either in a transmitter or receiver, be operated at the rated filament voltage. If the voltage is too low or too high, tube life is materially reduced. When in doubt, it is advisable to operate the filament at a slightly higher than normal voltage, rather than at lower voltage. The value of a filament resistor can be calculated by means of Ohm's Law, a very simple formula which indicates the relationship between voltage, current and resistance. If any two are known, the third can be determined. The three forms of this equation are:

$$E = IR \qquad R = \frac{E}{I} \qquad I = \frac{E}{R}$$

Where E = the voltage; I, current (amperes); R, resistance (ohms).

For example, assume the two type 30 tubes are being operated with their filaments in parallel and a 3 volt battery is to supply the filament power. But, since 3 volts is too high, it must be dropped to 2 volts through a series dropping resistor, which will give the normal operating voltage. To calculate the value of the series resistor, it is first necessary to determine the current drawn by the two tubes. The current in this case is 120 milliamperes, or .12 amperes. From the equation R = E/I, the resistance is computed by dividing the desired voltage drop by one volt (which is "I" in this case) by 12/100, which is the same as multiplying 100/12. The equation then is 1/1 × 100/12, which equals 8.3 ohms. Therefore, 8 ohms is the proper value of resistor to use, because fractional value resistors are not obtainable. When connecting two tubes in series, it becomes necessary to provide twice as much heating voltage as when only one tube is used; however, there is no increase in heating current. When the filaments of two type 30 tubes are connected in series, it is necessary to provide 4 volts at 60 milliamperes (0.06 amperes). Either a 4½ volt "C" battery or three 1½ volt dry cells connected in series provide a convenient means for operating the two tubes in series. The dropping resistor should be 8 ohms, which is determined by dividing the voltage drop of ½ volt by the total filament current of .06 amperes. Care should be taken to see that tubes which draw different values of filament current are not connected in series unless special precautions are taken, as shown in Figure 2. A shunt resistor must be connected across the filament of the tube drawing the least current, so that the sum of the current through the resistor, plus the current through the filament which it shunts, is equal to the current drawn by the other tube.

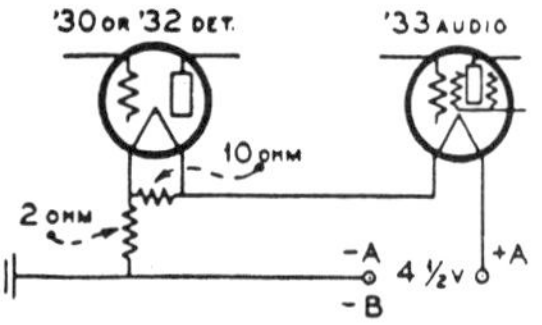

Fig. 2. Series connection for dissimilar filament currents.

CALCULATING VALUE OF SELF-BIASING RESISTORS: In practically all receivers utilizing either radio or audio frequency amplifying stages, some method of self-biasing the grids is employed. This bias is obtained by inserting a resistor in the cathode lead return wire and taking the necessary voltage drop across the resistor. The value of self-biasing resistors can be calculated by the formula:

$$\text{Ohms} = \frac{\text{grid bias} \times 1000}{\text{plate current}}$$

Thus, for a 45 tube which has a plate current of 34 ma. for which a grid bias of 50 volts is needed:

$$\frac{50 \text{ Volts} \times 1000}{34} = 1{,}470 \text{ Ohms}$$

The wattage or power consumed in the resistor equals E × I or .034 × 50 or 1.7 watts. For push-pull amplifiers combine the plate currents of each tube. For screen-grid and pentodes use the sum of the plate and screen currents.

Simple Receivers
2-Tube "DX-ER"

This receiver does not in any sense represent a new development in the short-wave construction field. Instead, it is one in which the designer combined well-known and accepted principles to produce a set that is simple and inexpensive to build.

From a casual examination of the schematic diagram it will be seen that the receiver is of the single-circuit regenerative type, with tickler feed-back. The placement of the parts is extremely important for effective results. As in all receiver designs where the maximum efficiency is desired, only the highest quality of parts should be used. Equipment of inferior design, carelessly assembled, will not bring the desired results.

For economical operation, two type 30 low-drain two-volt tubes are used. The first serves as a regenerative detector; the second as an audio amplifier. The tuning range of the receiver is 15 to 200 meters, covered by a set of four plug-in coils. Regular broadcast reception is optional, by adding a set of two plug-in coils to cover 200-500 meters.

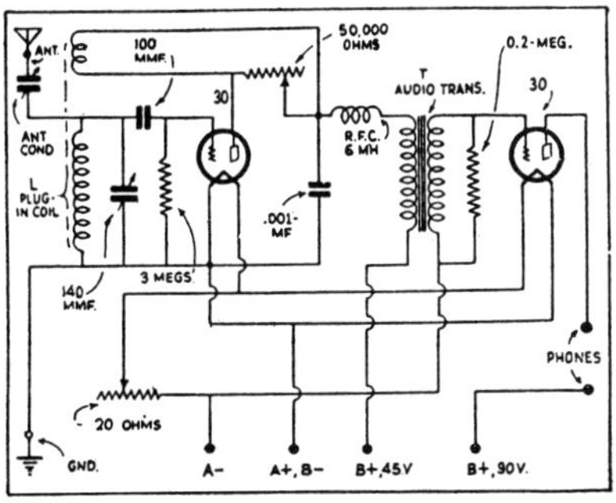

Simple 2-Tube Regenerative Receiver.

Only two dry-cells and two 45 volt "B" batteries are required for complete operation.

Regeneration is controlled by a 50,000 ohm variable resistor connected across the tickler leads. The output of the detector is transformer-coupled to the audio tube by a shielded transformer having a ratio of 1 to 5. A load resistor of 200,000 ohms is connected across the secondary of the audio-transformer to eliminate any possibility of "fringe howl."

The antenna is coupled to the tuning coil by a semi-variable "postage stamp" condenser having a maximum capacity of 80 uufds.

Tuning is accomplished by a 140 uufd. midget variable condenser mounted on the front panel. A smooth vernier-type dial is used to insure proper tuning.

OPERATING NOTES: Phone signals are loudest just below the oscillation point, and CW signals just above the oscillation point. When tuning the "DX-ER," set the regeneration control to the point where the detector just starts to oscillate; then the tuning dial should be carefully turned until a "whistle" is heard. Careful tuning at this point and further adjustment of the regeneration control will bring in the intelligible signal.

Simple Receiver With One Type 19 Tube

This receiver gives surprisingly good volume on DX signals; it is especially recommended for the beginner who is contemplating the design of a simple and inexpensive set.

The circuit diagram is self-explanatory; however, there are some details that need explanation. The grid and plate connections must be properly made, as shown in the circuit diagram. The grid bias is secured by means of the rheostat in the filament circuit. The constructor, therefore, is cautioned to connect the movable arm of the rheostat to the negative A, and also to the negative F on the audio transformer. Best results are secured with a 5 megohm gridleak; smaller values may cause the detector to regenerate with an unpleasant roar. Smoother regeneration is sometimes secured by connecting a 250,000 ohm ½ watt resistor across the secondary (GF) terminals of the audio transformer.

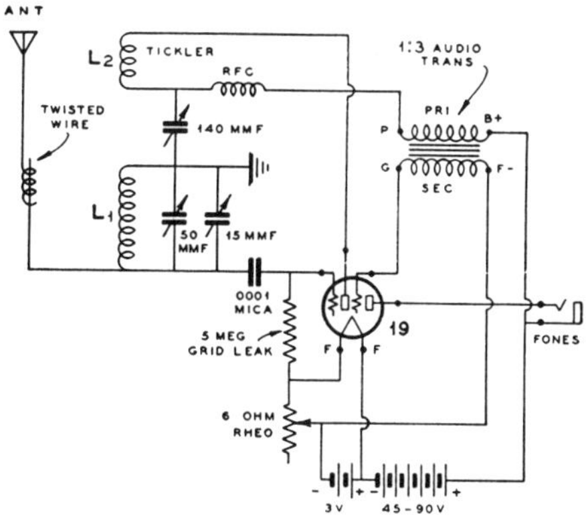

Schematic circuit diagram of the one-tube receiver. L1 is the secondary, or grid coil. L2 is the "tickler," or regeneration coil.

The band-spread tuning condenser is a 3-plate midget variable; the tank tuning condenser is a 50 uufd (or 100 uufd) midget variable. A 140 uufd. midget variable condenser is used for the regeneration control. The secondary and tickler coils are both wound on the same form, and both coils **must** be wound in the same direction; otherwise the detector will not oscillate.

General Construction: The front panel is made of a piece of No. 12 or No. 14 gauge aluminum, 7 in. x 9 in. The wood baseboard is 9 in. x 11 in. The band-spread, tank condenser and regeneration condenser are mounted directly on the panel and the rotors of these condensers are grounded to the panel. The rotors may be connected together, and the connecting wire bonded to the ground or panel. An inexpensive airplane dial enhances the symmetry of the front panel. This dial controls the 3-plate band-spread tuning condenser.

Ordinary Fahnestock battery connection-clips can be used for headphone connections in place of the phone jack; these connectors can be secured to the baseboard in any convenient location, preferably near the audio frequency transformer.

An on-off switch can be added, or the dry cells can be disconnected from the receiver when not in use. Two 1½-volt dry cells are required. These will give excellent service for a long period of time. The B-battery voltage may be as low as 22 volts, but at a sacrifice in audio volume; 45 to 90 volts is more suitable for normal operation, except when the receiver is used as a portable. With 22 volts the tickler coil must be placed very close to the secondary coil.

Antenna Connection: The antenna is coupled to the "high potential end" of the secondary coil by a few turns of lead-in wire twisted around the grid-lead of coil L1; a single turn loop wound around the top of L1 will give the same results. The small midget condenser shown in rear-view photograph of the receiver is connected in series with the antenna lead and the top lead of L1. It can be used as a substitute for the twisted-wire coupling arrangement.

Noise-Free Two-Tube Autodyne

The circuit of this receiver is conventional in every respect. It utilizes a 57 detector in an electron-coupled "Hartley" circuit which has proven so simple to make oscillate at high frequencies. Regeneration is controlled by varying the screen-grid voltage by a potentiometer across the power supply. A RF filter is incorporated in the plate lead from the detector as a precaution against spurious RF currents flowing through the audio impedance. The audio stage uses a 56 type vacuum tube, although a 27 may be substituted. In general, the circuit includes all refinements commonly found in standard practice, except for the filtering of the phone and power leads, and the link coupling to the antenna.

The receiver housing is made of aluminum, approximately 7½ inches deep. The actual panel dimensions are left to the discretion of the builder. Inside the housing is an aluminum sub-panel formed by making two rectangular, or flat "U," bends two inches deep. Another piece of aluminum is closely fitted and fastened to the bottom of the housing by tapping holes in ¼-inch "dural" corner posts which hold the assembly together. The top is fitted in the same manner as the bottom, with the exception that no drilling is necessary—the top merely rests on the corner posts.

RECEIVER ASSEMBLY: The tuning condenser is mounted on an aluminum bracket which rigidly supports it; the bracket also serves to shield the audio

COIL DATA

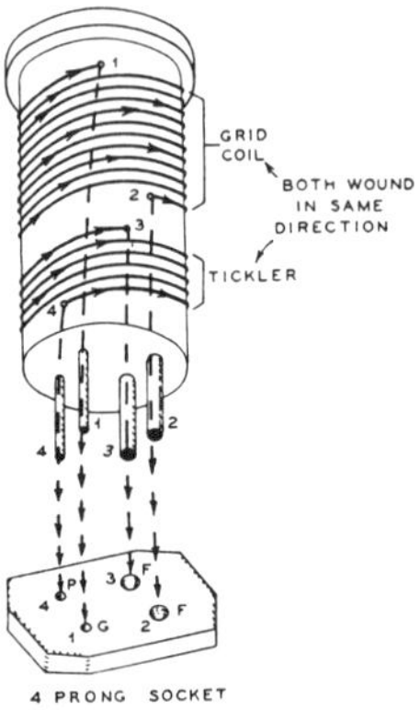

The upper coil is the grid (secondary) coil. Start the winding at point 1, make the connection to prong 1. The bottom of the grid coil (2) connects to prong 2. The top of the tickler coil (3) connects to prong 3; the bottom of the tickler (4) connects to prong 4. Mark the coil prongs and the coil socket contacts to correspond with these numbers. See the pictorial layout to show how the connections are made to the coil socket. Make certain that Connection No. 1 goes to the stators of both tuning condensers, and also to one side of the .0001 mfd. grid condenser. Likewise, take care to see that Connection No. 4 goes to the plate of the detector portion (P2) of the type 19 tube. If these connections are not properly made, the receiver will not function. The antenna lead-in wire can be looped around the No. 1 connecting lead.

COIL FORM LEGEND

Terminal No. 1 connects to one side of the .0001 mfd. mica fixed condenser and to the stator of the 100 mmf. (or 50 mmf.) condenser, as well as to the stator of the 3-plate midget variable tuning condenser. Likewise, the insulated antenna lead-in wire is twisted around the lead which connects to Terminal No. 1.

Terminal No. 2 connects to the **rotors** of all three variable condensers, and at the point where the three are connected together another lead is run to the **"ground"** terminal of the receiver.

Terminal No. 3 connects to the **stator** of the 140 mmf. variable condenser which is used for regeneration, and the same terminal also connects to one end of the 2.5 mh. RF choke.

Terminal No. 4 connects to the **P2** terminal on the type '19 tube.

COIL WINDING DATA

The secondary coil and the tickler coil are **both** wound in the same direction.

20-Meter Coil: Secondary winding—7 turns of No. 22 DSC wire, space-wound to cover a winding space of 1-in.

Tickler Winding—5 turns of No. 22 DSC wire, close-wound, and spaced about ⅛-in. from the secondary winding.

40-Meter Coil: Secondary Winding—14 turns of No. 22 DSC wire, space-wound to cover a winding space of 1-in.

Tickler Winding—11 turns of No. 22 DSC wire, close-wound, and spaced ⅛-in. from secondary winding.

80-Meter Coil: Secondary Winding—27 turns of No. 22 DSC wire, close wound.

Tickler Winding—11 turns of No. 22 DSC wire, close wound, and spaced ⅛-in. from secondary winding.

160-Meter Coil: Secondary Winding—60 turns of No. 22 DSC wire, close-wound.

Tickler Winding—17 turns of No. 32 Enameled wire, close-wound, and spaced ⅛-in. from secondary winding.

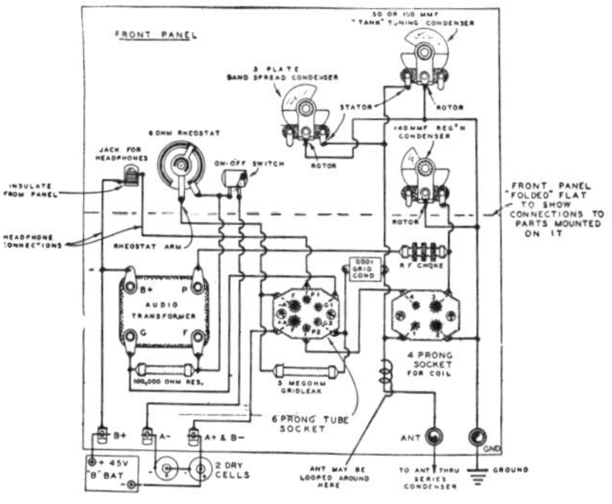

Pictorial layout of parts for 1-tube receiver. This arrangement should be closely adhered to.

Rear View of the Completed Receiver.

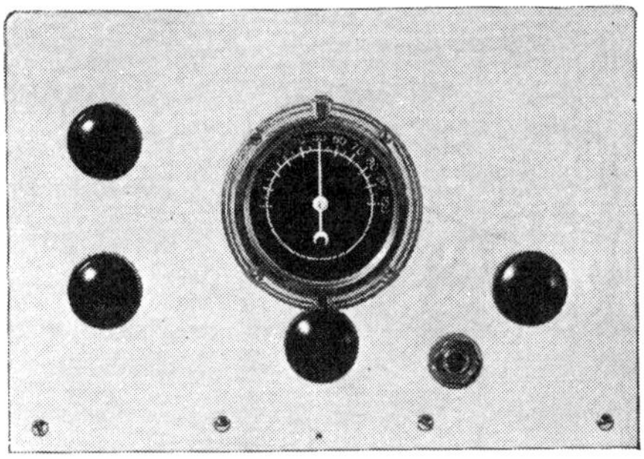

Front Panel Layout.

The Controls on the Front Panel Are:

Top, left—"Tank" tuning condenser.
Bottom, left—Regeneration condenser.
Center—Airplane tuning dial.
Extreme right—Rheostat control.

The headphone jack is mounted between the airplane dial and the rheostat control. This jack MUST be insulated from the metal front panel and a hole at least ⅛-inch larger in diameter than the outside diameter of the screw thread on the jack should be drilled in the panel.

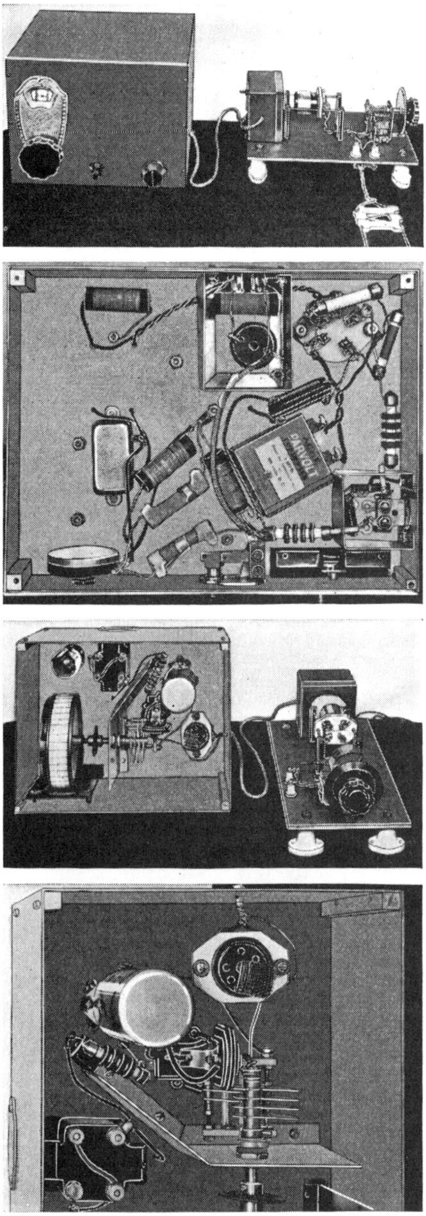

Four views of the Noise-Free Autodyne. The center picture shows how the small shield compartments are arranged under the chassis.

impedance (choke) from affecting the detector stage, thus eliminating possibility of "fringe howl." The grid leak and condenser are fastened to the tuning condenser, thereby making the leads very short to the detector tube. Only short and direct leads make

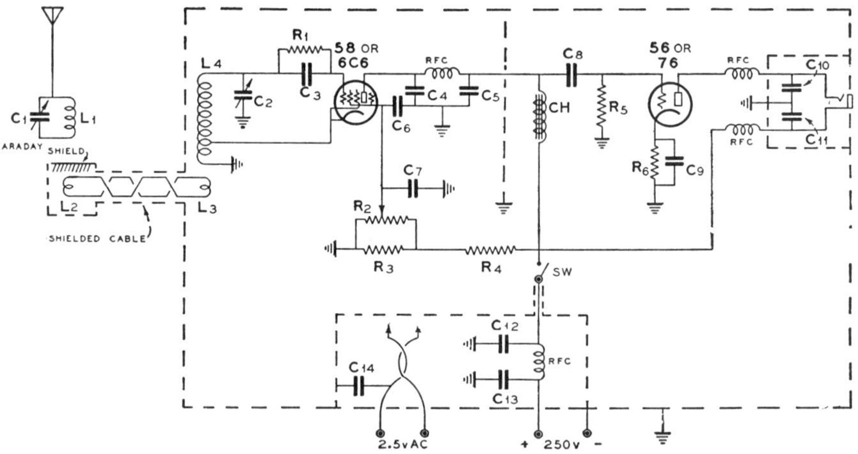

LEGEND FOR THE NOISE-FREE AUTODYNE

L1—Similar to L4, but with fewer turns, depending on type of antenna used. L2-L3—See coil table. L4—Described in Text. C1—100 mmf. midget variable. C2—20 mmf. National SEU-20. C3—100 mmf. Sangamo, with grid clip. C4, C5, C10, C11—250 mmf. mica Aerovox postage stamp type. C6, C12, C13—.01 mfd. mica condensers. C7—½ mfd. 400 volt non-inductive condenser. C8—.01 mica, Sangamo. C9—1 mfd. 200 volt paper condenser. C14—.01 mfd. non-inductive. R1—2 to 5 megohm grid leak (experiment for noiseless one). R2—50,000 ohm Centralab variable resistor. R3—4,000 ohm 10 watt. R4—15,000 ohm 10 watt. R5—½ megohm 1 watt. R6—3000 ohm 1 watt. RFC—Good short-wave choke. CH—Old A.F. Transformer or high inductance choke.

possible the ease by which this set oscillates on 28 MC; this, coupled to the fact that more coil turns are required in the circuit than is common in ordinary practice, make for a high LC ratio—a prerequisite for high sensitivity. The plate filter is mounted above the sub-panel to keep leads short and the RF from under the chassis.

Under the sub-panel, the wiring arrangement is completely conventional, with the exception of the RF filters and the number of by-pass condensers. Note, for example, that the screen-grid of the detector tube is by-passed twice—once at the socket of the 6C6, and again by a .05 ufd. condenser across the regeneration control. This latter condenser eliminates any noises that may be injected into the circuit by the sliding contact on the potentiometer. A simple output filter consists of two .00025 ufd. condensers and two RF chokes; the condensers and phone jack are included in a special shielded can, as may be seen in

COIL DATA FOR L4

3.5 MC—46 turns No. 30 enameled, close wound, tapped 1½ turns up.

7 MC—23 turns No. 18 enameled, spaced diameter of wire, tapped ⅞ turns up.

14 MC—11 turns No. 18 enameled, spaced 1½ diameters, tapped 2/3 turns up.

(Above coils wound on 1½-inch five-prong coil forms).

28 MC—9 turns No. 14 enameled wound ¾-in. diameter on air, tapped 1⅛ turns up. Turns spaced about ½ diameter.

Each link coupling loop consists of two turns interwound between the two bottom turns of each coil.

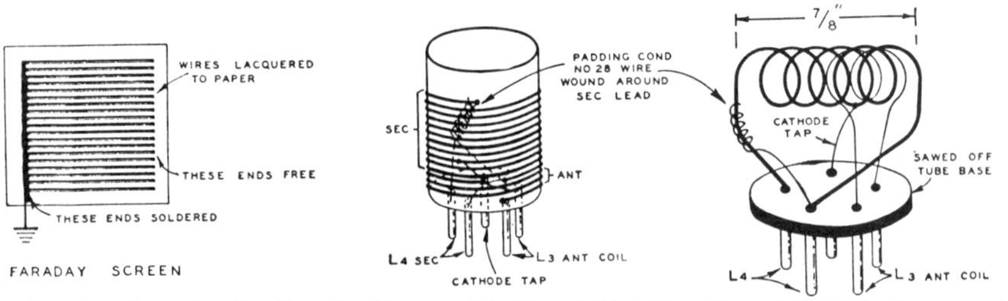

Showing how to make the Faraday Screen, 3.5, 7 and 14 MC coil, and (right) the special 28 MC coil.

the photographs. Another shield can encloses the power supply RF filter; a configuration made from a pi section filter of two .01 ufd. condensers shunted across an RF choke inserted in series with the positive terminal of the "B" battery or power supply. The two shields are made from small pieces of aluminum bent to form three sides of a box, and another piece, to serve as top, is made by bending the edges to fit snugly over the cans.

Proper band-spreading is achieved by placing padding condensers across the tuning condenser. These condensers are not shown in the photographs or wiring diagrams. Each coil contains its own padding condenser. A piece of No. 28 enameled wire is soldered to the ground side of the coil, right in the coil itself, and this is wrapped around the lead that goes up to the grid end of the coil. This permits accurate spotting of each coil right into the band, and the more turns, the more capacity; consequently the more band-spread. When the coil has once been adjusted, the extra wire is cut off and the connections made permanent. NOTE: If the coil does not cover the band for which it has been designed, it is only necessary to repeat the spotting adjustments by either lengthening or shortening the wire wrapped around the grid lead; only by experimentation can the best setting be found.

A-C Operated Gainer
An Ideal Amateur Receiver

Of the many two-tube circuits developed for amateur reception, the improved circuit shown in the accompanying diagram will be found superior to others of similar design. Although series band-spread tuning is shown, the constructor can substitute parallel band-spread tuning, the latter being a more simple method for the beginner to use. If the constructor embodies the parallel band-spread system in the circuit design, the variable condenser C_1 should have a capacity of 100 mmfds.; this condenser is shunted across coil L_1. The band-spread condenser C_2 may be a 3 plate midget, 15-25 mmfds., shunted across C_1, the tank condenser.

The receiver may be mounted on a metal chassis, 9x7 inches, with a "U" supporting bend 2 inches high. The space under the chassis is used for mounting resistors R3, R4, R5, R7 and condensers C3, C4, C5, C6, C7, and C8. The regeneration control is brought out to the front of the panel, as are controls R6 (gain) and the band-spread tuning dial for condenser C_2. The tank tuning condenser knob C_1 should also be on front of the panel. The grid condenser C9 and grid leak R1 are air-supported above the chassis, close to the grid cap of the 57 detector. The lead from R2 to the screen of the 57, and the lead from R5 to the phone jack are run through shielded braid. Plug-in coils are used in this receiver. L1 is the secondary coil; L2, the cathode regeneration coil. Both of these coils are wound on ordinary 4-prong tube bases or on standard plug-in coil forms, 1¼ or 1½ inches in diameter. The coils are wound as shown in the table under the List of Parts.

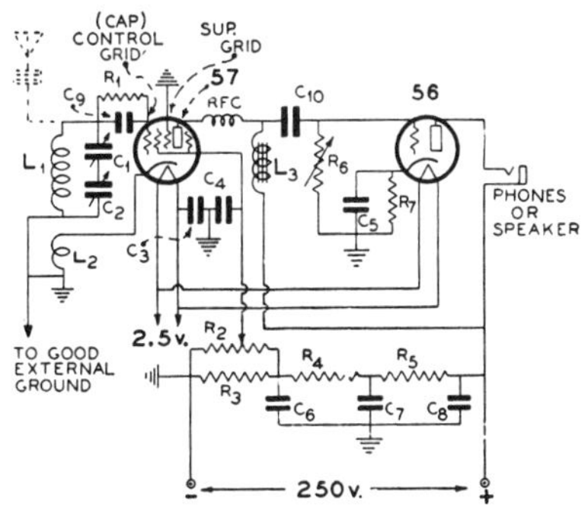

AC "Gainer" Circuit Diagram

L1—Secondary winding. L2—Tickler winding. C1, C2—Band-spread condensers, each 100 mmf., for series-band-spread tuning. C3—.01 mfd. C4—.5 mfd. C5—.1 mfd. C6, C7, C8—Each .5 mfd. C9—.0001 or .00025 mfd. C10—.002 mfd. R1—2 megs. R2—50,000 ohm potentiometer. R3, R4—Each 10,000 ohms, 10 watt. R5—5,000 ohms, 10 watt. R6—500,000 ohm potentiometer. R7—2500 ohms, 1 watt. L3—Iron-core choke (or impedance) 100 henry, or larger. An ordinary audio transformer, with primary and secondary windings connected in series, can also be used at L3. If parallel band-spread is to be used, C1 and C2 are connected in parallel, instead of in series, as shown above, and C1 should then be a 100 mmf. variable condenser, C2 a 3-plate (approx. 15 mmf.) band-spread condenser of the midget type.

L1—20 meters— 8 turns of No. 22 DCC.
40 meters—16 turns of No. 22 DCC.
80 meters—32 turns of No. 22 DCC.

L2—(Wound on the same form as L1, spaced about 3/16 inch away from L1) 4 turns of No. 22 DCC. (L2 is the same for all coils.)

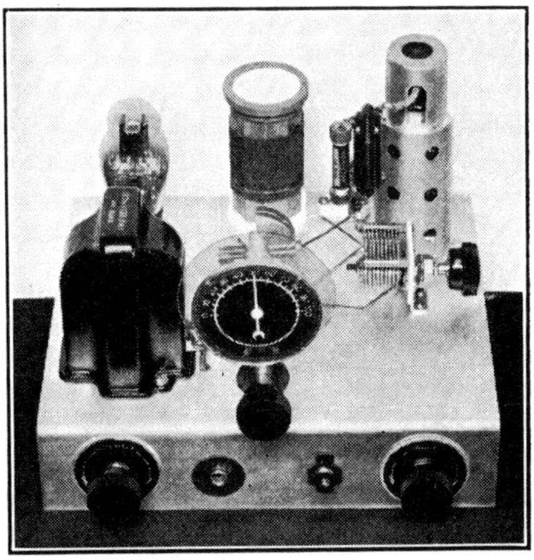

The AC "Gainer" with front panel removed to show correct arrangement of parts.

Superheterodyne Receivers

The highest grade receiver for general amateur use is the superheterodyne. The circuit design is more complex than a regenerative autodyne. One of the salient points about a superheterodyne is that it has a remarkable ability to reject undesired signals, and is less susceptible to overload from powerful local transmitters.

In general, superheterodynes may be classified according to their uses, because the ideal superheterodyne for CW reception differs in many respects from the type used for phone reception. For both classes of work a superheterodyne must necessarily be a compromise between the two ideals.

In a superheterodyne exclusively used for CW reception, the two most important points are: (1) extreme selectivity; (2), freedom from noise. To date no superheterodyne has been designed which is too selective or too free from noise. A superheterodyne for CW must also have particular attention given to the high-frequency and beat-frequency oscillators, because a frequency drift in either oscillator of only a few cycles, can make the received signal entirely disappear. This point is less important in a receiver used for phone on account of the signal being considerably broader than a CW signal, and the oscillator drift, if any, is of little importance.

Conventional automatic volume control systems have no place in a CW receiver, which is primarily designed to operate from the variations in a continuous carrier. Hence, the variations in sensitivity caused by a CW signal merely make the signal difficult to read. Likewise, high-fidelity has no place in a CW receiver. In fact, many of the best CW superheterodynes utilize intentionally-poor audio fidelity by means of a peaked audio filter which passes the audio-beat note being received, and suppresses all others.

Automatic volume control belongs in a receiver for phone use, as does good fidelity in the audio channel. A receiver for phone generally has more audio amplification than a receiver for strict CW reception, in order to satisfactorily drive a loudspeaker. This is because the majority of phone operators prefer loud-speaker reception, while most of the CW men prefer the use of headphones. A receiver designed exclusively for phone reception probably would not require a beat-frequency oscillator, while in a CW receiver it is an essential device in order to produce an audible beat tone in the headphones.

A superheterodyne designed for phone use need not have the extreme selectivity required for CW as the modulation sidebands as well as the carrier coming from the transmitter must be passed through the receiver for detection. Thus, the conventional type of "series crystal filter" (explained later) is undesirable in a "phone" receiver on account of its extreme selectivity impairing the intelligibility of the received voice signal.

Circuit Pre-selection

The question of pre-selection arises in the design of both CW and phone superheterodynes. Pre-selection ahead of the first detector minimizes "image interference." An explanation of image interference requires a brief outline of how a superheterodyne operates.

In superheterodynes, it is important to note that instead of tuning the major receiver elements to the incoming signal, it remains fixed on one frequency and the received signal is then changed in frequency to the frequency of the intermediate amplifier, which is the real heart of the superheterodyne. This portion of the receiver provides 90 per cent of the selectivity and amplification achieved. The undesired image response is a characteristic of the frequency-changer in the front end, which consists of the first detector and high-frequency oscillator. An incoming signal from

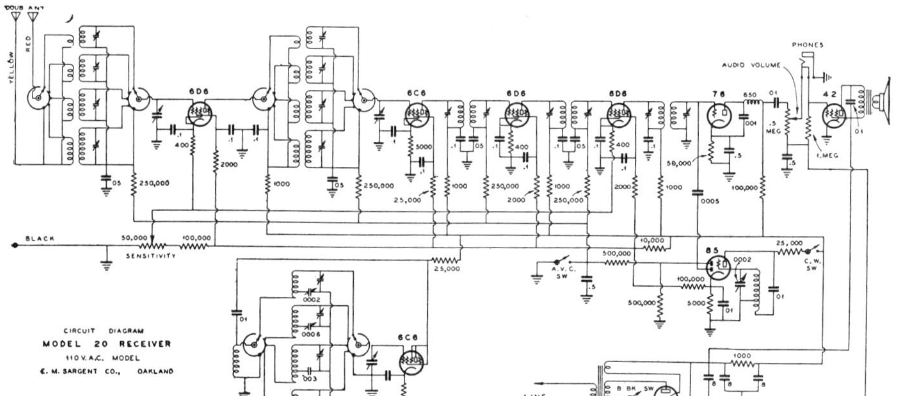

Typical All-Wave Superheterodyne with Coil Switching.

the antenna is applied to the first detector or mixing tube, then a second signal, locally generated by a high-frequency oscillator, is likewise applied to the mixing tube. The presence of the two signals combine in the tube and cause the generation of sum and difference beat notes to appear in the mixing tube plate circuit. For example: Suppose the signal coming from the antenna is exactly 7,000 kilocycles, and the signal coming from the local oscillator is 7,460 KC. In the plate circuit of the mixing tube there will be, therefore, the sum and difference of these two frequencies, namely 14,460 KC and 460 KC. It is the 460 KC frequency that is wanted in this particular case, on account of the intermediate frequency amplifier being tuned to this frequency. The sum frequency (14,460 KC) would be bypassed to ground in the first intermediate amplifier transformer, while the difference frequency is the one usually chosen for amplification.

While the desired signal was 7,000 KC, and the local oscillator frequency was 7,460 KC, it will be seen that if there is a signal of 7920 KC present in the antenna and the first detector circuits, this 7920 KC frequency will also "heterodyne" or "beat" with the local oscillator frequency to produce a difference frequency of 460 KC. Because one of 460 KC signal is just like any other 460 KC signal, the intermediate frequency amplifier has no way of rejecting the undesired beat produced by the 7920 KC interfering signal. It is this interfering signal that has been termed the "image," and the frequency of the image signal is almost always two times the intermediate amplifier frequency higher in frequency than the signal which the operator is trying to receive. Therefore, the only method by which the image response can be minimized is to provide enough tuned circuits, or selectivity, AHEAD of the first detector in order to pre-select the desired signal and at the same time to reject the image.

Image interference is not always present. It only occurs when there is a powerful transmitter in operation on a frequency twice the intermediate frequency away from the desired signal being received. Because the intermediate frequencies chosen in most amateur work are in the neighborhood of 450 KC, the image interference is largely from stations approximately 900 KC higher in frequency than the signal being received. This means that the image cannot be produced by other amateur stations, because none of the commonly-used amateur bands are 900 KC wide. Thus the interference most often heard originates from either commercial or government stations. A selective pre-selector interposed between the antenna and first detector will eliminate, or at least minimize, this form of interference.

The Super-Gainer

This three tube superheterodyne circuit has a regenerative first and second detector, no intermediate-frequency stage, and is selective and sensitive; it answers the problem which has long confronted the experimenter of limited means. It does not "block" on strong, undesired signals.

Technical Details: By employing detector regeneration at two frequencies, three tubes do the work of six, as shown in the unique circuit accompanying this description. On account of regeneration, a separate '76 tube oscillator is necessary to prevent interlock or reaction between the first detector and oscillator. The front-end of this receiver is smiliar to the "222" described elsewhere in this section.

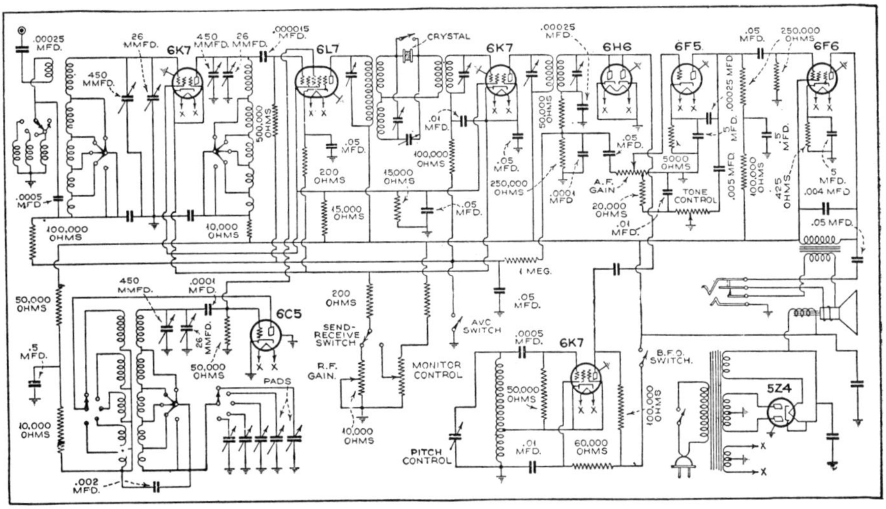

All-Wave Metal Tube Superheterodyne, Hallicrafter's Super-Skyrider. Illustration courtesy "Radio News."

Professional construction characterizes this model of the Jones "Super-Gainer." Note the short, direct leads. The antenna is coupled to the grid by twisting a few turns of the antenna lead around the grid lead.

The second detector, a '79 twin-triode, is the most important component in this new receiver. The tube functions as a regenerative second detector, beat-frequency oscillator, and as an additional stage of audio amplification. Regeneration in the second detector, even when oscillating for CW reception, eliminates the need of an IF stage. By the same token, a separate BFO tube is eliminated. The second triode only functions as a stage of resistance coupled audio amplification.

Cathode regeneration is used in the first section of the 79 tube. The cathode coil consists of an old BCL receiver coil of about 90 turns of No. 30 wire, wound on a 1¼-inch diameter form. The regeneration is controlled by means of a tapered

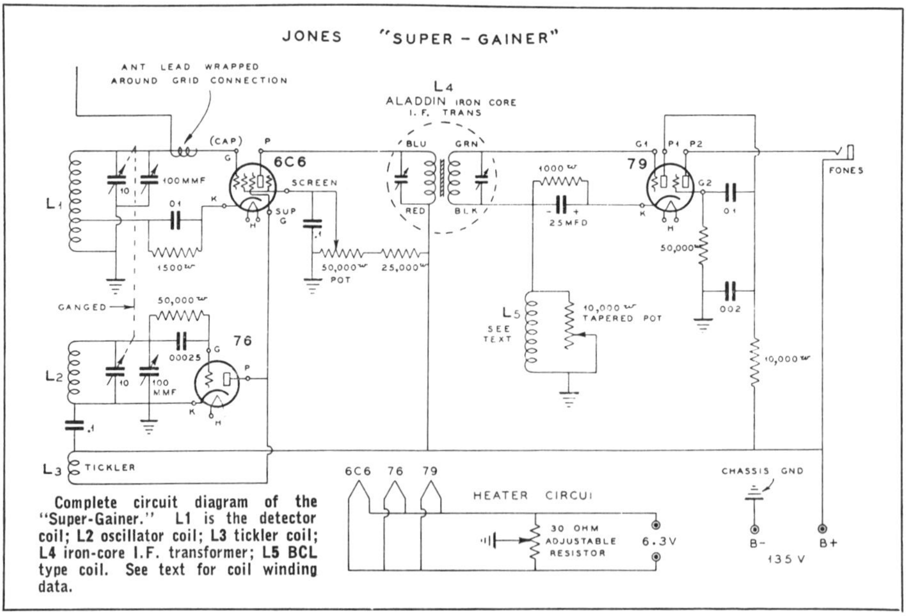

Complete circuit diagram of the "Super-Gainer." L1 is the detector coil; L2 oscillator coil; L3 tickler coil; L4 iron-core I.F. transformer; L5 BCL type coil. See text for coil winding data.

Front view of the Jones "Super-Gainer."

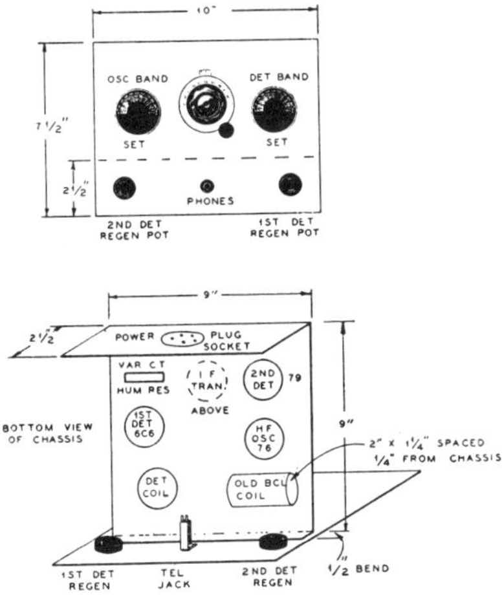

Front panel view and under-chassis layout of alternate design using standard front panel and "U"-bend chassis.

10,000 ohm variable resistor shunted across the BCL coil. This latter component is not directly a part of the 456 KC tuned circuit, and therefore no trouble is encountered from a detuning effect on CW for various settings of the regeneration or oscillation control. A 1000 ohm control may give smoother control.

A single Aladdin iron-core IF transformer (465 KC) provides sufficient selectivity for this receiver. This unit has a screw adjustment on the side of the shield-can which varies the coupling between the two tuned coils. When the second detector is made to regenerate it is necessary that very loose coupling between the circuits be maintained. For this reason only such types of IF transformers should be used which will allow adjustment of coupling.

The main tuning is accomplished by means of a two-gang double-spaced condenser, originally having 35 mmfd. max. capacity per section. To prevent interlock effect on 20 meters, an aluminum shield is placed around the oscillator section of the condenser. By removing one stator plate from each of the inside ends of the stators, space is made available for the ground shield. The oscillator section of the condenser also has its front plate removed; thus, this section has 7 dielectric spaces between rotor and stator, while the detector has eight spaces. The detector band-setting condenser is adjusted for maximum signal or noise pick-up by advancing the first detector regeneration control; that is, increasing the screen-grid voltage. The cathode-tap on the first detector coil allows regeneration at the signal frequency; variation of screen-voltage provides a convenient adjustment of regeneration. The tube should never be permitted to oscillate; otherwise it will bring in undesired stations which will differ in frequency from the desired station by the value of the intermediate frequency.

The antenna is capacitively coupled to the grid of the 6C6 by twisting a few turns of the lead-in wire around the grid lead of the first detector. If the antenna is inductively coupled to the receiver, too much coupling, as when using a resonant antenna, will prevent sufficient regeneration.

Receiver Adjustments: The second detector must oscillate when its regeneration control is adjusted. The IF transformer tuning can then be adjusted to resonance with the secondary by noting the spot at which it tends to pull this detector out of oscillation.

After the second detector is operating properly, the 76 oscillator can be aligned on some strong signal, or by a calibrated modulated oscillator. The first detector

RECEIVER COIL DATA

All in 1½" Diameter Forms

Wavelength	L_1	L_2	L_3
160 Meters	1¾" winding of #24E. Tapped at 1½ turns. Close wound.	1¼" winding of #24E. Close wound. Grid on top end.	12t #24E. Close wound ⅛" from L2. Same direction as L2 with plate on far end.
80 Meters	40t #20 DSC, spaced to cover 1¾". Tap at ¾ turn.	33t #20DSC, spaced to cover 1¾".	8t #24E. Close wound 1/16" from L2.
40 Meters	12t #20DSC, spaced to cover 1½". Tap at ½ turn.	11t #20DSC, spaced to cover 1¼".	5t #24E, spaced ¼" from L2.
20 Meters	5t #20DSC, spaced to cover ⅞". Tap at ½ turn.	5t #20DSC, spaced to cover ⅞".	2½t #20DSC, spaced ¼" from L2.
10 Meters	3½t #20DSC, spaced to cover 1". Tap at ⅓ turn.	3½t #20DSC, spaced to cover 1".	2½t #20 DSC ¾" from L2, and 1/16" between turns.

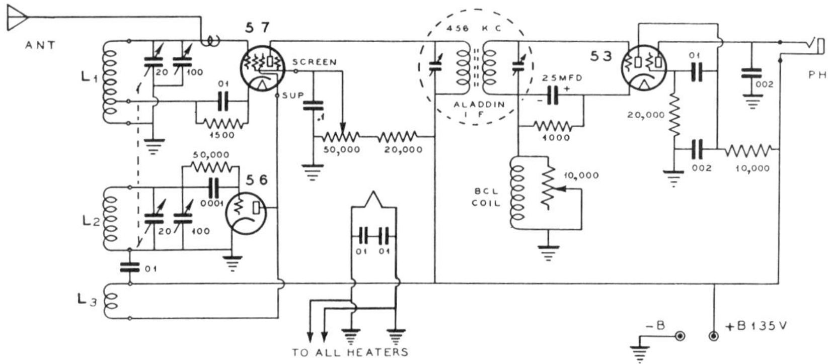

3-tube "Super-Gainer" with 2.5 volt heater tubes.

control must not be advanced to the point of actual oscillation. The antenna coupling can be adjusted so that it will allow the first detector to actually oscillate. All tests can be made by listening with a headset plugged into the telephone jack. The audio volume is not sufficient for operating a loudspeaker.

IMPORTANT DATA:

When more than 135 volts plate supply is used, the H-F oscillator voltage must be reduced by means of a 25,000 or 50,000 ohm, 1 watt resistor, then by-passed to ground with a 0.1 mfd. condenser. The value of the second detector cathode resistor should be reduced to approximately 250 ohms. Smoother second detector regeneration can be obtained by using either a 400 ohm or 1,000 ohm variable wire-wound resistor instead of the 10,000 ohm resistor across the BCL coil. Sometimes a few turns must be added to the BCL coil when a lower value of variable resistor is used.

Two - Tube Super - Gainer: Multi-purpose tubes are used in this receiver producing results comparable to 6- or 7-tube superheterodynes. The inherent selectivity of this set is greater than that of a tuned RF receiver and the sensitivity is comparative.

Technical Considerations: A 6F7 dual-purpose tube serves as a regenerative first detector and separate oscillator. A 6A6 double triode performs the functions of regenerative second detector, beat-oscillator and audio amplifier. The receiver sensitivity is apparently higher than the three-tube super-gainer, but has a slight interlock effect which is encountered on 10 and 20 meters. This effect is practically unnoticeable after the two band-setting 100-mmfd. condensers have been properly adjusted for any given band. Turning over any portion of the communication spectrum between 10 and 160 meters is accomplished by means of a two-gang 20-mmfd. condenser.

Selectivity is obtained from regeneration in the iron-core intermediate-frequency transformer. In general, the circuit is a simplified superheterodyne. The triode portion of the 6F7 is the H.F. oscillator, tuned to about 456KC higher in frequency than

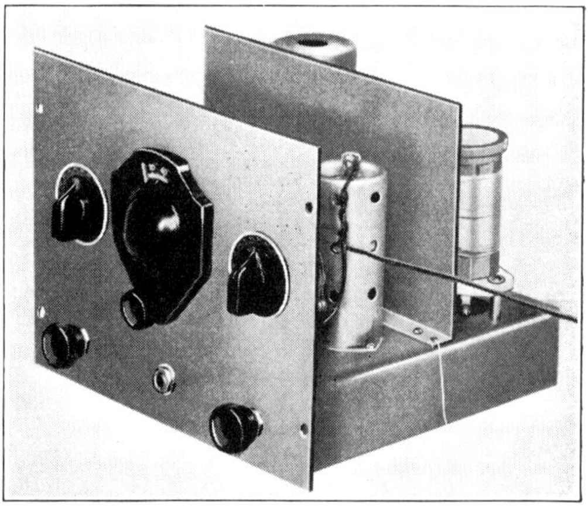

Front view of the 2-tube "Super-Gainer," showing shield partition and antenna "condenser" (twisted lead around grid connection).

the first detector input. The pentode portion of the 6F7 is a regenerative first detector with cathode-tap for regeneration and H.F. oscillator coupling. Screen-grid voltage variation serves for both volume and regeneration control.

The I.F. transformer coupling is set to a value which will allow regeneration and oscillation within the range of the tapered variable resistor control. This control shunts the 6A6 cathode-coil which consists of 100 turns of No. 32 DSC wire "jumble-wound" on a ½-in. diameter rod. The second detector is by-passed with a .004 mfd. by-pass condenser to ground while the grid and cathode are above ground poten-

2-Tube "Super-Gainer" Layout

2-tube "Super-Gainer" Layout, 6A6 tube shield removed.

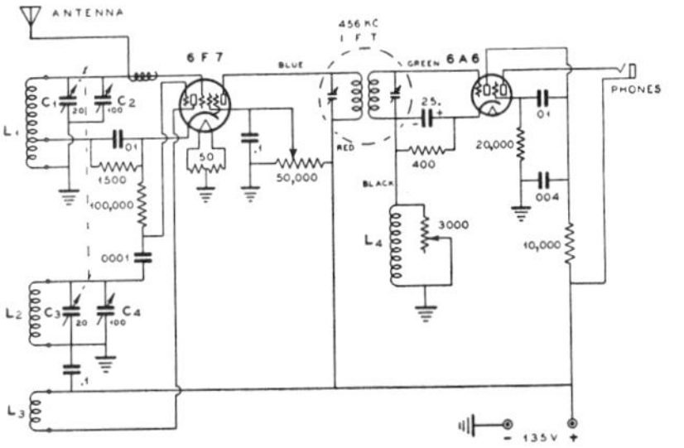

The circuit diagram. See table on page 42 for coil winding data.

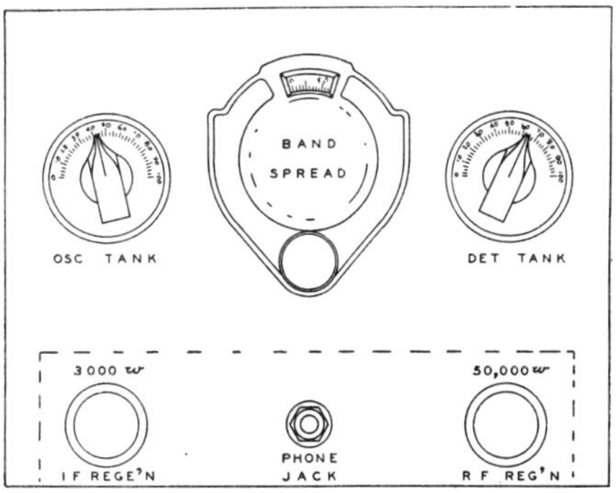

FRONT PANEL LAYOUT FOR 2-TUBE SUPER-GAINER
METAL PANEL 9" WIDE, 7" HIGH

The front panel is 9" wide, aluminum or steel.

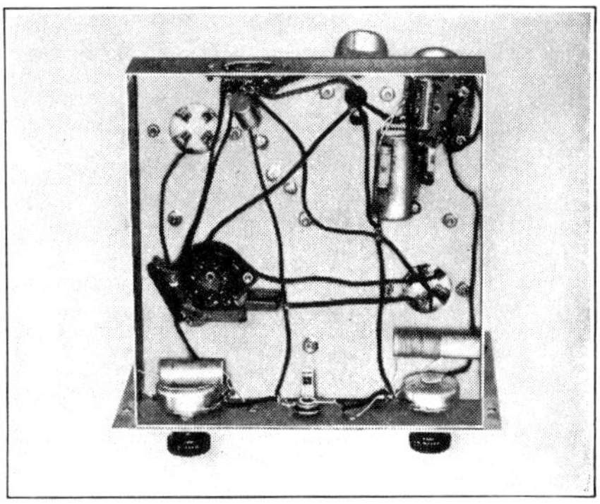

Under-chassis view, showing the BCL coil, L4.

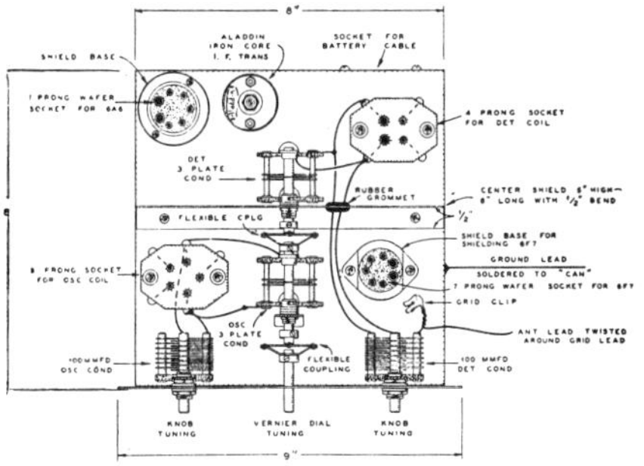

Pictorial arrangement for correct parts placement.

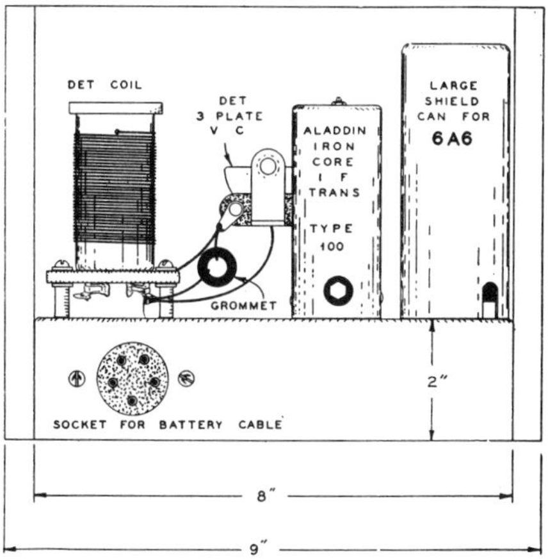

Rear view showing shield can for 6A6 tube, iron-core I. F. transformer, detector coil and detector condenser.

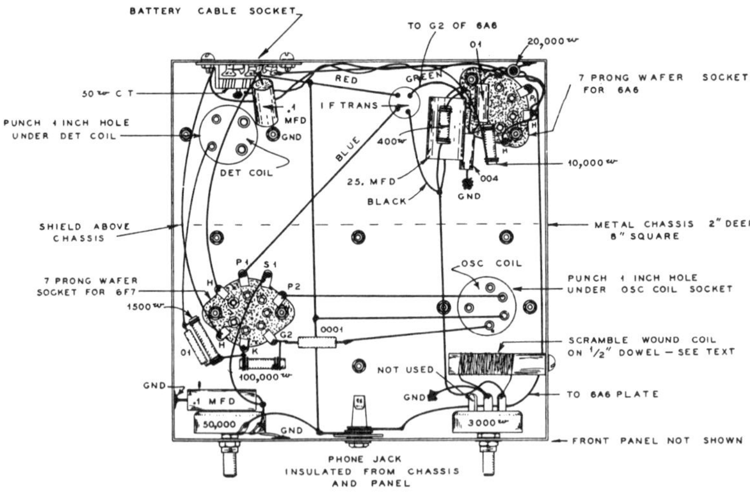

Construction of the 2-tube "Super-Gainer" is simple, if this under-chassis pictorial wiring plan is carefully followed.

tial for RF, or rather I.F. This forms a regenerative or oscillating circuit controlled by the 3000-ohm variable resistor. The value of the tapered resistor may have a maximum as high as 5000 or 10,000 ohms; control, however, taking place in the region between 0 and 2000 ohms.

The 400-ohm cathode-resistor must be by-passed with a large low-voltage, electrolytic condenser in order to prevent degenerative amplification (motor-boating). The detector is resistively coupled into the audio amplifier part of the 6A6 by low ohmic resistors.

Antenna coupling is varied by twisting more or less insulated hook-up wire around the 6F7 detector grid-lead until smooth regeneration is obtained up to the point of oscillation. Note: A modulated test oscillator will simplify all preliminary adjustments.

The chassis is about 8 x 8 x 1¾th inches with a front panel 8 x 7 inches. A shield 5 inches high separates the first detector and the H.F. oscillator coils and tuning condensers. The latter are ganged by means of a flexible shaft coupling, and tuned by a vernier dial. The two 100-mmfd. band-setting condensers should be controlled from the front panel in order to accurately resonate the detector circuit when using regeneration. The coil turns may be compressed or expanded before cementing in place, so as to obtain circuit tracking across each amateur band. Both tubes should be shielded.

2 TUBE SUPER-GAINER COIL DATA

All Coils Wound on 1½" Diameter Forms

Wavelength	L_1 Detector	L_2 Oscillator	L_3 Tickler
160 Meters	1¾" of #24 E. Tapped at 4 turns. Closewound.	1¼" of #24 E. Closewound. Grid on top end.	20t #24 E. Closewound ⅛" from L2. Same direction as L2 with plate on far end.
80 Meters	40t #20 DSC., Spaced to cover 1¾". Tap at 2 turns.	33t #20 DSC., Spaced to cover 1¾".	10t #28 DSC. Closewound 1/16" from L2.
40 Meters	12t #20 DSC., Spaced to cover 1½". Tap at 1½ turn.	11t #20 DSC., Spaced to cover 1¼".	7t #24 E. Spaced ⅛" from L2.
20 Meters	7t #20 DSC., Spaced to cover 1⅛". Tapped at one turn.	7t #20 DSC., Spaced to cover 1⅛".	4t #20 DSC., Spaced ⅛" from L2.
10 Meters	3½t #20 DSC., Spaced to cover 1". Tap at ⅓ turn.	3½t #20 DSC., Spaced to cover 1".	3t #20 DSC., ⅛" from L2 and 1/16" between turns.

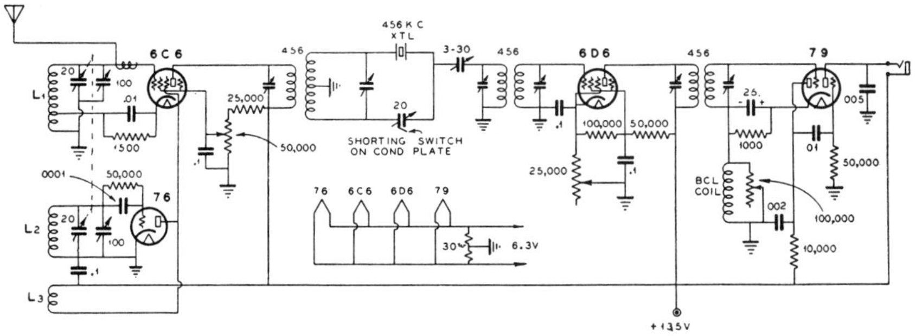

Experimental "Super-Gainer" with Crystal Filter.

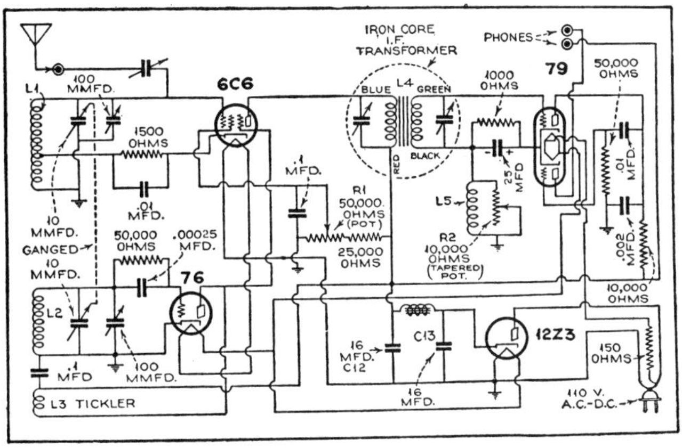

A-C operated "Super-Gainer" with 12Z3 rectifier tube in the power supply. This is a diagram of the McMurdo-Silver factory-built "Super-Gainer."

Metal Tube Super-Gainer: This receiver has four of the new metal tubes in the Super-Gainer circuit, the characteristics of which are similar to the receiver previously described except that with the inclusion of the 6L7, special mixer tube, the receiver has a higher degree of sensitivity. The 6L7 tube has a higher plate impedance as a first detector so that I.F. gain is as high with a small Aladdin iron-core I.F. unit as with a larger unit and 6C6 tube. The 6L7 also makes a very effective regenerative first detector with variable screen-voltage control. A cathode-tap on the detector grid coil serves as a means of obtaining regeneration at the signal frequency.

Miscellaneous Notes: Second detector regeneration and oscillation is controlled by a 5000-ohm tapered variable resistor shunted across a cathode coil. The latter is made of 100 turns of No. 26 or No. 28 DSC wire "scramble-wound" on a short section of ⅝th-inch diameter dowel rod. There is no magnetic coupling between this coil and the second detector grid coil. A 6F5 high-mu tube functions as the detector of the grid-leak or bias type. Grid-leak detection is shown, but generally cathode-bias detection will allow the circuit to regenerate smoother.

A 6C5 tube similar to a 76 serves as an audio amplifier, resistance-coupled to the detector circuit. Another 6C5 tube functions as a H.F. oscillator with cathode-tap for oscillation. The grid-leak and condenser bias this tube as well as the special injection grid of the 6L7 tube.

The set is assembled on a 7 x 7 x 1¾-inch metal chassis with a small shield placed between the coils and ganged-condenser sections. The sections are made from 35-35 mmfd. midget condenser having only four stator plates per section (the others being removed). The 100 mmfd. condensers are band-setting controls which are manipulated by small dials on the front panel, the latter is of aluminum 7 x 8 inches 12-gauge. The vernier dial is insulated from the tuning condenser shaft in order to eliminate multiple ground leads and resulting noise when tuning. A power-plug and socket are mounted at the rear of the chassis for connection to a 6.3 volt filament transformer and 135-volts of B-battery. or to similar values of voltage from

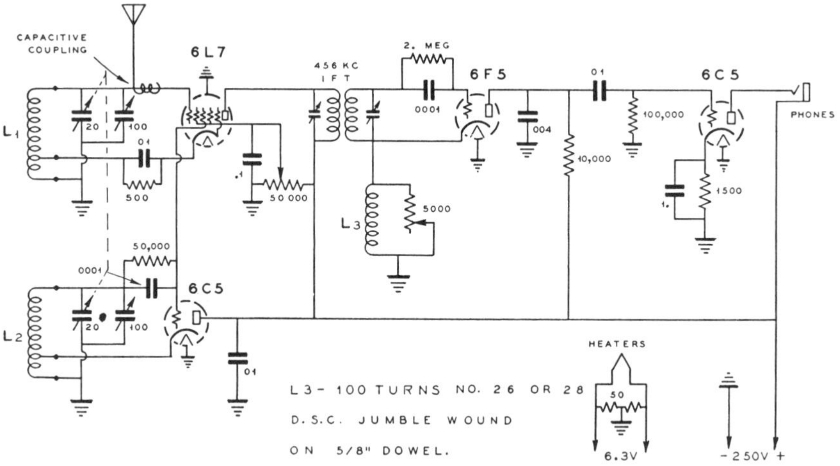

METAL TUBE SUPER-GAINER

Looking into the Metal-Tube "Super-Gainer."

The airplane tuning dial adds beauty and convenience.

an AC power supply. With a power-pack, the DC voltage should not be over 180-volts and an 8mfd. condenser must be connected across the voltage divider at this point.

The coils are similar to those listed under the three tube Super-Gainer except that no tickler is needed on the oscillator coils. The cathode-tap in this case is from ¼th to ⅓rd of the total turns up from the grounded end of each oscillator coil. The antenna coupling should be semi-variable because of the effects of antenna resonance on the first detector regeneration.

METAL TUBE SUPER-GAINER COIL TABLE

All Coils Wound on 1½" Diameter Forms

Wavelength	Detector Coil	Oscillator Coil
160 Meters	1¾" of #24 E., closewound. Tap at 1¼ turns.	1¼" of #24 E., closewound. Tap at 1/3 of total turns.
80 Meters	38t #22 DSC., 1¾" long. Tap at ¾ turn.	32t 22 DSC., 1¾" long. Tap at 10 turns.
40 Meters	12t #22 DSC., 1½" long. Tap at ½ turn.	11t #22 DSC., 1¼" long. Tap at 3½ turns.
20 Meters	6t #22 DSC., 1" long. Tap at ¼ turn.	6t #22 DSC., 1" long. Tap at 1½ turns.
10 Meters	3½t #22 DSC., 1" long. Tap at ¼ turn.	3½t #22 DSC., 1" long. Tap at 1 turn.

Amateur Superheterodyne Receivers The "222" Radio Series

A splendid ultra-sensitive amateur communications receiver featuring the superheterodyne principle together with many engineering refinements is given herewith. The receiver will cover both the 20 and 40 meter wave bands without coil changing; for 80 meter operation a separate set of coils are required.

In describing the circuit complement, reference should be made to the circuit diagram from which the more salient points can be taken into consideration.

which would cause steady beat-note whistles in certain band-settings in the short-wave range. The oscillator strength is adjusted by simply twisting the wire-coupling capacity to the second detector. This type of coupling allows maximum signal to BFO noise ratio. The high value given to the plate and screen resistors limit the harmonic output, in addition, simplifies the shielding problem for the BFO.

IF Amplifier: The IF amplifier has only one stage, as two stages complicate the set and tends to increase the noise to signal ratio. With one high-gain IF stage operating in the neighborhood of 500 KC, no iso-

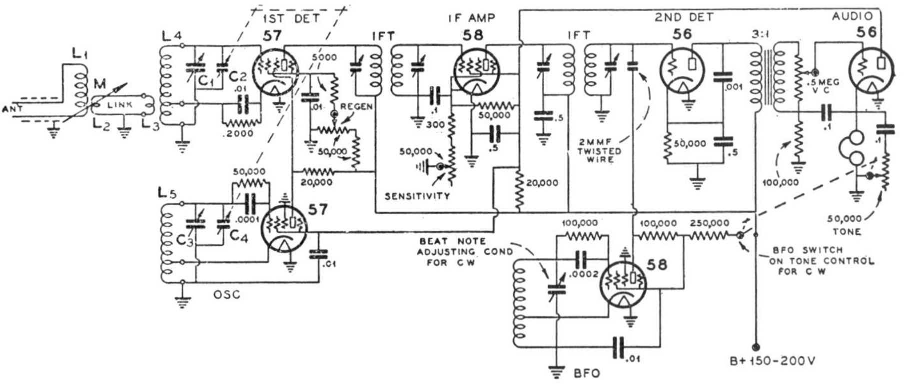

6-Tube Jones "222" Superheterodyne, ideal for amateur operation.

Antenna and Coupling: Regeneration is used and a variable antenna coupling allows maximum effect from the regeneration. The antenna coupling is the same as shown for the "pre-selector" on page 50. The antenna and first detector coils are connected by link coupling; one of the link coils sliding backward or forward to vary the degree of coupling. The advantage of link coupling minimizes capacity coupling to the antenna without using a Faraday electro-static screen, and at the same time minimizes man-made static.

First Detector: Note that the regenerative effect is obtained by means of a cathode tap on the detector coil which gives a more uniform effect to the regeneration for certain sets of coils. In addition, the detector conversion gain is increased many fold due to regeneration and to the method of oscillator coupling. A careful study of the circuit will show that the suppressor-grid is connected directly to the plate of the oscillator; this connection practically eliminates oscillator radiation into the antenna due to the screen-grid being by-passed to ground which electrostatically shields the suppressor-grid from the control-grid circuit. The positive potential placed on the suppressor-grid augments the sensitivity of the first detector.

Electron-coupled Oscillators: The first oscillator is made to oscillate strongly for good conversion gain, while the second oscillates weakly to minimize harmonics

lating condensers and resistors are needed in the plate, screen-grid and cathode circuits. Flexibility of control is provided by an IF and volume control, each operating independently of the other.

Detector and Audio Circuits: The detector circuit is conventional, while the audio amplifier has an interesting modification which utilizes the telephone headset as a bias resistor for the tube with the tone control across the phones. This connection allows the telephone jack to be grounded to the aluminum chassis or panel. The grid circuit is confined to the grid and cathode by means of a 1 megohm resistor and a 0.1 mfd. by-pass from the audio transformer to the cathode. This scheme prevents audio degeneration and the loss of signal; the output, therefore, is the same as if the cathode resistor and a large by-pass condenser were used and the headset placed in the plate circuit.

Power Supply: The power supply is isolated to keep stray capacity, hum and other sources of spurious noises at a distance. If "A" and "B" batteries supply the necessary power, it will be necessary to provide some means of cutting off both A and B leads by a switch when disconnecting the power supply from the receiver.

CONSTRUCTION: In the original design, a pair of Aladdin iron-core IF transformers were used as they had better selectivity and higher gain than ordinary air-core IF transformers. If these transformers are

Front view of the "222" Super.

Looking into the "222."

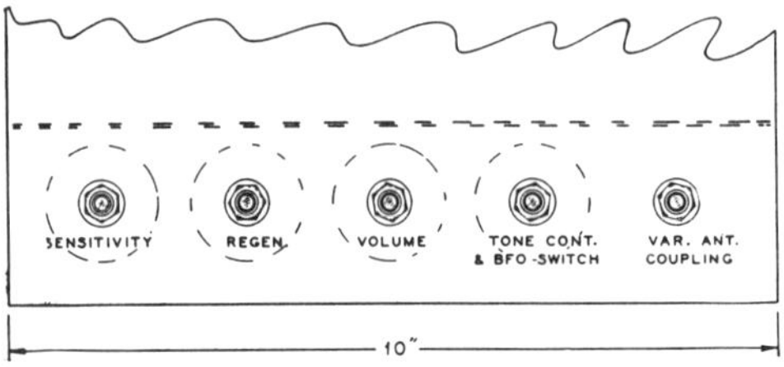

Front Panel Control Arrangement for the "222."

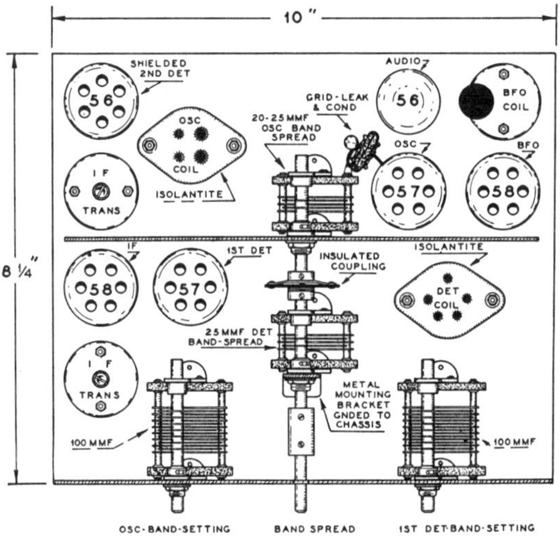

The constructor is advised to use the exact layout of parts, as shown above. All tubes are shielded, other than the type 56 audio tube. Isolantite sockets should be used for the Detector and Oscillator coils.

not available, air-core transformers may be substituted with entire satisfaction. In most all IF units, the coupling has been adjusted at the factory for best broadcast reception gain and band-width. This is generally to close for best short-wave practice where greatest selectivity and good gain are desirable. The two coils should be at least 1¼ inches apart for most all air-core types. Some makes can be adjusted by warming the supporting tube with a soldering iron tip until the wax softens, then sliding the coils apart. The iron-core transformers have a pair of coils mounted at right angles to each other on short molded straight cores. Coupling is adjusted by a screw adjustment on the lower coil which slowly moves it along its axis.

As previously stated, a single stage of IF will give ample gain if the front-end of a "super" is functioning properly. A stage of RF ahead of the first detector is sometimes desirable, but it does not compare with a "super" having a regenerative first detector unless regeneration is used in the RF stage.

The oscillator tuning condenser consists of a double-spaced midget condenser of eight plates, while the detector condenser has nine plates double-spaced. These condensers are made from "Cardwell 100 mmfd. Trim-Air" normally spaced midget condensers, similar to those used for band-setting. By winding the oscillator coil to cover a greater winding space of 1¼ inches as against 1⅛ inches for the detector coil, the oscillator and detector will track throughout the narrow amateur bands. With the number of plates left in these double-spaced condensers, the 20 meter band covers approximately 15 divisions on the airplane type of dial, and the 40-meter band about 60. Greater band-spread may be had by removing plates from each of these condensers. A flexible coupling **must** be used to gang the oscillator condenser to the front detector condenser to eliminate torsion detuning effects on the beat-note of a CW station. This effect always occurs with all types of dial and condenser mountings.

A pair of shielded lead-in wires are connected directly from the antenna system to the fixed antenna coil underneath the chassis. (See photo of under-chassis view.) The antenna coil consists of 12 turns of No. 24 DSC wire closely wound on a 1¼ inch diameter bakelite tube, approximately 1¼ inches long. The sliding coil is made by closely winding 4 turns of No. 24 DSC on a 1 inch diameter tube. Flexible leads form the remainder of the link coupling device to the isolantite coil-socket above the chassis. Four turns of this same wire were wound on the detector coil about one-eighth inch from the ground end. This 1 inch bakelite tube is controlled from the front panel by means of a plunger action knob over a distance of approximately 1 inch. The knob is fitted with a ¼ inch diameter brass rod extending through the front panel and fastened to the one-inch tubing with two machine screws. The bearing, retaining and pressure spring is simplicity itself, being an ordinary telephone jack. The rear tip connection acts as a pressure spring against the brass rod, making it remain in whatever position it is adjusted to by merely manipulating the knob.

The antenna coupling device allows adjustment of the resonant antenna coupling to obtain optimum value of first detector regeneration. This scheme is applicable to any type of antenna system, the latter being externally adjusted or tuned to resonance until the optimum coupling is found. The results are very gratifying. The image interference on 40 meters measures 60 DB units down in level from the desired signal, using a signal generator for these measurements. 60DB means an image rejectivity of 1000-to-1 which is extremely good for sets using a well designed stage of RF. The image measures 50 DB down on 20 meters, which is more than most superheterodyne receivers can even approach at that wave length. The receiver has practically no image whistles of "phantom" commercial signals in the amateur bands, unless the commercial signal is of very high field intensity. The signal generator gives an audible signal in the headset with an input of 130 DB down from 1 volt, which is less than 1 micro-volt input. This is ample sensitivity, with low internal noise level, to reach down into the atmospheric noise level in any locality.

The receiver is built into a metal cabinet measuring 8½ inches deep, 7 inches high, and 11 inches long. The front panel is 7x11 inches and is made of No. 12 gauge aluminum. The chassis is also made from the same gauge aluminum, bent in the form of a U, two inches deep and 8¼ inches wide by 10 inches long. All of the necessary tube socket and dial holes can be punched, or cut out with a circle cutter and drill press. The shield partition between the oscillator and first detector is also made from No. 12 gauge aluminum, 7 inches long, 4¾ inches high with a ½ inch lip along the bottom for fastening to the chassis with three machine screws.

In building this set, it is a good plan to take all the largest parts and set them on the chassis so as to get the proper chassis

COIL WINDING TABLE FOR "222" COMMUNICATIONS RECEIVER

Coils L1, L2 and L3 are the same for 20, 40 and 80 meter operation. L1—12 turns No. 24 DSC wire, close wound, on 1¼-in. dia. tubing.

L2—4 turns No. 24 DSC wire, close wound, on 1-in. dia. tubing. This coil slides into coil L1; the coupling is made variable by sliding L2 into and out of L1.

L3—4 turns, No. 24 DSC wire, wound on 1½-in. dia. tubing, separated ⅛ in. from L4.

For 20 and 40 meters: (same coils used for both bands).

L4—11 turns, No. 18 DCC wire, space-wound on 1½-in. dia. tubing, to cover a winding space of 1⅛ in. long, and tapped at one and one-third turns from bottom.

L5—11 turns, No. 18 DCC wire, space wound on 1½-in. dia. tubing, to cover a winding space of 1¼ inches, and tapped at 2½ turns from bottom.

C1-C3—100 uufd. midget variable condenser.

C2—9-plate double-spaced midget condenser to give approx. 25 uufd.

C4—7-plate double-spaced midget condenser to give approx. 20 uufd.

(Use 8 plates for C2 and 6 plates for C4 if more bandspread is desired.)

Condensers C2 and C4 are standard Cardwell 100 uufd. "Trim-Air" midgets, with alternate plates removed so as to double-space the plates.

L1, L2, L3 same as for 20 and 40 meter operation.

L4—30 turns, No. 24 DSC wire, wound to cover a space of 1½ in. on a 1½-in. dia. form, with cathode tap taken at one turn from bottom.

L5—26 turns, No. 24 DSC wire, wound to cover a space of 1⅜ in. on a 1½-in. dia. form, with cathode tap taken at 4¼ turns from bottom.

NOTE—The cathode tap on the oscillator coil must not be too high, otherwise image interference will become serious.

TUBES—Instead of using type 56, 57 and 58 tubes, this receiver will give equal satisfaction if the types 6C6, 6D6 and 76 are used for 6.3 volt operation.

160-METER BAND—This receiver will not operate successfully on the 160-meter band unless large variable condensers are used in place of the small midgets. The receiver was primarily designed for 20, 40 and 80 meter operation.

CONDENSER SETTINGS

Band	Oscillator Band-Setting Condenser	Detector Band-Setting Condenser	Coverage on Main Tuning Dial
20 Meters	8°	10°	12° to 15°
40 Meters	80°	95°	50° to 60°
75 Meter Phone Band	45°	50°	25°
80 Meter C.W. Band	50°	55°	100°

layout before drilling. The accompanying pictures and the plan drawing should enable anyone to duplicate the design without trouble. The lower knobs on the front panels, from left to right, are sensitivity, regeneration, audio volume, tone control and BFO switch combination, and antenna coupling. The upper row: Oscillator band-setting adjustment with knob and small 0-100 metal escutcheon plate, main tuning control, and last, the detector band-setting control and a 0-100 division plate. The antenna leads, power-pack cable plug, and telephone jack are at the rear of the chassis with large holes drilled around them through the metal cabinet. The cabinet has a metal hinged lid.

The "222" Receiver with Improved Crystal-Filter and BFO

This receiver is exceptionally sensitive and selective, and is capable of remarkable signal-to-noise ratio. The receiver incorporates a new crystal-filter arrangement. One of the features of the new unit are that no loss in sensitivity occurs when switching from the "off" to the "series" position, because the impedances remain matched whether the crystal is "in" or "out." Another interesting feature of the improved crystal-filter is that it can be put into any existing superheterodyne receiver without disturbing the IF amplifier in any way, except to disconnect the detector plate leads.

Circuit Data: The circuit of this receiver is similar in many respects to the "222" receiver previously described. The circuit modifications in this design are new, and have only been incorporated after having proved meritable in laboratory and field tests.

The receiver consists of a stage of semi-tuned RF using plug-in resonant chokes, a regenerative first detector, a single stage of IF, second detector, audio and BFO. The HF oscillator, detector and RF are exactly the same as the original "222" with only minor deviations. Here, it was found that tuning condensers using bakelite instead of isolantite insulation require about ¼ more coil turn in the first detector cathode-tap. It was also found desirable to use a separate set of coils for 20 meters to obtain more bandspread. Five turns one-inch long on the 1½ inch diameter plug-in coils, are satisfactory for this band. All other coil data are given in the coil table for the "222" receiver with RF (see preceding descriptive articles).

The plate circuit of the first detector must be tuned for maximum signal gain so that the plate tuning condenser acts as an effective RF by-pass to increase detector efficiency. The center-tapped coil and neutralizing or phasing condenser form a Wheatstone bridge to balance out the crystal holder capacity. At resonance the succeeding tuned IF circuit would be over-coupled to the first detector tuned-plate circuit, because effectively there is only a resistance of a few thousand ohms between the "live" ends of the tuned circuits. To prevent any bad effects caused by over-coupling, a small 3-30 mmfd. condenser is placed in series with the crystal. This allows the use of tuned circuits between the crystal and first IF amplifier grid without loss in signal. By this matching device there is no appreciable loss in the crystal filter when it is cut into the circuit. The noise level decreases because of the very narrow band passed through the IF amplifier.

With an efficient circuit of this type, only one stage of high-gain IF is necessary. In general, superheterodyne circuits should have high gain in the front end, but should not depend too much upon the IF amplifier for gain. The main function of the IF amplifier is to increase selectivity.

Crystal Filter: The crystal filter is made by removing the center universal wound coil of a Hammarlund 2.1 mh. RF choke, thereby providing a center-tapped plate coil which is tuned by a 7-70 mmfd. trimmer condenser. A 25 mmfd. variable condenser is employed for phasing; the value of the condenser depends upon the plate-to-plate capacity of the crystal holder. The condenser is mounted on the front panel with insulating bushings; by resorting to plugging, the crystal may be placed "in" or "out" of the circuit. The stator plate of the phasing condenser is bent to cause a short-circuit in

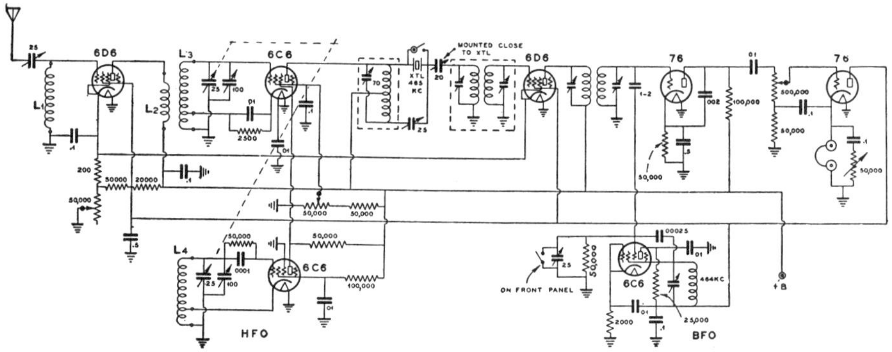

"222" Superheterodyne with Jones Crystal Filter and Improved B.F.O.

the condenser at minimum capacity setting for phone operation. The idea may also be included to turn the BFO "on" or "off" for CW or "phone" reception.

Beat-Frequency Oscillator: The oscillator is of the relaxation type. The advantages are in simplicity, since no tickler or cathode-tap are necessary in the tuned circuit; in addition, the circuit is highly stable, and the harmonic content is less than in an electron-coupled circuit. Unless the oscillator is **completely** shielded, harmonics will be heard in the form of steady carrier signals at various points throughout the short-wave spectrum.

The function of the circuit depends upon feedback in phase to the suppressor-grid through condenser C_4 of the BFO circuit diagram. The screen is more positive than the plate. The plate voltage is adjusted to approximately +22½ volts, the screen from +75 to +100, the usual control grid at zero potential, and the suppressor-grid at about 6 to 10 volts negative with respect to the cathode. The various potentials are reduced to the proper value by means of resistors.

The BFO coil L_1 and condenser C_1 must tune to the IF; the combination can be made from an old IF coil unit by simply removing coil turns until it resonates at the desired frequency by manipulating the shunt condenser and the trimmer condenser mounted on the front panel. As an alternative the combination can be made from a "jumble wound" coil with a fixed .001 mfd. and a semi-variable 70 mmfd. condenser. Front panel control of the BFO frequency can be obtained by C_2 which acts as a vernier adjustment for C_1. On account of the rotor plates C_1 being grounded, the condenser can be mounted on the metal front panel. Output from the BFO is taken from the suppressor-grid in the form of a short length of hookup wire with its free end twisted once or twice around the second detector grid lead.

Operating Notes: Lack of good single-signal effect can usually be traced to extraneous capacity coupling, lack of proper setting of neutralizing or BFO condensers, or insufficient circuit isolation. In the receiver shown, it was found necessary to shield the grid lead to the IF amplifier to prevent direct capacity coupling past the crystal-filter. This decreases the undesired signal from R9 to R5 ratio up to R9 to R3 ratio. Even better ratio could probably be obtained by better cathode, screen and plate return-lead isolation resistors and condensers.

To properly line-up this receiver, reference should be made to the sub-topic "Receiver Adjustments."

Coil Winding Table for Frank C. Jones' 222 Communications Receiver With R.F. Stage

	L1 (R. F. Grid Coil)	L2 (Plate Winding)	L3 (Detector Coil)	L4 (Oscillator Coil)
For 10 Meters	20 Turns No. 18 DCC Wire. Winding space 1 inch long on a 7/16 inch dia. tube.	3 Turns No. 36 DSC Wire, interwound with L3.	4¼ Turns No. 22 DSC Wire, space wound to cover a winding space 1 inch long on a 1½ inch dia. coil form. Tapped at ⅛ turn.	4 Turns No. 22 DSC wire, space wound to cover a winding space 1 inch long on a 1½ inch dia. coil form. Tapped at 1¼ turns.
For 20 Meters	35 Turns No. 22 DSC Wire. Winding space 1 inch long on a 7/16 inch dia. tube.	7 Turns No. 36 DSC Wire, interwound with L3.	10 Turns No. 22 DSC Wire, space wound to cover a winding space 1 inch long on a 1½ inch dia. coil form. Tapped at ½ turn.	8¾ Turns No. 22 DSC wire, space wound over a winding space 1 inch long on a 1½ inch dia. coil form. Tapped at 2¼ turns.
For 40 Meters	60 Turns No. 26 Enameled Wire. Winding space 1 inch long on a 7/16 inch dia. tube.	7 Turns No. 36 DSC Wire, interwound with L3.	10 Turns No. 22 DSC Wire, space wound to cover a winding space of 2 inches long on a 1½ inch dia. coil form. Tapped at ½ turn.	8¾ Turns No. 22 DSC Wire, space wound to cover a winding space 1 inch long on a 1½ inch dia. coil form. Tapped at 2¼ turns.
For 80 Meters	160 Turns No. 36 DSC Wire. Scramble wound on a 7/16 inch dia. tube, 1 in. long.	16 Turns No. 36 DSC Wire, interwound with L3.	30 Turns No. 22 DSC Wire over a winding space of 1⅜ inches long on a 1½ inch dia. coil form. Tapped at ¾ turn.	26¾ Turns No. 22 DSC wire over a winding space of 1⅜ inches long on a 1½ inch dia. coil form. Tapped at 4¼ turns.
For 160 Meters	300 Turns No. 36 DSC Wire. Scramble wound on a 7/16 inch dia. tube, 1 in. long.	30 Turns No. 36 DSC Wire, interwound with L3.	60 Turns No. 28 DSC Wire over a winding space of 1½ inches long on a 1½ inch dia. coil form. Tapped at 1¼ turns.	53 Turns No. 28 DSC wire over a winding space of 1½ inches long on a 1½ inch dia. coil form. Tapped at 7 turns.

Regenerative Pre-selector With Variable Antenna Coupling

This pre-selector consists of a single RF amplifier stage placed ahead of any short-wave superheterodyne receiver. By the use of variable antenna coupling and cathode regeneration, this single stage can be made equivalent to the usual two stage RF pre-selector. The function of this class of apparatus is to increase the signal-to-noise ratio and to reduce image interference.

The variable antenna coupling is obtained by means of a sliding coil whose electrical constants need not be changed for different amateur bands. An efficient plug-in coil is used in the tuned circuit inductance to insure the correct placement of the cathode tap for each band. Regeneration is controlled by means of 50,000 ohm potentiometer which varies the screen voltage. The

Under-chassis assembly, showing variable antenna coupler.

Side view of Jones Regenerative Pre-Selector.

screen-grid series resistor of 5,000 ohms tends to prevent the regeneration control from introducing noise as the latter is varied. The plate voltage is fed through a small Hammarlund multi-section RF choke which is effective over all the amateur bands.

The plate circuit is connected through a coupling condenser to the receiver so this can connect to the antenna post on the main receiver, or this lead can be twisted around the first detector grid lead a few times to obtain capacity coupling. In the latter case the trimmer condenser must be re-set for best results.

The regeneration is slightly affected by the plate circuit load, requiring in some cases, a trial adjustment of the cathode-tap or changes in coupling to the receiver. The RF tube will smoothly slide into oscillation when the pre-selector is functioning properly. The point just below oscillation gives the greatest gain and selectivity.

The antenna coupler is made of two pieces of bakelite tubing, each 1½ inches long. The larger one is 1⅛ inch outside diameter, and the smaller one ⅞ inch diameter, so that the latter with its winding of 8 turns will slide readily inside the other tube. The large tube has 20 turns of No. 28 DSC wire, close wound which is connected to a doublet antenna for maximum outside noise reduction. The link coupling system employed here is similar to that used in the "222" receiver.

The tuning condenser is of the midget type, well insulated and having a maximum capacity of about 100 mmfd. A small aluminum bracket supports the condenser (see photo) at the proper level for the dial shaft connecting bushing. All parts are

Shield housing for Jones Pre-Selector.

mounted on a piece of 12-gauge aluminum bent in the shape of an inverted U. The original piece should be 8½ inches long and 7 inches wide, 1½ inches on the front edge and ¾ inch on the rear edge are bent down, so the top of the chassis is 8½x4¾ inches. The antenna coupler mounts underneath on one side and the regeneration

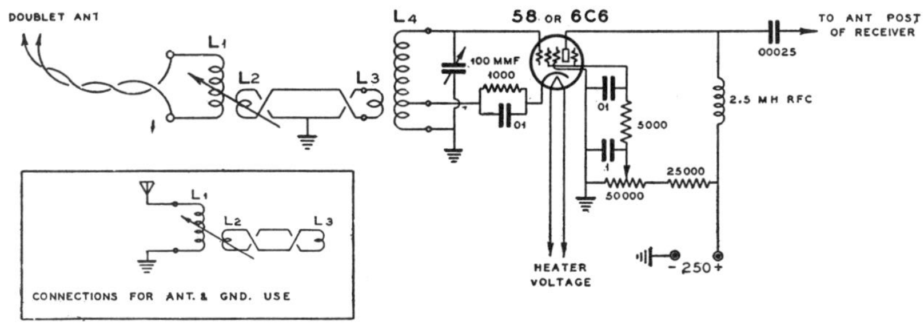

Regenerative Pre-Selector Circuit.

control on the other; the entire unit mounts in a can which comes equipped with a dial. The approximate dimensions of this can are 9½ inches long, 5 inches deep, and 6 inches high. The front and back are removable so the coil can be changed by snapping off the rear cover or by means of an opening in the rear. The dial is fastened to the chassis by a right-angle bend in the dial-mounting strap, the latter being fastened down by a machine screw. The chassis is fitted to the front cover or panel.

It is desirable to twist the antenna leads together for the two leads into the pre-selector. The plate coupling lead should come out at the other side of the rear cover and be as short as possible in its connection to the radio receiver. Coupling between this plate lead and the antenna would cause undesirable effects. Power for the tube can be obtained from the receiver. If a doublet antenna is not used, one of the antenna leads must be grounded.

Coil winding table for Pre-Selector.

L1—Same for all bands. 20 turns, No. 28 DSC, close wound on 1⅛-in. dia. tubing.

L2—Same for all bands. 8 turns, No. 28 DSC, close wound on ⅞-in. dia. tubing.

Coupling between L1 and L2 variable. L2 slides into and out of L1.

RF COIL FOR 160 METERS

L3—10 turns, No. 22 DSC, close wound on 1½-in. dia. low-loss coil form.

L4—60 turns, No. 22 DSC, close wound, and tapped 1¼ turns up from ground end. L4 is wound on same coil form as L3, and is spaced ⅛ in. from L3.

RF COIL FOR 80 METERS

L3—7 turns, No. 22 DSC, close wound, on 1½-in. dia. form.

L4—35 turns, No. 22 DSC, close wound, and tapped ½ turn up from ground end.

Spacing between L3 and L4 to be ⅛ in.

RF COIL FOR 40 AND 20 METERS

L3—5 turns No. 22 DSC, close wound, on 1½-in. dia. form.

L4—12 turns, No. 18 DSC, space-wound over a winding space of 1¼ in., and tapped ¼ turn from ground end.

NOTE—The ground end of the L4 is the bottom of the coil. The top end of L4 connects to the grid of the 58 or 6C6 tube in the Pre-Selector.

Dual Band 20-40 Meter Receiver

This is a receiver for the DX operator who devotes the greater portion of his time between the 20 and 40 meter wavebands. The circuit, as will be seen in the accompanying figure, has two front ends, one for 20 meters and the other for 40 meters, with a common IF amplifier, crystal filter circuit, detector and 800 cycle audio amplifier. The circuit is quite similar to that used in the "222 Receiver" in that a fixed tuned RF stage is placed ahead of the regenerative first detector.

Circuit Details: The HF oscillator employs a twin-triode type 79; one portion oscillating for the 20 meter band and the other for 40 meters. The oscillator circuits are stabilized with a combination grid-leak and cathode bias polarizing the grids. The cathode resistor is not by-passed; consequently it forms part of the oscillating circuit with an automatic regulating effect. The result is a high degree of frequency stability for changes in plate and filament voltages comparable to an electron-coupled oscillator.

The second detector employs a twin-triode, type 79; one portion acts as a bias detector and the other as an audio amplifier. The audio amplifier is tuned to series resonance at 800 cycles. The resonant reactor consists of a 4 henry audio choke-coil made from an old 250 mh RF choke with an "A metal" core from a small audio frequency transformer. The audio amplifier is tuned to the desired AF by adjusting the air-gap in the core. The coil of a small filter choke, with a few straight pieces of iron-core inserted in the coil form, will provide a 4 henry choke suitable for this purpose.

The first detector 2-plate main tuning condensers are ganged with flexible couplings to their respective 2-plate oscillator tuning condensers. A 2-gang 35 mmfd. per section condenser provides a tank condenser capacity, plus front panel trimmer adjustment, which is neded when using regeneration.

Coil Data: The RF coils are wound on ½-inch tubing to minimize the external field. The 20-meter coil consists of 40 turns of No. 22 DSC wire, with a primary of No. 36 DSC of 8 turns center-tapped. These primaries are wound over the

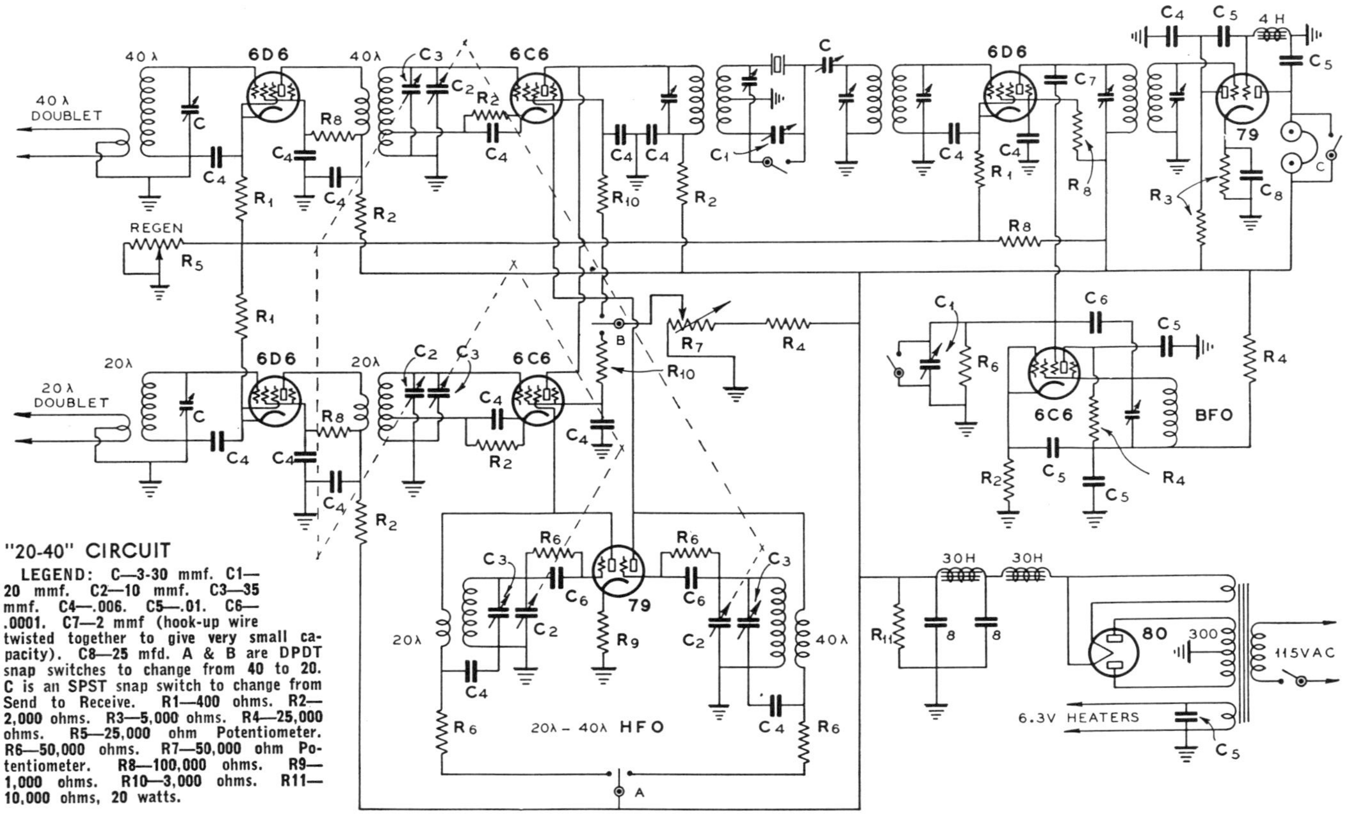

"20-40" CIRCUIT

LEGEND: C—3-30 mmf. C1—20 mmf. C2—10 mmf. C3—35 mmf. C4—.006. C5—.01. C6—.0001. C7—2 mmf (hook-up wire twisted together to give very small capacity). C8—25 mfd. A & B are DPDT snap switches to change from 40 to 20. C is an SPST snap switch to change from Send to Receive. R1—400 ohms. R2—2,000 ohms. R3—5,000 ohms. R4—25,000 ohms. R5—25,000 ohm Potentiometer. R6—50,000 ohms. R7—50,000 ohm Potentiometer. R8—100,000 ohms. R9—1,000 ohms. R10—3,000 ohms. R11—10,000 ohms, 20 watts.

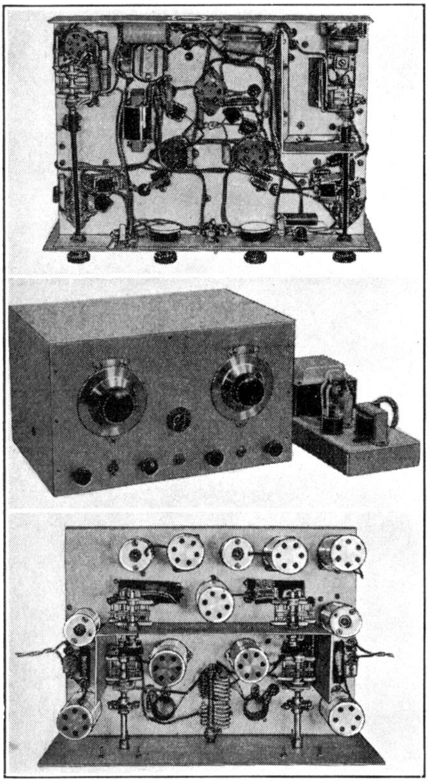

Three views of the "20-40" receiver. See page 52 for complete circuit diagram.

grounded end of the secondary in a small bunch winding with center grounded. The 40-meter RF coil has 66 turns of No. 26 enameled wire with a center-tapped 10-turn primary of No. 36 DSC wire.

The 20-meter detector coil consists of 10 turns of No. 22 DSC, 1-in. diameter, ¾ in. long, wound on celluloid strips. The wire is cemented to the strips with Duco cement. The primary consists of 7 turns of No. 36 DSC interwound with the secondary with the RF PLUS "B" connection to the "ground" end of the coil. The cathode tap is made of ¼ turn from the ground end. This tap should only be high enough to allow the first detector to spill into oscillation with the regeneration control well advanced.

The 40-meter detector coil is made in the same manner as the 20-meter coil, but with 24 turns, wound on a form one inch long and one inch in diameter. The cathode tap is made one-third of a turn up from ground and the primary is interwound for 14 turns; No. 36 DSC wire is used. For mechanical rigidity, the ends of the celluloid strips are cemented to bakelite tubing which is fastened to the chassis with a machine screw.

The oscillator coils are wound on one-inch bakelite tubing to provide great rigidity to the coil. The 20-meter coil has 10 turns of No. 22 wire wound on a form ¾ inch long, with a three-turn tickler interwound at the ground end of the secondary. The 40-meter oscillator coil has 22 turns of No. 22 DSC, wound on a form one inch long, one inch in diameter, with a 6 turn tickler of No. 36 DSC interwound. Duco cement is applied to the coils at various points to firmly secure the wires in place. All coils are mounted at right angles to each other, and an aluminum shield of No. 12 gauge is used between them. The RF coils are tuned by means of small compression-type 3-30 mfd. condenser soldered across the ends of the RF coils.

The change from 20 to 40 meters is accomplished by switching the detector screen-grids and oscillator plate-returns through a small DPDT snapswitch. There is no RF on these leads.

Antenna Connection: A 20 and separate 40 meter doublet with twisted-pair lead-ins should be used with this receiver in order to minimize auto ignition and power line noise pick-up. There is practically no antenna coupling capacity to the RF grid coil because a balanced primary is used. This prevents pick-up from the antenna feeders nearly as effectively as a very elaborate Faraday screen system.

The Perrine Superheterodyne

An amateur receiver setting up new standards and unexcelled for DX reception is here shown. The parts have been arranged so that no lead in the entire receiver is over one inch in length. A very high degree of shielding separates all the

Acorn Tube R-F Plug-In Unit for Perrine Super.

major electrical components. A minute study of the details will reveal a number of unique features. Some of these are: (a) the double by-passing of heater circuits; (b) coupling the oscillator plate to the detector suppressor; (c) the crystal filter circuit with a split-stator condenser which places twice as much capacity between the first detector plate and ground, thus by-passing more effectively the high-frequency components in the first detector plate circuit; (d) the air-tuned beat frequency

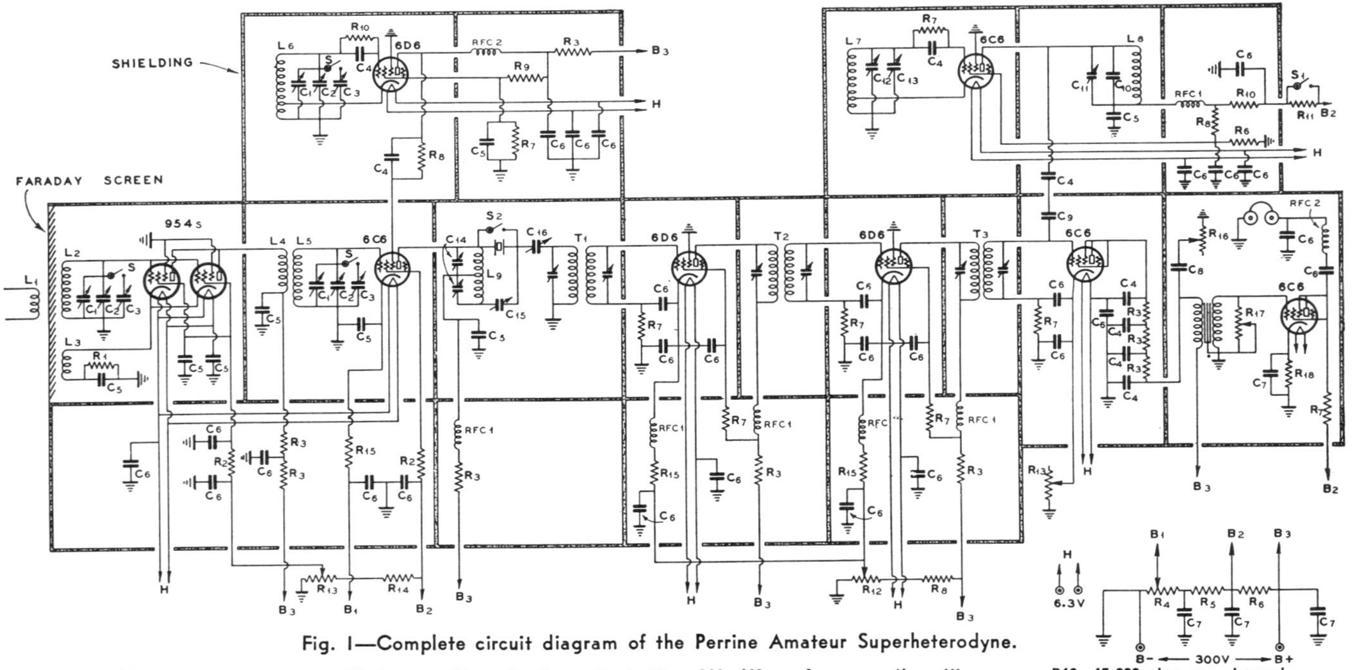

Fig. 1—Complete circuit diagram of the Perrine Amateur Superheterodyne.

LEGEND

L1—5 turns, 1-in. dia., No. 28 enameled.
L2—13 turns, 1-in. dia., No. 20 bare.
L3—2 turns, 1-in. dia., No. 32 SSC.
L4—9 turns, 1-in. dia., interwound with L5—No. 30 DSC.
L5—13 turns, 1-in. dia., No. 20 bare.
L6—12 turns, 1-in. dia., No. 20 bare tapped 4 turns from ground.
L7—95 turns, 1-in. dia., tapped at 30 turns, No. 32 DSC.
L8—30 turns, 1-in. dia., No. 32 DSC.
L9—Hammarlund choke with center pie removed.

C1—Approx. 20 mmf.—Hammarlund 50 mmf. with two stator plates removed.
C2—20 mmf. Cardwell Trimair.
C3—100 mmf. Cardwell Trimair.
C4—.0001 mica.
C5—.01 mica.
C6—.1 non-inductive paper (500 V.).
C7—1 mfd. paper (500 V.).
C8—.04 paper.
C9—3 mmf.—See text.
C10—.004 mica fixed.
C11—.001 mica variable (trimmer).
C12—.001 air variable.
C13—50 mmf. air midget (Hammarlund).
C14—140 mmf. per section (Hammarlund) midget.
C15—Phasing control—10 mmf.
C16—30 mmf. midget.
R1—400 ohm, 1 W.
R2—10,000 ohm, 1 W.
R3—1,000 ohm, 2 W.
R4—1,000 ohm, var. pot.
R5, R6—CT 30,000 ohm, 50 W. voltage divider.
R7—100,000 ohm, 1 W.
R8—50,000 ohm, 3W.
R9—40,000 ohm, 3 W.
R10—50,000 ohm, 3 W.
R11—500,000 ohm, 3 W.
R12—15,000 ohm, var. tapered
R13—50,000 ohm, var. pot.
R14—10,000 ohm, 3 W.
R15—200 ohm, 3 W.
R16—10,000 ohm, tone control.
R17—500,000 ohm, AF Gain.
R18—5,000 ohm, 3 W.
T1, T2, T3—Hammarlund 465 KC IF trans.
T4—UV712 RCA audio trans.—1,000 ohm peaked type.
RFC1—85 mh.
RFC2—2 mh.
S—Band changers.
S1—BFO strength control.
S2—Xtal shorting SW.

The beautiful Perrine Superheterodyne. Note placement of tube shields and individual shield housings.

oscillator which assures freedom from frequency drift and, in addition, has a high-C tuned-plate circuit which definitely reduces any strong harmonics in the output and thus reduces oscillation hiss—a switch is also provided to reduce the BFO plate voltage so that more readable signals are delivered to the output at low microvolt inputs.

A legend of the various parts is given on page 54. In constructing this receiver all coils should be rigidly mounted to prevent frequency changes due to vibration externally transmitted to the receiver chassis. The design is otherwise conventional in all respects.

Receiver Measurements

Satisfactory results can only be obtained from a radio receiver when it is properly aligned and adjusted. The most practical technique for making these adjustments is given in the following discussion.

The simplest type of regenerative receiver requires little adjustment other than those necessary to insure correct tuning and smooth regeneration over some desired range. Receivers of the tuned radio-frequency type and superheterodynes require almost precision alignment to obtain the highest possible degree of selectivity and sensitivity.

Testing Instruments: Only a very small number of instruments are necessary to check and align any multi-tube receiver. The most important of these testing units being a modulated oscillator and a DC and AC voltmeter. The meters are essential in checking the voltages applied at each circuit point from the power supply. NOTE: If the AC voltmeter is of the oxide-rectifier type, it can be used, in addition, as an output meter when connected across the receiver output when tuning to a modulated signal. If the signal is a steady tone, such as from a test oscillator, the output meter will indicate the value of the detected signal. In this manner line-up adjustments may be visually noted on the meter rather than by increases or decreases of sound intensity as detected by ear.

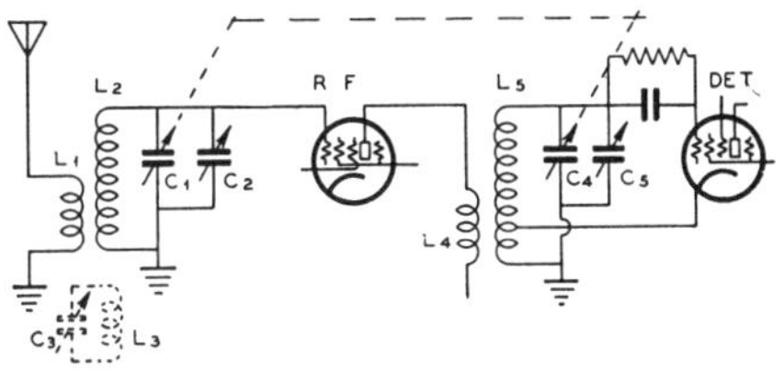

FIG. 1

R-F stage and regenerative detector.

Tuned RF and Regenerative Detector: In Figure 1 is shown a single stage of tuned RF and a regenerative detector. For proper performance, these two tuned circuits must resonate to the same frequency throughout the desired tuning range. It is required, therefore, that L_2 and L_5 have equal values of inductance and equal values of effective shunt capacity at each point on the tuning dial. The inductances may be closely matched by using similar coil forms and windings. If one coil is closer to some metal object, such as the chassis or shield, it will be difficult to obtain a good match unless coil turns are removed or shifted along the coil-form to change the effective coil length. A resonant antenna will unbalance the RF stage unless L_1 is loosely coupled to L_2.

Circuit Capacities: The shunt capacities are due to coil distributed capacity, wiring capacity, shunt condensers and tube capacity. Usually trimmer condensers C_2 and C_5 are needed to equalize the fixed circuit capacities. These should be adjusted for maximum signal sensitivity towards the high-frequency end of the tuning dial, that is, minimum capacity position for C_1 and C_4. After making this adjustment (usually with a screw driver) the alignment can be checked throughout the tuning range by bending "in" or "out" one of the outside rotor plates of tuning condenser C_1. Some receivers have condensers with slotted end-plates to facilitate bending to correct circuit alignment over the whole tuning range after C_2 and C_5 have been correctly set. The RF tube and primary L_4 reflect a capacity across L_5 which can be exactly balanced by having a duplicate primary winding L_3 on the RF grid coil. A small trimmer condenser simulates the RF tube plate circuit —this refinement is seldom used in receivers, but is well merited.

Multi-Stage Tuned RF Receivers: The alignment procedure in a multi-stage RF receiver is exactly the same as aligning a single stage. If the detector is regenerative, each preceding stage is successively aligned while keeping the detector circuit tuned to the test signal, the latter being a station signal or one locally generated by a test oscillator loosely coupled to the antenna lead. During these adjustments the RF amplifier gain control is adjusted for maximum sensitivity, assuming that the RF amplifier is stable and does not oscillate. Oscillation is indicative of improper by-passing or shielding. Often a sensitive receiver can be roughly aligned by tuning for maximum noise-pick-up, such as parasitic oscillations originating from static or electrical machinery.

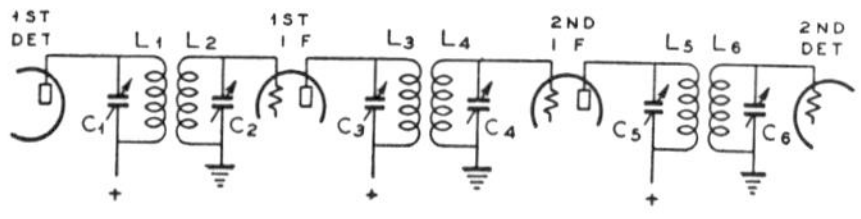

FIG. 2

I. F. Amplifier.

Superheterodynes: A superheterodyne presents an involved alignment procedure since it is necessary to align both the oscillator and first detector as well as the intermediate frequency amplifier. In this case, the latter should be aligned first. METHOD: A calibrated modulated oscillator is set to the frequency of the IF amplifier; this is usually between 175 KC and 500 KC. A lead from the oscillator is connected to the grid of the last IF stage, and C_5 and C_6 of Figure 2, varied until maximum signal strength is obtained in the output of the 2nd detector or audio amplifier. The adjustment can be simplified if the receiver has AVC, the tuning meter being used to indicate the maximum signal strength. Since the coupling inductances L_5 and L_6 are generally fixed, the only possible adjustment will be by varying the trimmer condensers. After C_5 and C_6 are properly set, the oscillator power is decreased, then coupled to the grid of the first IF amplifier tube. C_3 and C_4 may then be adjusted for maximum signal strength. The RF input to the receiver must be kept at an optimum value to insure signal readability. The procedure is repeated to align C_1 and C_3, providing the receiver has two IF stages. Sometimes it is necessary to disconnect the first detector grid lead from the coil, it then being grounded in series with a 1000 or 5000 ohm grid leak, and the test oscillator coupled through a small capacity to the grid. The oscillator should have some form of attenuator; however, the coupling may be varied by moving the oscillator lead further away from the tube grid into which it is coupled. For test purposes, the 1000 ohm resistor prevents the RF coil from short-circuiting the IF of the test oscillator so the first detector acts as an amplifier. After the IF is aligned, the first detector grid lead is connected back to its RF coil.

The technique of lining-up the first detector and RF stages, if any, is precisely the same as that described in aligning a tuned RF receiver. However, the line-up with the RF oscillator is slightly modified. METHOD: The HF oscillator is used to provide a signal in the first detector which will beat with the desired signal to form a new signal at the frequency to which the IF amplifier is tuned. If this is 450 KC, the HF oscillator should tune to 450 KC higher frequency than that of the first detector and RF stage. Figure 3 illustrates this circuit. In general, coil L_2 must have less inductance than L_1, and C_4 must have less tuning range than C_1. These requirements necessitate that L_2 have less turns than L_1, and less capacity in C_4 than in C_1. If C_1 and C_4 are of the same capacity and are coupled in tandem, a fixed or variable condenser C_3 is placed in series with C_4 to reduce its maximum capacity. C_2 and C_5 may be either trimmer or band-setting condensers. C_3 is required at longer wavelengths where the ratio of the oscillator to detector frequency is not approaching unity of equality. For example: at 14,000 KC with the oscillator at 14,450 KC no series condenser is necessary, but one would be required at frequencies of 2,000 KC and 2,450 KC if the tuning condensers C_1 and C_4 were very large.

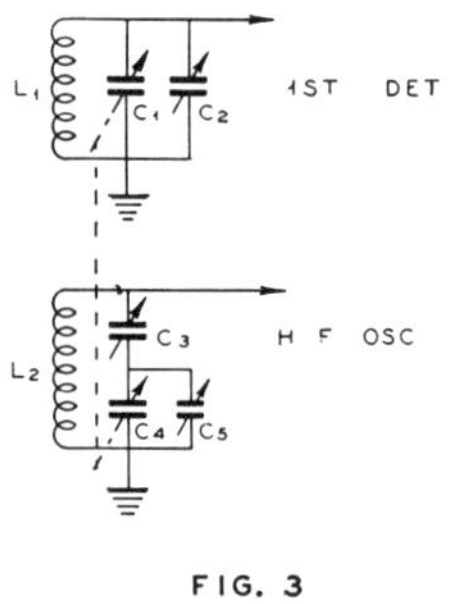

FIG. 3

Front-End of Superheterodyne.

Alignment Procedure: Actual alignment of the front end of a "superhet," such as shown in Figure 3, follows: The test oscillator is set at the highest frequency which can be tuned-in with a given set of coils.

This may require a little manipulation, but if the tuning range is known or can be estimated, an approximate frequency setting of the test oscillator can be made. The test signal is increased in value until it is heard or can be measured at the output of the receiver. C_2 is then adjusted to bring the dial reading to the desired point for a given frequency, that is, providing the dial is calibrated. C_1 and C_4, of course, being tuned simultaneously; afterwards, C_2 is adjusted for maximum sensitivity. Next, the tuning dial is rotated through to nearly full capacity setting of C_1 and C_4, of Figure 3, and the oscillator set for this lower frequency. These circuits can be aligned by moving the tuning dial while adjusting C_3 with a screwdriver or plate bending of C_1. A middle dial setting can be checked by means of a third setting of the test oscillator and plate bending of C_1. If alignment cannot be obtained by plate bending adjustments, a new value of trimmer condenser settings of C_5 and C_2 will have to be used and the whole procedure repeated. Sometimes L_2 has to have considerably less turns than L_1, and a few turns added or subtracted to allow the HF oscillator to tune through the whole range at precisely 450 KC higher in frequency than the detector and RF stages.

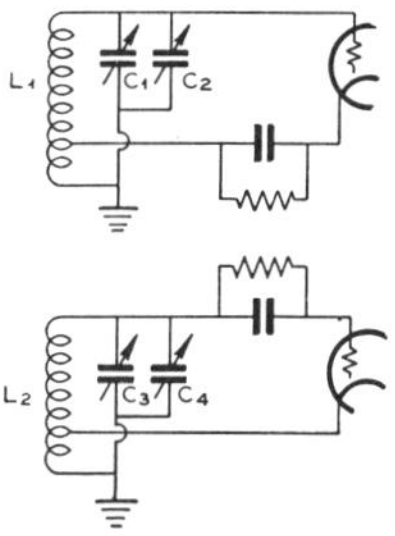

FIG. 4

Another type of front-end.

Multi-band Receivers: Individual coils in multi-band receivers with coil switching arrangements must have small trimmer condensers shunted across the inductive circuits, as shown in Figure 5. This allows fairly accurate alignment in each band by following the procedure previously outlined. In assembling a superheterodyne, the labor of checking is rather long and tedious since each coil must have exactly the correct number of turns because bending the main tuning condenser plates would unbalance or misalign all other coils. Unfortunately in receivers incorporating coil switching arrangements, it is impossible to obtain accurate circuit alignment. Many commercially built receivers use two stages of RF ahead of the first detector, tuned rather broadly to overcome this defect and obtain better signal-to-noise and image ratios.

If either the circuits of the RF stage are regenerative, they must track exactly with the HF oscillator. This type of circuit is shown in Figure 4, where C_1 and C_3 are

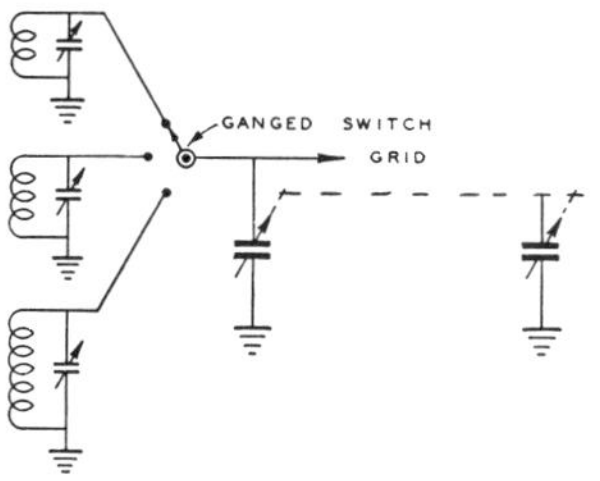

FIG. 5

Tuned circuits for coil switching.

approximately 20 to 30 mmfds. ganged tuning condensers on the main tuning dial, and C_2 and C_4 are band setting condensers of 100 to 140 mmfds. In this instance, C_2 can be used as a panel operated trimmer condenser to hold the circuits exactly in line at high degrees of regeneration. The series condensers C_3 of Figure 3 are not required in this class of receiver due to the very narrow band tuning-range of C_1 and C_3. The coil turns on L_1 and L_2 can be adjusted so that at random settings of C_2 and C_4 they will give practically perfect alignment. In practice, the adjustment occurs at slightly greater capacity settings of C_2 than for C_4, together with a small increase in inductance L_1. Varying the coil turns and spacing between turns will insure good tracking throughout all the amateur bands with the possible exception of the 160 meter band. This form of receiver invariably uses plug-in coils which must be adjusted properly, the turns being cemented in place with celluloidal cement.

Beat-frequency Oscillator: A beat-frequency oscillator, BFO, is lined up by tuning it so that its hiss is loudest in the receiver output; later, a signal is impressed to give a 1000 or 800 cycle beat-note. For example: If the IF amplifier is lined up to 450 KC, the BFO must be tuned to either 499 or 451 KC. If a crystal filter forms part of the IF amplifier complement, a vernier adjustment for the BFO should be available on the front panel in order to exactly set the beat-note for best results. The BFO input to the second detector need only be sufficient to give a good beat-note on a fairly strong signal. Too much coupling to the second detector will mean excessive hiss level with loss of very weak signals in the noise background. The BFO must be well shielded to prevent harmonics of the circuit from radiating and setting up unwanted signals. The oscillating circuit must have a high C to L ratio in order to generate oscillatory currents of high stability.

Crystal Filters: In lining up the IF amplifier for use with a crystal-filter, it is necessary to employ the crystal itself as an oscillator, providing a calibrated test oscillator is unavailable and the exact frequency of the crystal unknown. When the crystal itself functions as the oscillating medium, the circuit shown in Figure 6 should be used. In the diagram, the crystal is connected as a conventional crystal-oscillator in a transmitter, with the exception that a

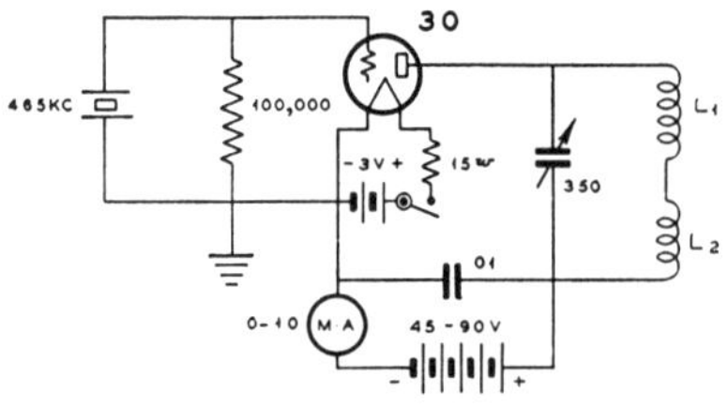

TEST OSCILLATOR CIRCUIT
FIG. 6

Crystal Filter Aligning.

small air-gap is used and the grid-leak and choke combination eliminated. A winding from an IF transformer for the plate inductance with the trimmer attached are all that are required for tuning. For lining-up purposes, a type 30 tube with 2 volts AC on the filament will suffice; the AC modulates the signal and simplifies the adjusting procedure. Plate voltage (180 volts) is secured from a tap on the voltage divider. A milliammeter inserted in the plate circuit will indicate oscillation, the plate current dipping as the trimmer condenser tunes the inductance to the resonant frequency of the crystal. A piece of insulated wire is brought near the inductance and the far end of the wire hooked over the grid input to the first IF. Tuning the IF to exact resonance with the crystal then becomes a simple matter. Unless the IF amplifier is lined up to the exact crystal frequency, the crystal will introduce a very decided loss in sensitivity when it is switched into operation.

In adjusting the crystal filter, the phasing condenser and input tuning condenser should be adjusted simultaneously for maximum signal response, then a slight readjustment of the phasing condenser will allow elimination of the other sideband.

Notes: In lining up a receiver which has automatic volume control (AVC), it is considered good practice to keep the test-oscillator signal near the threshold sensitivity at all times to give the effect of a very weak signal relative to the audio amplifier output with the audio gain control on maximum setting.

In checking over a receiver certain troubles are often difficult to locate. In general, by making voltage or continuity tests, blown-out condensers, or burned-out resistors, coils or transformers may be easily located. Oscillators are usually checked by means of a DC voltmeter connected from ground to screen or plate-return circuits. Short-circuiting the tuning condenser plates should usually produce a change in voltmeter reading. A vacuum-tube voltmeter is also very handy for the purpose of measuring the correct amount of oscillator RF voltage supplied to the first detector circuit. The value of the RF voltage is approximately one volt less than the fixed grid bias on the first detector when the voltage is introduced into either the grid or the cathode circuit.

Incorrect voltages, poor resistors or leaky bypass or blocking condensers will ruin the audio tone of the receiver. Defective tubes can be checked in a tube tester. Loudspeaker rattle is not always the defect in the voice coil or spider support, or metallic filings in its air-gap; more often the distortion is caused by overloading the audio amplifier. An IF amplifier can also impair splendid tone due to a defective tube or overloading the final IF tube. In some AVC circuits, the last IF tube will easily overload if too much bias is fed back on strong carrier signals. Diode detectors give best fidelity when operated at fairly high input levels which means that there must be ample voltage swing delivered to the output of the last IF tube.

Quartz Crystal Filters: The subject of quartz-crystals is confusing to many users, which may be attributed to the complexities underlying the technical nature of the device.

Briefly, a quartz-crystal cut on certain axes and with parallel faces, has the property of mechanically oscillating in alternating-current electric fields of certain frequency. In addition, it has the very unique property of functioning as a resonator. In CW reception, the self-resonant feature is utilized in a filter circuit to limit the received signal to a band of approximately 100 cycles wide, such an electrical combination improves the signal-to-noise ratio as well as assures the highest selectivity obtainable for CW radio telegraphic reception.

General Details: To generally illustrate the function of the crystal and filter circuit, assume that the latter is replaced with its electrical equivalent in inductance and capacity. A crystal of 451.5 has an equivalent inductance of 3.5 henries and a capacity of less than 0.1 micromicrofarad. The effective "Q" of such a circuit ranges from 1000 to 10,000. Since the "Q" is the property which governs the shape of the resonance curve, the circuit would have a very narrow shoulder with a sharply peaked characteristic. Apparently no combination of inductance and capacity could eclipse these effects. Similarly, to an electrical equivalent circuit, the crystal has also properties of a series-resonant circuit. A circuit of this type offers very low impedance to the resonant frequency (that frequency where the inductive reactance and capacitive reactance are equal), while at the same time presents very high impedance to all other frequencies. A series-resonant circuit will pass the resonant frequency (in this case the frequency to which the receiver is tuned) and reject all other adjacent signals. In general, resonance curves do not have vertical sides, they slope. The "steepness" of the slope is dependent, among other things, upon the "Q" of the circuit. With a circuit having a resonance curve with gradual sloping sides, an interfering signal removing 10 KC from the desired signal may only be ten points down in strength from the desired signal at the output of the receiver. In contrast, a quartz-filter circuit with extremely steep sides can cause interfering signals to be cut down from the unwanted signal 10,000 times. These figures are merely illustrative of the effect of the extreme discrimination of such circuits as compared with ordinary tuned-parallel resonant circuits used in an IF amplifier.

Technical Discussion: The impedance of a quartz-crystal oscillator to an AC electrical current is exceptionally low and it can therefore be used as a series element of an electrical filter for CW reception. The quartz-crystal may be compared to the electrical equivalent circuit shown in Figure 1, where C1 is the capacity across the quartz plate when not vibrating; R, the resistance equivalent to the frictional effects of the vibrating crystal; L, the inductance corresponding to inertia; and C, the capacity corresponding to the elasticity. One side of resonance the circuit has capacitive reactance due to the elastic forces which control the crystal virbrations, while on the other side of resonance the reactance is inductive on account of the inertia effects. At resonance, the crystal vibrates freely, its amplitude being limited by the frictional

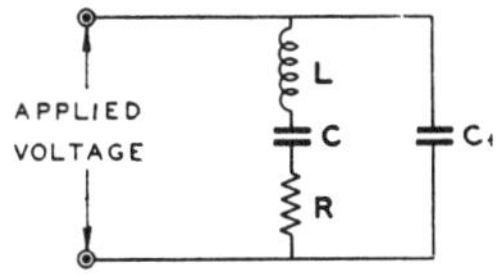

FIG. 1

Equivalent circuit of a quartz crystal.

effects; in the resonating state, L and C are equal in reactance and the resonant frequency is the same as the mechanical vibratory mode.

If the impressed voltage is at the resonant frequency, the current through it will be large, limited only by the resistance R. There is also a leading component due to C1 which can be balanced out by means of a "phasing" condenser. (Note: A phasing condenser is used in all single-signal receiver circuits to eliminate the by-passing effect of C1, of Figure 1, or to use it as a means of eliminating one sideband.) C1 combined with L and C have a sufficient inductive effect to provide a parallel circuit at a frequency slightly different from series resonance.

By placing the phasing condenser in the circuit so that the voltage across it is out of phase with that across the crystal, the parallel resonance can be shifted above or below crystal-resonance. Thus, the phasing condenser can be adjusted so that the parallel resonance causes a sharp dip in the response curve at some desired point, such as 2 KC away from the desired signal peak. This means that the other sideband 1 KC away from zero-beat can be practically eliminated with a beat-frequency oscillator. The series-resonant effect is used to pass the desired signal through an IF amplifier for further amplification.

Quartz-Filter Circuits: In reception, it is required that the noise-to-signal ratio be kept at a very low value; to obtain the optimum noise ratio requires circuits having selective and highly-peaked response curves. Thus, it is desirable to have a band-width only about 100 cycles wide, down to a point at where the gain of the receiver will discriminate against undesired signals audible in the output. A well-designed crystal filter will provide an attenuation of about 60 DB to signals more than 5 KC off resonance with, of course, that much more attenuation of the opposite sideband, 1 KC from zero-beat on the opposite side from the peak response.

Quartz-crystals have a greater "Q" at lower frequencies. For this reason most filters are designed for operation at 500 to 450 KC, and used in an IF amplifier resonating at the crystal frequency. From a selectivity standpoint, frequencies lower than 450 KC would be desirable because the crystal "Q" would be greater; however, in the lower ranges image interference becomes a problem.

In quartz-crystal filter circuits, the R value ranges between 2,500 and 10,000 ohms which requires that the circuit be designed to minimize its loading effect on any tuned circuits, otherwise the impedance irregularity will cause an excessive loss at the desired signal frequency. This latter condition occurs in the popular circuit shown in Figure 2.

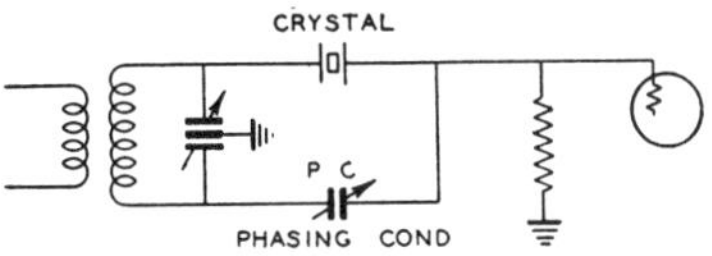

FIG. 2

Lamb's crystal filter circuit.

Some of the undesirable effects of the circuit shown in Figure 2 are eliminated in the circuit of Figure 3. Here, the grid-leak is replaced by a tapped resonant RF choke. The resonant effect, plus the mid-point connection, gives a step-up in impedance from the series element (the quartz-crystal) with only a slight loss in signal strength. To realize the full possibilities of this system requires that the resonant choke be properly designed; unfortunately, the design is difficult.

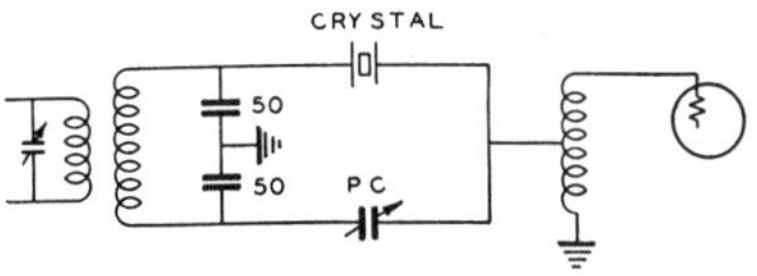

FIG. 3

McMurdo Silver's crystal filter.

The difference between the circuits of Figures 2 and 3 is in the manner of obtaining an out-of-phase voltage across the crystal. The coil can be center-tapped to ground, or the center point of the two condensers may be used. In either case, the crystal-input circuit tuning condenser and phasing condenser are simultaneously adjusted for maximum signal response and greatest single signal effect.

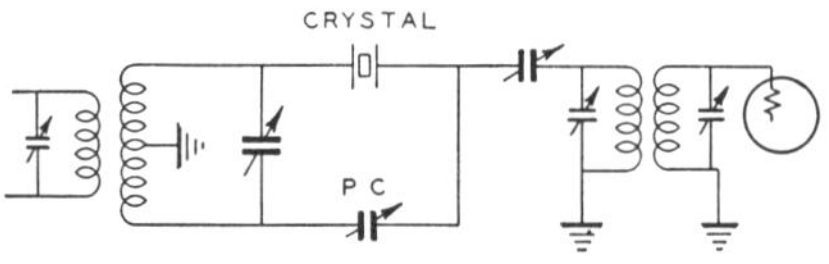

FIG. 4

Frank C. Jones' crystal filter.

In the circuit of Figure 4 the crystal is used as a series element, connecting two parallel resonant circuits together in a band-pass circuit. The small condenser C of 20 to 30 uufds. is necessary to prevent over-coupling between the tuned IF transformers, because at series resonance, only a few thousand ohms is offered as impedance. The small condenser C does not appreciably decrease the signal strength, its function is that of coupling the two tuned circuit together. The extra tuned circuits, which cause only an effective loss, eliminate the usual spurious side-band responses of most quartz crystals. The side-band responses are a few kilocycles away from resonance, but by careful tuning of the IF transformers, these effects can be attenuated to practically zero value.

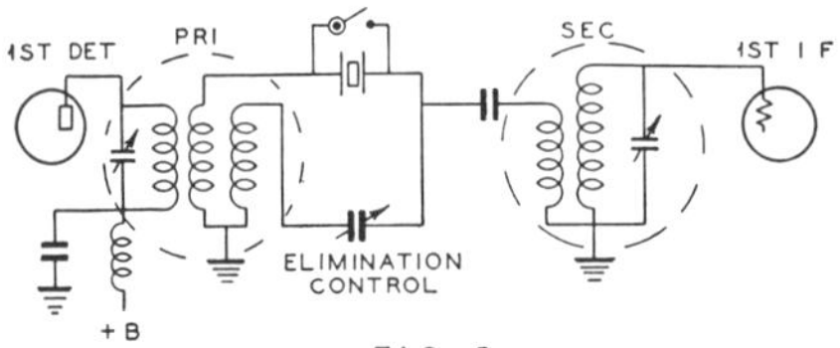

FIG. 5

Comet "Pro" crystal filter

Another method for matching impedances is shown in Figure 5. Here the low impedance of the crystal at resonance does not over-couple the two parallel tuned circuits. A 30-1 step-down ratio of impedance works into the crystal, and a similar step-up ratio couples it into the tuned-grid circuit. In this circuit, as well as in the one above, a small series condenser prevents over-coupling. Laboratory and field tests show that very little, if anything, is gained by the step-down transformers as compared with the system shown in Figure 4. The circuits shown in Figures 3, 4 and 5 are better than that of Figure 2.

The illustrations to the right show modern designs for home-built crystal filter I-F amplifier and B.F.O. circuits.

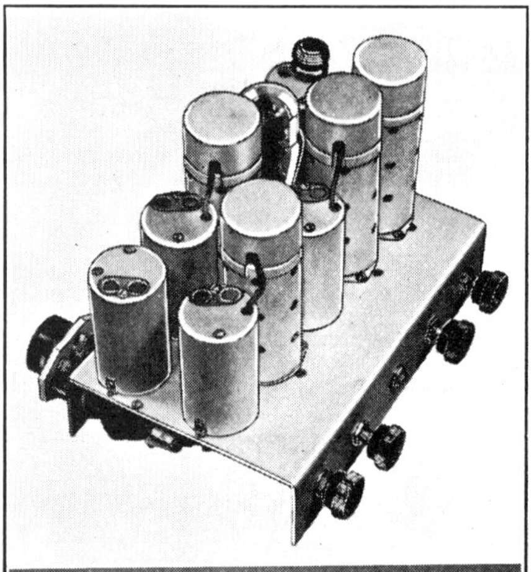

The under-chassis view shows the placement of the crystal phasing condenser, which has one of its rotor plates bent over slightly so that the condenser will be short-circuited when it is in the full "in" position.

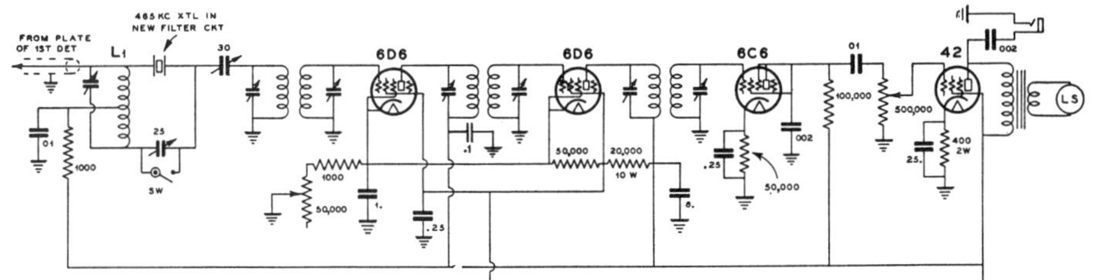

Intermediate frequency amplifier with Jones crystal filter and R.C.A. BFO circuit. The photographs on the preceding page show two types of units in which these modern circuits are incorporated. They can be used for converting a regenerative detector and R-F into a superheterodyne.

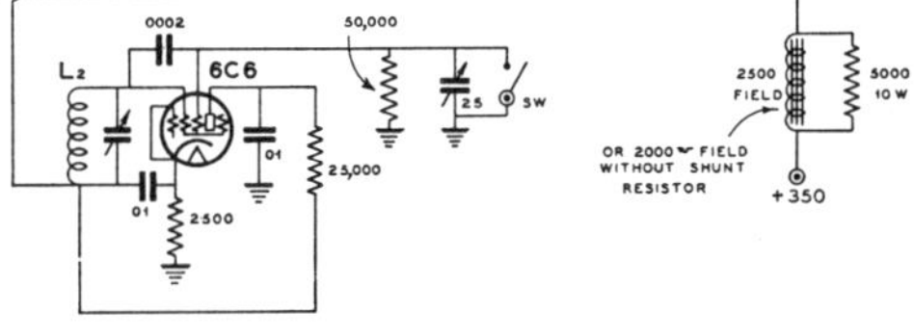

IMPROVED BFO CIRCUIT

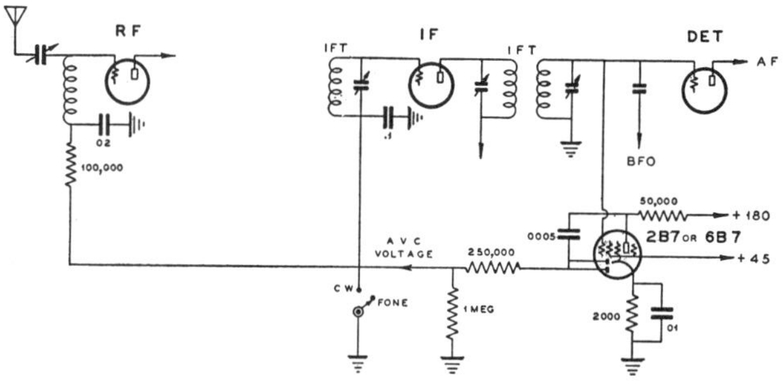

AMPLIFIED A V C FOR 222 RECEIVER OR ANY OTHER SHORT WAVE SUPERHET

Automatic volume control is a distinct advantage when the receiver is used for phone reception. It automatically compensates for signal fading. Briefly, its action depends upon the carrier signal strength which causes a control voltage to vary in proportion to the actual carrier intensity.

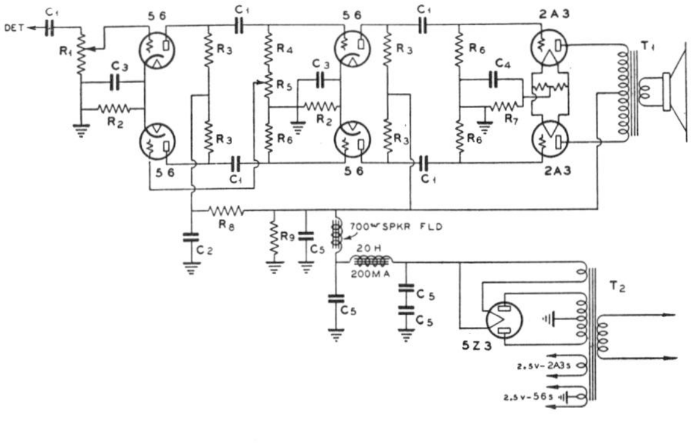

High quality audio amplifier with a maximum output of 15 watts. Legend—

C1—0.1. C2—4 mfd. C3—4 mfd. C4—10 mfd. C5—8 mfd. R1—500,000 ohm Potentiometer. R2—1250 ohms, 1 watt. R3—100,000 ohms, 1 watt. R4—250,000 ohms, 1 watt. R5—250,000 ohm Potentiometer. R6—500,000 ohms, 1 watt. R7—750 ohms, 10 watts. R8—20,000 ohms, 1 watt. R9—25,000 ohms, 20 watts. T1—Output Transformer from 2A3s Push-Pull to Dynamic Speaker. T2—Power Transformer, 300 volts each side of center-tap at 150 MA.

C_1—.01. C_2—.1. C_3—.0001. C_4—.00025. C_5—.00025. C_6—8. C_7—12. C_8—25. C_9—.002. R_1—2,000 ohms, ½ watt. R_2—100,000 ohms, ½ watt. R_3—100,000 ohms, ½ watt. R_4—100,000 ohms, 1 watt. R_5—10,000 ohms, ½ watt. R_6—1,000 ohms, ½ watt. R_7—200 ohms variable wire wound. R_8—150 ohms, ½ watt. R_9—5,000 ohms, ½ watt. R_{10}—50,000 ohms, ½ watt. R_{14}—30,000 ohms variable. R_{15}—1 megohm variable. R_{16}—1 megohm, ½ watt. R_{17}—20,000 ohms, 1 watt. R_{18}—500 ohms, ½ watt. R_{19}—250,000 ohms variable. R_{20}—250,000 ohms, ½ watt. T_1—Power Transformer. T_2—Audio Output Transformer. CH—30 H. Filter Choke.

Voltage Divider: R_{11}—6,800 ohms; R_{12}—7,200 ohms; R_{13}—410 ohms.

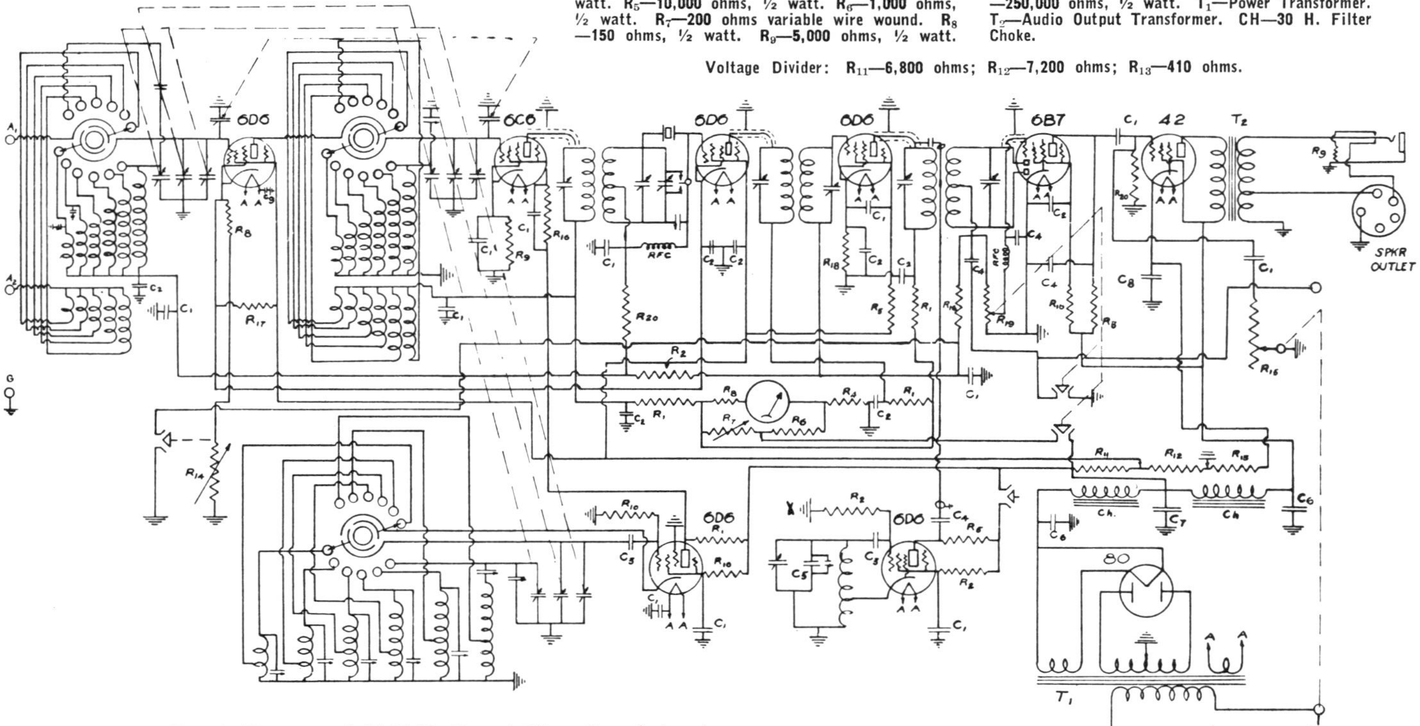

Circuit Diagram of RME-69 Crystal Filter Superheterodyne.

Band Pass Crystal Filters

An ideal characteristic for an I.F. amplifier in a c-w receiver would be a band width 500 cycles broad at the top, and practically straight-sided. The total attenuation would be down at least 120 D.B. at approximately 100 cycles either side of this band-pass. The attenuation should extend down to 120 D.B. in order to eliminate "slop-over" from very powerful local stations.

A multiple quartz crystal filter, combined with a number of tuned I.F. circuits, would approach this ideal condition for phone reception; on the other hand its use would not be desirable for c-w reception. Series crystal filter circuits as used in single signal superheterodynes give a very narrow width, but the shape of the curve resembles the outline of a volcano. It is too sharp

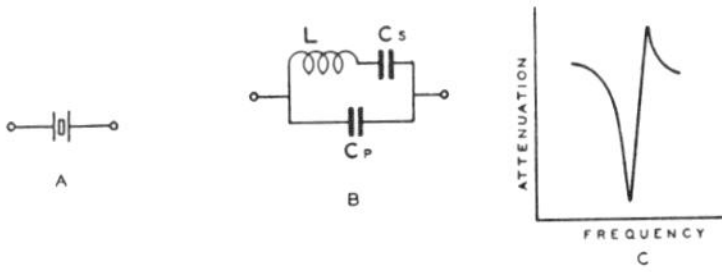

FIG. 1

for easy tuning on the peak, and altogether too wide at the base; therefore the strong local signals cannot be eliminated. The peak portion of the curve is too selective for phone reception, and for this reason the series crystal circuits will eventually be discarded.

The equivalent circuit of a quartz crystal is shown in Fig. 1, wherein both series and parallel resonance occur. Series resonance is due to the equivalent inductance and series capacity:

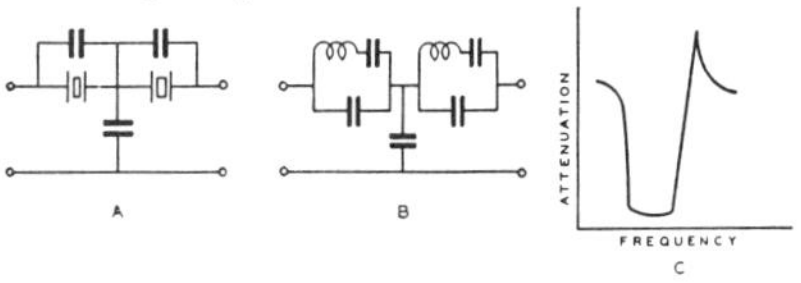

FIG. 2

$$F_s = \frac{1}{2\pi\sqrt{LC_s}}$$

The crystal holder introduces a shunt or parallel capacity C_p across the crystal, and parallel resonance occurs at:

$$F_p = \frac{1}{2\pi}\sqrt{\frac{C_s + C_p}{LC_sC_p}}$$

The parallel resonance effect can be varied by means of a "phasing" condenser in a single signal receiver in such a manner that it will nearly eliminate the second beat note of a c-w signal which is tuned-in on the peak of resonance. The parallel resonance is too sharp to make possible the elimination of the entire undesired beat note, except over a certain range, such as from 800 to 900 or 1,000 cycles. Thus a weak, undesired signal of higher or lower beat note can still be heard, especially if the lower beat note signal is of sufficient intensity.

Fig. 2 shows two crystals in a band-pass circuit. The crystals used in band-pass circuits are slightly different in frequency.

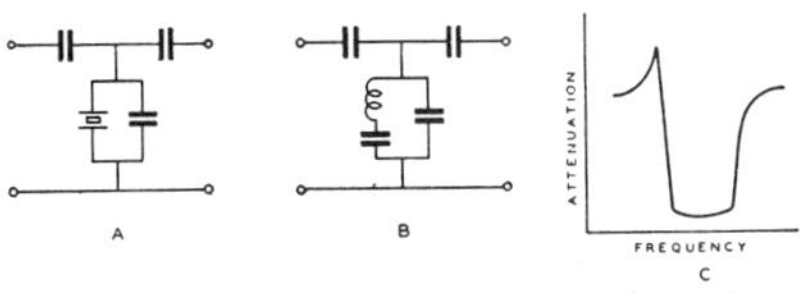

FIG. 3

In Fig. 2 the response curve is wider at the base, which is the point of least attenuation (the peak of response in a receiver) than for the single series crystal shown in Fig. 1c.

Fig. 3 shows a shunt single crystal filter circuit with series condensers. The circuit is similar to that of Fig. 2, except for the reversal in the point of greatest attenuation. The curve of (c) depends upon the proper impedance terminations, as well as the correct values of shunt and series condensers.

Fig. 4 shows a system with three crystals for better band-pass characteristic. The band-pass width is less than 0.4% of the series resonance frequency of the crystals; consequently for a 465 KC crystal the band width would not be greater than 1750 cycles.

These band-pass filters have a low impedance, depending upon their band widths. The narrower the band, the lower is the value of impedance to match. This impedance ranges from a few hundred ohms, downward. Impedance matching can be accomplished with tuned I.F. coils which have low inductance untuned secondary and primary windings.

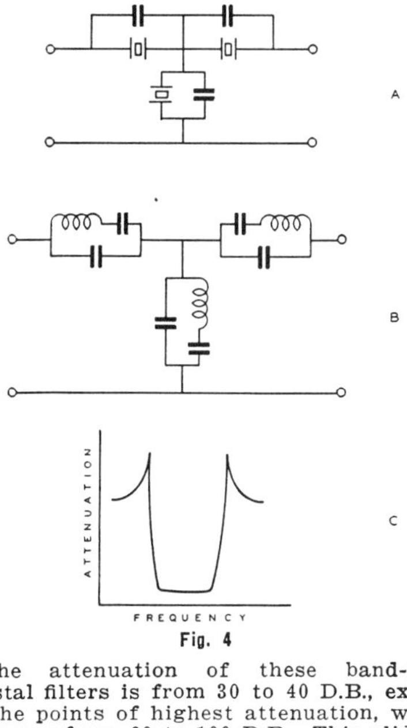

Fig. 4

The attenuation of these band-pass crystal filters is from 30 to 40 D.B., except at the points of highest attenuation, which may run from 60 to 100 D.B. This sliding-off effect on the sides beyond the parallel resonant cut-off points means that additional attenuation in the I.F. amplifier is required, or more than one section of crystal filter must be used between stages.

Tube Symbols and Bottom Views of Socket Connections

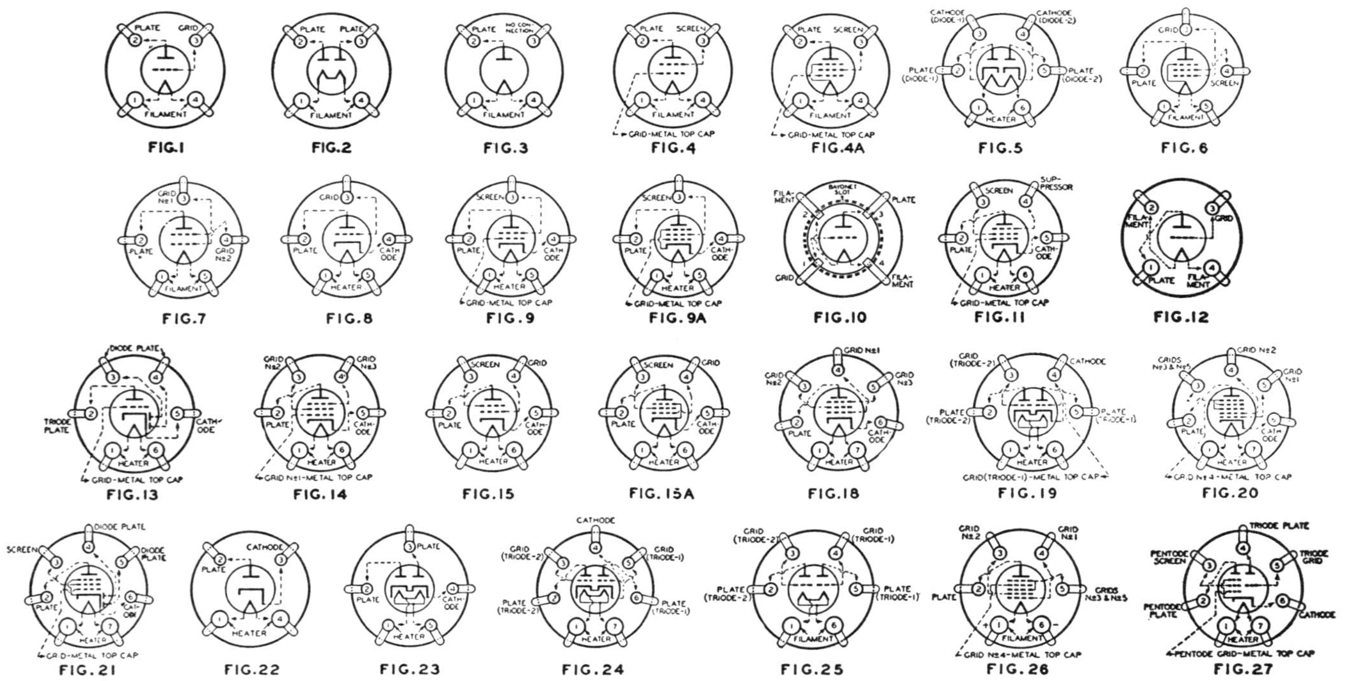

Tube Characteristics and Socket Connections courtesy RCA Cunningham Radiotron Co., Inc.

Characteristics of Receiving Tubes

TYPE	NAME	BASE	SOCKET CONNECTIONS	DIMENSIONS MAXIMUM OVERALL LENGTH x DIAMETER	CATHODE TYPE ■	RATING: FILAMENT OR HEATER VOLTS	RATING: FILAMENT OR HEATER AMPERES	RATING: PLATE MAX. VOLTS	RATING: SCREEN MAX. VOLTS	USE Values to right give operating conditions and characteristics for indicated typical use	PLATE SUPPLY VOLTS	GRID VOLTS ■	SCREEN VOLTS	SCREEN MILLI-AMP.	PLATE MILLI-AMP.	A-C PLATE RESISTANCE OHMS	MUTUAL CONDUCTANCE MICROMHOS	VOLTAGE AMPLIFICATION FACTOR	LOAD FOR STATED POWER OUTPUT OHMS	POWER OUTPUT WATTS	TYPE
1A6	PENTAGRID CONVERTER ◘	SMALL 6-PIN	FIG. 26	4 17/32" x 1 9/16"	D-C FILAMENT	2.0	0.06	180	67.5	CONVERTER	180	−3.0 min.	67.5	2.4	1.3	500000	Anode-Grid (#2) 135 max. volts, 2.3 ma. Oscillator Grid (#1) Resistor, 50000 ohms. Conversion conductance, 300 micromhos.				1A6
1C6	PENTAGRID CONVERTER ◘	SMALL 6-PIN	FIG. 26	4 17/32" x 1 9/16"	D-C FILAMENT	2.0	0.12	180	67.5	CONVERTER	180	−3.0 min.	67.5	2.0	1.5	750000	Anode Grid (#2) 135 max. volts, 3.3 ma. Oscillator Grid (#1) Resistor, 50000 ohms. Conversion conductance, 325 micromhos.				1C6
2A3	POWER AMPLIFIER TRIODE	MEDIUM 4-PIN	FIG. 1	5 3/8" x 2 1/16"	FILAMENT	2.5	2.5	250	—	CLASS A AMPLIFIER	250	−45	—	—	60.0	800	5250	4.2	2500	3.5	2A3
								300	—	PUSH-PULL AMPLIFIER	300 300	−62 −62	Self-bias Fixed-bias		40.0 40.0	Power Output is for 2 tubes at stated load, plate-to-plate			5000 3000	10.0 15.0	
2A5	POWER AMPLIFIER PENTODE	MEDIUM 6-PIN	FIG. 15A	4 11/16" x 1 13/16"	HEATER	2.5	1.75	250	250	CLASS A AMPLIFIER	250	−16.5	250	6.5	34.0	100000	2200	220	7000	3.0	2A5
2A6	DUPLEX-DIODE HIGH-MU TRIODE	SMALL 6-PIN	FIG. 13	4 17/32" x 1 9/16"	HEATER	2.5	0.8	250	—	TRIODE UNIT AS CLASS A AMPLIFIER	250✖	−1.35	—	—	0.4	—	—	Gain per stage = 50-60			2A6
2A7	PENTAGRID CONVERTER ◘	SMALL 7-PIN	FIG. 20	4 17/32" x 1 9/16"	HEATER	2.5	0.8	250	100	CONVERTER	250	−3.0 min.	100	2.2	3.5	360000	Anode Grid (#2) 200 max. volts, 4.0 ma. Oscillator Grid (#1) Resistor, 50000 ohms. Conversion conductance, 520 micromhos.				2A7
2B7	DUPLEX-DIODE PENTODE	SMALL 7-PIN	FIG. 21	4 17/32" x 1 9/16"	HEATER	2.5	0.8	250	125	PENTODE UNIT AS R-F AMPLIFIER	100 250	−3.0 −3.0	100 125	1.7 2.3	5.8 9.0	300000 650000	950 1125	285 730	—	—	2B7
										PENTODE UNIT AS A-F AMPLIFIER	250‡	−4.5	50	—	0.65	—	—	—	—	—	
6A4 also LA	POWER AMPLIFIER PENTODE	MEDIUM 5-PIN	FIG. 6	4 11/16" x 1 13/16"	FILAMENT	6.3	0.3	180	180	CLASS A AMPLIFIER	100 180	−6.5 −12.0	100 180	1.6 3.9	9.0 22.0	83250 45500	1200 2200	100 100	11000 8000	0.31 1.40	6A4 also LA
6A7	PENTAGRID CONVERTER ◘	SMALL 7-PIN	FIG. 20	4 17/32" x 1 9/16"	HEATER	6.3	0.3	250	100	CONVERTER	250	−3.0 min.	100	2.2	3.5	360000	Anode Grid (#2) 200 max. volts, 4.0 ma. Oscillator Grid (#1) Resistor, 50000 ohms. Conversion conductance, 520 micromhos.				6A7
6B7	DUPLEX-DIODE PENTODE	SMALL 7-PIN	FIG. 21	4 17/32" x 1 9/16"	HEATER	6.3	0.3	250	125	PENTODE UNIT AS R-F AMPLIFIER	100 250	−3.0 −3.0	100 125	1.7 2.3	5.8 9.0	300000 650000	950 1125	285 730	—	—	6B7
										PENTODE UNIT AS A-F AMPLIFIER	250‡	−4.5	50	—	0.65	—	—	—	—	—	
6C6	TRIPLE-GRID DETECTOR AMPLIFIER	SMALL 6-PIN	FIG. 11	4 13/16" x 1 9/16"	HEATER	6.3	0.3	250	100	SCREEN GRID R-F AMPLIFIER	250	−3.0	100	0.5	2.0	exceeds 1.5 meg.	1225	exceeds 1500	—	—	6C6
										BIAS DETECTOR	250	−1.95	50	Cathode current 0.65 ma.		—	Plate coupling resistor 250000 ohms. Grid coupling resistor 250000 ohms.**				
6D6	TRIPLE-GRID SUPER-CONTROL AMPLIFIER	SMALL 6-PIN	FIG. 11	4 13/16" x 1 9/16"	HEATER	6.3	0.3	250	100	SCREEN GRID R-F AMPLIFIER	250	−3.0 min.	100	2.0	8.2	800000	1600	1280	—	—	6D6
										MIXER IN SUPERHETERODYNE	250	−10.0	100	—	—		Oscillator peak volts = 7.0.				

◘ Grids #3 and #5 are screen. Grid #4 is signal-input control-grid.

‡ Applied through plate coupling resistor of 200000 ohms. **For grid of following tube.
✖ Applied through plate coupling resistor of 250000 ohms.

TYPE	NAME	BASE	SOCKET CONNECTIONS	DIMENSIONS MAXIMUM OVERALL LENGTH x DIAMETER	CATHODE TYPE ■	RATING: FILAMENT OR HEATER VOLTS	RATING: FILAMENT OR HEATER AMPERES	RATING: PLATE MAX. VOLTS	RATING: SCREEN MAX. VOLTS	USE	PLATE SUPPLY VOLTS	GRID VOLTS ■	SCREEN VOLTS	SCREEN MILLI-AMP.	PLATE MILLI-AMP.	A-C PLATE RESISTANCE OHMS	MUTUAL CONDUCTANCE MICROMHOS	VOLTAGE AMPLIFICATION FACTOR	LOAD FOR STATED POWER OUTPUT OHMS	POWER OUTPUT WATTS	TYPE
6F7	TRIODE-PENTODE	SMALL 7-PIN	FIG. 27	4 17/32" x 1 9/16"	HEATER	6.3	0.3	100	—	TRIODE UNIT AS AMPLIFIER	100	−3.0	—	—	3.5	17800	450	8	—	—	6F7
								250	100	PENTODE UNIT AS AMPLIFIER	250	−3.0 min.	100	1.5	6.5	850000	1100	900	—	—	
								250	100	PENTODE UNIT AS MIXER	250	−10.0	100	0.6	2.8	Oscillator peak volts = 7.0. Conversion conductance = 300 micromhos.					
'00-A	DETECTOR TRIODE	MEDIUM 4-PIN	FIG. 1	4 11/16" x 1 13/16"	D-C FILAMENT	5.0	0.25	45	—	GRID LEAK DETECTOR	45	Grid Return to (−) Filament			1.5	30000	666	20	—	—	'00-A
01-A	DETECTOR★ AMPLIFIER	MEDIUM 4-PIN	FIG. 1	4 11/16" x 1 13/16"	D-C FILAMENT	5.0	0.25	135	—	CLASS A AMPLIFIER	90 135	−4.5 −9.0	—	—	2.5 3.0	11000 10000	725 800	8.0 8.0	—	—	01-A
10	POWER AMPLIFIER TRIODE	MEDIUM 4-PIN	FIG. 1	5 5/8" x 2 3/16"	FILAMENT	7.5	1.25	425	—	CLASS A AMPLIFIER	350 425	−31.0 −39.0	—	—	16.0 18.0	5150 5000	1550 1600	8.0 8.0	11000 10200	0.9 1.6	10
11 12	DETECTOR★ AMPLIFIER TRIODE	WD 4-PIN MEDIUM 4-PIN	FIG. 12 FIG. 1	4 1/8" x 1 3/16" 4 11/16" x 1 7/16"	D-C FILAMENT	1.1	0.25	135	—	CLASS A AMPLIFIER	90 135	−4.5 −10.5	—	—	2.5 3.0	15500 15000	425 440	6.6 6.6	—	—	11 12
19	TWIN-TRIODE AMPLIFIER	SMALL 6-PIN	FIG. 25	4 1/4" x 1 9/16"	D-C FILAMENT	2.0	0.26	135	—	CLASS B AMPLIFIER	135 135	0 −3.0	—	—	Power output value is for one tube at stated load, plate-to-plate.				10000 10000	2.1 1.9	19
'20	POWER AMPLIFIER TRIODE	SMALL 4-PIN	FIG. 1	4 1/8" x 1 3/16"	D-C FILAMENT	3.3	0.132	135	—	CLASS A AMPLIFIER	90 135	−16.5 −22.5	—	—	3.0 6.5	8000 6300	415 525	3.3 3.3	9600 6500	0.045 0.110	'20
22	R-F AMPLIFIER TETRODE	MEDIUM 4-PIN	FIG. 4	5 1/32" x 1 13/16"	D-C FILAMENT	3.3	0.132	135	67.5	SCREEN GRID R-F AMPLIFIER	135 135	−1.5 −1.5	45 67.5	0.6* 1.3*	1.7 3.7	725000 325000	375 500	270 160	—	—	22
24-A	R-F AMPLIFIER TETRODE	MEDIUM 5-PIN	FIG. 9	5 1/32" x 1 13/16"	HEATER	2.5	1.75	275	90	SCREEN GRID R-F AMPLIFIER	180 250	−3.0 −3.0	90 90	1.7* 1.7*	4.0 4.0	400000 600000	1000 1050	400 630	—	—	24-A
										BIAS DETECTOR	250●	−5.0 approx.	20 to 45	—	Plate current to be adjusted to 0.1 milliampere with no signal.						
26	AMPLIFIER TRIODE	MEDIUM 4-PIN	FIG. 1	4 11/16" x 1 13/16"	FILAMENT	1.5	1.05	180	—	CLASS A AMPLIFIER	90 180	−7.0 −14.5	—	—	2.9 6.2	8900 7300	935 1150	8.3 8.3	—	—	26
27	DETECTOR★ AMPLIFIER TRIODE	MEDIUM 5-PIN	FIG. 8	4 1/4" x 1 9/16"	HEATER	2.5	1.75	275	—	CLASS A AMPLIFIER	135 250	−9.0 −21.0	—	—	4.5 5.2	9000 9250	1000 975	9.0 9.0	—	—	27
										BIAS DETECTOR	250	−30.0 approx.	—	—	Plate current to be adjusted to 0.2 milliampere with no signal.						
30	DETECTOR★ AMPLIFIER TRIODE	SMALL 4-PIN	FIG. 1	4 1/4" x 1 9/16"	D-C FILAMENT	2.0	0.06	180	—	CLASS A AMPLIFIER	90 135 180	−4.5 −9.0 −13.5	—	—	2.5 3.0 3.1	11000 10300 10300	850 900 900	9.3 9.3 9.3	—	—	30

Characteristics of Receiving Tubes

(Continued)

TYPE	NAME	BASE	SOCKET CONNECTIONS	DIMENSIONS MAXIMUM OVERALL (LENGTH X DIAMETER)	CATHODE TYPE ■	RATING: FILAMENT OR HEATER VOLTS	RATING: FILAMENT OR HEATER AMPERES	RATING: PLATE MAX. VOLTS	RATING: SCREEN MAX. VOLTS	USE (Values to right give operating conditions and characteristics for indicated typical use)	PLATE SUPPLY VOLTS	GRID VOLTS ■	SCREEN VOLTS	SCREEN MILLI-AMP.	PLATE MILLI-AMP.	A-C PLATE RESISTANCE OHMS	MUTUAL CONDUCTANCE MICROMHOS	VOLTAGE AMPLIFICATION FACTOR	LOAD FOR STATED POWER OUTPUT OHMS	POWER OUTPUT WATTS	TYPE
31	POWER AMPLIFIER TRIODE	SMALL 4-PIN	FIG. 1	4 1/4" x 1 9/16"	D-C FILAMENT	2.0	0.13	180	—	CLASS A AMPLIFIER	135 180	−22.5 −30.0	—	—	8.0 12.3	4100 3600	925 1050	3.8 3.8	7000 5700	0.185 0.375	31
32	R-F AMPLIFIER TETRODE	MEDIUM 4-PIN	FIG. 4	5 1/32" x 1 13/16"	D-C FILAMENT	2.0	0.06	180	67.5	SCREEN GRID R-F AMPLIFIER	135 180	− 3.0 − 3.0	67.5 67.5	0.4* 0.4*	1.7 1.7	950000 1200000	640 650	610 780	—	—	32
										BIAS DETECTOR	180♥	− 6.0 approx.	67.5	—	Plate current to be adjusted to 0.2 milliampere with no signal.						
33	POWER AMPLIFIER PENTODE	MEDIUM 5-PIN	FIG. 6	4 11/16" x 1 13/16"	D-C FILAMENT	2.0	0.26	180	180	CLASS A AMPLIFIER	180	−18.0	180	5.0	22.0	55000	1700	90	6000	1.4	33
34	SUPER-CONTROL R-F AMPLIFIER PENTODE	MEDIUM 4-PIN	FIG. 4A	5 1/32" x 1 13/16"	D-C FILAMENT	2.0	0.06	180	67.5	SCREEN GRID R-F AMPLIFIER	135 180	− 3.0 min.	67.5 67.5	1.0 1.0	2.8 2.8	600000 1000000	600 620	360 620	—	—	34
35	SUPER-CONTROL R-F AMPLIFIER TETRODE	MEDIUM 5-PIN	FIG. 9	5 1/32" x 1 13/16"	HEATER	2.5	1.75	275	90	SCREEN GRID R-F AMPLIFIER	180 250	− 3.0 min.	90 90	2.5* 2.5*	6.3 6.5	300000 400000	1020 1050	305 420	—	—	35
36	R-F AMPLIFIER TETRODE	SMALL 5-PIN	FIG. 9	4 17/32" x 1 9/16"	HEATER	6.3	0.3	250	90	SCREEN GRID R-F AMPLIFIER	100 180 250	− 1.5 − 3.0 − 3.0	55 90 90	— — 1.7*	1.8 3.1 3.2	550000 500000 550000	850 1050 1080	470 525 595	—	—	36
										BIAS DETECTOR	100● 250●	− 5.0 − 8.0	55 90	—	Plate current to be adjusted to 0.1 milliampere with no signal.						
37	DETECTOR★ AMPLIFIER TRIODE	SMALL 5-PIN	FIG. 8	4 1/4" x 1 9/16"	HEATER	6.3	0.3	250	—	CLASS A AMPLIFIER	90 180 250	− 6.0 −13.5 −18.0	—	—	2.5 4.3 7.5	11500 10200 8400	800 900 1100	9.2 9.2 9.2	—	—	37
										BIAS DETECTOR	90 250	−10.0 −28.0	—	—	Plate current to be adjusted to 0.2 milliampere with no signal.						
38	POWER AMPLIFIER PENTODE	SMALL 5-PIN	FIG. 9A	4 17/32" x 1 9/16"	HEATER	6.3	0.3	250	250	CLASS A AMPLIFIER	100 180 250	− 9.0 −18.0 −25.0	100 180 250	1.2 2.4 3.8	7.0 14.0 22.0	140000 115000 100000	875 1050 1200	120 120 120	15000 11600 10000	0.27 1.00 2.50	38
39-44	SUPER-CONTROL R-F AMPLIFIER PENTODE	SMALL 5-PIN	FIG. 9A	4 17/32" x 1 9/16"	HEATER	6.3	0.3	250	90	SCREEN GRID R-F AMPLIFIER	90 180 250	− 3.0 min.	90 90 90	1.6 1.4 1.4	5.6 5.8 5.8	375000 750000 1000000	960 1000 1050	360 750 1050	—	—	39-44

★For Grid-leak Detection—plate volts 45, grid return to + filament or to cathode.
■Either A. C. or D. C. may be used on filament or heater, except as specifically noted. For use of D. C. on A-C filament types, decrease stated grid volts by ½ (approx.) of filament voltage.
● Applied through plate coupling resistor of 250000 ohms or 500-henry choke shunted by 0.25 megohm resistor.
♥Applied through plate coupling resistor of 100000 ohms.
*Maximum.

TYPE	NAME	BASE	SOCKET CONNECTIONS	DIMENSIONS MAXIMUM OVERALL (LENGTH X DIAMETER)	CATHODE TYPE ■	RATING: FILAMENT OR HEATER VOLTS	RATING: FILAMENT OR HEATER AMPERES	RATING: PLATE MAX. VOLTS	RATING: SCREEN MAX. VOLTS	USE	PLATE SUPPLY VOLTS	GRID VOLTS ■	SCREEN VOLTS	SCREEN MILLI-AMP.	PLATE MILLI-AMP.	A-C PLATE RESISTANCE OHMS	MUTUAL CONDUCTANCE MICROMHOS	VOLTAGE AMPLIFICATION FACTOR	LOAD FOR STATED POWER OUTPUT OHMS	POWER OUTPUT WATTS	TYPE
40	VOLTAGE AMPLIFIER TRIODE	MEDIUM 4-PIN	FIG. 1	4 11/16" x 1 13/16"	D-C FILAMENT	5.0	0.25	180	—	CLASS A AMPLIFIER	135✖ 180✖	− 1.5 − 3.0	—	—	0.2 0.2	150000 150000	200 200	30 30	—	—	40
41	POWER AMPLIFIER PENTODE	SMALL 6-PIN	FIG. 15A	4 1/4" x 1 9/16"	HEATER	6.3	0.4	250	250	CLASS A AMPLIFIER	100 180 250	− 7.0 −13.5 −18.0	100 180 250	1.6 3.0 5.5	9.0 18.5 32.0	103500 81000 68000	1450 1850 2200	150 150 150	12000 9000 7600	0.33 1.50 3.40	41
42	POWER AMPLIFIER PENTODE	MEDIUM 6-PIN	FIG. 15A	4 11/16" x 1 13/16"	HEATER	6.3	0.7	250	250	CLASS A AMPLIFIER	250	−16.5	250	6.5	34.0	100000	2200	220	7000	3.00	42
43	POWER AMPLIFIER PENTODE	MEDIUM 6-PIN	FIG. 15A	4 11/16" x 1 13/16"	HEATER	25.0	0.3	135	135	CLASS A AMPLIFIER	95 135	−15.0 −20.0	95 135	4.0 7.0	20.0 34.0	45000 35000	2000 2300	90 80	4500 4000	0.90 2.00	43
45	POWER AMPLIFIER TRIODE	MEDIUM 4-PIN	FIG. 1	4 11/16" x 1 13/16"	FILAMENT	2.5	1.5	275	—	CLASS A AMPLIFIER	180 250 275	−31.5 −50.0 −56.0	180 250 275	—	31.0 34.0 36.0	1650 1610 1700	2125 2175 2050	3.5 3.5 3.5	2700 3900 4600	0.82 1.60 2.00	45
46	DUAL-GRID POWER AMPLIFIER	MEDIUM 5-PIN	FIG. 7	5 5/8" x 2 3/16"	FILAMENT	2.5	1.75	250	—	CLASS A AMPLIFIER □	250	−33.0	—	—	22.0	2380	2350	5.6	6400	1.25	46
								400	—	CLASS B AMPLIFIER ◆	300 400	0 0	—	—	Power output values are for 2 tubes at indicated plate-to-plate load.				5200 5800	16.0 20.0	
47	POWER AMPLIFIER PENTODE	MEDIUM 5-PIN	FIG. 6	5 3/8" x 2 1/16"	FILAMENT	2.5	1.75	250	250	CLASS A AMPLIFIER	250	−16.5	250	6.0	31.0	60000	2500	150	7000	2.7	47
48	POWER AMPLIFIER TETRODE	MEDIUM 6-PIN	FIG. 15	5 3/8" x 2 1/16"	D-C HEATER	30.0	0.4	125	100	CLASS A AMPLIFIER	96 125	−19.0 −20.0	96 100	9.0 9.5	52.0 56.0	— —	3800 3900	— —	1500 1500	2.0 2.5	48
49	DUAL-GRID POWER AMPLIFIER	MEDIUM 5-PIN	FIG. 7	4 11/16" x 1 13/16"	D-C FILAMENT	2.0	0.12	135	—	CLASS A AMPLIFIER □	135	−20.0	—	—	6.0	4175	1125	4.7	11000	0.17	49
								180	—	CLASS B AMPLIFIER ◆	180	0	—	—	Power output values are for 2 tubes at indicated plate-to-plate load.				12000	3.5	
50	POWER AMPLIFIER TRIODE	MEDIUM 4-PIN	FIG. 1	6 1/4" x 2 11/16"	FILAMENT	7.5	1.25	450	—	CLASS A AMPLIFIER	300 400 450	−54.0 −70.0 −84.0	—	—	35.0 55.0 55.0	2000 1800 1800	1900 2100 2100	3.8 3.8 3.8	4600 3670 4350	1.6 3.4 4.6	50
53	TWIN-TRIODE AMPLIFIER	MEDIUM 7-PIN#	FIG. 24	4 11/16" x 1 13/16"	HEATER	2.5	2.0	300	—	CLASS B AMPLIFIER	250 300	0 0	—	—	Power output value is for one tube at stated load, plate-to-plate.				8000 10000	8.0 10.0	53
55	DUPLEX-DIODE TRIODE	SMALL 6-PIN	FIG. 13	4 17/32" x 1 9/16"	HEATER	2.5	1.0	250	—	TRIODE UNIT AS CLASS A AMPLIFIER	135 180 250	−10.5 −13.5 −20.0	—	—	3.7 6.0 8.0	11000 8500 7500	750 975 1100	8.3 8.3 8.3	25000 20000 20000	0.075 0.160 0.350	55
56	SUPER-TRIODE AMPLIFIER DETECTOR★	SMALL 5-PIN	FIG. 8	4 1/4" x 1 9/16"	HEATER	2.5	1.0	250	—	CLASS A AMPLIFIER	250	−13.5	—	—	5.0	9500	1450	13.8	—	—	56
										BIAS DETECTOR	250	−20.0 approx.	—	—	Plate current to be adjusted to 0.2 milliampere with no signal.						
57	TRIPLE-GRID DETECTOR AMPLIFIER	SMALL 6-PIN	FIG. 11	4 15/16" x 1 9/16"	HEATER	2.5	1.0	250	100	SCREEN GRID R-F AMPLIFIER	250	− 3.0	100	0.5	2.0	exceeds 1.5 meg.	1225	exceeds 1500	—	—	57
										BIAS DETECTOR	250	− 1.95	50	Cathode current 0.65 ma.		—	Plate coupling resistor 250000 ohms. Grid coupling resistor 250000 ohms**				

★For Grid-leak Detection—plate volts 45, grid return to + filament or to cathode. ●Applied through plate coupling resistor of 250000 ohms or 500-henry choke shunted by 0.25 megohm resistor. *Maximum.

Characteristics of Receiving Tubes

(Continued)

TYPE	NAME	BASE	SOCKET CONNECTIONS	DIMENSIONS MAXIMUM OVERALL LENGTH x DIAMETER	CATHODE TYPE ■	RATING: FILAMENT OR HEATER VOLTS	RATING: FILAMENT OR HEATER AMPERES	RATING: PLATE MAX. VOLTS	RATING: SCREEN MAX. VOLTS	USE Values to right give operating conditions and characteristics for indicated typical use	PLATE SUPPLY VOLTS	GRID VOLTS ■	SCREEN VOLTS	SCREEN MILLI-AMP.	PLATE MILLI-AMP.	A-C PLATE RESISTANCE OHMS	MUTUAL CONDUCTANCE MICRO-MHOS	VOLTAGE AMPLIFICATION FACTOR	LOAD FOR STATED POWER OUTPUT OHMS	POWER OUTPUT WATTS	TYPE
58	TRIPLE-GRID SUPER-CONTROL AMPLIFIER	SMALL 6-PIN	FIG. 11	4 15/16" x 1 9/16"	HEATER	2.5	1.0	250	100	SCREEN GRID R-F AMPLIFIER	250	−3.0 min.	100	2.0	8.2	800000	1600	1280	—	—	58
										MIXER IN SUPERHETERODYNE	250	−10.0	100	—	—	Oscillator peak volts = 7.0.					
59	TRIPLE-GRID POWER AMPLIFIER	MEDIUM 7-PIN♯	FIG. 18	5 3/8" x 2 1/16"	HEATER	2.5	2.0	250	—	AS TRIODE ¶ CLASS A AMPLIFIER	250	−28.0	—	—	26.0	2300	2600	6.0	5000	1.25	59
								250	250	AS PENTODE ■■ CLASS A AMPLIFIER	250	−18.0	250	9.0	35.0	40000	2500	100	6000	3.00	
								400	—	AS TRIODE ● CLASS B AMPLIFIER	300 400	0 0	—	—	Power output values are for 2 tubes at indicated plate-to-plate load.				4600 6000	15.0 20.0	
71-A	POWER AMPLIFIER TRIODE	MEDIUM 4-PIN	FIG. 1	4 11/16" x 1 13/16"	FILAMENT	5.0	0.25	180	—	CLASS A AMPLIFIER	90 180	−19.0 −43.0	—	—	10.0 20.0	2170 1750	1400 1700	3.0 3.0	3000 4800	0.125 0.790	71-A
75	DUPLEX-DIODE HIGH-MU TRIODE	SMALL 6-PIN	FIG. 13	4 17/32" x 1 9/16"	HEATER	6.3	0.3	250	—	TRIODE UNIT AS CLASS A AMPLIFIER	250✖	−1.35	—	—	0.4	—	—	Gain per stage = 50-60			75
76	SUPER-TRIODE AMPLIFIER DETECTOR★	SMALL 5-PIN	FIG. 8	4 1/4" x 1 9/16"	HEATER	6.3	0.3	250	—	CLASS A AMPLIFIER	250	−13.5	—	—	5.0	9500	1450	13.8	—	—	76
										BIAS DETECTOR	250	−20.0	—	—	Plate current to be adjusted to 0.2 milliampere with no signal.						
77	TRIPLE-GRID DETECTOR AMPLIFIER	SMALL 6-PIN	FIG. 11	4 17/32" x 1 9/16"	HEATER	6.3	0.3	250	100	SCREEN GRID R-F AMPLIFIER	100 250	−1.5 −3.0	60 100	0.4 0.5	1.7 2.3	650000 1500000	1100 1250	715 1500	—	—	77
										BIAS DETECTOR	250	−1.95	50	Cathode current 0.65 ma.		—	Plate coupling resistor 250000 ohms. Grid coupling resistor 250000 ohms**				
78	TRIPLE-GRID SUPER-CONTROL AMPLIFIER	SMALL 6-PIN	FIG. 11	4 17/32" x 1 9/16"	HEATER	6.3	0.3	250	125	SCREEN GRID R-F AMPLIFIER	90 180 250 250	−3.0 min.	90 75 100 125	1.3 1.0 1.7 2.6	5.4 4.0 7.0 10.5	315000 1000000 800000 600000	1275 1100 1450 1650	400 1100 1160 990	—	—	78
79	TWIN-TRIODE AMPLIFIER	SMALL 6-PIN	FIG. 19	4 17/32" x 1 9/16"	HEATER	6.3	0.6	250	—	CLASS B AMPLIFIER	180 250	0 0	—	—	Power output value is for one tube at stated load, plate-to-plate.				7000 14000	5.5 8.0	79
85	DUPLEX-DIODE TRIODE	SMALL 6-PIN	FIG. 13	4 17/32" x 1 9/16"	HEATER	6.3	0.3	250	—	TRIODE UNIT AS CLASS A AMPLIFIER	135 180 250	−10.5 −13.5 −20.0	—	—	3.7 6.0 8.0	11000 8500 7500	750 975 1100	8.3 8.3 8.3	25000 20000 20000	0.075 0.160 0.350	85
89	TRIPLE-GRID POWER AMPLIFIER	SMALL 6-PIN	FIG. 14	4 17/32" x 1 9/16"	HEATER	6.3	0.4	250	250	AS TRIODE ¶ CLASS A AMPLIFIER	160 180 250	−20.0 −22.5 −31.0	—	—	17.0 20.0 32.0	3300 3000 2600	1425 1550 1800	4.7 4.7 4.7	7000 6500 5500	0.300 0.400 0.900	89
										AS PENTODE ■■ CLASS A AMPLIFIER	100 180 250	−10.0 −18.0 −25.0	100 180 250	1.6 3.0 5.5	9.5 20.0 32.0	104000 80000 70000	1200 1550 1800	125 125 125	10700 8000 6750	0.33 1.50 3.40	
										AS TRIODE ● CLASS B AMPLIFIER	180	0	—	—	Power output values are for 2 tubes at indicated plate-to-plate load.				13600 9400	2.50 3.50	
V-'99 X-'99	DETECTOR★ AMPLIFIER TRIODE	SMALL 4-NUB SMALL 4-PIN	FIG. 10 FIG. 1	3 1/2" x 1 1/16" 4 1/8" x 1 3/16"	D-C FILAMENT	3.3	0.063	90	—	CLASS A AMPLIFIER	90	−4.5	—	—	2.5	15500	425	6.6	—	—	V-'99 X-'99
112-A	DETECTOR★ AMPLIFIER TRIODE	MEDIUM 4-PIN	FIG. 1	4 11/16" x 1 13/16"	D-C FILAMENT	5.0	0.25	180	—	CLASS A AMPLIFIER	90 180	−4.5 −13.5	—	—	5.0 7.7	5400 4700	1575 1800	8.5 8.5	—	—	112-A

★For Grid-leak Detection—plate volts 45, grid return to + filament or to cathode.
■Either A. C. or D. C. may be used on filament or heater, except as specifically noted. For use of D. C. on A-C filament types, decrease stated grid volts by ½ (approx.) of filament voltage.
♯Requires different socket from small 7-pin.

■■Grid #1 is control grid. Grid #2 is screen. Grid #3 tied to cathode.
¶Grid #1 is control grid. Grids #2 and #3 tied to plate. ✖Applied through plate coupling resistor of 250000 ohms.
●Grids #1 and #2 connected together. Grid #3 tied to plate. **For grid of following tube.

RECTIFIERS

TYPE	NAME	BASE	SOCKET CONNECTIONS	DIMENSIONS	CATHODE TYPE	FILAMENT VOLTS	FILAMENT AMPERES	PLATE MAX. VOLTS	SCREEN MAX. VOLTS	RATINGS	TYPE
5Z3	FULL-WAVE RECTIFIER	MEDIUM 4-PIN	FIG. 2	5 3/8" x 2 1/16"	FILAMENT	5.0	3.0	—	—	Maximum A-C Voltage per Plate .... 500 Volts, RMS Maximum D-C Output Current .... 250 Milliamperes	5Z3
12Z3	HALF-WAVE RECTIFIER	SMALL 4-PIN	FIG. 22	4 1/4" x 1 9/16"	HEATER	12.6	0.3	—	—	Maximum A-C Plate Voltage .... 250 Volts, RMS Maximum D-C Output Current .... 60 Milliamperes	12Z3
25Z5	RECTIFIER-DOUBLER	SMALL 6-PIN	FIG. 5	4 1/4" x 1 9/16"	HEATER	25.0	0.3	—	—	Maximum A-C Voltage per Plate .... 125 Volts, RMS Maximum D-C Output Current .... 100 Milliamperes	25Z5
1-V°	HALF-WAVE RECTIFIER	SMALL 4-PIN	FIG. 22	4 1/4" x 1 9/16"	HEATER	6.3	0.3	—	—	Maximum A-C Plate Voltage .... 350 Volts, RMS Maximum D-C Output Current .... 50 Milliamperes	1-V°
80	FULL-WAVE RECTIFIER	MEDIUM 4-PIN	FIG. 2	4 11/16" x 1 13/16"	FILAMENT	5.0	2.0	—	—	A-C Voltage per Plate (Volts RMS).. 350 400 550 D-C Output Current (Maximum MA.) 125 110 135 The 550 volt rating applies to filter circuits having an input choke of at least 20 henries.	80
'81	HALF-WAVE RECTIFIER	MEDIUM 4-PIN	FIG. 3	6 1/4" x 2 7/16"	FILAMENT	7.5	1.25	—	—	Maximum A-C Plate Voltage .... 700 Volts, RMS Maximum D-C Output Current .... 85 Milliamperes	'81
82	FULL-WAVE➤ RECTIFIER	MEDIUM 4-PIN	FIG. 2	4 11/16" x 1 13/16"	FILAMENT	2.5	3.0	—	—	Maximum A-C Voltage per Plate ..... 500 Volts, RMS Maximum D-C Output Current .... 125 Milliamperes Maximum Peak Inverse Voltage ..... 1400 Volts Maximum Peak Plate Current .... 400 Milliamperes	82
83	FULL-WAVE➤ RECTIFIER	MEDIUM 4-PIN	FIG. 2	5 3/8" x 2 1/16"	FILAMENT	5.0	3.0	—	—	Maximum A-C Voltage per Plate ..... 500 Volts, RMS Maximum D-C Output Current .... 250 Milliamperes Maximum Peak Inverse Voltage ..... 1400 Volts Maximum Peak Plate Current .... 800 Milliamperes	83
84 also 6Z4	FULL-WAVE RECTIFIER	SMALL 5-PIN	FIG. 23	4 1/4" x 1 9/16"	HEATER	6.3	0.5	—	—	Maximum A-C Voltage per Plate .... 350 Volts, RMS Maximum D-C Output Current .... 50 Milliamperes	84 also 6Z4

➤ Mercury Vapor Type. ° Interchangeable with Type 1.

★For Grid-leak Detection—plate volts 45, grid return to + filament or to cathode. □ Grid next to plate tied to plate. ◇Two grids tied together. ●Requires different socket from small 7-pin. ❖Applied through plate coupling resistor of 250000 ohms. **For grid of following tube.

Characteristics of Metal Tubes

TUBE TYPE	Fil. or Heater V.	Fil. or Heater A.	Max. Pl. V.	Max. S.-G. V.	Grid V. Neg.	Pl. Ma.	Cath. Ma	Plate Resis.	Mut-ual Cond.	Amp. Factor	Plate Load	Out-Put Watts*	Equiv Types	No. of Pins	Function
6A8 *RK*	6.3	0.3	250	100	3.0	4.0	14	300M	520				6A7	8	Pent. Converter
6A8 *A*	6.3	0.3	250	100	3.0	2.6	12.8						6A7	8	Pent. Converter
6A8 *TNS*	6.3	0.3	250	100	3.0	3.5		360M					6A7	7	Pent. Converter
6C5 *RATNKS*	6.3	0.3	250		8.0	8.0		10M	2,000	20			76	6	Triode Amply.
6D5 *RATNKS*	6.3	0.7	275		40	31		2,250	2,100	4.7	7,200	1.4	45	6	Triode Amp., Class A
6D5 *NKA*	6.3	0.7	300		50	23					5,300	5.0	45	6	Triode Amp., Class AB
6F6 *RKS*	6.3	0.7	250	250	16.5	34		100M	2,300	200	7,000	3.0	42	7	Pentode Output, Class A
6F6 *TAN*	6.3	0.7	250	250	16.5	34	40.5	100M	2,200	220	7,000	3.0	42	7	Pentode Output, Class A
6F6 *KS*	6.3	0.7	250		20.0	31	31	2,600	2,700	7.0	4,000	.85	42	7	Triode Output, Class A
6F6 *K*	6.3	0.7	250	250	26.0	17	19.5				10,000	19.0	42	7	Pentode Output, Class AB
6F6 *K*	6.3	0.7	350		38.0	22.5					6,000	18.0	42	7	Triode Output, Class AB
6H6 *RATNKS*	6.3	0.3	100	Direct Current 2 Ma. (max.)									none	7	Duodiode Detector
6J7 *RTKANS*	6.3	0.3	250	100	3.0	2.0	2.5	1.5 meg. +	1,225	1,500+			6C6	7	Pentode Det.-Amp. (Non-var. Mu.)
6K7 *RTANKS*	6.3	0.3	250	100	3.0	7.0	8.7	800M	1,450	1,160			6D6	7	Var. Mu. Amplifier
6L7 *RNKS*	6.3	0.3	250	150	6.0	3.5		2.0 meg +	325				none	7	Pentagrid Mixer-Amplifier
6L7-G *A*	6.3	0.3	250	100	3.0	5.3		800M	1,100				none	7	Pentagrid Mixer-Amplifier
5Z4 *RKNTS*	5.0	2.0	400			125 Max							5Z3	5	Full-wave—H.-V Amplifier
6P7 *A* (*Pent. section*)	6.3	0.3	250	100	3.0	6.5	8.0	850M	1,100	900			6F7	8	Pentode and
(*Triode sections*)			100		3.0	3.5	3.5	17,800	450	8			6F7	8	Triode Amp in one Bulb
43-MG *T*	25.0	0.3	135	135	20	34	41	35,000	2,300	80	4,000	2.0	43	7	AC-DC Power Amp. Pentode
6B6	6.3	0.3	250		2.0	0.8		91,000	1,100	100			75	7	Duodiode-Triode
6F5 *NATKS*	6.3	0.3	250		2.0	0.9	0.9	66,000	1,500	100			none	5	High-Mu. Triode
25Z5-MG *T*	25.0	0.3	125	100									25Z5	7	Full-Wave Rectifier
5Y3 *A*	5.0	2.0	400	125									80	5	Full-Wave Rectifier
50A2-MG *T*	50 V. total drop; 0.3-A.;												none	4	Ballast tube
50B2-MG *T*	50 V. total drop; 0.3-A.;												none	4	Ballast tube

Courtesy "Radio Craft"

R-RCA and Raytheon; *K*-Ken-Rad; *A*-Arcturus; *T*-Triad; *N*-National Union; *S*-Sylvania. These letters appearing after the tube types above mean that data was available from the makers on these particular types. Some manufacturers do not as yet make all the types at present available. Arcturus tube designations are all terminated by "G," meaning glass-"metal"; the Triad termination is "MG", meaning metal-glass. Where manufacturers differ somewhat in their tube characteristics, the tube is listed twice, as is the case with the 6A8.

The power tubes, 6D5 and 6F6 appear more than once because they are used under different operating conditions. The 6H6 is equivalent to the two diodes of a 75, while the 6F5 resembles the triode section of a 75. The Triad 50A2-MG and 50B2-MG are ballast tubes, both having a voltage drop of 50, the former for use with one Type No. 40 pilot lamp and the latter for use with two. They are to be used in A.C.-D.C. sets, in place of the usual series resistors.

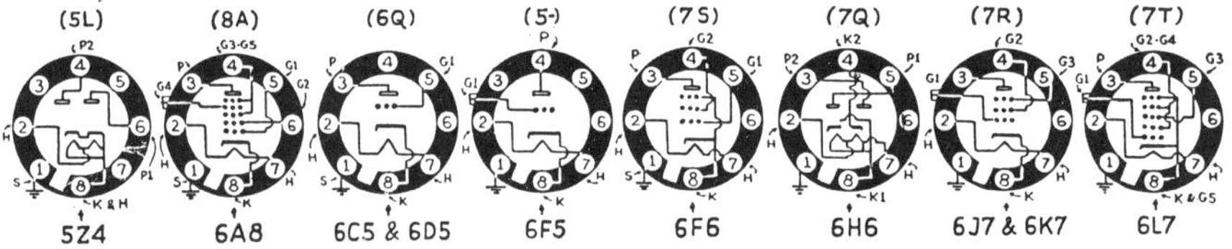

Metal Tubes Released After Above Chart Was Compiled:

6X5—Full-wave rectifier for automobile service.

6Q7—Duplex Diode, high mu (70) triode. 6.3v heater.

25A6—Power-Amplifier-Pentode. 18v heater.

25Z6—Rectifier, voltage doubler. 85 m.a. Heater 0.3 amp.

0Z4—Gas-filled filamentless rectifier (Raytheon).

Compact I-F Amplifier with metal tubes and Aladdin midget iron-core I-F transformers.

The Eimac Tube Family

Front row, left to right: 50T, 150T, 35T. In the rear are the larger tubes; left to right: 300T, 500T and a special-duty high-voltage rectifier.

TRANSMITTING TUBES
Static Characteristics

Type	Max. Continuous Plate Loss In Watts	Cathode Voltage	Cathode Current (Amps.)	Max. Plate Voltage	Max. Continuous D-C Plate Current M. A.	Amp. Factor	Inter-Electrode Capacities: Grid to Plate uufds.	Grid to Fil. uufds.	Plate to Fil. uufds.	Base	Purpose
45	10	2.5	1.5	300	40	3.5	8	5	3	M4	Low MU General Purpose Triode, Class A Audio, Self-Excited Oscillator, etc.
46	10	2.5	1.75	300	45	30	......	......	......	M5	High MU Dual Grid Triode. Class B Audio and Doubler-Buffer
59	10	2.5	2	300	45	30	......	......	......	M7	General Purpose Pentode. Class B and A Audio, Crystal Oscillator and Doubler
47	10	2.5	1.75	300	45	150	1.25	8.7	13.2	M5	General Purpose Pentode. Class A Audio, and Crystal Oscillator
6B5	10	6.3	.8	400	......	......	......	......	......	......	Audio Amplifier and Crystal Oscillator
RK15	10	2.5	1.75	400	......	......	7.5	2.7	5	M4C	Class B Modulator and R. F. Doubler
RK16	10	2.5	2	400	......	6	7.5	3.8	6	M5	Class B Driver or R. F. Amplifier
RK17	10	2.5	2	400	......	220	1	7.5	16	M 5G	Crystal Oscillator and Doubler
53 6A6	10	2.5 6.3	2 .8	300	40	27	......	......	......	M7	Dual High MU Triodes. Class B Audio and Crystal Osc.-Doubler (Jones Exciter)
RK34	10	6.3	.8	300	40	27	2.7	4.2	2.1	M7AA	Similar to 53-6A6, except Plate leads out of top of envelope
RK23 RK25	12	2.5 6.3	2 .8	500 500	50 50	* *	.02 .02	10 10	10 10	M7A M7A	Shielded R. F. Pentode. For Crystal Osc., Doubler and Suppressor Mod. Amp.
802	15	6.3	......	500	50	*	......	......	......	M7A	Somewhat similar to RK 23
WE307A	15	5.5	1	500	50	*	.55	15	12	M5A	Similar to RK 23, but has filament instead of heater
2A3 6A3	15	2.5 6.3	2.5 1	300	60	4.2	13	9	4	M4	Low MU General Purpose Triode. Class A Audio, Self-Excited Osc., etc.
210	15	7.5	1.25	600	75	8	8	5	4	M4	General Purpose Triode, Medium MU. Class B & C Primarily
841	15	7.5	1.25	600	75	30	8	5	3	M4	General Purpose High MU Triode. Class B & Class C Primarily
865	15	7.5	2	750	75	*	.05	10	7.5	M4A	R. F. Tetrode. Mainly useful as Buffer-Doubler below 10 Megacycles

TRANSMITTING TUBES
Static Characteristics

Type	Max. Continuous Plate Loss In Watts	Cathode Voltage	Cathode Current (Amps.)	Max. Plate Voltage	Max. Continuous D-C Plate Current M. A.	Amp. Factor	Inter-Electrode Capacities Grid to Plate uufds.	Grid to Fil. uufds.	Plate to Fil. uufds.	Base	Purpose
801	20	7.5	1.25	750	75	8	6	4.5	1.5	M4	General Purpose Medium MU Triode. Carbon Plate. Similar to 210
250	25	7.5	1.25	550	60	3.8	9	5	3	M4	Low MU Triode for Class A and Class AB Audio
WE254B	25	7.5	3.25	900	90	*	.085	11.2	5.4	M4A	R. F. Tetrode similar to 865, but larger
800	35	7.5	3.25	1,250	100	15	2.5	2.75	1	M4AC	General Purpose Medium MU High Frequency Triode. Useful to 200 MC
RK20	35	7.5	3	1,250	85	*	.012	11	10	M4A	R-F Pentode. Used as Crystal Osc., Buffer-Doubler, Suppressor Mod. Amp.
RK18	40	7.5	3	1,250	85	18	4.8	4.6	2.9	M4A	General Purpose Medium MU Triode. Useful up to approx. 50 MC.
RK30	40	7.5	3.25	1,250	90	14	2.5	2.75	2.75	M4AC	General Purpose Medium MU High Frequency Triode. Similar to 800
RK31	40	7.5	3	1,250	85	27	......	......	......	M4	Zero Bias High MU Triode. Mainly for Class B Audio
RK32	40	7.5	3.25	1,250	90	14	3	2	1	M4AC	Ultra-High Frequency Triode
830	40	10	2	800	75	8	9.9	4.9	2.2	M4	General Purpose Medium MU Triode. Somewhat similar to 801
830B 841A	50	10	2	1,250	100	30	10	6	2	M4A	High MU Triode. Class B Audio and Class C Amplifier below 15 Megacycles
WE304A	50	7.5	3.25	1,500	90	11	2.5	2	.67	M4AC	Ultra High Frequency Triode. Useful up to about 250 Megacycles
WE282A	70	10	3	1,500	125	*	.2	12.2	6.8	M4A	R-F Tetrode. Buffer Doubler useful below 15 Megacycles
50T	75	5	6	3,000	125	12	2	2	.2	M4AD	General Purpose Medium MU Low C Triode. Useful up to 250 Mc.
WE242A	100	10	3.25	1,250	175	12	13	6.5	4	J4	General Purpose Medium MU High C Triode. Class A, B & C up to 10 Mc.
WE261A	100	10	3.25	1,250	175	12	9	6.5	4	J4	Similar to 211-242A
WE276A	100	10	3	1,250	160	12	9	6	4	J4	Similar to 211-242A

TRANSMITTING TUBES
Static Characteristics

Type	Max. Continuous Plate Loss In Watts	Cathode		Max. Plate Voltage	Max. Continuous D-C Plate Current M. A.	Amp. Factor	Inter-Electrode Capacities			Base	Purpose
		Voltage	Current (Amps.)				Grid to Plate uufds.	Grid to Fil. uufds.	Plate to Fil. uufds.		
211	100	10	3.25	1,250	175	12	15	8	7	J4	Similar to WE-242A
203A	100	10	3.25	1,250	175	25	15	8	7	J4	General Purpose Triode, High MU. Class B & C. Useful up to 10 Mc.
838	100	10	3.25	1,250	175	27	8	6.5	5	J4	High MU, High C Triode. Zero Bias Class B; Class C, up to 15 Mc.
845	100	10	3.25	1,250	175	5	15	8	7	J4	Low MU, High C Triode. Class A and AB Audio. Also Self Excited Osc.
850	100	10	3.25	1,250	165	*	.2	17	26	J4C	R-F Tetrode. Buffer-Doubler. Useful below 5 Megacycles
RK28	100	10	5	2,000	160	.....	.02	15.5	5.5	J5A	R-F Pentode. Similar to 803
WE284A	100	10	3.25	1,250	160	4.7	8.2	7	7.8	M4A	Low MU Triode. Similar to 845
852	100	10	3.25	3,000	125	12	3	2	1	M4BC	Medium MU Triode. Mainly for High and Ultra-High Frequencies
860	100	10	3.25	3,000	125	*	.05	8.5	9	M4BC	R-F Tetrode. Doubler-Buffer. Useful below 15 Megacycles
211H	120	10	3.25	1,500	175	12	8	5.5	2	J4A	General Purpose High Frequency Triode. Useful also for Ultra-HF.
803	125	10	3.25	2,000	160	*	.15	15.5	28.5	J5A	R-F Pentode. Used as Buffer-Doubler and Suppressor or Plate Modulated Amplifier
HF200	150	10.5	3.4	2,000	200	18	5.8	5.2	1.2	J4AD	Medium MU Low C Triode. Class B and Class C. Use to 60 Megacycles
HK354	150	5	10	4,000	300	14	4	9	.2	J4A	General Purpose Triode. H.F. and U.H.F. Class B & C to 250 Mc.
150T	150	5	10	3,000	200	12	3.2	3	1	J4AD	General Purpose Medium MU Low C. Triode. Class B & C up to 250 Mc.
F108A	175	10	11	3,000	175	12	7	3	2	J4BC	Tungsten Filament General Purpose Triode. Class C up to 100 Mc.
HF300	200	11.5	4	2,500	250	23	6.5	6	1.4	J4AD	Medium High MU, Low C Triode. Class B & C up to 60 Megacycles

TRANSMITTING TUBES
Static Characteristics

Type	Max. Continuous Plate Loss In Watts	Cathode Voltage	Cathode Current (Amps.)	Max. Plate Voltage	Max. Continuous D-C Plate Current M. A.	Amp. Factor	Inter-Electrode Capacities Grid to Plate uufds.	Grid to Fil. uufds.	Plate to Fil. uufds.	Base	Purpose
WE212D	200	14	6	1,500	250	16	19	19	12	Spec.	W. E. 4 Pin Base, Gen. Purpose Medium MU, High C Triode. Not useful above 2 Mc.
204A	250	11	3.85	2,500	250	24	17	8	3	J3A	High MU, High C Triode. Class B & C up to about 10 Mc.
300T	300	7.5	11	4,000	300	16	4	3.5	1.5	J4AD	General Purpose Medium MU, Low C Triode. Class B & C, up to 250 Mc.
WE270A	300	10	9.75	2,500	300	16	21	18	2	WE2BC	General Purpose Triode. Class B & C up to 10 Megacycles
849	350	11	5	2,500	300	19	33.5	17	3	J3A	General Purpose Triode. Class B & C up to 5 Megacycles
831	400	11	10	3,000	300	14.5	4	3.8	1.5	J3BC	Medium MU, Low C Triode. Class C up to 100 Mc.
861	400	11	10	3,000	300	*	.1	17	13	J3BC	R-F Tetrode. Buffer-Doubler up to 15 Mc.
HK255	500	14	30	5,000	1,000	3	5	12	7	Spec. H-K	General Purpose Gridless Triode with Gamma Plate
500T	500	7.5	20	4,000	500	14	4.5	4	1.5	Spec. † 4AD	General Purpose Medium MU, Low C Triode. Class A, B, or C up to 150 Mc.
F100A	500	11	25	4,000	300	14	10	4	2	J3BC	Tungsten Filament Triode. Low C, and useful Class C up to 100 Megacycles
851	750	11	15.5	2,000	500	20	55	30	7	J3A	High MU, High C Triode. Class A, B & C, up to 2 Megacycles
WE251A	750	10	16	3,000	500	10.5	8	10	6	WE2BC	Medium MU Triode. Useful Class B & C below 15 Megacycles
WE279A	1000	10	21	3,000	650	10	18	15	8	WE2BC	Similar Triode toWE-251A, but larger
HK254	1500	14	45	5,000	2,000	20	15	25	2.5	Spec. H-K	High Power Air-Cooled Triode

A Plate Lead Brought Out of Top of Envelope. B Plate Lead Brought Out of Side of Envelope.
C Grid Lead Brought Out of Top of Envelope. D Grid Lead Brought Out of Side of Envelope.
M-4, 5, 6, 7:—Medium Receiving Type Socket. J3—RCA 250-Watt Socket.
J4—RCA 50-Watt Socket.
WE2—Western Electric 2-Pin Socket.
† Special EIMAC 4-Pin Base.
* μ (MU) of all Tetrodes and Pentodes varies with screen voltage. The amplification factor is lowered as the screen voltage is raised, but the transconductance is increased.

Limits of plate loss, plate voltage and plate current, given in the above table, are all independent limits and should not be exceeded. Rarely will the circuit efficiency permit the above values of plate voltage and plate current to be used at the same time. These limiting values are not to be considered the actual operating conditions. In general, as the frequency of operation is raised the limits must be reduced in order to keep the radio-frequency grid and plate currents from heating the tube seals. See Manufacturers' Application Notes for limits of r-f grid and plate current.

TRANSMITTING TRIODES
Class C Operating Constants

Type	Max. Plate Loss in Watts	U (MU)	Inter-Electrode Capacity Identification	Neut. Capa-city uufds.	Frequency Above Which Ratings Must Be Reduced	Plate Voltage	Plate Current in M. A.	Grid Current in M. A.	Negative Grid Bias Voltage	Approx. Recom-mended Grid-Leak Resistance in Ohms	Approx. R. F. Grid Driving Power in Watts	Approx. Power Output in Watts	Type
45	10	3.5	Med. C	8	7,500	400	40	2 to 5	−175 to −250	50,000 to 100,000	1 to 3	10	45
46	10	30	High C	......	7,500	400	45	5 to 10	−20 to −50	2,500 to 5,000	2	12.5	46
46 As Doubler	10	30	High C	......	7,500	400	55	5 to 10	−20 to −50	2,500 to 5,000	2 to 4	6 to 10	46 As Doubler
210	15	8	Med. C	8	7,500	600	60	12	−120 to −160	10,000	3 to 5	20 to 25	210
841	15	30	Med. C	8	7,500	600	55	16	−40 to −67	3,000 to 5,000	....	15 to 20	841
801	20	8	Med. C	6	15,000	700	65	12	−120 to −160	10,000	4 to 6	25 to 35	801
800	35	15	Low C	2.5	30,000	1,250	85	20	−150 to −220	7,500 to 10,000	4 to 6	65 to 75	800
RK-18	40	18	Low C	5	30,000	1,250	85	15	−200 to −250	15,000	5	60 to 75	RK-18
RK-18 As Doubler	40	18	Low C	5	30,000	1,000	85	15	−325 to −400	25,000	5 to 8	25 to 50	RK-18 As Doubler
WE-304A	50	11	Low C	2.5	60,000	1,250	85	20	−180 to −250	10,000	....	65 to 75	WE-304A
830-B 930-B	50	30	Med. C	10	15,000	1,250	85	20	−75 to −135	5,000	5 to 8	65 to 75	830-B 930-B

TRANSMITTING TRIODES
Class C Operating Constants

Type	Max. Plate Loss in Watts	U (MU)	Inter-Electrode Capacity Identification	Neut. Capacity uufds.	Frequency Above Which Ratings Must Be Reduced	Plate Voltage	Plate Current in M. A.	Grid Current in M. A.	Negative Grid Bias Voltage	Approx. Recommended Grid-Leak Resistance in Ohms	Approx. R. F. Grid Driving Power in Watts	Approx. Power Output in Watts	Type
50T	75	12	Low C	2	16,000	1,000	125	20	−150 −200	10,000	5 to 10	70 to 90	50T
						2,000	125	20	−300 −400	20,000	15	175	
						3,000	100	25	−450 −600	25,000	25	225	
211*	100	12	High C	13 15	5,000	1,250	175	25	−150 to −200	7,000 to 10,000	10 to 20	150	211*
203 A	100	25	High C	13 15	5,000	1,250	175	35	−90 to −135	5,000	10 to 25	150	203 A
838	100	27	High C	8	7,500	1,250	165	65	−75 to −135	2,000	....	140	838
852	100	12	Low C	3	60,000	2,500	100	35	−650 to −950	12,500	....	150	852
						3,500	100	50		20,000	....	250	
354	150	14	Low C	4	60,000	1,500	285	50	−400	8,000	30	320	354
						2,500	285	40	−800	20,000	40 to 50	580	
						4,000	277	40	−1300	30,000	60 to 80	950	
150 T	150	12	Low C	3.5	60,000	1,000	200	20	−135 to −180	7,500	5 to 10	140	150 T
						2,000	200	25	−270 to −360	12,500	15 to 20	300	
						3,000	200	30	−400 to −600	20,000	30 to 40	450	
HF 200	150	18	Low C	6	45,000	1,500	175	30	−150 to −190	6,000	25	175	HF 200
						2,500	175	30	−230 to −280	8,500	40	300	
HF 300	200	23	Low C	6.5	45,000	1,500	250	50	−120 to −150	3,000	40	275	HF 300
						2,500	225	50	−180 to −250	5,000	60	450	

*Data on the 211 also applies to Types WE-242A, 261A, 276A.

TRANSMITTING TRIODES
Class C Operating Constants

Type	Max. Plate Loss in Watts	U (MU)	Inter-Electrode Capacity Identification	Neut. Capa-city uufds.	Frequency Above Which Ratings Must Be Reduced	Plate Voltage	Plate Current in M. A.	Grid Current in M. A.	Negative Grid Bias Voltage	Approx. Recom-mended Grid-Leak Resistance in Ohms	Approx. R. F. Grid Driving Power in Watts	Approx. Power Output in Watts	Type
204-A	250	24	High C	17	5,000	2,000	250	60	−167 to −220	2,500	60	300	**204-A**
						3,000	250	75	−250 to −325	5,000	75	500	
300 T	300	16	Low C	4	60,000	2,000	300	30	−270 to −333	10,000	60	450	**300 T**
						3,000	300	30	−400 to −500	15,000	75	700	
						4,000	300	35	−540 to −666	20,000	90	950	
849	350	19	High C	34	5,000	2,000	300	65	−180 to −220	3,500	....	400	**849**
						3,000	300	80	−270 to −350	4,000	....	650	
831	400	14.5	Low C	4	30,000	3,000	300	60	−360 to −500	7,500	....	700	**831**
						4,000	300	75	−600 to −850	10,000	....	900	
500 T	500	14	Low C	4.5	60,000	1,000	500	50	−150 to −185	3,500	....	350	**500 T**
						2,000	500	50	−300 to −370	7,000	....	750	
						3,000	500	50	−450 to −550	10,000	....	1150	
851	750	20	High C	55	2,500	2,000	500	100	−180 to −220	2,000	....	650	**851**
WE-251A	750	10.5	Med. C	8	20,000	2,000	500	100	−360 to −440	4,000	....	700	**WE-251A**
						3,000	500	100	−550 to −700	6,500	....	1050	

TRANSMITTING TRIODES

CLASS BC OPERATING CONDITIONS

Operating As a Linear Amplifier Or As a Grid-Bias Modulated Amplifier

TYPE	Max. Plate Dissip. In Watts	μ (MU)	Net Plate Voltage (Gross, Less Cathode Bias)	Plate Current In M. A.	Fixed "Cut-Off" Bias (Volts)	Cathode Bias Resistor (Ohms)	Approximate Power Output: Peak 100% Mod. (Watts)	Approximate Power Output: Carrier Zero Mod. (Watts)	TYPE
210	15	8	650	34	— 80	2,350	30	7.5	210
801	20	8	750	40	— 90	2,250	40	10	801
800	35	15	1,250	44	— 80	1,800	80	20	800
WE 304 A	50	11	1,250	60	—110	1,800	100	25	WE 304 A
50T	75	12	1,500	75	—120	1,600	160	40	50T
			2,500	50	—200	5,000	200	50	
211*	100	12	1,000	150	— 80	530	200	50	211*
			1,500	110	—120	1,000	260	65	
852	100	12	2,000	75	—160	2,100	200	50	852
			3,000	55	—240	4,300	260	65	
HF 200	150	18	1,500	150	— 85	570	300	75	HF 200
			2,500	100	—130	1,300	400	100	
HK-354 OR 150T	150	12	1,500	150	—120	800	300	75	HK-354 OR 150T
			2,500	100	—200	2,000	400	100	
HF 300	200	23	1,500	200	— 67	350	400	100	HF 300
			2,500	133	—100	800	532	133	
212 D & E	200	16	1,750	180	—100	600	500	125	212 D & E
204 A	250	24	2,000	187	— 80	425	500	125	204 A
			3,000	133	—120	1,000	600	150	
300 T	300	16	1,500	300	— 90	300	600	150	300 T
			2,500	200	—150	750	800	200	
849	350	19	1,500	300	— 75	250	700	150	849
			2,500	225	—120	550	840	210	
831	400	14.5	2,500	240	—165	700	800	200	831
			3,500	190	—225	1,200	1,060	265	
500 T	500	14	2,000	375	—130	350	1,000	250	500 T
			3,000	277	—200	700	1,333	335	
851	750	20	2,500	500	—120	250	2,000	500	851
WE 251 A	750	10.5	2,000	500	—185	400	1,500	333	WE 251 A
			3,000	415	—280	700	2,000	500	

*Data on the 211 Also Applies to WE-242A, 261A, 276A.

TRANSMITTING TETRODES AND PENTODES
— OPERATING CONDITIONS —

Type	Max. Plate Loss In Watts	Max. Screen Loss In Watts	Use	Plate Voltage	Screen Voltage	Suppr. Voltage	Bias Voltage	Plate Current In M. A.	Screen Current In M. A.	Control Grid Current In M. A.	Grid Driving Power In Watts	Approx. Carrier Power Output In Watts	Type
802	10	6	C	500	250	40	—100	45	12	2	.25 to .75	16	802
			SM	500	200	—45	—90	22	28	4.5	.5	3.5	
RK-23-25	12	8	C	500	200	45	—90	55	35	6.0	.33 to .5	24	RK-23-25
			SM	500	200	—45	—90	32	40	6.0	.33 to .5	5.5	
WE-307A	15	5	C	500	250	45	—90	45	14	2.5	.....	16	WE-307A
			SM	500	200	—40	—75	25	22	6.0	.....	3.5	
865	15	3	C	750	125	.....	—90	50	.....	10	.....	25	865
WE-254B	25	5	C	1,000	150	.....	—167	75	.....	10	.....	50	WE-254B
RK-20	40	15	C	1,250	300	45	—100	92	32	5.0	1	80	RK-20
			SM	1,250	300	—45	—100	43	36	5.0	1	18	
WE-282A	70	7.5	C	1,250	250	.....	—167	125	.....	15	.....	100	WE-282A
850	100	10	C	1,250	175	.....	—167	160	.....	35	.....	130	850
860	100	10	C	2,500	300	.....	—200	125	.....	35	.....	220	860
			C	3,500	350	.....	—300	100	.....	45	.....	270	
803	125	30	C	2,000	500	40	—30	160	42	16	1.75	210	803
			SM	2,000	500	—135	—50	80	55	15	1.6	53	
RK-28	100	35	C	2,000	400	45	—100	140	60	10	1.8	200	RK-28
			SM	2,000	400	—50	—100	80	85	11	2	60	
861	400	35	C	3,000	600	.....	—250	275	.....	50	.....	550	861
			C	4,000	600	.....	—300	250	.....	60	.....	700	

C=Class C R-F Oscillator, Buffer or Amplifier.
SM=Suppressor-Grid Modulated Amplifier.
Note—The small variations from any indicated inner-electrode voltage in a multi-element tube can cause a material change in all of the electrode currents.

CONTINUOUS WAVE (C-W) TELEGRAPHY

The analysis of the circuit and component parts of so complex a device as a radio transmitter is not an easy task, but the development of the subject material can be made plain so that the beginner as well as the advanced amateur can develop as well as enrichen their scope of the subject.

Definition: A radio transmitter consists of some form of a high-frequency oscillator, and buffer amplifier stages which serve the dual purpose of amplifying the relatively weak output of the oscillator, and isolating the oscillator from the keying or modulation surges usually applied to the final amplifier.

In addition to the above, certain types of frequency stabilizing equipment (such as piezo-crystal stabilization) are employed to maintain the frequency at one value. The use of buffer amplifiers may be necessary when doubling the frequency generated by a crystal oscillator; doublers are required because mechanical limitations prevent the stable operation of piezo-crystals at frequencies higher than about 8 megacycles, whereas the final amplifier may be required to operate on much higher frequencies. The various buffers and doublers drive the grid or grids of the tubes used in the final amplifier stage. The final stage functions as a converter of direct-current plate current into radio-frequency alternating current, which is supplied to the radiating portion of the antenna system through some form of coupling device.

The Oscillator: The function of each portion of the parts in a transmitter, and the effect of varying the characteristics follow in a step-by-step analysis. Figure 1 shows the fundamental circuit of a typical transmitter using a 47 crystal oscillator and a 46 buffer-doubler.

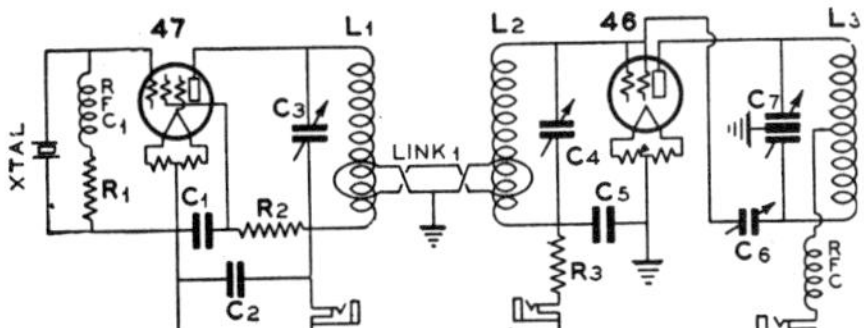

FIG. 1

Pentode crystal oscillator, link coupled to buffer stage.

The first component, from left to right, is the piezo-crystal. It usually consists of a thin, flat quartz plate whose physical dimensions permit it to resonate mechanically at the frequency of the oscillator. The crystal is mounted between two flat metal plates which rest very lightly on the crystal in order to avoid the dampening effect of any pressure which would tend to retard and make difficult mechanical oscillation. Some of the better type of crystal holders have an air gap between the top plate and crystal to minimize any dampening effects. The two metal plates are lapped perfectly flat, and the flat piece of quartz functions as the dielectric. When the crystal is set into a state of mechanical vibration an alternating-current voltage is developed across the condenser which is impressed across the grid circuit. The crystal will continue to vibrate so long as there is some kind of electrical vibrating stimulus applied to it. Hence, this stimuli can come from a separate source, such as an oscillator and tube circuit in which a small portion of energy is taken from the tuned plate circuit and fed back to the crystal circuit to sustain its oscillation.

When the crystal is maintained in oscillation, it acts as a very sharply tuned series resonant circuit, consisting of high inductance, low capacity and low resistance. The actual frequency is slightly higher or lower than exact resonance so as to give an inductive or capacitive reactance depending on the character of the oscillating circuit. A crystal-oscillator circuit has a very high "Q," which is an index of its resistance to changes in resonant frequency with variation in external constants. In this manner, the crystal acts as a tuned grid circuit whose resonant frequency is quite free from changes caused by load or voltage variations. However, the frequency may vary slightly with changes in the temperature of the plate, but in amateur practice the temperature effect need not be seriously considered (at least for the present).

The low-frequency crystals which operate upwards to 4000 KC usually start easier and develop more output energy than the higher frequency types, which are more fragile and rather difficult to handle.

The Radio-Frequency Choke: The next components in the oscillator circuit are the radio-frequency choke, RFC1, and the resistor, R1, which are connected in series and shunted across the crystal. The purpose of the resistor is to provide a DC return for the grid of the oscillator tube. In addition to the DC bias on the grid of the tube, there is also present an AC voltage which is caused by the plate-to-grid feedback in the tube. This AC voltage exceeds the DC bias and causes the grid to periodically go slightly positive with respect to the filament. When the grid is positive it attracts some of the electrons emitted by the filament which are rectified into a uni-directional current (half-wave rectifier). This small rectified DC current flows back through resistor R1 to the filament; during this flow a voltage drop occurs across R1 which is impressed on the grid and therefore becomes the source of DC bias voltage. The purpose of choke RFC1 is to impede the flow of AC current while at the same time offering little or no resistance to the passage of DC to ground through R1.

In general, lowering the ohmic value of R1 down to about 10,000 ohms will increase the RF output from the oscillator, although the use of high resistive values up to about 50,000 ohms will permit the crystal to start easier. It is has been found that the better made crystals start with a 10,000 ohm resistor, while poorer or inferior makes require higher ohmages.

In crystal oscillator circuits where harmonic generation is utilized in the crystal stage itself, such as in the "Tri-tet," "Dow Crystal Doubler," and in the "Jones All-Band Exciter," a high value of grid leak is used for an altogether different purpose. The distortion in the output of a vacuum-tube amplifier increases as the bias is increased, and it is the harmonic distortion which produces the second or fourth harmonic selected by the output tank circuit.

Center-tapped Resistor: These are used to divide the DC and RF currents equally across both halves of the filament. If these returns were connected to only one-half or to one side of the filament the 60 cycle AC hum would increase in the output, because one-half of the filament heating voltage is periodically added to and subtracted from the grid voltage, which effectively modulates it with the hum frequency.

Oscillator Tube: The oscillator tube requires little mention. Vacuum tube theory and operation are completely covered elsewhere (see Index). The ideal crystal oscillator tube should have a high amplification factor, medium-to-low plate resistance, as well as low inter-electrode capacities. The screening need not be perfect as some feedback is necessary for self-oscillation, but it must be kept at a very low value to keep the RF current at a minimum. In some transmitting pentodes, such as the RK20 and 802, the screening is so perfect that a small external capacity must be used to provide the necessary feedback. This is advantageous in that it allows some adjustment of the feedback so that the best possible compromise between power output and RF current through the crystal can be obtained.

By-pass Condensers: At all points where radio-frequency energy is by-passed in an amateur transmitter, non-inductive condensers of the mica dielectric type should be used.

Resistor R2: This resistance drops the plate voltage for the screen circuit to approximately 100 volts. The value of R2 can be between 25,000 and 50,000 ohms because the screen current varies enough to offset variation in this resistor, thus varying the drop through the resistor so that the screen voltage is normal.

The Buffer Doubler: The grid circuit is ALWAYS tuned to the same frequency to which the plate tank circuit which feeds it energy is tuned, even if the stage operates as a frequency multiplier, because frequency multiplication manifests itself in a plate circuit. The resistor R3 acts as the grid leak for the 46 stage and places a DC bias on the grid, due to the rectified current which flows through it causing the usual voltage drop. Whether the stage is to operate as a straight buffer amplifier, as a frequency multiplier, or doubler, are also factors which determine the value of R3. If the available excitation is low (less than 10 milliamperes of DC grid current, measured at J2) the grid leak can be eliminated and the lower end of L2 connected directly to ground. In this case, condenser C5 would also be eliminated. However, more excitation than 10 milliamperes is generally available and thus the grid leak is desirable. Its value is not critical up to 2,000 ohms, and values as high as 5,000 ohms are sometimes desirable for best doubling efficiency.

The grid by-pass condenser C5 provides a path for the RF return so that the grid circuit is completed back to the filament.

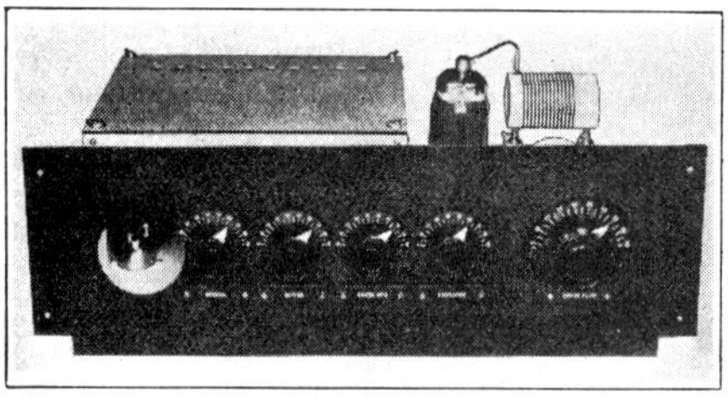

A typical oscillator and pentode buffer driver for a medium power pentode doubler.

In other words, the DC grid path goes to ground through R3, while the RF grid path to ground flows through C5 and not through R3.

In the first buffer stage maximum power amplification is desired, not maximum plate efficiency.

Neutralizing: As was previously shown in the oscillator stage, the plate and grid of an ordinary vacuum tube act as two plates of a small condenser, so that a measurable amount of RF voltage present in the plate circuit is by-passed back to the grid circuit, where it adds to the voltages already present in that circuit in again increasing the amplitude of the RF voltage in the plate circuit. Thus there is a cumulative rise in the AC plate and grid voltages which continues and rises even after the excitation voltage from the oscillator is removed. This condition is called **Self-oscillation;** it is the frequency at which oscillation is not controlled by the quartz crystal. This state of oscillation is avoided by the process of neutralization.

The fundamentals of resonant circuits shows that the voltages at the opposite ends of a parallel resonant tank circuit are equal, though opposite in polarity at any given instant, when the center of the coil is the reference point. In the case of the plate tank coil L3 and the condenser C7, the reference point is established at the center of the coil and in the condenser by grounding the split-stator rotor. So if the capacity of C6 is equal to the plate-to-grid capacity of the 46 tube, the voltage drop across this condenser will be equal to the voltage drop across the small condenser consisting of the tube, thereby balancing out the AC voltage. If this voltage was not neutralized, the tube would go into a state of self-oscillation.

Neutralization to prevent self-oscillation is necessary only when the stage is operated as a straight buffer-amplifier. When the stage is employed to function as a doubler, there is little tendency for self-oscillation because of the plate tank circuit being tuned to a different frequency than that of the grid tank. However, the neutralizing circuit becomes a regeneration circuit and actually aids in doubling by increasing the grid drive at the output frequency due to capacity C6. In a doubling circuit this capacity should be greater than the capacity necessary to properly neutralize a stage which operates as a straight amplifier.

The RF power in L3 can be employed to excite an antenna by means of any of the diverse antenna coupling methods, or to excite another RF amplifier stage by means of a coupling link, similar to link 1 between the oscillator and the 46 stage.

Notes on Mechanical Design and Construction

The factors entering into the mechanical design and construction of an amateur transmitter are those which govern the efficiency and the results obtainable from a circuit specification. It is important, therefore, that a great deal of consideration be given to the constructional details. Practical notes are given herewith.

Before constructing a transmitter, all of the various parts should be laid out on a board (commonly called a "breadboard") or chassis and moved about until all of the RF leads are as short and direct as it is possible to make them. It is not necessary to strive for a symmetrical layout in order to improve the appearance to an onlooker; short and direct leads are important if the transmitter is to operate efficiently.

It matters little what type of base the apparatus is mounted on. A metal chassis is preferred by some constructors, while others prefer wood; however, the metal is a little more difficult to work. If the chassis is of metal, aluminum or copper should be used, especially if radio-fréquencies are present. Cadmium or copper plating on steel is often satisfactory.

Boards have certain losses on account of most soft woods being poor dielectrics, which absorb energy in strong electro-static fields, such as those surrounding a transmitter stage. The losses may be minimized by selecting dry hardwood for the base.

An excellent breadboard base can be made of ordinary white pine covered with a thin sheet of No. 20 or 30 gauge aluminum, the sheet being neatly fastened by bending over the edges and tacking the underside down with small wire brads. This type of base allows condensers and coils to be mounted with ordinary wood screws. The metal acts as a shield and also keeps the capacity-to-ground constant from the various parts of the transmitter. Shielding is a necessary requirement, although it represents a small loss. Natural shielding; that is, the greatest permissible space between stages, is better than metal shields.

Since link coupling has been universally accepted as being a superior method for interlinking stages, there is now no valid reason for placing more than one stage on a single breadboard, as coupling links can be lengthened upwards to ten feet. If transmitter stages closely adjoin each other, it is necessary to resort to interstage shielding to prevent feedback. Complete shielding, as is specified in receiver construction, is not a necessary requirement in transmitter design; a double baffle separated by at least one-fourth inch and six inches high will suffice in practically all cases. The plates must not touch except where they are supported to the common ground connection, or screwed to the metal chassis.

Typical final amplifier with two W. E. 211-D tubes.

Recently a movement has been introduced to more or less standardize the size of breadboards and chassis. This practice ought to be encouraged by all amateurs, as it facilitates the quick exchange and replacement of component parts. In general, chassis sizes more or less follow the standard rack construction specifications originally adopted by the Bell System. The front panel of a standard rack is 19 inches wide and is some even multiple of one and three-quarter inches high. Three common sizes are: seven, eight and three-quarters, and ten and one-half inches high.

Jones Exciter on aluminum or steel chassis.

The breadboards or chassis that are mounted on or behind these front panels cannot be wider than 17 inches, due to the clearance limits between the side members of the standard rack. Most chassis are eight and one-half to twelve inches deep. No movement toward standardizing the depth has been started on account of the limitations in the strength of the supporting structure.

A neat way to lay out a transmitter is to obtain several pieces of five-ply veneer, eight and one-half inches square, and then covering these pieces with No. 28 gauge

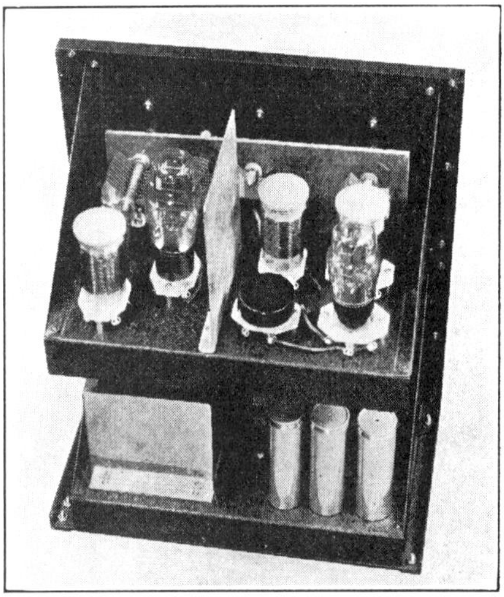

Relay rack mounting for exciter and buffer. Power supply on lower deck

aluminum. Such breadboards are of the correct size for a single low-power stage, and each can be quickly removed from the completed transmitter when rebuilding or when changes are necessary. For standard rack mounting two of these small breadboards may be mounted behind each panel. A plug and jack arrangement conveniently allows almost any stage to be taken out and replaced with another, especially in transmitters having a 47 oscillator, Jones Exciter, or electron-coupled oscillator stage. Here the buffer and doubler stages are identical and use type 210 tubes; hence, a 50

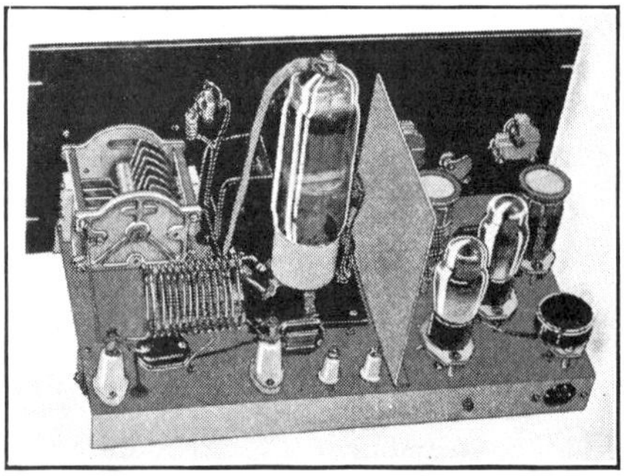

803 pentode, driven by Jones Exciter, all on a single relay rack.

1 k-w Audio Products amplifier, wood framework construction. Two HK-354 triodes in push-pull. Grid condenser and coil at bottom.

watt stage may replace any of the push-pull 210s.

Nothing in the transmitter should be nailed down. Each coil, condenser, etc., need only be fastened firmly to some support; however, it is a prerequisite that all the apparatus be placed in such a manner as to permit quick replacement should any part burn out or fail.

Not even a radio engineer can lay out a radio transmitter with the hopes of expecting it to operate perfectly the first time it is tested. Often it is found that there is insufficient excitation to some particular stage, requiring, of course, the addition of another buffer stage; this is a relatively simple problem if each stage is mounted on its own little board. Difficulty experienced from parasitic oscillation can be more easily corrected when individual breadboards are used.

When laying out a stage on a breadboard or chassis, the grid coil must be placed as far from the plate coil as possible—at least five times the diameter of the plate coil away from it. If the two coils are in close proximity, difficulties may be encountered in neutralizing the stage. In some cases, especially in high-power stages, it is desirable to orient the coils so that they are at right angles to each other so that the fields around the two coils will have the minimum of interaction between them.

High power Gammatron driver.

choke is to connect it across the tank condenser and depress the key. If the presence of the RF choke across the tank condenser materially detunes the circuit from resonance, the choke was functioning inefficiently. Few RF chokes can withstand this test. One of the disadvantages of shunt-feed is that no choke has infinite impedance, and therefore a finite amount of RF power is lost to ground. It is difficult to design and build a RF choke that is effective when used on more than one of the amateur bands. These bands are even harmonics of the lowest frequency band, whereas RF chokes operate best on the odd harmonics of the lowest frequency for which they are designed; hence, a multiband choke is only a compromise on all bands and is theoretically perfect on none.

Single-ended HK-354 Gammatron amplifier.

Tank coils for low power stages may be wound on receiving-type plug-in coil forms, providing that they are made of ceramic or Isolantite insulating materials. Most amateurs prefer five prong plug-in coils to simplify coil change and exchange between stages.

Isolantite sockets for tubes and coil forms are desirable. Some of the newer wafer-type sockets are satisfactory for stages operating with less than 500 volts plate voltage. The latest type midget condensers give splendid results when used in the grid and plate circuit of low power stages. Good practice dictates that a closed-circuit jack be placed in every grid and plate circuit, even though some meters may always be in the circuit. When the stage is removed from the transmitter for test or rebuilding, it is always convenient to be able to quickly check the grid and plate current while the stage is being tested on the workbench.

Notes on Electrical Design

Shunt-Feed and Series-Feed Tank Circuits: Two methods are employed to supply plate power to the transmitting tube; one of these is known as "Shunt-Feed," which delivers the DC from the power supply directly to the plate of the tube. This method prohibits the passage of any radio-frequency voltage present on the plate of the tube from being by-passed back to ground through the power supply. The RF currents are retarded from seeking this path by the inclusion of a RF choke shunted directly across the plate tank coil. Thus, a good test for a radio-frequency

The only advantage of shunt-feeding plate voltage through an RF choke is that it allows the plate tank coil and condenser to operate at ground potential with respect to the DC plate voltage. This condition is sometimes desirable in the design of transmitters in which the connecting leads must be kept at a minimum to permit quick band changing.

Series-feed applies the DC voltage at the bottom, or low potential end (middle of the coil in a split-tank circuit) of the plate tank coil; no radio-frequency difference voltages exist between this point and ground, and practically no RF finds its way back into the power supply. In some cases where the grounding of the transmitter is somewhat uncertain it is advisable to use an RF choke at the ground end of the coil to prevent the passage of any small RF potential differences from one part of the transmitter to another. The choke has very little work to do and can be small in size.

Eliminating Key Clicks: The transmission of intelligence by means of radiotelegraphy involves the variation of the RF carrier output between the full "on" and the full "off" position. "Mark" and "Space" are defined by the presence and absence of radiated output, respectively.

53 exciter, 803 doubler, 803 buffer for driving 1 k-w final amplifier.

The carrier wave is usually cut "on" and "off" during the keying process by opening and closing the supply circuit which delivers plate power to one or more stages in the transmitter. If the change from the "no output" condition to the "full output" condition occurs too quickly, an undesired key click will be produced. This click will be radiated over a very wide range of frequencies on each side of the carrier frequency, causing a particularly annoying form of interference to other radio services. Key clicks are often audible within a 100 mile radius, but usually cause aggravating interference to radio reception nearby.

There are two distinct types of key clicks; the most common occur at the start of an impulse, or when the key is closed. If the voltage builds up too rapidly, a discontinuous wave will be produced, and its amplitude may be several hundred times the amplitude of the signal wave. This type of click is usually damped-out by providing some form of time-lag in the circuit which forces the DC current to build up relatively slowly. By "slowly" is meant that the time required for the current from the power supply to go from zero to maximum be about one-one-hundredth second. If the time is less than approximately one-five-hundredth second, annoying clicks will be produced.

The most common form of lag circuit is one that uses a variable series inductance in series with the key, or keying relay. Often no variable inductance is available, but the inductance of any choke coil can be readily varied by connecting a variable resistor across it, as shown in Figure 4. The required value of the inductance depends on several factors, such as the amount of current flowing, the plate voltage, the voltage regulation of the source of supply, the characteristics of the filter on the supply, etc. Thus, no definite value can be specified in advance. Eliminating keying interference resolves itself into trying every known remedial measure until a satisfactory one is found.

The second type of key click is that which occurs at the end of each impulse when the key is opened. This click is a combination of the spark produced at the key contacts and the sudden change in voltage applied to the RF amplifier. The use of a series inductance increases this type of key click due to the large inductive back EMF when the circuit is opened, and the spark across the key contacts. Ordinarily, the click produced when the key is opened is considerably less bothersome than that produced when the keying contact is closed. However, a series inductance can often eliminate the "make" click at the expense of doubling or tripling the amplitude of the "break" click. The latter type of click is best eliminated by connecting a condenser in series with a variable resistor

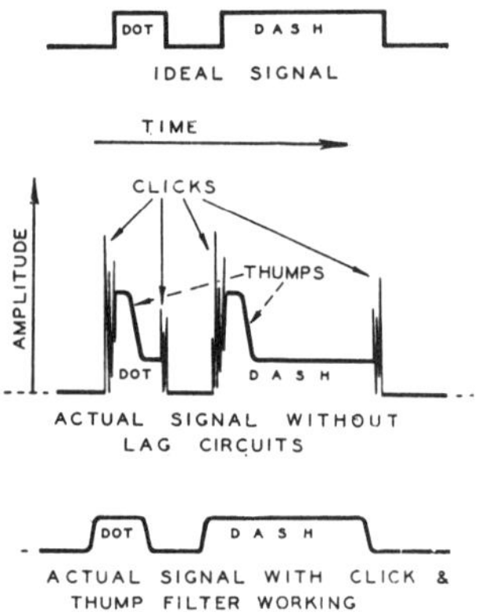

How the three types of c-w signals appear on the screen of an oscilloscope.

Vacuum tube keying unit.

across the keying contacts. The condenser-resistor circuit represents a compromise between a minimum of clicks and good keying characteristics. The value of the condenser is not critical, it may be between ½ and 2 microfarads. However, the resistor must be carefully adjusted for best results. If the value of the resistor is too large, it will put "tails" on the dots, making the signal difficult to read. If the ohmic value is too low, the plate voltage will diminish too rapidly and clicks will be produced. A time-constant of approximately one-one-hundredth second will, in most cases, allow satisfactory keying without bothersome clicks, although a fast operator who manipulates an automatic key may find that the dots are accentuated to a higher degree if the time constant is reduced to 70-to-80-thousandths second.

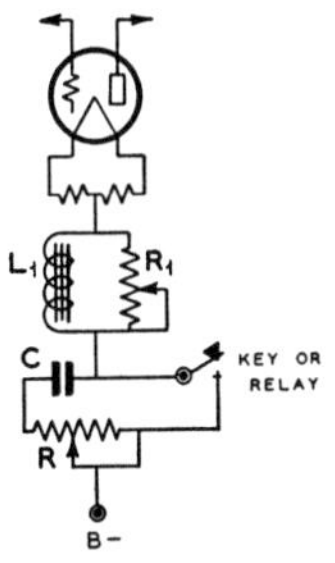

FIG. 4.

Conventional center-tap keying with an adjustable key-click filter. This system gives very good results. The actual amount of inductance and capacity in the circuit depends on the amount of current being keyed, and also on the voltage regulation of the plate power supply. L1 should be of a value between 1 and 5 henrys; R1, 20,000 ohms; C, between ¼ and 2 mfds.; R, 2,000 ohms.

To minimize the effects of the above compromise, it is desirable to key in some circuit that draws negligible power. The grid-block method of keying is useful because the key is required to open a circuit that carries no current at that instant. When keying the oscillator stage to obtain perfect break-in operation, the screen can be employed; on the other hand, the center-tap method of keying the stage is also satisfactory. Most of the high-powered commercial transmitters that key many kilowatts of power at speeds up to 500 words per minute use some variation of the vacuum-tube keying system of which representative examples are given in Figures 2 and 5.

A click at the "make" means that some form of series inductance must be added in the plate or grid circuit of one or more of the amplifiers. A click at the "break" indicates that a condenser is required across the keyed circuit to enable the voltage to diminish slowly and evenly. The adjustment of the series resistor is by far the most important in eliminating clicks.

Key Thumps and their Prevention: The deep keying thump which causes considerable interference is largely due to the plate voltage power supply building-up when the key is open, thereby causing a sudden surge of output at the instant the key is closed. The transient may rise to several times the average amplitude of the steady carrier. The thump may be eliminated by improving the voltage regulation so that it is not over 15 per cent higher when the key is open than when it is closed, during which times a power demand is being made. The best way to improve the voltage regulation is to connect a bleeder resistor across the output of the filter; the bleeder should draw enough power to sustain the voltage when the key is up. The exact value of the bleeder can best be determined by experiment because the regulation of most power supplies varies quite widely.

To design a scheme for the prevention of key thumps and clicks, it is only necessary to place a sufficient amount of inductance in series with the key to prevent too-sudden building-up of oscillations. By selecting the proper value of inductance the desired degree of "lag" can be introduced. The effect of the inductance in the circuit is satisfactory when the key is closed, but when the contact is broken, an arc occurs,

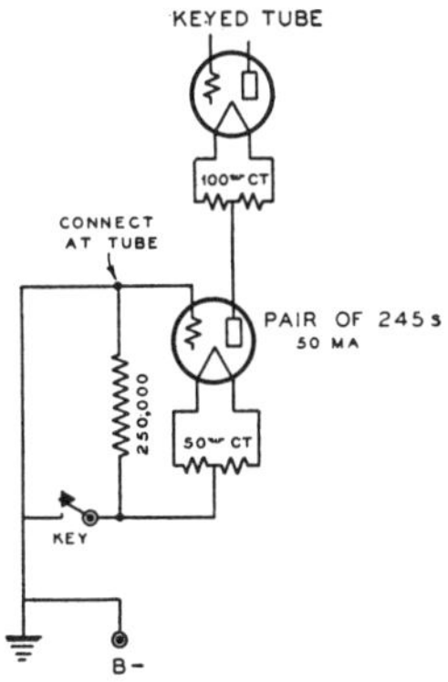

FIG. 2.

Vacuum Tube Keying. This circuit shows one of the more simple vacuum tube keying circuits. Some current flows through the key and this system sometimes produces clicks when the key is opened. Both filament transformers must be insulated from each other and also from ground. This circuit will not completely cut off the plate current to the keyed stage, but will reduce it to a very small value.

ORDINARY CENTER TAP KEYING

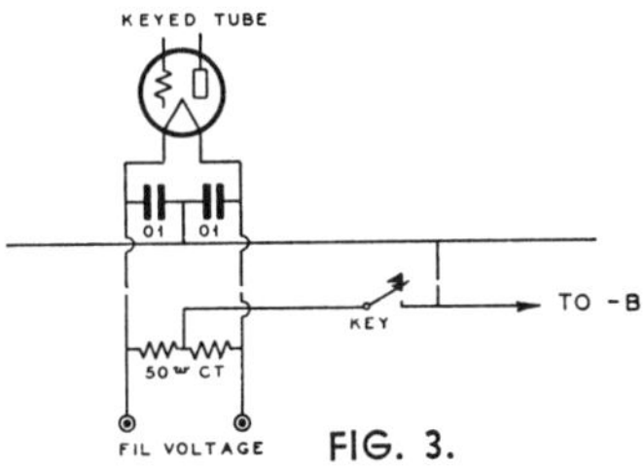

FIG. 3.

Ordinary center-tap keying. The center-tap of the filament transformer must not be grounded. As a general rule, the filament transformer which supplies the keyed stage will not be used to supply any of the other stages. The B minus lead from the power supply should be grounded.

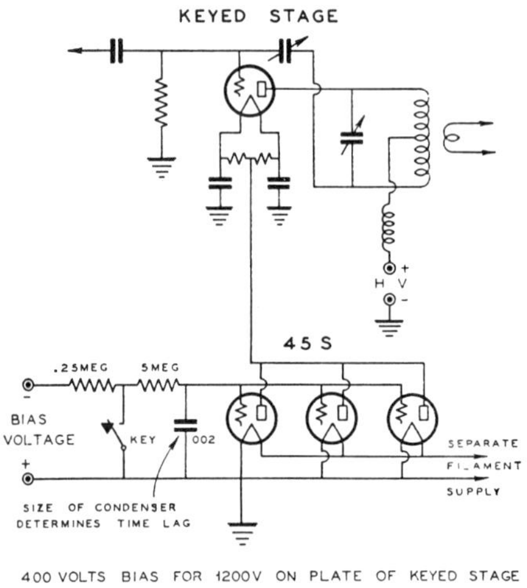

FIG. 5.

High speed commercial vacuum-tube keying system.

which has the tendency to burn the contacts. To offset this effect, it is only necessary to connect a condenser across the contacts of the key to absorb the "inductive kick-back." Now, when the key is again closed, the condenser gives up its charge, causing a spot-welding effect on the key contacts, which gives rise to parasitic interference from impact excitation of associated circuits. To remedy this condition, a resistor is placed in series with the condenser to prevent sudden discharge. Unfortunately, the resistor impairs the ability of the condenser to take on a sudden charge, absorbing the self-induced voltage of the inductance at the opening of the key and to some extent defeats the original purpose of the condenser. To compromise between the small arc occurring at the opening of the key and the small welding effect on the contacts at the closing of the key necessitates some sort of ingenious remedial measure; Figure 6 shows a scheme of great practicability. There, L1 and L2 are in series with the key and provide the necessary "lag." The "kick" from L1 is cushioned by C2. In turn, L2 prevents C2 from spot-welding the contacts on discharge. Similarly, the self-induced voltage in L2 at the opening of the circuit is taken care of by C1 and L1 and prevents a sudden discharge of C1. The correct values of L and C can be determined experimentally. The combined capacity of C1 plus C2 should be 1 microfarad or less. The value of the chokes are similar to those used in power packs.

Primary Keying: This is a type of keying which permits a grid-leak bias to be used on the keyed stages. This method prevents clicks and safeguards the filter condensers in the keyed stages, and in addition, does away with the necessity of using a high-

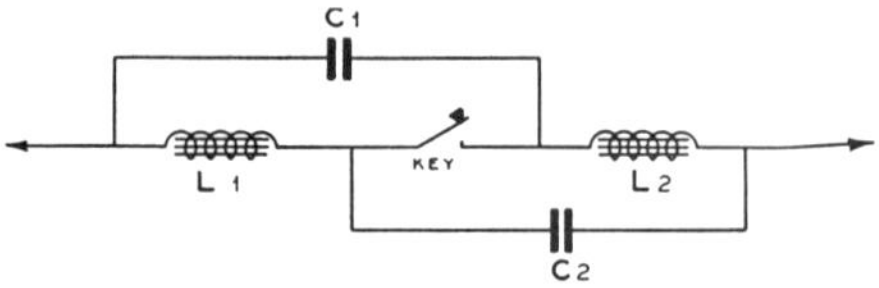

FIG. 6.

Key thump filter.

voltage bleeder and eliminates back wave 100%, if more than one stage is keyed.

The disadvantages of primary keying are:

(a) requires a heavy current relay that can break an inductive AC circuit.

(b) Tends to blink the lights when used on high power.

(c) Sometimes creates band thumps in BCL sets on the same line, caused by 60 cycle surges.

(d) makes perfect keying at high speeds difficult due to the tendency of the filter condensers to add "tails" to the dots in some cases.

Center-Tap Keying: This is another widely used method of keying which allows the use of grid-leak bias on the keyed stage, but separate bias must be used on all succeeding stages.

The advantages of center-tap keying are:

(a) will follow an automatic key ("bug") perfectly.

(b) improves the readability of received signals.

(c) permits the use of high voltage DC relays which are relatively modest in price.

The disadvantages of this system of keying are: causes bad clicks unless a well-designed click filter is used. Thumping is increased if a heavy bleeder is not placed across the high voltage; in addition, the bleeder is a necessary accessory to protect the filter condensers from failure when the key is open.

Keying the Oscillator: This is not a type of keying but a place to key. It justifies special mention because it sems to give best results at the present time.

The outstanding features of this keying arrangement permit complete break-in and completely eliminate back-wave; also, practically eliminate clicks and thumps, and will key at high speed. Unfortunately, the plan requires that a fixed bias be supplied to all the amplifier stages; also, the keying may give rise to a "chirping effect" unless the screen voltage for the crystal oscillator tube is taken from a voltage divider, rather than from a series resistor.

Blocked-Grid Keying: This method can be satisfactorily used to eliminate key clicks in low or medium power transmitters. In Fig. 7, R1 is the usual grid leak; fixed bias is applied through the 100,000 ohm resistor R2 in order to block the grid

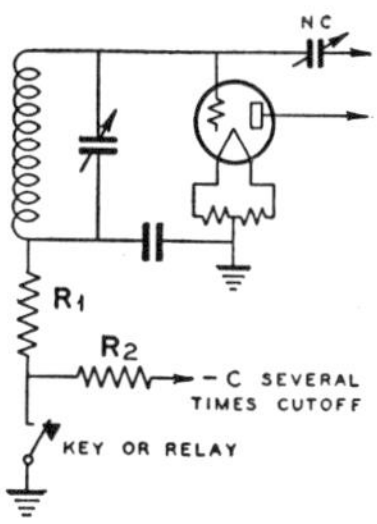

BLOCKED GRID KEYING

FIG. 7.

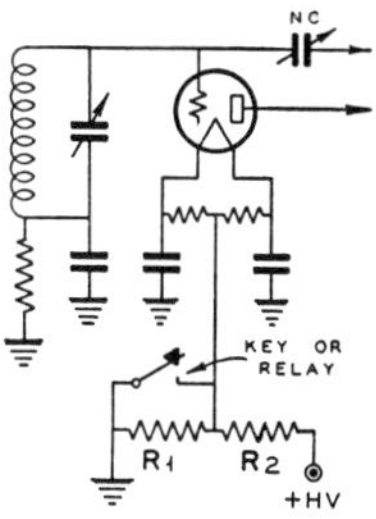

BLOCKED GRID KEYING

FIG. 8.

current. As a general rule, 200 to 400 volts of bias from a small C bias supply will reduce the output to zero. In Fig. 8, the value of R1 is from two to three times as high in value as R2, the combination being connected across the high voltage supply. The keying relay shorts out the additional bias obtained in this manner when transmitting.

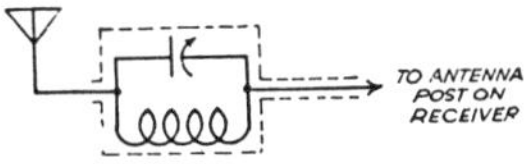

FIG. 9. Wave trap.

Interference Elimination by Wave Trap: Interference caused by amateur transmitters in the neighborhood of broadcast receivers is usually due to the fact that the **first RF stage** in the set does not possess sufficient selectivity. Thus the high-frequency signal from the amateur transmitter rides through into the grid of the first tube in the receiver. Usually no amount of selectivity beyond this point will eliminate the interference. The amateur signal causes detection and cross modulation in the first tube. One method of reducing this type of interference is to place a tuned wave-trap in series with the antenna lead to the broadcast receiver. The trap is tuned for the weakest response to the interfering signal; the device should be placed as close as possible to the antenna post on the set. It is also essential that the receiver be provided with a short low-resistance connection to ground to prevent the AC power line from bringing in the interfering signal, in spite of the wave-trap.

Eliminating the Chirps when Keying the Crystals: In the conventional pentode crystal oscillator circuit, the screen voltage is obtained from the plate power supply by means of a series dropping resistor. When an attempt is made to key in the center-tap of such a circuit, a bothersome and spurious-like chirp is manifested in the signal tone. When the key is up, the screen voltage rises to the same value as the plate voltage, which is from 350 to 450 volts. With the key open, no space current flows through the tube because there is no current through the screen dropping resistor; hence, there is no voltage drop, and the high voltage is thus applied to the screen. When the key is closed, and space current starts to flow in the tube, the screen current causes a voltage drop across the usual series dropping resistor and the screen voltage than drops back to its normal 100 volts. However, it does not drop back instantaneously; during the time the screen voltage is dropping there is often a very noticeable change in the frequency, which

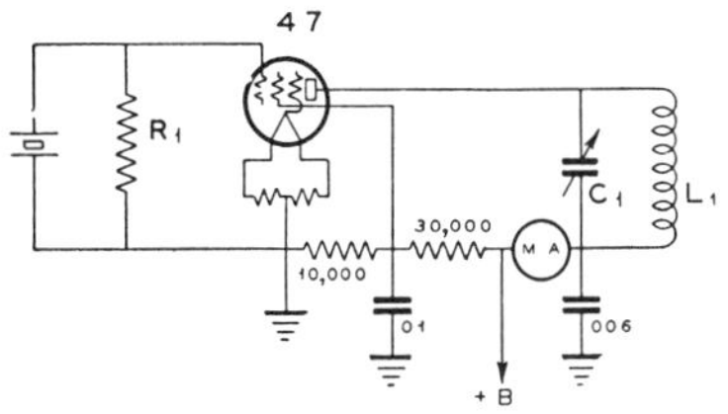

PENTODE OSCILLATOR

FIG. 10. Key chirp eliminator.

causes the chirp. This effect can be eliminated by keeping the screen voltage approximately constant, whether the key is open or closed. The remedial measure requires the use of a voltage divider, instead of a series dropping resistor as a source of screen voltage, as shown in the circuit above. The value of the resistance R should be chosen so that the voltage on the screen, when the key is closed, is 100 volts when measured by a high resistance voltmeter.

Miscellaneous Notes on Transmitter Adjustments

A transmitter for either phone or C.W. requires a proper adjustment of all the circuit components for the attainment of satisfactory operation. Below are a few practical notes which are of inestimable value in making transmitter adjustments.

Crystal Oscillators: In oscillatory circuits employing pentode tubes such as the 47, 2A5, 42, or 59, the plate circuit should have a low ratio of tuning capacity to inductance since the plate circuit is tuned for maximum output consistent with stability. Condenser C_1, of Figure 11, is tuned for a dip in plate current and then re-adjusted for slightly greater capacity for maintaining stability. A 6.3 volt pilot lamp connected in series with a turn of wire and coupled in the proximity of the oscillator coil makes a good oscillation indicator. Another type of indicator consists of taking a small neon tube and touching the tip connection to the stator plates of C_1, a pink glow indicates oscillation. The plate current of the crystal oscillator will be between 10 and 30 MA, depending upon the applied plate voltage. Potentials below 350 volts will exert less strain on the crystal and will tend to prevent it from fracturing, in addition, will tend to minimize the heating effect, which is one of the causes of frequency drift. The screen voltage should seldom be over 125 volts. The value of the grid resistor

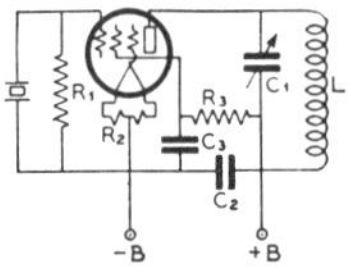

FIG. 11. Pentode oscillator.

R_1, in Figure 11, will vary with different crystals; generally, 10,000 to 50,000 ohms sufficing—the stability increases and the output decreases for higher resistive values.

With ordinary low-mu tubes in the oscillatory circuit, such as a 27, 56 or 10 type tubes, the procedure is the same. Plate voltage must never exceed 250 volts maximum for this type of oscillator tube.

In the Tritet or Dow oscillators, oscillation occurs by cathode regeneration at the frequency control, hence, harmonics of this frequency may be selected from the plate circuit by means of a tuned circuit C_2L_2, in Figure 12. The cathode coil and condenser are not tuned to resonance with the crystal, but to a frequency approximately twice as high. This circuit scheme exerts less strain on the crystal with high C than for low C in the cathode circuit, so at least 100 uufd of operating capacity is needed. The coils must be of such dimensions to allow operation with at least 100 uufds. Oscillation will take place over a rather wide range of cathode tuning; certain settings, however, will give greatest power output. The plate circuit must have

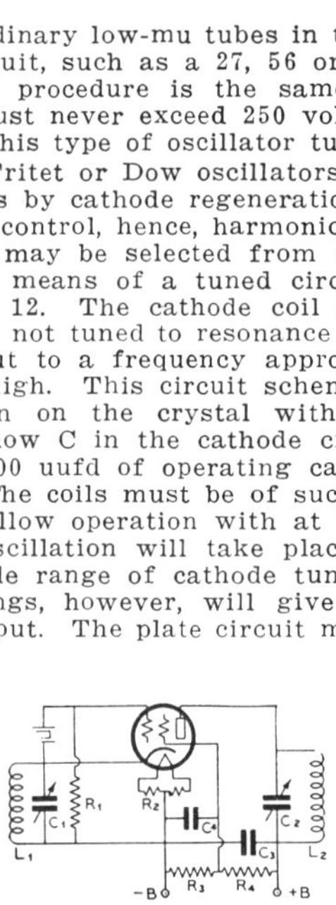

FIG. 12. Tritet oscillator.

low C and high inductance at the harmonic chosen. When tuning to the second harmonic, the resonance condition is indicated by a decrease in plate current; the decrease will be less pronounced as the load is increased by the succeeding stage. A lamp indicator used here is adaptable to making the best output adjustments comparable with oscillator stability. If a thermo-galvanometer is placed in series with the crystal, it will be found in many cases that the Tritet oscillator has more crystal current than the pentode oscillator; high current means increased heating of the crystal and frequency creepage.

In a few cases, the above oscillator is used with vacuum-tubes having large screen-grids such as the RK20 or 803 pentodes. Here the cathode coil consists of a double winding wound in series with the filament. The tuning procedure is exactly as described for the smaller Tritet oscillator.

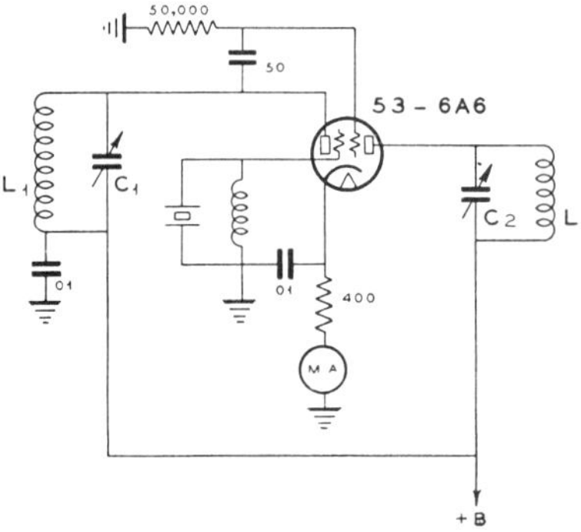

FIG. 13. Jones 53-6A6 oscillator-doubler with cathode bias resistor.

Oscillator Doublers: The 53 or 6A6 oscillator-doubler circuit shown in Figure 13 is adjusted as follows: C_1 is tuned to the crystal frequency and its capacity increased until the circuit approaches the point where oscillation is about to cease; this point is indicated by maximum output; the total plate current or cathode current will be between 50 and 75 MA, depending upon the plate voltage. The second triode acts as a doubler, and C_2L_2 are tuned to the harmonic as in a Tritet oscillator. The plate current will dip at resonance and a lamp indicator will glow to indicate maximum RF output. The adjustment of C_1 for greatest output gives about 20 per cent less cathode current than the maximum obtainable while tuning C_1 through oscillation. With cathode bias, the plate current will drop off to 20 or 30 MA. when the tube is not oscillating.

A crystal oscillator normally drives a buffer or doubler stage for greater output or frequency multiplication. In Figure 14 is shown a very simple form of frequency doubler to operate in the 40-meter band with an 80-meter crystal. The grid bias, due to the grid current flowing through the grid leak, should be higher than for a buffer stage, since the doubler is functioning as a distorting device.

A well-designed oscillator-doubler circuit is shown in Figure 15, operating as either a neutralized buffer or regenerative doubler stage. As a buffer it is neutralized in the conventional way, but as a doubler, a small coil is used together with a larger capacity value in C_2. If the preceding tuning circuit is of very low C, its impedance to the second harmonic will be high, so C_2 acts as a regeneration con-

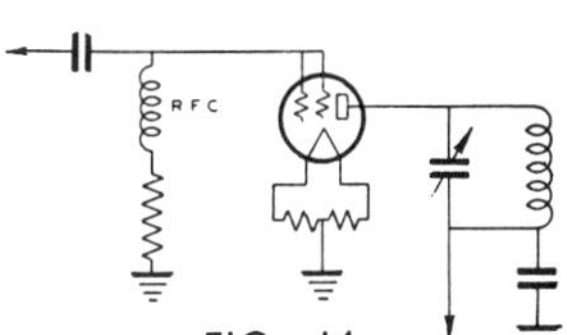

FIG. 14.

Simple 46 doubler circuit.

denser feeding back second harmonic power to the grid circuit. If C_2 is too large the tube will oscillate at the plate circuit resonant frequency, but if properly adjusted, the output is from 50 to 100 per cent higher than in a non-regenerative doubler. A RF indicator, such as a test lamp, will glow when adjusting C_1 and C_2 for maximum output without actual oscillation. No oscillations should be detectable without the crystal oscillator functioning.

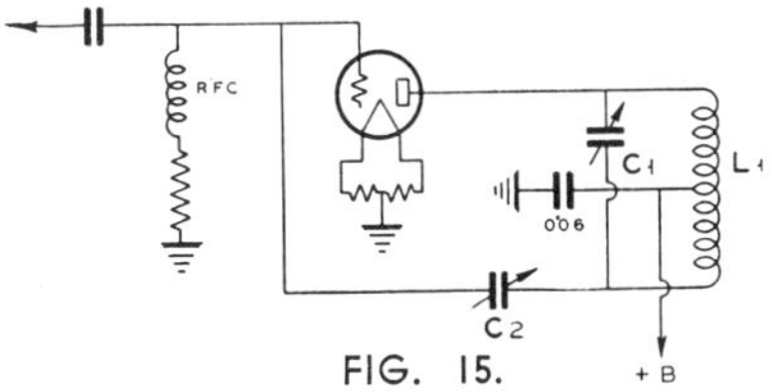

FIG. 15.

Regenerative doubler or neutralized buffer.

In any low powered doubler stage, a fixed C bias from C batteries or C bias supply is more desirable than that of grid-leak bias. The bias voltage should nearly be 3½ times the cut-off bias (plate voltage divided by the mu of the tube) as a minimum value, and practically no DC grid current need flow. If greater RF excitation if available, higher C bias can be applied with greater output and efficiency. The reason for low grid current is that the doubler tube only gives a surge of power to the tuned plate circuit every other cycle, since the frequency of the latter is twice as high as that of the grid circuit driving power. When grid current is flowing, the plate circuit receives a surge of power with an actual loss in efficiency; with grid leak bias, some grid current must flow in order to create the polarizing voltage on the grid, in this case the loss must be tolerated.

Figure 16 shows a popular doubler circuit which gives a surge of power or "push" every cycle to the second harmonic tuned plate circuit. The efficiency of this circuit is as high as some amplifier circuits and is easily adjusted. The grid circuit is tuned to the fundamental frequency with link coupling to the preceding stage and the plate circuit to twice that frequency. The C bias is made at least 3½ times cut-off and the RF excitation sufficient to allow some grid current to flow. A 53 or 6A6 tube makes an excellent "push-push" doubler, or a pair of tubes such as the high-mu type 46s, 59s or 42s can be used. A split-stator grid condenser is needed to provide capacitive reactance to the second harmonic which prevents spurious oscillation in the doubler circuit, that is, similar to a TNT oscillator. A single tun-

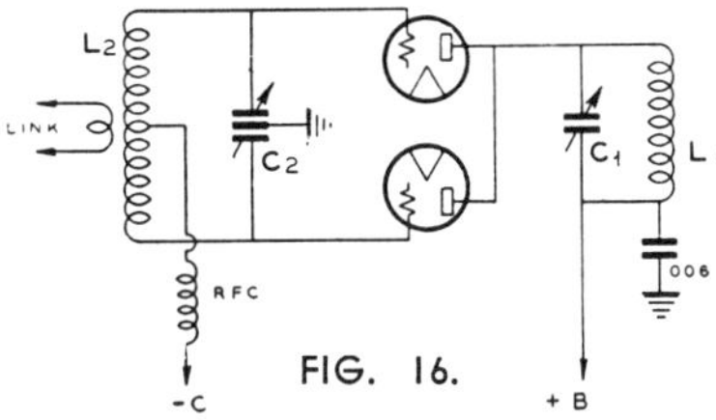

FIG. 16.

Push-push doubler

ing condenser with a bypass condenser from the center of the grid coil to the ground, allows half of the grid coil to act as the untuned grid coil of a TNT oscillator. This same circuit is also applicable for high-powered output doublers on 10 or 20 meters; with efficiencies from 60 to 70 per cent.

Figure 17 shows another form of regenerative doubler which works effectively at high frequencies such as 14 or 28MC. Here, the cathode circuit is by-passed with only a small condenser which causes it to have an impedance common to both grid and plate circuits. If this impedance is made very high, such as by placing an RF choke in the cathode circuit, the tube will oscillate. With the values shown, the circuit will regenerate on 14 or 28MC when the grid is excited with 7 or 14MC of power, In all these doubler circuits, the DC grid bias must always be as high as can be used for the available amount of RF excitation. The plate circuit can be loaded fairly heavily by the following stage and the highest allowable potential applied to the plate for a given plate-heating effect.

The circuit of Figure 15 is a popular form

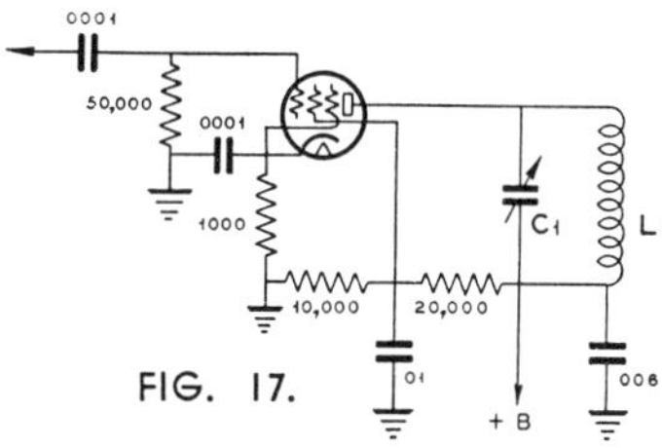

FIG. 17.

Regenerative doubler

of buffer or amplifier stage. It can be capacitively or link coupled to the preceding stage and either fixed bias, grid-leak bias or combination of both can be applied. The C bias is made at least twice cut-off value for class C operation, but often this can not be used in practice. Quite frequently a buffer stage is employed for maximum power gain rather then for maximum efficiency. Class B operation with cut-off bias will give the greatest power gain; but, unless a large tube is used, the low plate efficiency will cause excessive plate heating. Generally, a compromise between class B and class C operation will give the greatest output for driving the following stage. In a high powered transmitter this means balancing the cost of several low power tube class C intermediate stages against fewer large tube class B to C stages. For example, a single 211 tube buffer stage might be more economical than two stages of 800s or 801s in push-pull.

Figure 18 shows a grid neutralized buffer or final amplifier stage; plate or grid neutralization being optional, though the former, as shown in Figure 19, has certain advantages. This form with a split-stator plate tuning condenser will remain in neutralization for multi-band operation provided the coils are designed to permit operation of the split-stator condenser at a medium high scale setting. This is much more important for phone operations than for C W.

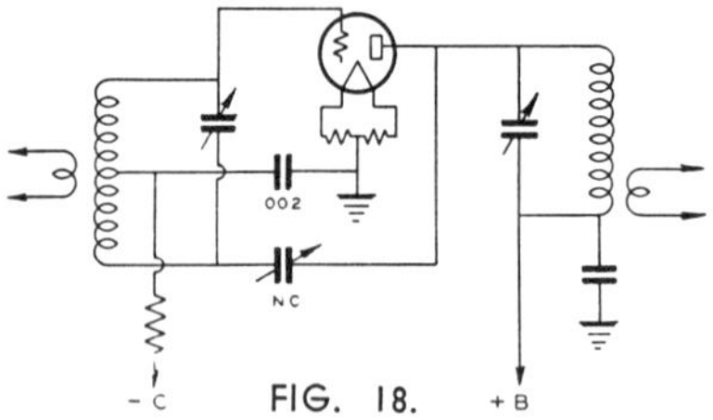

FIG. 18.

Grid neutralized stage.

Neutralizing in either case may be accomplished as follows: The plate voltage is disconnected and the grid circuit tuned for maximum grid current as indicated by a DC grid milliammeter, or neon lamp. Then the plate circuit is tuned for maximum RF excitation as indicated by means of a neon or flash-light lamp, or by a thermo-galvanometer with a turn of wire. The neutralizing condenser N_c is adjusted to the point of minimum RF current in the amplifier plate circuit, keeping the grid and plate circuit tuned to resonance. At neutralization, the effect of tuning the plate circuit through resonance will be negligible on the grid circuit DC milliammeter. If the circuit is improperly neutralized, there will be a sharp deflection of the meter pointer.

Neutralizing High-Power Stages: In high power stages where the grid driving power is 50 watts or more, the plate r. f. current cannot always be brought to zero on account of the presence of RF in the plate circuit caused by inter-circuit flow of high-frequency currents through the N_c condenser

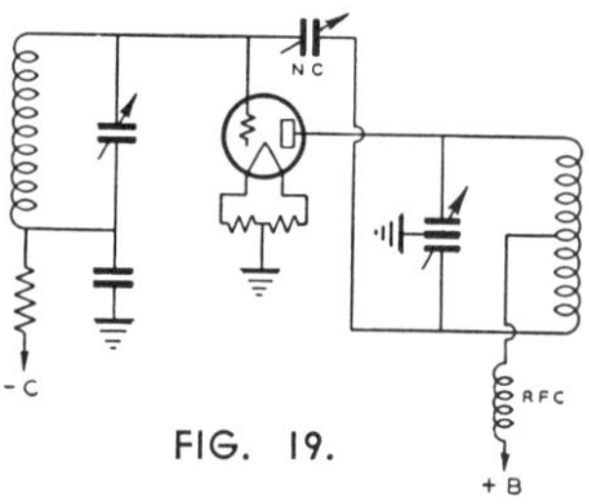

FIG. 19.

PLATE NEUTRALIZED STAGE

and through the tube element capacities. Radio-frequency current will always be detected by the RF indicator unless it is coupled to the exact nodal point or center of the plate coil. The DC grid meter is the most reliable RF indicator for neutralization; the measurements are always made with the plate voltage disconnected. After neutralizing, the voltage is applied and the plate circuit tuned for minimum plate current, preferably at reduced power. This stage is then loaded for the desired output and plate current at full plate voltage.

Push-pull amplifiers of the type shown in Figure 20 are neutralized by adjusting both N_c condensers as nearly simultaneously as possible. In this circuit the tube leads must be very short to prevent parasitic oscillation at ultra-high frequencies. Sometimes grid suppressors are needed for either push-pull or parallel operation of tubes; they are made by winding about 10 turns of No. 14 wire on a form one-half inch in diameter, the coil form is then

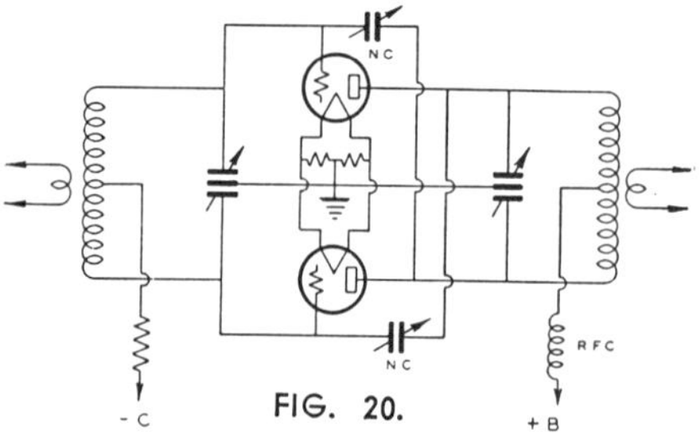

FIG. 20.

PUSH-PULL STAGE

extracted from the core and the coil shunted with a 200 ohm carbon resistor having a 1 or 2 watt dissipation rating. When suppressors are required, they are connected in series with the grid lead as near as possible to the grid terminal of the tube.

Tuning the final amplifier stage of a C.W. transmitter is similar to that of a buffer stage except that sufficient RF excitation must be available to allow the stage to operate class C; that is, with at

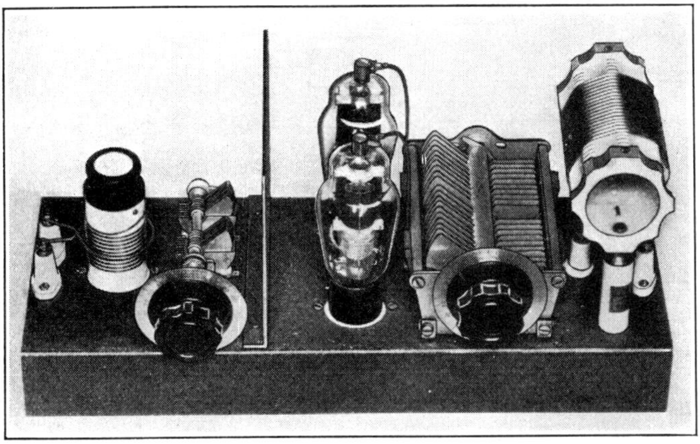

FIG. 21. Push-pull amplifier with type 865 screen-grid tubes.

least twice cut-off grid bias. The plate load is then increased so that normal plate current is drawn. The antenna adjustments are covered in the section on "Antennas." These adjustments must always be made for maximum power into the actual antenna (not the dummy antenna) for a given value of plate current or tube heating effect.

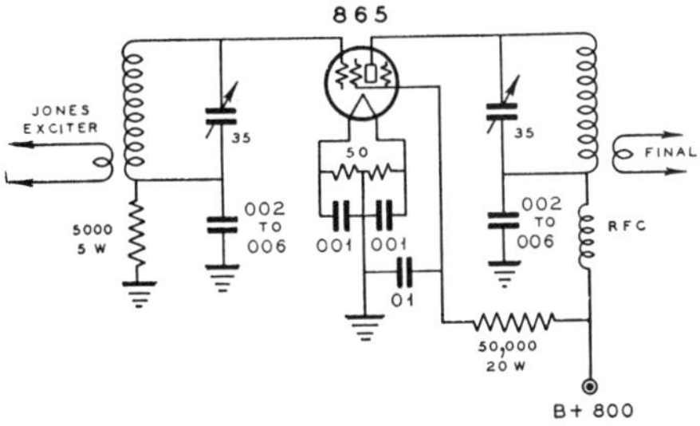

FIG. 22. Single 865 buffer or doubler circuit.

For phone operation, the modulated stage must be exactly neutralized and the plate circuit shielded from the grid or any preceding stages. The latter also applies to screen-grid tube stages in either phone or C.W. transmitters. The grid excitation for a modulated phone stage is about twice that actually needed for C.W. operation unless the latter happen to have an excess of excitation. This can be checked by means of a V.T. voltmeter, linear rectifier or an oscilloscope for studying both positive and negative values of modulated waves. Another check is to draw a curve of the output r-f. current versus the plate voltage, since there must be a constant increase for similar increases of plate voltage; too little excitation will cause a droop in this curve. The plate load on a modulated stage must be constant and of the proper value to allow good modulation. An oscilloscope and sine wave audio oscillator are necessary instruments for adjusting all classes of phone transmitters. The section on "Electrical and Radio Measurements" cover this subject. The grid bias, grid excitation, plate load on the modulator, plate RF load, over-modulation, tube ageing and other variables are made visual for quantitative analysis by means of the cathode-ray oscilloscope.

Test for High-RF Efficiency: To test the high efficiency of an RF stage apply some excitation to the grid circuit and apply the plate voltage to the plate (after neutralizing). A grid leak bias is used temporarily. Now, with no load coupled to the plate tank, the plate current should drop below 10 milliamperes. The plate current in an efficient stage reads about one-twentieth the normal operating plate current, when no load is connected. If the current does not fall, it is an indication that the tube is not functioning correctly or that there is an undesired loss somewhere in the stage. A high plate current reading with the stage unloaded may be indicative of a high-resistance connection in either the grid or plate tank circuits. It may also be due to the use of inferior materials for the grid and plate coil forms.

As soon as the plate tank is detuned from resonance (unloaded) the plate current suddenly rises to a point high above the normal operating plate current. In other words, the most efficient stage will show the greatest dip in plate current as the plate tank is tuned through resonance with no load coupled to the plate circuit.

Notes on Grid Circuit Excitation: The problem of obtaining sufficient grid excitation is of utmost importance to the amateur. There is a saving in buffer power, or the possible elimination of one buffer stage, if the RF amplifier can be equally well driven with less grid current.

The circuits shown in Figure 1a and 1b, are both plate neutralized; that of 1b has been the standard circuit because of its ability to maintain neutralization for coil changes—but recent tests show that it takes approximately twice as much grid excitation power as the circuit of Figure 1a.

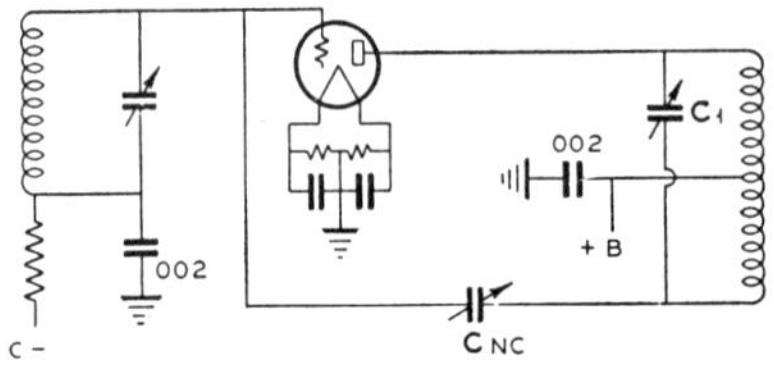

FIG. 1A

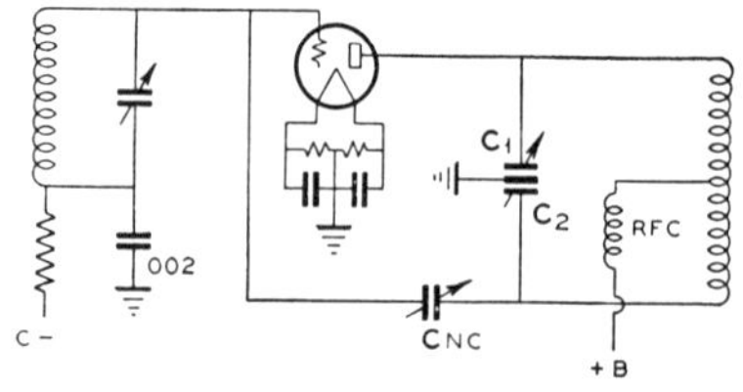

FIG. 1B

The tests were conducted with a 50T in the final amplifier, it being driven by an 801 in the buffer stage. In a given case with the best link coupling adjustments in both circuits between the 801 and 50T, the grid current was always 1½ to 2 times as great in Figure 1a as in Figure 1b which means that there was a greater output from the 50T up to the point of grid current saturation. A practical saving would, therefore, result from the use of a 2A3 at 400 volts supply as against an 801 at 600 volt supply as a buffer stage for the 50T operating at about 1400 volts plate potential. Or, the 50T could be operated at 2500 or 3000 volts with an 801 buffer in Figure 1a but with only 1500 volts in Figure 1b, due to lack of grid excitation in the latter case.

In Figure 1a, coil changes can be made for different bands without having to re-neutralize by inserting a small condenser of a few uufds (depending upon the tube used) from the plate coil side of the neutralizing condenser (C_{nc}) to ground. (Hint: a small grounded aluminum sheet bent up near the rotor end-mounting plate can be used as added capacity to ground.)

Use of the 45 and 46 Tube in Low-Powered R. F. Amplifiers

The 45 tube provides better "buffing" action than a 46 when functioning as a buffer to isolate the final amplifier from the oscillator. Even slight changes in plate voltage or plate load cause a noticeable change in the grid impedance of a 46. With a 45, changes in the output circuit react but little upon the grid impedance. As an RF amplifier, the 45 eclipses the 46 in performance. The two tubes can be compared further; thus, while the 45 has a somewhat lower wattage filament, it also has a higher mutual conductance (measured at zero bias) than the 46 (grids tied together and considered as a single grid). The lower mutual conductance of the 46 is largely due to the greater "shadow" effect of the grids, which becomes quite appreciable in multiple-grid tubes. Because of its higher mutual conductance, the 45 actually requires fewer watts excitation than a 46 to drive it **to a given output with a given efficiency.** Though it takes more **voltage swing**, it can be said that the 45 is the easier to excite, because driving **power,** not voltage, is the criterion of ease of excitation.

The plate impedance of the 46 is several times that of the 45. Thus, for a given efficiency in the output circuit (ratio of load impedance to plate impedance), much looser coupling must be used to the plate tank of the 46 (raise the load impedance). Then, to regain the output, the plate voltage must be increased beyond a safe operating limit. Although the inter-electrode spacing and the spacing of the plate lead coming through the stem is much greater in the 46, it will not stand any more plate voltage than a 45. The gas content, not the spacing, limits the plate voltage that can be safely applied to a 46. Paradoxically, the residual gas in many 46s will ionize at a given plate voltage and input quicker than a 45 of the same make operated under the same conditions! The 45 permits **greater efficiency** than is possible with a 46, both adjusted to a given output at a given plate voltage.

Because of its high grid impedance the 45 can be more advantageously capacitively coupled to the preceding stage than a 46 (presuming it is desired to connect from the high-potential end of the plate tank of the preceding stage to avoid parasitics). The 46, with its very low grid impedance, requires an extremely small coupling capacity to give the preceding stage a sufficiently high load impedance, and most of the excitation is being wasted. The grid impedance of a 45 offers a very respectable load for most tubes, and the grid of a 45 can be capacitively coupled on the lower frequency bands with almost as much efficiency and as great a transfer of energy as can be obtained with link coupling.

The optimum ohmic value for the grid resistor is very high (between 50,000 and 75,000 ohms for a single tube); hence, it is permissible to dispense with the grid choke in capacity-coupled circuit using a 45. The only precaution necessary is that the grid resistor be either of the carbon or metallized types, these being non-inductive.

Frequency Multiplication

Quartz-crystal oscillators have, unfortunately, a vibratory limit of about 8 megacycles; hence, to operate on a frequency higher than this value, one or more stages of frequency multiplying amplification must be added between the crystal oscillator and final amplifier. In almost every vacuum-tube amplifier there is a certain amount of distortion which represents the generation of new frequencies that are integral multiples of the exciting grid frequency. By tuning the plate circuit to the frequency of the desired harmonic, the fundamental and all undesired frequencies are by-passed to ground, while the selected harmonic (usually the second, third, or fourth) is transferred to the succeeding grid circuit.

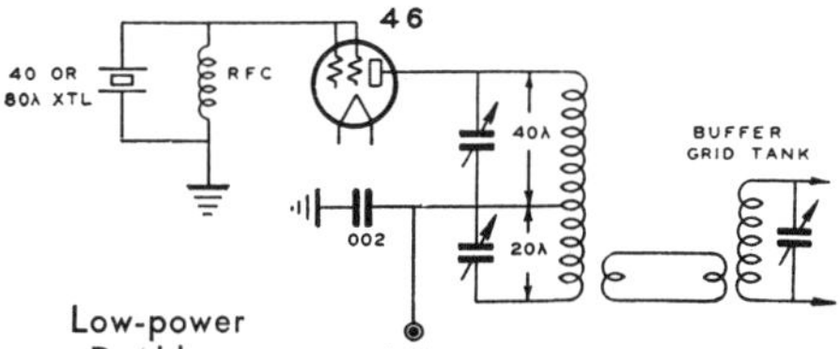

Low-power Doubler.

Two-band operation from a single crystal is secured by tapping the plate coil and tuning each section of the coil with a separate condenser. Moderate power output is obtained.

For efficient doubling, it is essential that the doubler amplifier be carefully adjusted. For every tube there is one particular value of grid excitation and grid bias that will give maximum output; thus, a means must be provided to smoothly adjust these factors. It will be found that more bias is necessary for plate doubling than for straight class-C operation. Pentodes and high-mu triodes such as the 53, 46, 59, 841, 203A, RK21 and 838 function well as doublers, although there is some question as to whether or not high-mu tubes are better than those having medium-mu, such as the 210, 211, 852, 50T, 354, and 150T, all with regeneration. The latter can be applied to any single-ended doubler stage by using any of the conventional neutralizing circuits. When the plate is tuned to a harmonic of the grid circuit, the neutralizing circuit becomes a feedback circuit.

Push-Pull Doubling: The push-pull circuit in Figure 2 differs from most doubler circuits in that doubling is not dependent on distortion, but on the fact that each RF impulse applied to the grid circuit results in two plate current impulses being applied to the plate tank circuit. This is because the grids are excited in push-pull and the plates excite the plate tank in parallel; thus, there are twice as many current impulses in the plate circuit as there are cycles in the grid circuit—in other words, the frequency of the plate tank is twice that of the grid tank.

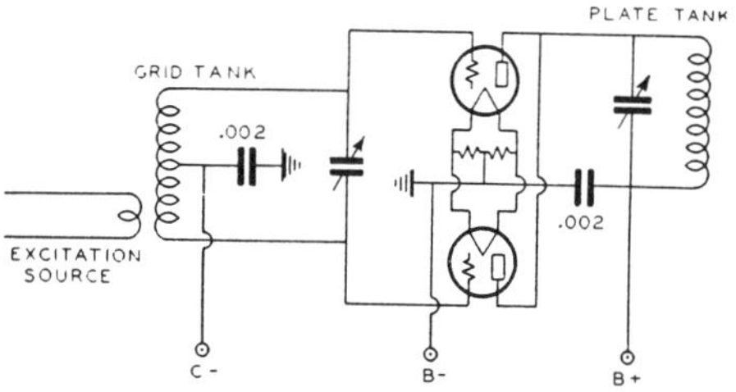

FIG. 1

Wrong way.

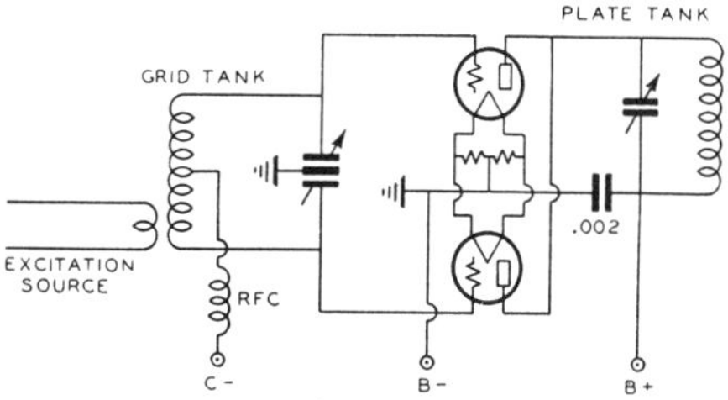

FIG. 2

Right way.

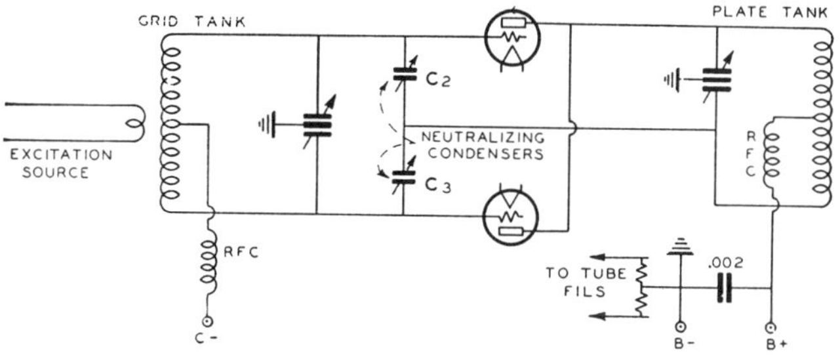

FIG. 3

The simpler forms of the push-pull doubler sometimes show a marked tendency to oscillate. The circuit in Figure 1 (Page 93) is particularly noted for this: the better circuit appears in Figure 2, particularly if a shielding baffle is provided between the grid and plate tanks. When using the higher-C tubes as push-pull doublers, it is often desirable to utilize the KH type of doubler shown in Figure 3. Here, oscillation is effectively prevented by separately neutralizing each tube. High grid bias is necessary for efficient operation.

The circuits shown in Figures 2 and 3 should be neutralized while connected as regular push-pull amplifiers, after which no further changes are necessary. To increase the frequency by a factor of two, requires changing the tank coil of the final amplifier to one that will tune to twice the frequency of the grid circuit. The circuit is then tested with reduced voltage while no load is coupled to the final during which times the tank condenser is varied until a pronounced dip in the plate current is found. The final is now ready for operation with a load.

Neutralizing the R. F. Amplifier

Neutralization of a radio-frequency power amplifier is necessary to prevent self-oscillation. The latter occurs in a power amplifier because of the electrostatic energy fed back through the plate-to-grid capacity of the tube. The energy in the plate circuit is many times that in the grid circuit and self-oscillation results when only a small fraction of the plate circuit energy is applied to the grid circuit. The capacity feedback through the tube is neutralized by dividing the plate or grid tank circuit so that the voltages at each end of whichever coil is divided are equal, but opposite in polarity with respect to the center of the split tank, which is at ground potential. Both ends of the split tank circuit are then connected to the high-potential end of the other tank circuit. In other words, when using plate neutralization both ends of the plate tank are connected to the grid of the tube (one through the tube capacity and the other through an external neutralizing capacity which is equal to the internal tube capacity). See Figure 1. Thus, two feedback voltages are applied to the grid, but because they are equal and opposing, the net

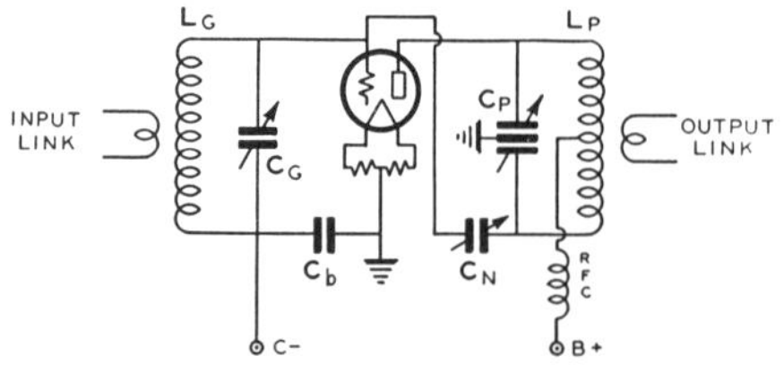

FIG. 1

Plate, or Hazeltine neutralization.

voltage is always zero, so the effective grid voltage (AC) is independent of the RF voltages in the plate circuit.

In the grid neutralized amplifier (Figures

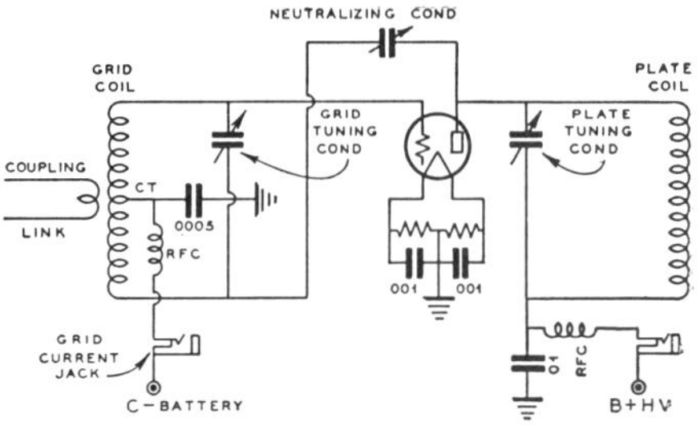

GRID NEUTRALIZING WITH SERIES FEED

FIG. 2.

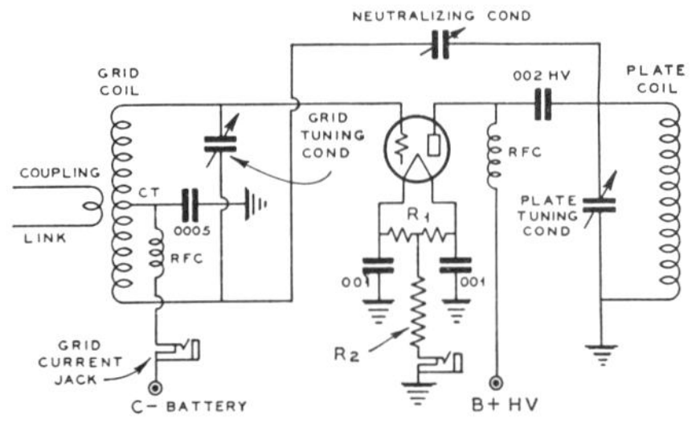

GRID NEUTRALIZING WITH PARALLEL FEED

FIG. 3.

2 and 3) the grid coil is split, the plate coil being continuous. Thus the RF plate voltage (AC) is applied simultaneously to both ends of the grid tank. For this reason there can be no potential difference between the two ends of the grid tank, caused by feedback from the plate tank, and the effective net grid voltage is again independent of that in the plate circuit. It will be seen that the two capacities which feedback the RF plate voltage to the grid must be exactly equal, if the two voltages are to exactly neutralize each other. For proper neutralization, the capacity of the neutralizing condenser must almost exactly equal the plate-to-grid capacity of the tube.

Grid neutralization may be preferable between stages that are joined by link coupling so that inexpensive plate tank and neutralizing condensers can be used; on the other hand, plate neutralization is more desirable with stages capacitively coupled.

How to Neutralize: In a perfectly neutralized RF amplifier there is no coupling from the plate circuit to the grid circuit. By the same token, there is no coupling from the grid circuit to the plate circuit. This characteristic is used in adjusting the neutralizing condenser during the neutralizing process.

Technique: With the plate voltage removed from the stage being neutralized, RF excitation is applied to the grid circuit. Some form of RF indicator, such as a thermo-galvanometer, neon bulb or flashlight globe with a single loop of wire, should then be coupled to the plate circuit. If the amplifier is **not** neutralized, there will

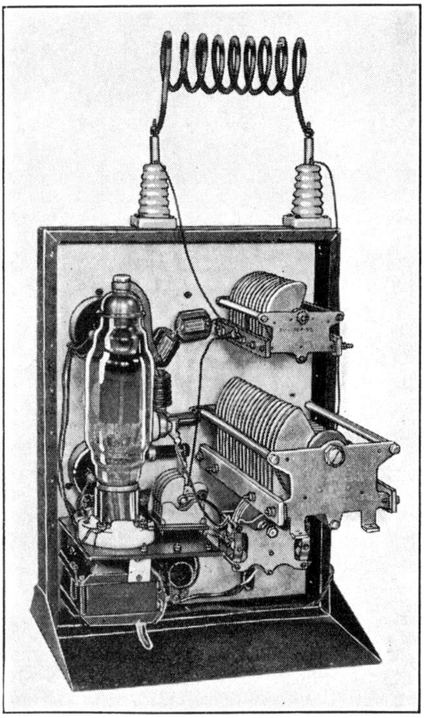

Grid-neutralized HF-300 c-w amplifier with simplified PI antenna network. 600 watts.

Shunt feed, as used in this amplifier, calls for an efficient RF choke which has no resonant dip near any of the bands on which the amplifier is operated. Many types of RF chokes fail in service when used on 80 meters, but stand up satisfactorily when used on 40 and 20 meters. If the RF choke becomes quite warm after a few minutes of operation, it is proof that power is being lost in the choke and replacement should be made with a more suitable type. The plate-blocking condenser should be mounted at least one-inch from the metal panel in order to minimize the capacity to ground.

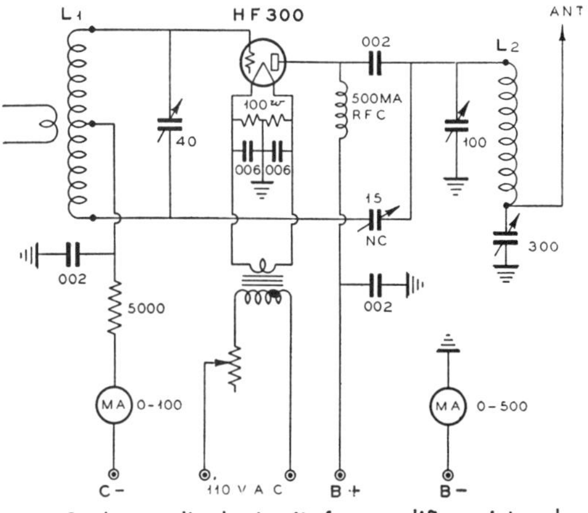

Grid-neutralized circuit for amplifier pictured above.

be an indication of RF in the plate tank circuit, when it is tuned to resonance. The neutralizing condenser should be slowly varied until **all** indications of RF in the plate tank circuit disappear. After each variation of the neutralizing condenser, it will be necessary to retune the grid and plate tank circuits in order to restore resonance in both these circuits.

The successfulness of the above procedure will depend upon the sensitivity of the RF indicator in the plate tank; incidentally, a neon bulb or flashlight globe is not particularly sensitive.

A better and very sensitive neutralizing indicator can be made by taking an 0-25 DC milliammeter and inserting it in the DC grid return of the stage being neutralized. Next, sufficient RF grid excitation must be applied to give a good current reading after tuning the grid circuit for maximum grid current.

If the amplifier stage is not perfectly neutralized, a variation in the DC grid current will be noted when the plate tank condenser is swung through resonance. The neutralizing condenser should be varied slowly until no variation in DC grid current is shown by the milliammeter (in the grid circuit) as the plate tank condenser is tuned through resonance.

If the amplifier which is being neutralized is NOT the final amplifier, another procedure can be followed: A DC grid current meter is placed in the grid circuit of the stage following the buffer stage which is being neutralized. Now, with no plate voltage on either stage, tune the stage being neutralized through resonance, then tune the next stage to resonance. A small grid current reading will be obtained as long as the buffer stage is not neutralized, but when it is in a neutralized state, the grid current on the following stage will entirely disappear. Note: The grid current of the stage which follows the one being neutralized acts as a diode vacuum-tube voltmeter and is a very sensitive indicator of RF which is present in the plate tank of the stage being neutralized.

Neutralization of a Push-Pull Stage: Push-pull RF amplifiers are neutralized by employing the same procedure as was used in neutralizing the single-ended amplifier. The neutralizing condensers are varied in small steps until all indication of RF disappears from the plate tank, or else there is no variation in DC grid current when the plate tank is tuned through resonance. Both neutralizing condensers should be varied simultaneously in the same direction; ganging the condensers will simplify the adjustment.

Most neutralizing troubles are caused by the RF return from the grid and plate tanks to ground. It is necessary that the low potential end of each tank coil (center of a split coil) have a short and direct RF path to the filament center-tap of the tube, or tubes. If a split-stator tank condenser is used, the rotor must be tied to the center-tap of the filament. If a single-section condenser is used with a split coil, the center of the coil must be by-passed back to the filament through a mica condenser of from .001 to .006 ufds., the value depending upon the frequency—low frequencies require higher capacities, in addition, high inter-electrode capacities require high capacities in the plate and grid returns.

It is an established fact that the lower-C tubes (such as the 800, WE304A, 50T, 852, 150T, 354 and 300T) are materially easier to neutralize, particularly at the higher frequencies, than those tubes which have higher inter-electrode capacities.

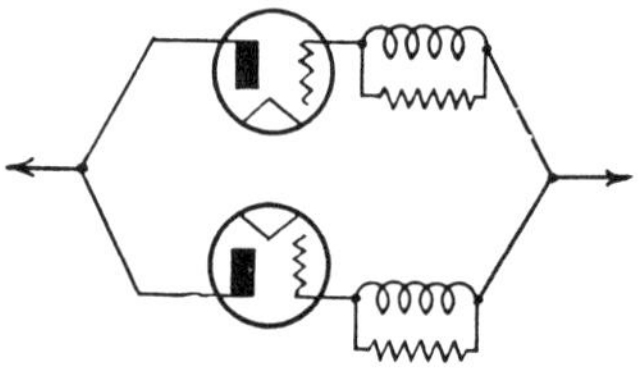

Grid chokes and resistors for parallel operation of tubes.

Operating Tubes in Parallel: Parasitic oscillations are sometimes introduced when operating vacuum-tubes in parallel at both audio and radio-frequencies. The most common type of parasitics occurring from parallel operation can be prevented by inserting small RF chokes, shunted with 50 to 200 ohm carbon resistors, in series with each grid lead. The choke need only consist of from 5 to 10 turns of No. 22 enameled wire wound on a form of about one-half inch diameter. Winding the chokes on the resistor will simplify the job as well as provide a convenient method of mounting

Grid Bias

The conditions under which practically all radio-frequency power amplifiers operate are such that plate current flows in the form of short, peaked impulses which last for less than one-half of the alternating current cycle. This means that plate current is "cut-off" during most of the RF cycle which makes for high efficiency and high power output from small tubes. To keep the plate current at zero during most of the RF cycle, it is necessary that the control grid of the amplifier tube be kept quite negative with respect to the filament by means of a DC voltage which is termed "negative bias." The AC grid excitation voltage, which usually comes from the plate circuit of the preceding amplifier stage, periodically overcomes the grid bias voltage and makes the grid slightly positive with respect to the filament causing a short impulse of plate current to flow.

If no grid bias were used, the tube would draw plate current all of the time. This would result in very inefficient operation because the plate would never have the opportunity to cool off.

Cut-Off Bias: Any value of grid bias which is just sufficient to reduce the plate current to zero is called the **"cut-off bias."** By taking a reading from milliammeter inserted in the plate circuit, with different varying values of negative bias on the control grid, it will be found that the plate current will decrease as the negative bias is increased. At a certain point the plate current will be reduced to zero, and any further increase in negative grid bias has no effect on the plate current which remains at zero. Thus, the lowest value of negative grid bias which reduces the plate current to zero is termed the "cut-off bias."

It is not necessary to experiment with bias batteries and different plate voltages to determine the cut-off bias for a given set of conditions. The required values can be calculated by simply dividing the voltage applied to the plate by the amplification factor; these data may be obtained from a table of tube characteristics. When estimating the cut-off bias add 5 to 10 per cent more bias to that calculated; this is required on account of the variable-mu tendency which is characteristic of all control grids as the cut-off point is approached.

Effect of Bias on Efficiency and Output: The amount of negative grid bias has a very definite effect on plate efficiency and power output. If the plate voltage and RF excitation voltage remain fixed, and if the bias voltage is increased beyond the cut-off point in a radio-frequency amplifier, the power output and input decline, although the plate efficiency rises. It is, therefore, necessary to make a compromise between power out and plate efficiency. The smallest amount of bias that allows the plate of the amplifier tube to run cool should be used beyond the cut-off point. This results in the maximum power output for a given tube, plate voltage, and RF excitation voltage. To increase the power output, it will be necessary to increase the plate voltage, loosen the antenna coupling, and in many cases increase the radio-frequency excitation voltage. With this procedure, the bias must be readjusted to the lowest value that allows the plate current to remain cool. The actual value of this bias, as measured in the number of times cut-off bias, will vary from about 1.25 times cut-off in a low-efficiency high-gain buffer stage, to about 4 times cut-off bias in an extremely high efficiency low-gain amplifier, operating with very high plate voltage and RF excitation. The bias voltage and the grid driving power are closely related, such that the higher the bias the more grid driving power is necessary to reach a given power output. For tubes of similar characteristics, the one with the highest zero bias mutual conductance (see tube tables) requires the least amount of bias and grid driving power for maximum power output and plate efficiency. As an example of the effect of mutual conductance on the required bias voltage (and therefore the amount of excitation power necessary), it is found that under a given set of conditions a type 852 must be biased to 3.5 times cut-off and excited with 106 watts of grid driving power to obtain 400 watts of radio-frequency power output at 80 per cent plate efficiency. On the other hand, a type 150T, or 354, which has considerably higher mutual conductance when used under the same conditions in the same stage, requires a bias of only 2.1 times cut-off and only 29 watts of grid driving power is necessary to obtain the same 400-watt output at the same plate efficiency (80 per cent).

When a radio-frequency power amplifier is plate modulated, the negative grid bias must be equal to or greater than twice cut-off. This is necessary in order that the peak power output can increase as the square of the plate voltage, which is essential for linear modulation.

Sources of Bias: In general, bias can be supplied from two distinct sources: (1) from within the amplifier circuit itself through a voltage drop taken across either a grid-leak resistor or cathode-bias resistor; (2), from an external source, such as batteries or a special rectified AC bias pack.

Grid-leak Bias: Whenever the control grid of an amplifier tube becomes positive with respect to the filament (as it does in all radio-frequency power amplifiers), the positive charge on the grid attracts some of the electrons emitted from the filament. The electrons flow back to the filament through the external DC grid-return and cause a current flow in the circuit. If a resistance is placed in series with the grid-return, a voltage drop will occur across it when the current flows; the end of the resistor closest to the grid will be negative with respect to the other end closest to the filament; thus, necessarily causing the grid to become negative with respect to the filament. The voltage drop across this grid-leak resistor consists of a varying AC voltage superimposed on a constant value of DC voltage, which is proportional to the effective value of the grid current impulses. The AC component is of no concern because it is by-passed by means of a condenser directly back to the filament, and thus by measuring the DC grid current with a DC milliammeter in series with the grid-leak, the grid bias can be calculated by multiplying the grid current by the ohmic resistance of the grid-leak.

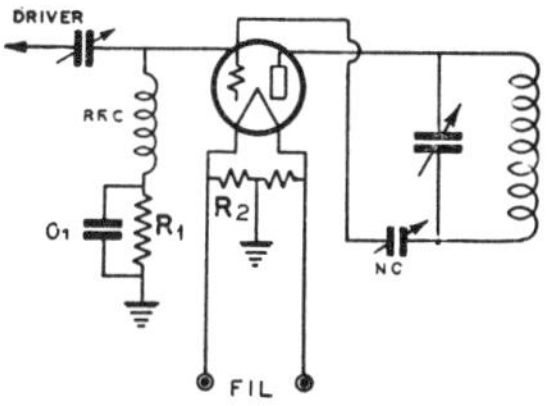

GRID LEAK BIAS

FIG. 1.

Grid-leak bias is quite flexible and more or less automatically adjusts itself with any variation in RF excitation. The value of grid-leak resistor is not particularly critical because the DC grid current usually decreases as the grid-leak resistance increases, thereby keeping the product of the two more or less constant for a given amount of RF excitation. Hence, the value of the grid-leak resistance can vary from one-half to two times the optimum value, a ratio of four to one, without materially affecting the negative DC bias voltage actually applied to the grid of the amplifier tube.

One of the disadvantages of grid-leak bias is that the bias voltage is proportional to the RF excitation, thus precluding its use in grid modulated or linear amplifiers, whose bias must be supplied from a well-regulated voltage source so that the bias voltage is independent of grid current. When grid-leak bias is used alone, it is evident that the bias disappears when the excitation fails, thereby allowing dangerously-high values of plate current to flow, with a consequent damage to the tube. It is always desirable to augment the grid-leak bias with either cathode or separate bias supplies to keep the plate current within safe limits whenever the excitation fails. The amount of the bias supplied in addition to the grid-leak bias should usually approximate cut-off bias so that the plate current will drop to zero if the crystal stage stops oscillating, or if the tuning elements are improperly adjusted in any of the other stages.

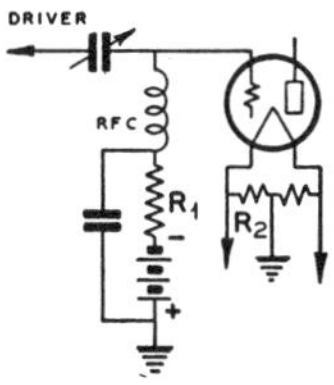

COMBINATION BATTERY & GRID LEAK BIAS

FIG. 2.

Cathode Bias: This form of bias utilizes the voltage drop across a resistor in the B-minus lead from the high-voltage power supply. Because the B-minus lead of most high-voltage power supplies is directly grounded, the bias resistor must be placed in the negative side of the DC plate circuit of the tube itself. The negative side of the DC plate circuit of a vacuum-tube amplifier is between the filament center-tap and ground, and a resistor placed between these two points will have the total plate current flowing through it. The voltage drop across the resistor will be equal to the product of the plate current in amperes, times the resistance in ohms. The grounded end of a cathode-bias resistor is more negative than the filament end by the amount of voltage drop across the resistor; hence, if the DC grid return is brought to the ground end of this resistor, the grid of the amplifier will be more negative with respect to the filament.

Cathode bias is probably the safest bias supply known, because the negative bias voltage is a function of the plate current and is largely independent of the RF grid excitation. With this type of bias the plate current can never reach a dangerously-high value if the excitation fails. Unfortunately, cathode-bias is generally unsuitable for class B linear amplifier, although its use is essential in the newer class BC linear or grid-modulated amplifiers.

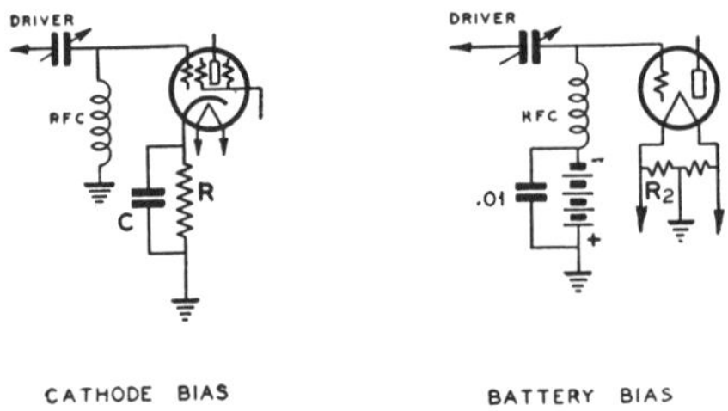

FIG. 3.

Cathode bias can be used in a plate-modulated class C amplifier provided a large audio by-pass condenser is connected across the bias resistor in addition to the usual mica radio-frequency by-pass condenser. The principal disadvantage of cathode bias is that the bias voltage must be subtracted from the total power supply in order to obtain the net plate voltage across the amplifier tube. In a high efficiency amplifier stage using a low-mu tube and biased to perhaps 3 times cut-off, it may require a 1600-volt power supply to actually realize 1000 volts on the plate of the amplifier tube, because 600 volts is deducted for negative bias. Cathode bias is sometimes called automatic bias because variations in plate current automatically change the bias to compensate for these variations.

Separate Bias Supplies: Negative grid bias may be supplied from any source of voltage external to the amplifier circuit itself, such as dry batteries or B eliminators. B batteries rarely fail without giving considerable warning to the operator and they represent one of the safest sources of negative bias. However, these batteries wear out rather rapidly due to the charging effect of the DC grid current which causes the voltage as well as the internal resistance of the batteries to rise. After a few months of service, the batteries often become noisy, especially when used for phone work. When charged-up they bulge and leak. It is not unusual to find a 45-volt battery which measures 60-volts after only a month or two of service in the grid circuit of a class C amplifier. This would be of no particular disadvantage if the 60 volts remained constant, but it often wavers and fluctuates causing the signal tone to be impaired.

Another form of separate bias supply consists of some form of rectifier and filter system whose positive terminal is grounded and whose negative terminal connects to the DC grid return of the amplifier stage. This bias supply often consists of an old B eliminator and is quite satisfactory if certain precautions are observed. If a high-power class C amplifier is biased by a B eliminator, some form of relay should be used and controlled by the bias supply so that the plate voltage to the amplifier is cut off if the bias supply fails. Most of the older B eliminators, and many of the newer types, have very poor voltage regulation which becomes troublesome when the eliminator is used to bias two or more separate amplifier stages. Poor voltage regulation merely means high internal resistance in the B eliminator. Any variation in grid current in any one of the amplifier stages will vary the voltage drop across this internal resistance and thereby affect the bias supplied to the other stages of the transmitter. If a B eliminator, or rectified AC bias supply furnishes bias to a class B or class BC linear or grid-modulated amplifier, it is essential that the DC bias voltage remain constant and independent of the DC grid current. This means that the bias supply must have extremely good voltage regulation. Low resistance transformers and filter chokes as well as a mercury-vapor rectifier tube and a low-resistance high-current bleeder should be used to minimize variation in output bias voltage with changes in grid current which normally occur in these types of amplifiers. To adjust the DC voltage output from a bias power supply, tap the primary of the transformer, or an autotransformer can be connected across the line to vary the voltage supplied to the bias transformer.

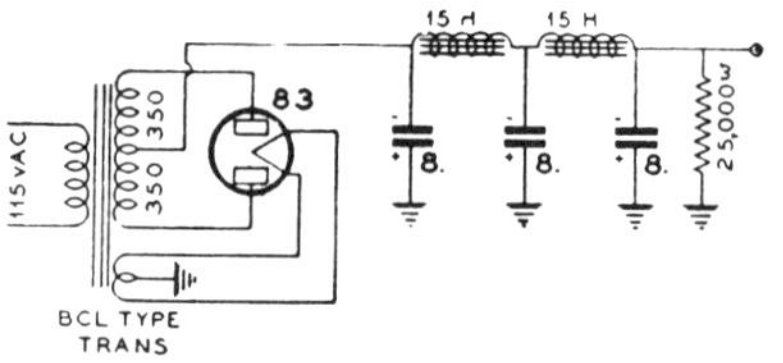

Here is a Bias Pack which uses a medium-to-high resistance bleeder. Voltage regulation is usually unimportant in biasing a class C amplifier and only enough bleeder is used to protect the filter condensers.

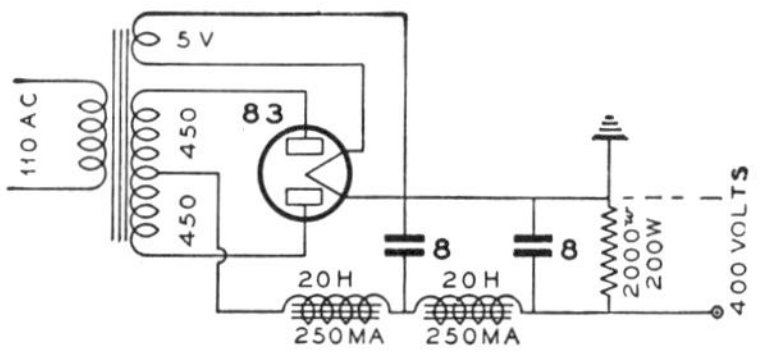

In the Bias Pack shown above, a low-resistance bleeder is used to provide a heavy, continuous current drain in order to stabilize the voltage output. A Bias Pack of this type is suitable for class B audio or class BC Linear Amplifier operation. The ungrounded side is the negative terminal.

More than one of the above bias supplies can be placed in series to bias a class C amplifier. In fact, it is recommended that a grid-leak be used to augment the cut-off value of bias which is best supplied by either cathode bias resistor, batteries or a separate bias pack.

A grid-leak common to more than one class C stage should be avoided, due to tremendous interaction caused by the two different grid currents in the respective stages.

To compute the wattage rating of the resistor, either as a grid-leak or to give cathode bias, multiply the square of the current in amperes flowing through the resistor by the resistance in ohms.

R. F. Impedances in Circuit Coupling

Low-C and High Efficiency: The difference between the DC plate input and the AC power output is the plate loss, and must be dissipated in the form of heat. Because the tube cost is almost related to plate dissipation, it pays to obtain high plate efficiency as it is then possible to secure high power output from small tubes. A vacuum-tube AC generator has a definite internal resistance to the flow of current. It varies with the applied voltage and the grid excitation.

Given a constant voltage generator, the generator efficiency increases as the ratio of the impedance mis-match increases—but the power output is maximum when the load impedance is matched to the internal impedance of the generator.

The Class C RF Amplifier: The most important application for impedance mismatching is found in the class C radio-frequency amplifier occupying the place of a final-amplifier in an amateur transmitter.

The greatest mis-match can be obtained from a tube with the lowest dynamic plate impedance, and with the highest voltage that the tube insulation and gas content will allow. The high plate voltage greatly reduces the internal impedance. The circuit should be adjusted so that the plate tank has a high L and low C. The antenna coupling is loosened as much as possible without reducing the input below that desired, and the bias is adjusted to several times cut-off. The excitation, as measured by the DC grid current, should be between 15 and 25 per cent of the DC plate current, and will vary for different types of tubes. In general, the higher the mutual conductance of the amplifier tube, the less excitation power is needed for a given load impedance.

Link Coupling

Advantages of Link Coupling Over other Types:

(1) Effectively establishes correct impedance relations between grid and plate circuits.

(2) Permits more efficient operation of circuits wherein low-mu tubes work into, or out of high-mu tubes, and vice-versa.

(3) Provides a flexible feed-line of several feet in length resulting in efficient operation between stages in "rack type" transmitters in which the stages are spaced quite far apart.

(4) Permits the use of series-feed in both grid and plate circuits.

(5) Makes possible maximum power output and minimizes oscillation difficulties.

(6) For a given amount of excitation on the grid of the first buffer, link coupling reduces plate current in the crystal oscillator stage and therefore reduces the RF current through the crystal itself.

(7) Eliminates the use of taps on coils, with their attendant losses.

(8) Because of the lack of capacitive coupling effect, neutralization is made easier.

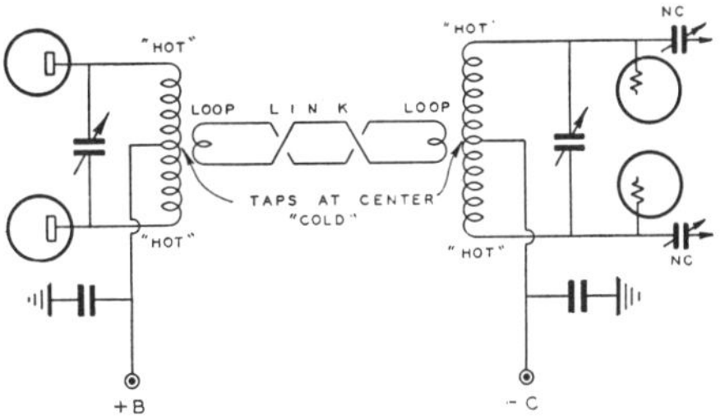

Link coupling between push-pull stages.

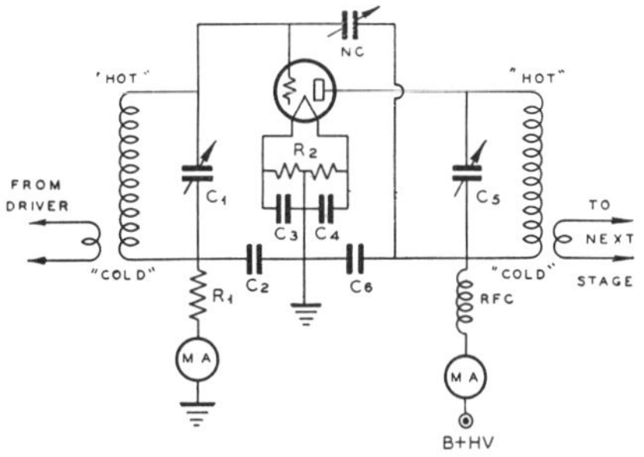

Link coupling between single-ended stages.

Link coupling provides a low impedance transmission line to transfer energy between two isolated tank coils, one of which is in the plate tank of the driver stage and the other the grid tank of the driven stage. This low impedance transmission line provides coupling of purely inductive nature, the capacitive loading effect of the coupling loop being negligible.

Feed lines, consisting of twisted pairs can be several feet in length. The wires can be of ordinary rubber-covered No. 18 to 14 wire.

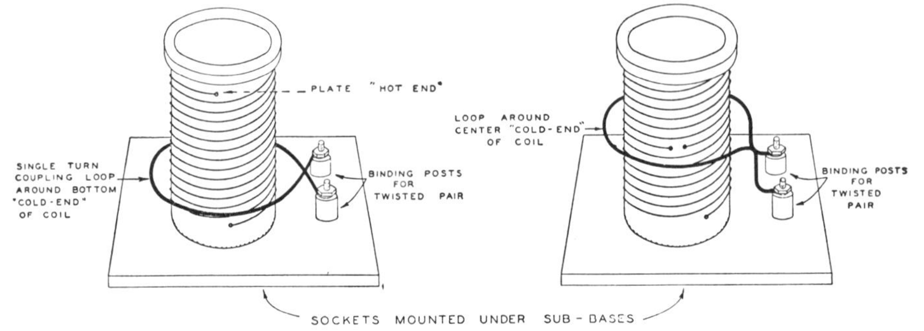

The link coupling loop is wound around the lower end of the coil when this end is by-passed to ground.

The link coupling loop is wound around the center of the coil when the coil is center-tapped.

A reference to the illustrations shows some of the mechanical arrangements suitable for link coupling. For low power stages one of the fixed-coupling-loop systems are recommended. One of the coil forms ideally suitable for the fixed coupling loop are the isolantite vertical plug-in type. This coupling scheme is shown in the illustrations as well as the system using one coupling loop adjustable from the baseboard.

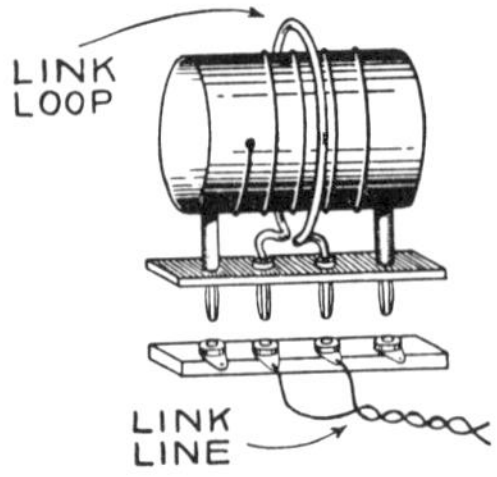

Coupling loop arrangement for large plug-in coil.

Link coupling can often be used between stages in a transmitter in a form which will give greater grid swing in each succeeding stage. Some tubes which have a high-mu, of the screen-grid type, have an extremely low grid impedance, especially under plate loaded condition. In such cases, it is difficult to obtain maximum grid swing or reasonable driver plate load with the usual form of one or two turns in the link coupling loop at each end of the link. To eliminate this difficulty, use one or two turns in the link coupling loop on the driver plate coil and two to six, or even seven turns at the grid coil when the driven tube is a high-mu or screen-grid type. The coupling between the grid coil and the link coil should be as close as possible, such as one winding directly over the other, or interwound. The latter is important for proper impedance matching.

Link coupling will give a certain amount of automatic impedance matching. This can be proved by noting that about 50 per cent or more grid swing can be obtained with the usual link coupling over the old form of capacitive coupling between a pair of type 210 tubes. Only a small part of this loss in capacitive coupling is due to the grid RF choke, since the latter can be made very effective, the loss is in the impedance mis-match when the grid of the following tube is across the entire tuned circuit. Link coupling gives an impedance

Air-wound tank coil with variable coupling loops for connection to twisted-pair feed line.

matching effect because the coupling is usually less than unity or maximum obtainable. The impedance reflected each way is not entirely dependent upon the ratio of tuned coil turns to link coil turns, since the effective coil coupling is relatively loose and resonant circuits are being used. Because the coupling is not unity between the coils, impedance matching takes place, if the ratio of impedances are not too great. When the impedances are greatly different, one being several times that of the other, then the low impedance circuit end should have more turns on the link coil and these two coils should be very closely coupled.

Unity Coupling

Unity coupling between stages of a transmitter can be used as an aid to eliminate RF choke troubles. This coupling is not as effective as link coupling, yet it does not require an additional tuned circuit. Unity coupling can be used advantageously in transmitters where space is lacking for link coupling. Care should be taken to see that the grid coil interwound turns are spaced sufficiently from the plate coil turns to prevent DC voltage flashovers. For low power operation, the grid coil is often wound inside of a copper tubing plate coil; wire used for this purpose must be well insulated.

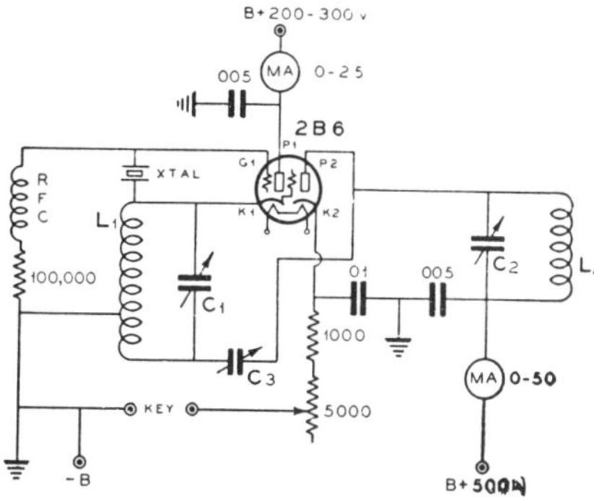

"LES-TET" 2B6 EXCITER

The "Les-Tet" Exciter provides high output with moderate plate voltage. When used as a crystal-oscillator amplifier the second section of the 2B6 tube must be neutralized. See page 138 for coil data.

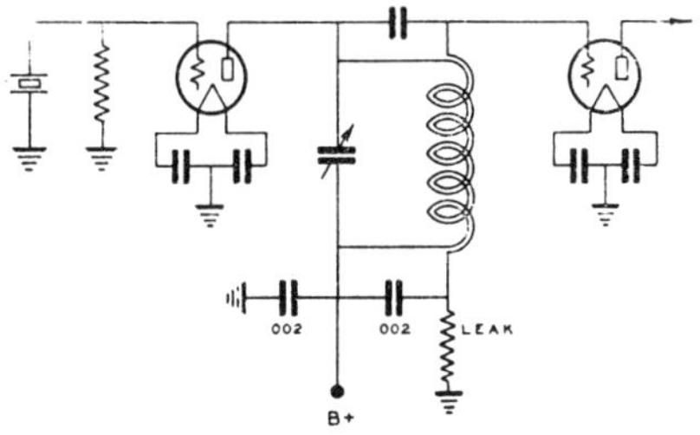

Unity coupling between crystal oscillator and buffer stages.

Final Amplifier Tuning Adjustment: The plate tuning of the amplifier should be set for a point at which minimum plate current occurs, rather than the point at which maximum grid current flows. Varying the plate tuning under operating conditions changes the plate impedance of the amplifier. The point of maximum plate impedance does not necessarily have to occur at the point of minimum grid impedance, as far as the DC current readings are concerned.

Grid Saturation: An excess of grid excitation will not increase the power output and may shorten tube life, due to overheating of the grid. The grid will become white hot when overloaded. In some tubes this will release gas retained in the metal and ruin the tube. Too little grid excitation will cause excessive plate heating and low output. More grid excitation is required as the plate voltage is increased, it also requires more grid excitation for closer antenna coupling than when the antenna is loosely coupled.

Parallel vs. Push-Pull Operation

The recent development of the low-C tubes practically eliminates the former difficulties encountered from parasitics and instability which accompany parallel operation of vacuum tubes. Parasitics are largely caused by the stray inductance and the capacity in the tubes themselves, as well as in the connecting leads between the tubes and the associated plate tank circuits. Parasitics are not confined to tubes in parallel, but are in fact nearly as common in push-pull circuits because the inductance of the leads when connected through the tank tuning condensers can form an ultra-high frequency tuned-plate-tuned-grid oscillator. This causes oscillations at a frequency other than that desired, with consequent low efficiency and reduced power output, as well as resulting in a poor note.

Parallel operation has many advantages over push-pull operation, even at the higher frequencies, provided low-C tubes are used. The plate tuning condenser can be one of a cheaper variety for a given tuning capacity and one neutralizing condenser is eliminated for parallel operation. Tubes are easier to drive to a given output when in parallel, due to the higher transconductance of the parallel circuit. The amplification factor is the same for a parallel connection as it is for one tube, whereas the plate resistance is cut in half.

High-C tubes, such as the 45, 2A3, 10, 211, 203A and 204A work best in push-pull below 40 meters because the high capacity shunted across the plate tank, when high-C tubes are used in parallel, makes the use of a low-C, low-loss tank circuit impossible.

R. F. Chokes: There are several varieties of Pie-wound RF chokes. Those who prefer to wind their own should choose the solenoid winding types. A RF choke designed for maximum impedance in one amateur band is not satisfactory for operation in other bands because of the effect of its distributed capacity and inductance. This choke would be satisfactory on three times the fundamental frequency, but worthless on twice that frequency.

The coil should be wound so that its fundamental operating frequency is about midway between its lowest and highest impedance. On even harmonic operation, therefore, the reactance will be fairly high. If the RF choke is to be used only on one band, it can be wound for resonance, as its effect across a tank circuit, when being tested, is negligible at that frequency. If, for example, it is to be used on both 80 and 40 meters, more or less turns are needed in order to keep the second harmonic impedance high. In all such cases the lower end of the RFC should be bypassed back to ground because some RF current will flow through it.

If the length of the solenoid is not great in comparison to its diameter, its fundamental resonance can be calculated from C in mmfd. = .24d, where d = diameter in inches, and C, the distributed capacity. The inductance can be obtained from L in micro-henrys:

$$L = \frac{a\,n}{9a + 10b}$$

where a, is the radius of the coil in inches; n, the number of turns; b, the length of the coil in inches.

$$f = \frac{1000}{2\pi\sqrt{LC}} \text{ in megacycles.}$$

The formulae give a starting point. Actual construction for short wavelengths, such as 5 to 40 meters, can well be made on small diameter rods or tubing with a long winding several times its diameter. The wire should be large enough to safely carry the DC plate or grid current, as well as an appreciable amount of RF current.

In high power circuits, RF chokes should be placed at points of low-RF potential, if possible, because even the pie-wound chokes, contrary to popular belief, are not efficient on all amateur bands. Care must also be taken to prevent a tuned-grid tuned-plate oscillation between the grid and plate RF chokes at the fundamental frequency of the RF chokes. If possible, the grid choke must have an inductance at least 10 to 20 times that of the plate choke.

Tank Circuits

The plate tank circuit of any transmitting radio frequency amplifier consists of a parallel resonant tuned circuit. The shunt impedance of any resonant tank circuit is the resultant of two factors: (1) the resistance of the tank circuit itself, and (2) the reflected resistance caused by coupling a load, such as an antenna, to the tank circuit. The output power dissipated in the resistance of the tank itself is entirely lost, so that for high output and efficiency it is desirable to make the tank losses as low as possible. The test for any tank circuit is to disconnect the load (antenna, etc.) and measure the DC plate current at normal plate voltage, bias, excitation, etc. This unloaded plate current should be less than 10 per cent of the normal loaded DC plate current in most circuits.

Tank "Q": The "Q" of a transmitting tank circuit is of importance only when determining the optimum ratio of L to C for a given frequency and load resistance. In general, a phone requires twice as much C as for a similar CW amplifier. In the plate tank of a self-excited oscillator, the C is required to be about three times greater than that for a given CW amplifier. Comparatively, the minimum "Q" of a single-ended amplifier should be kept about 5 for CW, 10 for phone, and approximately 15 for a self-excited oscillator.

A lower value of "Q" is permissible in the push-pull amplifier; a minimum Q of 3 for CW, 6 for phone, and 9 for a self-excited oscillators are satisfactory.

The accompanying table gives approximations of the optimum tank capacity for a single-ended CW amplifier (Q of 5) at different plate voltages, power and frequencies. Variations from the indicated values of capacity up to 20 per cent will not materially affect the operation of the amplifier. Larger capacities will increase the Q somewhat, but with an increase in the tank losses due to the increased circulating tank current, which reduces power output and efficiency. The use of less C than that shown will reduce the Q and may again reduce the efficiency and power output if minimum plate current does not coincide with maximum output current; that is, at the same point when the tank condenser is tuned. The capacities shown are those which should actually be applied, not just the maximum capacity of the tuning condenser.

The table shows most of the common combinations encountered in practice. However, for widely different frequencies or power inputs, the following formula will enable the approximate tank capacity to be directly determined. The following formula applies to a single-ended grid-neutralized (unsplit tank) amplifier for CW (Q of 5); for phone, (Q of 10) multiply the indicated capacity by 2.

For split-tank coils divide the indicated capacity by 4.

$$C = \frac{2{,}600{,}000}{fR_b}$$

Where C equals the tank capacity in mmf.; f, the frequency in megacycles; R_b, the DC resistance in ohms, of the plate to filament path of the amplifier (DC plate voltage divided by DC plate current, in amperes.

It will be seen that there will be relatively little difference between the cost of the tank condenser used with either grid or plate neutralization. With grid neutralization, the plate tank capacity must be four times as large as the capacity required in a plate neutralized amplifier. However, the

Table showing proper value of tuning capacity to use for a Circuit Q of 5 at the frequencies, plate voltages and plate currents indicated.

DC Plate Voltage E_B	DC Plate Current I_P	Plate Tank Capacity C_A	Plate Tank Capacity C_B	Frequency	DC Plate Voltage E_B	DC Plate Current I_P	Plate Tank Capacity C_A*	Plate Tank Capacity C_B§
375 V	.025 A	100	25	1750 KC	1500V	.100A	100 μμfds	25 μμfds
375	.050	200	50		1500	.200	200	50
375	.100	400	100		1500	.400	400	100
750	.050	100	25		3000	.100	50	12.5
750	.100	200	50		3000	.200	100	25
750	.200	400	100		3000	.400	200	50
375	.025	50	12.5	3500 KC	1500	.100	50	12.5
375	.050	100	25		1500	.200	100	25
375	.100	200	50		1500	.400	200	50
750	.050	50	12.5		3000	.100	25	6.25
750	.100	100	25		3000	.200	50	12.5
750	.200	200	50		3000	.400	100	25
375	.025	25	6.25	7000 KC	1500	.100	25	6.25
375	.050	50	12.5		1500	.200	50	12.5
375	.100	100	25		1500	.400	100	25
750	.050	25	6.25		3000	.100	12.5	3.12
750	.100	50	12.5		3000	.200	25	6.25
750	.200	100	25		3000	.400	50	12.5
375	.025	12.5	3.12	14,000 KC	1500	.100	12.5	3.12
375	.050	25	6.25		1500	.200	25	6.25
375	.100	50	12.5		1500	.400	50	12.5
750	.050	12.5	3.12		3000	.100	6.25	1.56
750	.100	25	6.25		3000	.200	12.5	3.12
750	.200	50	12.5		3000	.400	25	6.25
375	.025	6.25	1.56	28,000 KC	1500	.100	6.25	1.56
375	.050	12.5	3.12		1500	.200	12.5	3.12
375	.100	25	6.25		1500	.400	25	6.25
750	.050	6.25	1.56		3000	.100	3.12	.78
750	.100	12.5	3.12		3000	.200	6.25	1.56
750	.200	25	6.25		3000	.400	12.5	3.12

*C_A is the plate tank capacity to be used with all single-ended amplifiers when grid neutralization is used. This value is correct for a single tube (or tubes) in parallel, as long as the total DC plate current is as shown above.

§C_B is the total plate tank capacity to be used with all single-ended amplifiers which use plate neutralization. If a split-stator tank condenser is used, the capacity per section should be twice C_B in order that the total capacity will equal C_B.

If push-pull is used, a minimum circuit Q of 3 is permissible for CW use. Thus only 60% of the capacities shown in the column headed C_B should be used in a push-pull amplifier. As with the single-ended amplifier, multiply the indicated capacity by 2 for phone, and by 3 for a self-excited oscillator.

---

condensers in the plate neutralized tank circuit will have twice the peak RF voltage across them as well as twice the spacing as the condensers which are used in grid neutralized amplifiers.

Characteristics of Plate Tank Circuits: There are eight different arrangements of plate tank circuits for radio-frequency amplifiers. Fundamentally, there are two basic circuits; these are, (a) the split-tank with plate neutralization, shown in Figures 3, 4, 5, and 7; and (b), the unsplit-tank with grid neutralization, shown in Figures 1 and 2. Of course, the push-pull circuits, shown in Figures 6 and 8, also have a split plate-tank as well as a split grid-tank, because the neutralization of a push-pull stage may be considered to be both grid and plate neutralization.

From the standpoint of the optimum ratio between inductance and capacity in the plate tank circuit of a RF amplifier the circuit arrangement affects the required tuning capacity for a given tube, plate voltage, power output and frequency.

For a given set of conditions, the impedance in ohms, measured across the ends of a split tank coil, will be exactly four times the impedance across the unsplit plate tank coil. In the grid neutralized tank circuit shown in Figure 1 the plate circuit of the amplifier tube is connected across the entire circuit so that the required reflected load impedance appears across the entire tank circuit. When the same amplifier is changed to plate neutralization with either the split coil circuit shown in Figure 3, or the split-stator condenser circuit shown in Figure 4, the plate circuit of the tube is then tapped across only half of the tank circuit. Thus the impedance measured across either half of the plate tank must be the same in order that the tube will operate under exactly similar conditions as encountered in the grid neutralized circuit. Because this is an autotransformer arrangement, the impedances across part or all of the inductance will vary as the square of the turns ratio; and since there are twice as many turns across the entire tank coil as

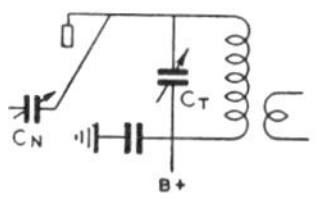

FIG 1

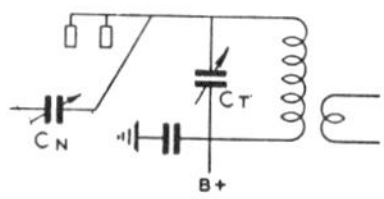

FIG 2

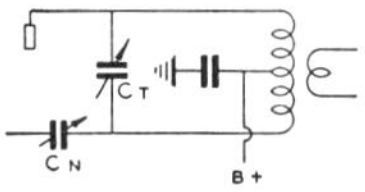

FIG 3

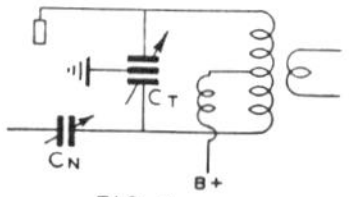

FIG 4

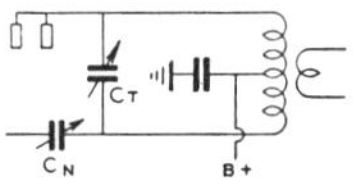

FIG 5

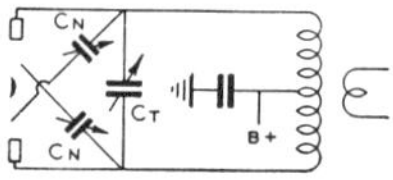

FIG 6

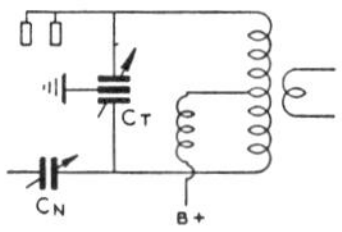

FIG 7

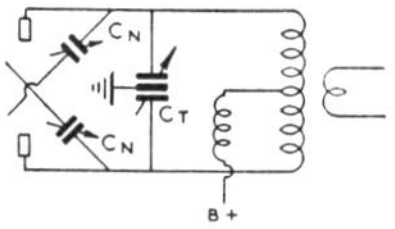

FIG 8

there are across either half, the impedance across the entire tank will be two squared, or four times the impedance that one-half of the tank reflects back into the tube. For a given power, tube and plate voltage, there is twice the peak RF voltage across the split tank as there is across an unsplit tank. This higher RF voltage means that dielectric losses in the plate tank circuit are four times as high in the split circuit as in the unsplit arrangement, but because the circulating RF current is twice as high in the unsplit tank the resistance losses in that circuit are four times as large.

In a single-ended grid neutralized high efficiency amplifier operating at less than 4000 volts DC plate voltage, the circulating current losses can be minimized by tapping the plate down on the plate coil in order to reduce the amount of C necessary for a given "Q" (see Figure 1A). This technique is more desirable than employing plate neutralization; furthermore, it allows the use of a single-section condenser.

Parallel Operation: The circuit shown in Figure 2 is exactly the same as that in Figure 1, with the exception that the two tubes are in parallel in Figure 2. If the two paralleled tubes draw the same plate current under the same conditions of operation as the one-tube circuit of Figure 1, then the load impedance across the two tank circuits will be equal, and the same tuning capacity will give the same circuit "Q"; but, if a second tube is added to an already existing amplifier to double the output, the bias must remain unchanged, even though the DC grid current will double. The neutralizing capacity must be doubled, and the antenna coupling must also be increased in order to make the amplifier draw twice the plate current as it did before. In addition, it will be found that the tank tuning capacity must be doubled to preserve the same circuit "Q."

Push-Pull: All push-pull circuits, such as those shown in Figures 6 and 8, have split tank coils. In these circuits there are no unbalances due to plate-to-ground capacities; the arrangement shown in Figure 8 is preferable to others, incidentally, the total plate tuning capacity is the same in either Figure 6 or 8.

With reference to Figure 8, if the two tubes together draw the same plate current as the one tube circuit in Figure 4 (assuming identical operating parameters), the load impedance across the entire circuit will be the same in both cases, and the required condenser capacities will be equal in value.

The push-pull circuit makes possible the use of a lower value of "Q" for the same circuit merit; the "Q" of a push-pull circuit need only be approximately 60 per cent of the "Q" of an equivalent single-ended amplifier. The purpose of "Q" in any tank circuit is to preserve the waveform of the alternating current. Thus, the particular advantage of the push-pull circuit is that it produces very few even harmonics and thus preserves the shape of the wave better than a single-ended circuit of the same "Q". The presence of harmonics in the distorted wave output of a low "Q" amplifier is precisely the reason why a high-C (meaning high "Q") tank circuit minimizes the radiation of undesirable radio-frequency harmonics.

Tank Circuit Relationships: The impedance across any tuned circuit is related to the series resistance of the tank. The higher the series resistance, the lower the shunt resistance. (Resistance and impedance are identical at resonance.) The shunt resistance is always "Q" squared, times the series resistance.

The reactance of either the coil or condenser of any resonant circuit is always equal to "Q" times the series resistance, or the shunt resistance divided by the "Q". Thus a tank loaded so that it has a shunt resistance of 5000 ohms at resonance would be said to have a series resistance of 50 ohms if the LC ratio were such that the circuit "Q" were 10. In order to have a "Q" of 10, the coil and the condenser reactance would have to be "Q" times the series resistance, or 10 times 50, or 500 ohms. The reactance is also shunt resistance divided by "Q", or 5000/10 = 500. The capacity required to equal a 500 ohm reactance can be calculated if the operating frequency is known by the following formula:

$$X_c = \frac{1{,}000{,}000}{2 \times \pi \times f \times C}$$

where X_c equals the reactance in ohms; **f**, the frequency in cycles per second; and C, the capacity in microfarads.

Antenna Tank Circuits: The use of link-coupling between the plate tank of the final amplifier and a separate antenna tank circuit to which the antenna or feeders are coupled has been universally popular. This type of coupling reduces harmonic radiation, preserves better balance on a push-pull stage, prevents the feeder radiation from altering stability of the various amplifiers in the transmitter, and tends to improve the effective "Q" of the plate tank circuit of the final amplifier.

The higher the "Q" of the antenna tank the more the harmonic radiation will be reduced. The "Q" of the antenna tank should not be less than 3, but preferably higher than 5. The "Q" is calculated or estimated in exactly the same manner as that of the plate tank.

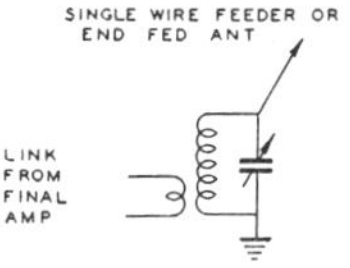

FIG 9

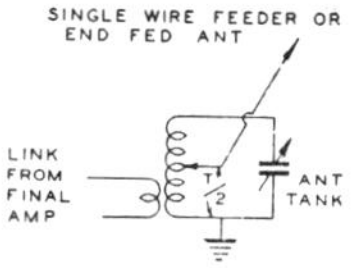

FIG 10

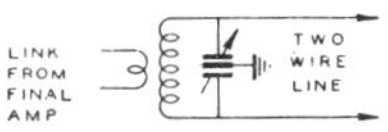

FIG 11

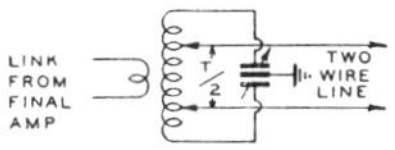

FIG 12

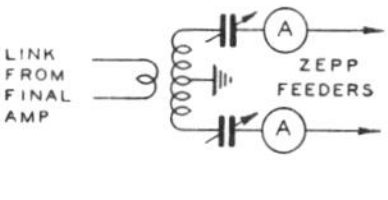

FIG 13

Various methods for coupling final amplifier to the antenna are here shown.

One of the simplest antenna tank arrangements appears in the schematic of Figure 9. If the tank is feeding an off-center Hertz antenna the shunt impedance across the tank will be the same as the characteristic impedance of the feeder, which is in the neighborhood of 600 ohms (Note: see the "ANTENNA" section for other details). Thus, to obtain a "Q" of 5, the condenser reactance at the operating frequency would be 120 ohms. At 7000 KC this would require a condenser capacity of 190 uufds. At 3500 KC, twice this capacity would be necessary. The values of capacity are larger than can be conveniently handled and therefore the arrangement shown in Figure 10 reduces the required capacity to one-fourth, although the RF voltage (for any given power output) is doubled; consequently the twice spacing must be provided. The feeder is tapped across one-half of the total turns, making the impedance across the entire tank four times the impedance from feeder to ground, or 2400 ohms across the tank for a 600 ohm feeder. The condenser reactance for a "Q" of 5 is 480 ohms; therefore only 48 uufds. of capacity is necessary at 7 MC. The capacity is independent of the power output of the transmitter, which is a point of difference between the antenna tank and a plate tank, because the power output of a transmitting tube is very closely related with the reflected load impedance into which the tube works. Therefore a 1 KW transmitter would require no more capacity in a given antenna tank than a 5 watt transmitter, but the voltage spacing would have to be much greater. The effective RF voltage across any tuned circuit is always equal to the square-root of the product of the power in watts, times the shunt impedance, in ohms; or writing

$$E = \sqrt{PZ}$$

where E equals the volts; P, watts; and Z, ohms.

Thus 1 KW of power across a 600 ohm feeder represents an effective voltage of 775 volts. The voltage across 2400 ohms for the same power is twice this value, or 1550 volts. The peak voltage can be about twice the effective voltage, particularly if harmonics are present or if the carrier output is voice modulated, and thus the antenna tank tuning condenser must be rated at from two to three times the peak voltage which is present.

If it is desired to use a still smaller condenser to tune the antenna tank, the feeder can be tapped farther down the tank coil. This steps-up the impedance across the entire tank circuit, according to the law of impedance transformation, wherein the impedance ratio is equal to the square of the turns ratio.

If an end-fed antenna is tapped directly to the antenna tank coil, the circuit of Figure 9 should be used, as it is not advisable to tap down on the coil. Figures 11 and 12 show split antenna tanks for feeding two-wire non-resonant transmission lines. Figure 13 describes how a Zepp antenna can be fed by means of a link from the final amplifier.

Power Transfer: In all transmitters, care must be taken to properly transfer the power into each succeeding stage; otherwise the output from the final stage will be low. The coupling link must be adjusted so that maximum grid current is obtained in the driven stage.

With capacity coupling between stages, the grid coupling condensers must have sufficient capacity to provide a normal load on the preceding tube with maximum grid current. Lower frequencies, such as 3,500 KC, require a .00025 grid condenser between the doubler circuit and the buffer grid (for an 801 tube) for the same loading effect on the doubler plate (6A6 in this example). With low impedance tubes, such as 6A6 or 53 types, the input to a buffer stage may be capacitively coupled with nearly as much grid drive as with link coupling, provided a high or medium grid impedance is offered. A low-mu tube offers a higher grid impedance load than does a high-mu tube, such as a 203A or 46. Capacity coupling between an 801 and 50T, both medium-mu tubes, gives only a little more than half as much grid current as is obtained with link-coupling. These important points must be carefully weighed when a transmitter is to be put into operation on 10 or 20 meters, as the margin of available grid excitation is much less than on 40 or 80 meters. Probably 90 per cent of the trouble with 20-meter transmitters is lack of sufficient excitation on one or more grid stages.

Filament By-passing: Each side of the filament must be by-passed with a .002 ufd. condenser to its particular RF stage ground-bus to provide low impedance paths for neutralizing purposes. Too low-C in the final tank circuit makes neutralization difficult, and does not give any more output on the fundamental frequency.

Self-Excited Oscillators

The self-excited oscillator (SEO) is one of the outstanding developments in the progress of radio transmitting apparatus. When properly designed, it is one of the best forms of frequency generation, for its use permits any desired frequency to be obtained with few adjustments. In amateur band operation this advantage results in selecting "clear spots" in which to operate. But it is a rather dangerous circuit for beginners to design. Few amateurs, especially the novice, have wavemeters and frequency meters to check the desired frequency with a self-excited oscillator. With these precepts, it is suggested that the SEO circuits be set aside until one has become well-grounded in radio knowledge and in practice.

Good design of the SEO necessitates a choice of good parts. solid connections, freedom from vibration, and a power supply with excellent voltage regulation.

tube (push-pull) affairs. They can be shunt or series fed.

Design and Technique: The push-pull circuit is to be recommended over the singled ended circuits, for there is a greater voltage swing, and the even harmonics are eliminated by circuit action. The rule to observe in construction of push-pull sets is symmetry—both mechanical and electrical. Exact electrical and mechanical symmetry cannot be obtained until left-handed and right-handed tubes are manufactured, because the grid and plate prongs of the tubes are reversed on the left-handed tubes. However, with the exception of the filament leads, a high degree of symmetry is obtainable. It is required that the leads to each inductance from each grid and plate socket be of the some length. The condensers can be connected to these leads in almost any manner without disturbing the constants. In many instances in which inductances are mounted on top of the condensers, unequal length of leads may result, even though they appear to be correct to the eye. Figure 7 illustrates the fact even though the grid and plate leads to the condensers are both of equal length, the condenser frame makes one of the leads

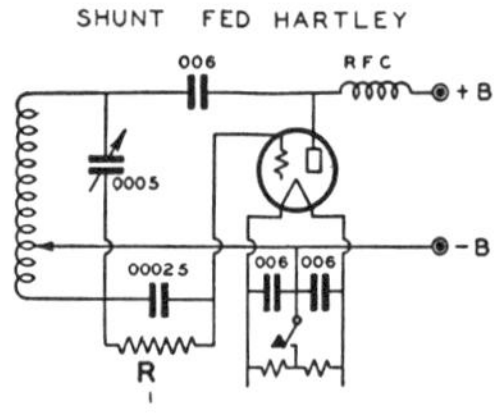

FIG. 1

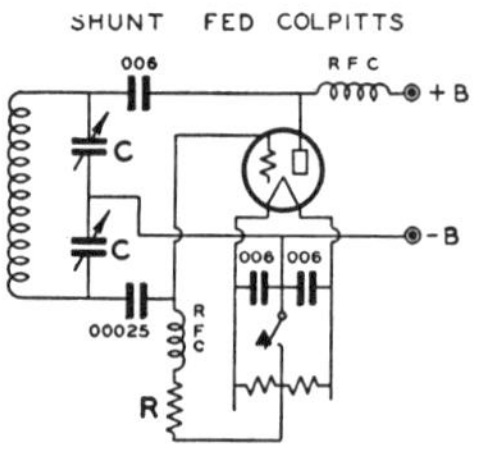

FIG. 2

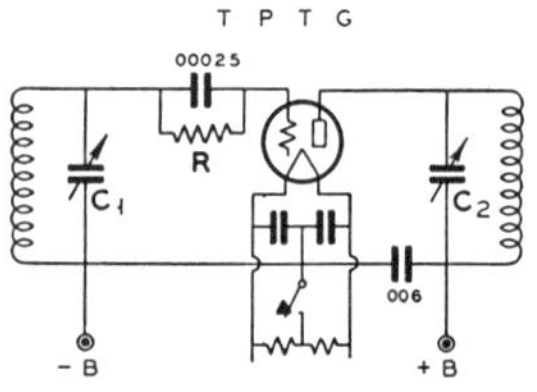

FIG. 3

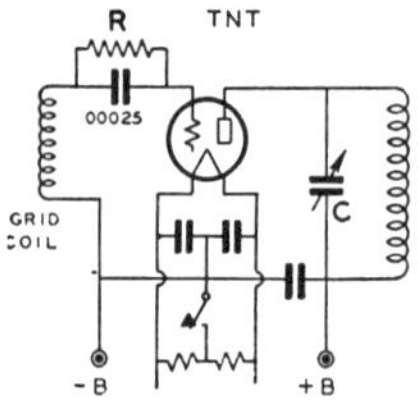

FIG. 4

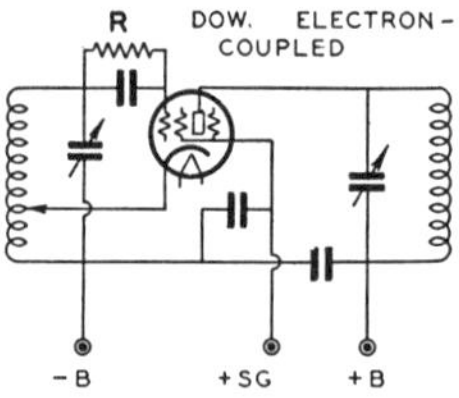

FIG. 5

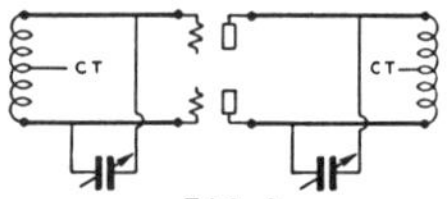

FIG. 6

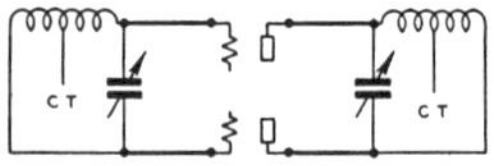

FIG. 7

Types of SEO Circuits: The common types are: The Hartley—Figure 1; the Colpitts—Figure 2; the tuned-plate tuned grid (TPTG)—Figure 3; the TNT—Figure 4, and the electron-coupled—Figure 5. These circuits need little explanation with possibly a reference to the TNT. Its name is correct; it is TNT in the hands of beginners and, therefore, is not a circuit for any newcomer to use.

SEO circuits can be single tube or two

longer than the other. To overcome this difficulty, mount the coils separately with their equal length connection to the sockets, and then connect the condensers to the leads. This might slightly throw off the balance, but odd length condenser leads still constitute capacity—and not inductance, if the leads are short and close together. This is illustrated in Figure 6. Note: Keep the condensers at least a coil's diameter away from the coil.

Piezo Quartz Crystals

Quartz and tourmaline plates are minerals having a crystalline structure which, when cut and ground on certain crystalographic (optical) axes, possess piezo-electric properties in the influence of an oscillating electrical field. The mechanical activity or frequency of a piezo-electric element depends upon its physical dimensions (the frequency being inversely proportional to the thickness). The stability of the oscillatory properties depends mainly upon the optical cut and the crystal-temperature coefficient.

Definition: Piezo-electric Oscillator (after the U.S.N. Conference in 1929): A circuit containing a resonator (crystal) and possessing too little regeneration to oscillate itself, but which oscillates through the reaction of the crystal when the latter is vibrating near one of its normal frequencies with energy derived from the circuit. Such a circuit is often called a "crystal controlled" or "piezo-oscillator."

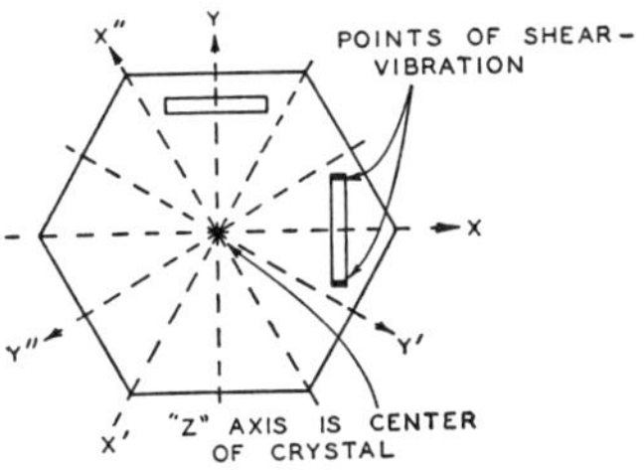

FIG. 1. Quartz crystal axes.

A quartz crystal plate (as used in amateur transmitters) is cut from the edges of a quartz crystal parallel to the optical axes known as X, Y and Z, see Figure 1. In general, crystals are cut with their faces either parallel or perpendicular to the Z or electric axis of the crystal. An X-cut is one that is parallel to the X-axis, while a Y-cut is parallel to the Y-axis. Y-cuts are sometimes referred to as 30-degree cuts. The thickness dimensions of the plate are parallel to the X- and Y-axes, respectively, while the rectangular length or elongation of the plate is perpendicular to the optical or Z-axis. An X-cut crystal vibrates in the direction of the Y-axis, and the chief mode of vibration for a Y-cut is that of a shearing vibrational-strain taking place about the Z-axis; with this latter cut, the crystal actually becomes elastic and waves are produced parallel to the Y-axis. A crystal cannot oscillate along the Z-axis, as the forces which hold the atoms of the crystal together are so great that there is relatively little expansion along this axis.

In general, quartz plates are most widely used for controlling frequencies below 10 megacycles, because of their relative cheapness as compared to tourmaline plates. On the higher frequencies, tourmaline is to be preferred for fundamental control, as quartz plates oscillating above 7 megacycles have a slight tendency toward sidetone oscillation. Tourmaline crystals are mechanically stronger than quartz, and are also easier to grind on account of their smaller diameter and greater thickness for a given frequency.

In amateur practice, X-cut crystals are sometimes ground with trick contours to boost the power output, but if the process is carried beyond a certain stage, the crystal will oscillate at more than one frequency unless special precautions are taken with the oscillator to prevent it. The temperature coefficient of a Y-cut plate is twice that of an X-cut (and in the opposite or negative direction), but if the oscillator is run underloaded, the drift will be negligible with either cut. Because of the temperature characteristics, X-cut crystals have a negative temperature coefficient, and Y-cuts positive; for these reasons, an X-cut plate is preferable for use just inside the HF edge of a band, and a Y-cut for the low-frequency edge.

Frequency Drift and "Twin-Peaks": Crystals that oscillate at more than one frequency are commonly known as crystals with "twin peaks." This dual vibrational tendency is more pronounced with Y-cuts, and to a certain degree is exhibited by many X-cuts. The use of a well-designed, space wound, low "C" tank coil in an oscillator will prohibit the crystal from oscillating at two frequencies, and in addition will increase the output. Experiments have shown that the frequency stability is not improved by large tank capacities, which only tend to augment the double frequency phenomenon.

Y-cut crystals having perfectly parallel sides, lapped to a high precision, are the worst offenders in regard to twin frequencies, sometimes making it necessary to resort to a special form of clamp holder in addition to an extremely low capacity tank to confine the oscillations to one peak.

An X-cut crystal that has been accurately ground, with both sides absolutely flat and parallel, will oscillate at only one peak, provided the edges are free from imperfections or nicks. Good output from an X-cut crystal can only be obtained when the top electrode of the crystal-holder does not press too heavily against the crystal. By grinding a special contour into the crystal, a medium output is obtainable that under certain conditions may suffice; however, a crystal ground in this manner will have its output appreciably reduced by heavy electrode pressure. In grinding, if the convexity is carried too far, the crystal will have two oscillating peaks.

Twin frequencies appear in several ways: sometimes the crystal will have two frequencies several hundred cycles apart, oscillating on both frequencies at the same time, and producing an acoustically audible beat note. Other crystals will suddenly "jump" frequency as the tank tuning condenser is varied past a certain setting. Operation with the tank condenser adjusted near the point where the frequency shifts is very unstable, the crystal sometimes going into oscillation on one frequency and sometimes on the other as the plate voltage is cut "on" and "off." Still other crystals will jump frequency only when the temperature is varied over a certain range. And some plates will jump frequency with a change in either tank tuning or temperature, and produce an audible beat tone at the same time, showing actually two pairs of frequencies!

Crystals are often cut with their axes between the X and Y points, to reduce the temperature coefficient. Since X- and Y-cut plates have a frequency drift in opposite directions with increase in temperature, a plate cut between the two axes will have a negligible or near-zero drift.

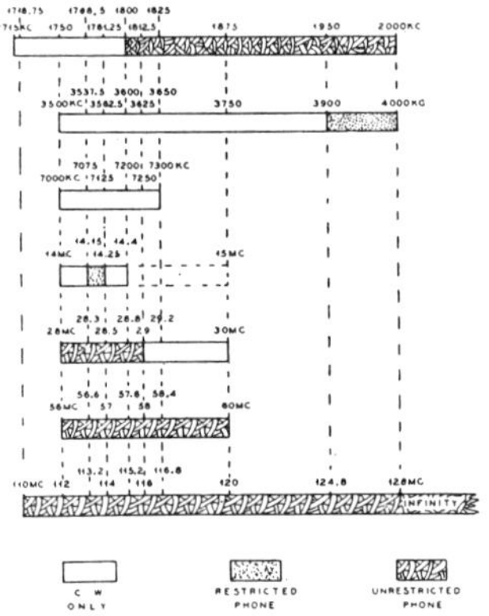

Bliley frequency chart of Amateur Bands and harmonic relationship.

Use and Care of Crystals: When operating close to the edge of one band, it is advisable to make sure that the crystal will respond to but one frequency in the holder and oscillator in which it is functioning; for a crystal with two peaks can suddenly leave a band and operate on another without giving any indications of the change on the meter readings of the transmitter. If the transmitter frequency is such that the operation takes place on the edge of the band at all times, under all conditions of room temperature, some form of temperature control will be required for the crystal. When working close to the edges of the 14 megacycle band it is essential that the crystal temperature be kept at a fairly constant value; the frequency shift in kilocycles per degree Centigrade increases in direct proportion to the operating frequency, regardless of whether the fundamental or harmonic is used. When a crystal shifts its frequency by two kilocycles, its second harmonic has shifted 4 kilocycles. Amateurs not operating on the edge of a band need not concern themselves about frequency drift due to changes in room temperature. If a pentode tube is used for the crystal oscillator having a plate potential of approximately 300 volts, the temperature of the crystal will not increase appreciably to cause any noticeable drift at even 14 kilocycles. When a crystal oscillator is keyed on 3.5 or 1.7 megacycles, the frequency drift is not of any consequence, even with much higher values of plate input, because of the keying and of the fact that the drift is not multiplied as it would be with harmonic operation of a final amplifier.

Crystal holders have a large effect on the frequency; for example, the frequency of an 80 meter crystal can vary as much as 3 kilocycles in different holders. Even greater variations are possible on account of the unevenness of some electrodes in various types of manufactured holders. Warped electrodes touching a crystal in two or three spots form, in effect, a sort of air-gap holder. Holders having a spring to provide tension on the top electrode appreciably affect the frequency, and their use is to be discouraged.

Periodic or weekly crystal cleaning done by rubbing the top electrode around on the crystal surface to dislodge dust particles that may have worked in between the electrode and crystal will, after a year or longer, increase the frequency of the crystal. CAUTION: Do not rub the crystal or electrodes; disassemble the holder and clean the parts with alcohol, ether, or carbon-tetrachloride (carbona). With polished crystals there is less tendency of wear; however, as a safety measure, all crystals should be placed in dust-proof holders.

40 Meter Crystals: A 40 meter crystal can be used in the conventional 47 crystal oscillator circuit and link coupled to an 841 doubler running at about 700 volts to excite a 210 to full output with high efficiency on 20 meters, provided the 841 is also link coupled to the 210. On the higher frequencies there is a worthwhile increase in efficiency and output when using the link form of inductive coupling, rather than capacitive coupling. Capacitive coupling is justified at the higher frequencies only for the sake of simplicity where reduction in efficiency can be tolerated.

Special precaution must be taken with 40 meter crystals, and more care given to the circuit details than with lower-frequency crystals. Here, a suitable crystal-holder is of prime importance, as many 40 meter crystals refuse to oscillate in any holder except the particular type in which the crystal was designed to operate. Because a holder works well with an 80- or 160-meter plate does not indicate that a 40-meter crystal will function likewise.

A 40-meter crystal requires a very light top electrode with no additional pressure spring for maximum output; spring pressure is not necessary for stability unless the transmitter is subject to severe vibration. The faces of a 40-meter crystal are practically flat, and if the surfaces of the holder electrodes are truly plane, the top electrode will not tend to "rock" on the crystal and cause frequency instability.

Before placing a 40-meter crystal in its holder, the edges of the crystal should be carefully examined for nicks and imperfections. A nick almost invisible to the naked eye will sometimes have an appreciable effect on the output. If the edges show that they have been chipped, the crystal should be returned to the manufacturer for refinishing.

Grinding Quartz-Crystals

What Cut to Grind: Amateurs who have had no previous grinding experience should first attempt grinding a Y-cut 80- or 160-meter crystal. Although it requires a much longer time to grind an X-cut after one has become proficient, X-cuts must be finished with a greater degree of precision and are therefore best avoided by the novice for the first attempt.

Grinding Materials: The necessary material and equipment required for grinding with a minimum of labor and difficulty are: a micrometer, several pieces of heavy plate glass, an oil can filled with water, a pan of clean water, several clean towels, a bottle of India ink, a test oscillator, and a frequency measuring device such as a calibrated receiver, and lastly, small quantities of No. 150, No. 280 and No. 400 carborundum. The latter grain is used only in finishing X-cut plates, and need not be procured if only Y-cuts are to be ground. Water is used in preference to kerosene, because it is necessary to remove all oily traces each time the crystal is tested in the oscillator. A one-half inch micrometer, reading to ten-thousandths, is best adapted for thickness measurements, but a one-inch instrument, reading to thousandths, will measure close enough for Y-cut places by estimating to ten-thousandths and with care can even be used for the lower-frequency X-cuts if nothing better is available. It is advantageous to grind down the movable face of the micrometer on a wheel so that the tip resembles a cone with a rounded point, rather than the end of a cylinder. This enables one to measure a point on a crystal instead of a section of the crystal.

The test oscillator must be equipped with a plate milliameter, a dummy load which can be cut out of the circuit, and plug-in inductances so that either low or high "C" can be used in the tank for test purposes. An RF meter in series with the crystal is optional.

Grinding Technique: Assuming that one has the necessary materials and a 160 or 80-meter Y-cut blank which it is desired to convert into a good finished crystal, it is first necessary to finish one side flat, to use as the reference side (some blanks already have the reference side finished and marked). This can be done by rubbing one side around with even pressure on a piece of plate glass that has been smeared with No. 150 carborundum grain and water until India ink marks which have been placed on the tip of each corner disappear. The crystal is then rinsed in the pan of water and rubbed on another piece of plate glass smeared with No. 280 carborundum and water for half a minute or longer—care being taken to see that the pressure is fairly even all over the crystal and that the grinding is being done on the same side. The crystal is then washed and dried, and one corner of the finished side marked with India ink. All subsequent grinding is done on the other side. Using a finer grain of abrasive for finishing Y-cut crystals is not advisable, because it does not increase the output, but only aggravates the tendency toward twin-frequency peaks. By using a medium grain of carborundum for finishing and by giving the right contour to the side that is not yet finished, the second peak can be eliminated.

A piece of raw quartz and several unfinished slabs from which oscillating crystal blanks are cut. The best quartz is mined in Brazil.

The crystal should now be roughened down with No. 150 grain carborundum until it is .002 or .003 inch thicker than the calculated finished thickness, which will be very close to .022 for 3500 KC and .0435 for 1750 KC. The finished thickness of a crystal of either cut can be predetermined for a given frequency within fairly close limits by applying the following formula:

$$\text{X-cut} \quad T = \frac{112.6}{F}$$

$$\text{Y-cut} \quad T = \frac{77}{F}$$

Where T is the thickness in inches, and F, the frequency in kilocycles.

The next step is to finish the crystal down with No. 280 abrasive to about .0004 inch greater than the calculated thickness (.001

for a 160 meter plate), frequent micrometer readings being taken to prevent any high or low spots from appearing. The crystal is then put in the oscillator. If its surfaces are reasonably parallel, it should now oscillate. If it oscillates at but one frequency as the tank tuning condenser is varied, it is a most unusual Y-cut crystal and is not acting in characteristic fashion. Making certain that it is oscillating at the low-frequency peak, the frequency should be checked to ascertain how closely it is agreeing with the formula.

The second peak, which is the highest in frequency, can be eliminated by giving the face now being ground a convex contour. The degree of convexity necessary to give one-frequency operation will vary with different crystals, but in every case the second peak will disappear before the process is carried far enough to affect the output to any great extent. In fact, a moderate curvature will actually increase the output slightly over that of a Y-cut crystal that has been ground with both sides perfectly flat.

At this point the corners should be slightly rounded and the edges finished up. It is best to finish these before putting the final touches on the crystal as a preliminary to grinding the crystal to an exact frequency, because grinding on the edges will sometimes affect the characteristics of the plate. Grinding on the edges has a minor effect on the frequency, and also will sometimes cause a crystal that checks at one frequency to develop a second peak.

The optimum amount of convexity can only be determined for each particular crystal by trial, but it is not critical as long as no spot is higher than the center of the crystal. A contour that has been found suitable for most 80 meter crystals of the Y-cut type is as follows:

Edges between corner .0001 inch lower than center; corners .0003 to .0005 inch lower than center. For 160 meter crystals the convexity can be slightly greater if necessary to eliminate twin peaks. A piece of glass that has been slightly worn down facilitates grinding a uniform convex contour, but until one has used a piece of glass for roughing-down several crystals it will not be hollowed out sufficiently for the requirements. If the glass is nearly flat, pressure on each of the edges and corners—one at a time—will be necessary to get the desired curvature.

A final check for twin peaks is made by using a tank coil in the test oscillator which requires slightly more capacity to tune to resonance than will ordinarily be required in the transmitter oscillator. No attempt should be made to keep the crystal from oscillating at two frequencies with an extremely high-C tank, because almost any crystal will show a second peak if the oscillator is made very high-C. If the "medium-C" tank shows two frequencies, it will be necessary to grind down the tips and the corners until the second one disappears. A soldering iron should then be held near the crystal as the beat note is monitored in the receiver, until the crystal frequency creeps 10 or 15 kilocycles. The shift should occur gradually without sudden "jumps." The tank is then tuned through resonance for only one frequency; if two appear, the tips of the corners of the plate must be ground further. Very few plates will be found to require such drastic treatment, the slight convexity usually being sufficient.

To test for output and freedom of oscillation, the original low-C tank coil is employed. It is helpful to have a crystal that is known to be a good oscillator for comparative purposes. A low value of minimum plate current is the criterion for freedom of oscillation. A low minimum plate current means nothing, however, if the crystal becomes unstable or goes out of oscillation the moment a load is coupled to the tank. The oscillator must stand a reasonable amount of loading without going out of oscillation; in addition, must be stable when loaded.

If the finished crystal gives good output and has only one frequency response, one is then justified in attempting to grind an X-cut plate.

Grinding an X-cut Plate: The reference side of an X-cut blank is ground down with No. 150 and No. 280 carborundum grain in the same manner as a Y-cut blank. It is then rubbed around in a circular motion for a half minute on a new piece of glass which is covered with No. 400 abrasive and water. It is imperative to use a **new** piece of glass for finishing the reference side of an X-cut blank, because maximum output will not be obtained if either side has the slightest amount of a convex curvature. One exception can be made: some manufacturers grind their X-cut plates with a special contour which calls for sections of the crystal being very slightly convex, but an amateur inexperienced in grinding will do well to keep away from such special trick contours. The output of an X-cut plate can be boosted by merely grinding it slightly concave on the finishing side. Paradoxically, while Y-cut plates have twin frequencies when the curvature is not great enough, X-cut plates exhibit double frequencies only when the curvature is too great. It is necessary, however, to remove a large section out of the center of an X-cut crystal before the second frequency appears, unless a very high-C tank is used in the oscillator. Grinding the center of an 80 meter X-cut plate .0001 or .0002 inch low will boost the output without encouraging a second frequency.

After inking the reference side, the blank is roughed down to about .03 or .004 inch over the calculated finished thickness with No. 150 carborundum, and then down a little further with No. 280 grain. The final grinding is done on a little-used piece of glass, covered with No. 400 grain and water. Enough pressure is exerted in the center of the crystal with one finger to bring the center .0001 or .0002 inch lower than the edges and corners. No spot should be lower than the center, otherwise the output will be disappointing. A new piece of glass should be used for finishing each X-cut crystal. The glass is then suitable for grinding Y-cut plates or roughing-down X-cut plates. Because 160 meter X-cut crystals are too thick to be hollowed out easily by exerting presure in the center, even if new pieces of plate glass are used, it is necessary to finish them on a special piece of convex glass.

Finishing the edges on X-cut plates is of greater importance than on Y-cuts. An X-cut plate with unfinished edges may even refuse to oscillate unless the edges and corners do not vary over .0001 inch, or unless about .0005 inch has been hollowed out of the center (which is sufficient to cause double response frequencies). A crystal with variations greater than .0001 inch between the different corners and edges may give full output after the edges are finished. It is important that every minute imperfection be removed from the edges when finishing X-cut plates. X-cut crystals that refuse to give full output can sometimes be made to oscillate more freely by grinding the edges so that the cross-section of the crystal is reduced; that is, to grind so that the dimensions of the crystal along the other axes are changed.

To finish the edges of crystals of either cut, all nicks are first ground out by using the same grade of abrasive as used to finish the faces, but applying less water. To complete the work, the corners and edges are then slightly rounded off.

The India ink reference mark can be removed with a moistened, soft rubber eraser. The plate should be washed with soap and warm water to remove any rubber gum which may adhere to the unpolished surfaces. X-cut crystals can be polished to high transparency with rouge; however, the output will not be increased over that which is obtained by grinding with a high grade abrasive.

One-Tube Push-Pull Transmitter

A simple push-pull crystal oscillator circuit employing a type 53 tube is an ideal means for securing 10 watts of output for a CW transmitter.

A complete unit, ready for connection to the power supply.

This illustration shows a more elaborate single-tube transmitter with large plate coil. A milliammeter, on-off switch, and a keying jack are mounted on a small Masonite panel. It gives no better results than the smaller unit.

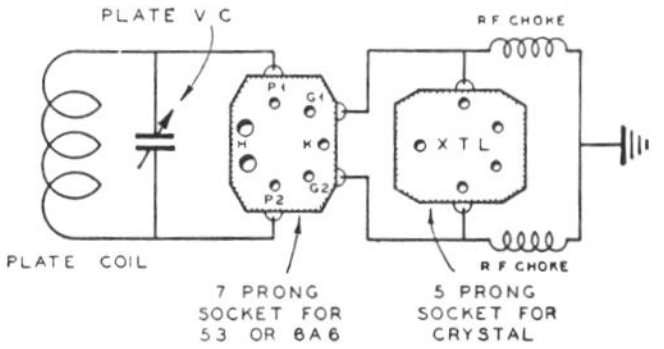

CORRECT PARTS LAYOUT

Single-tube 53 or 6A6 transmitter.

Details: The arrangement of the various parts should be followed as shown in the accompanying photographs. For 160-meter operation, a 140 uufd. midget variable condenser is used. This condenser is mounted about 6 inches back from the front panel and a ¼-inch round bakelite shaft is coupled to it, then extended to the tuning dial. The front panel is a piece of No. 12 or No. 14 gauge aluminum, 4 in. x 10 in. The baseboard is 11 in. x 15 in. x ¾ in. Wooden cleats, screwed to the two ends of the baseboard, raises it sufficiently off the table to permit mounting the bypass condensers and resistors underneath. The filament and plate supply wires are also under the baseboard.

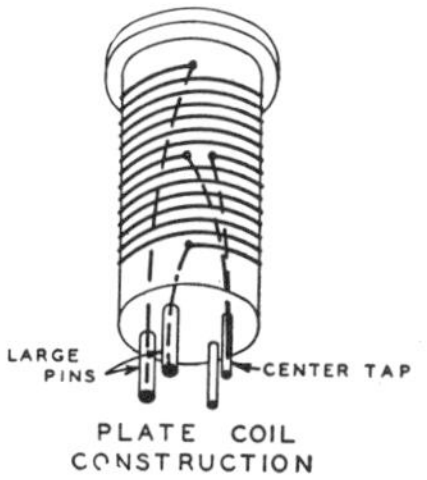

The top and bottom connections of the coil terminate in the LARGE pins of the coil form, where they are soldered in place. The center-tap (mid-point of the winding) is brought to one of the small prongs. This mid-point connection can be soldered where the tap is taken, and only one wire need then be brought to the small pin, instead of two wires, as shown in the accompanying sketch.

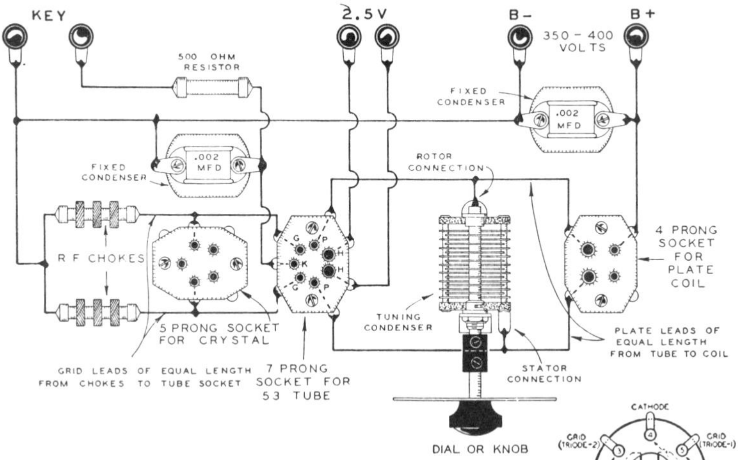

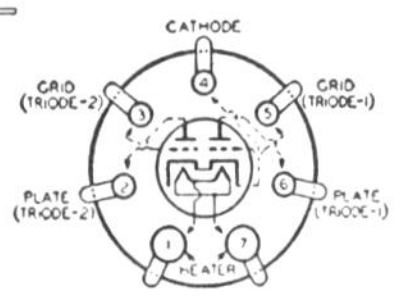

Pictorial wiring diagram, showing how the various components are connected in the circuit. The 53 tube base symbol in the extreme lower right hand corner shows BOTTOM OF SOCKET connections, whereas in the main pictorial circuit TOP OF SOCKET connections are shown.

Schematic wiring diagram of the complete one-tube push-pull transmitter and power supply. L1 is the oscillator plate coil, L2 and L3 are the coupling loops, L4 is the antenna coil. A simpler form of antenna coupling is shown below. The pictorial diagram above is the same as the schematic circuit here shown.

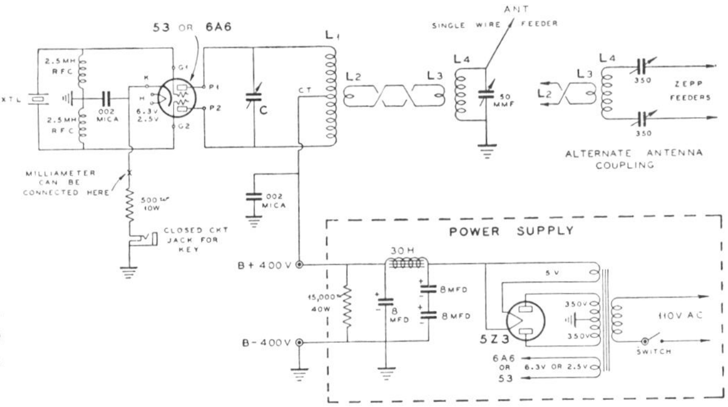

10 WATT PUSH-PULL CRYSTAL CONTROLLED BEGINNERS TRANSMITTER USING A SINGLE 53 OR 6A6 TUBE

The circuit diagram shows two methods of making the antenna connection.

If "air-wound," large diameter coils are used, the plate coil for 80-meter operation is wound with 37 turns of No. 14 enameled wire on a form 2¾ inches in diameter; the wire is wound over a space of 4½ inches and a tap is taken at the center of the coil for the plus "B" connection. Small diameter plate coil winding data is on page 118.

If the plate coil is not wound to specifications, a higher capacity condenser will be required for tuning the 80-meter plate coil. As an alternative, a single-spaced 100 μufd. midget variable condenser can be used

The plate coil also serves as a support for the antenna coil, as the illustration shows. The antenna coil is wound on a piece of tubing which can be slid over the plate coil, or supported on top of the coil form. The coupling is decreased by raising the coil.

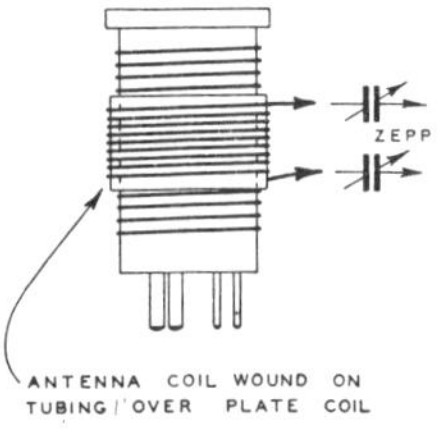

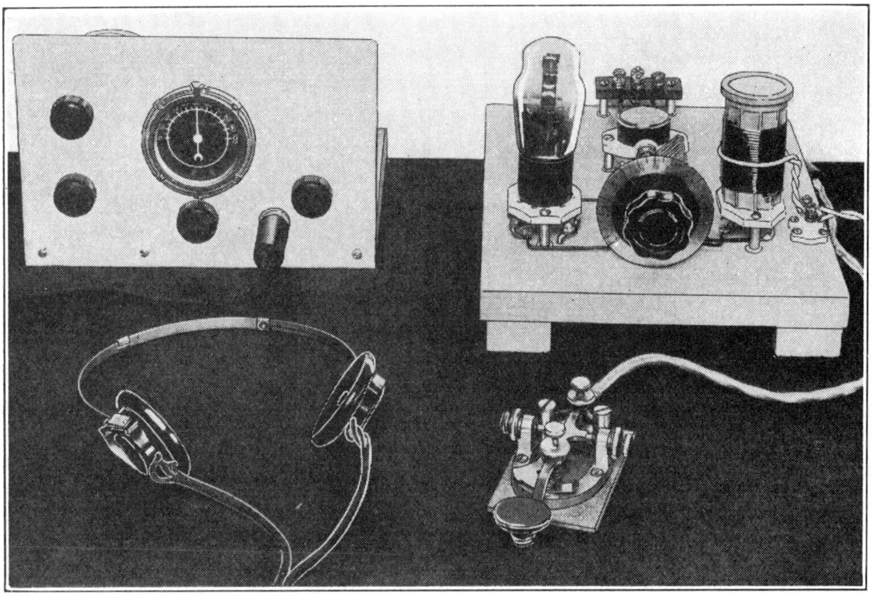

Complete newcomers' station, showing the one-tube push-pull transmitter, receiver with type 19 tube, telegraph key and headphones. This is an ideal arrangement—strictly modern.

for 80-meter operation if the plate supply is less than 400 volts. For 40 meters a 35 uufd. double-spaced midget variable is satisfactory. The plate coil for this band has 22 turns of No. 14 enameled wire wound on a form 2¾ inches in diameter in a space of 5 inches.

The two connecting leads from the plate coil to the plate condenser are of the same length.

47 Oscillator and 210 Amplifier Transmitter for Newcomers

With this transmitter, amateurs in San Francisco have contacted others in Australia and New Zealand on the 80 meter band. The construction given herein is such that additional stages can be added at any time in the future.

All coil-turns data is shown in the table on page 115. A turn or two can be added to, or removed from each coil, depending on the particular conditions under which the transmitter will be called to operate. However, the winding data is useful for coils tuned with small midget condensers of the capacity as specified.

The illustration shows the R-F portion of the transmitter. The 47 stage is at the extreme left. The 47 tube has given good results. The oscillator stage, from front to rear, consists of a 100 uufd. tuning condenser, plate coil, 47 tube, and quartz crystal. The resistors and condensers for the oscillator stage are mounted beneath the baseboard, the latter being raised from the operating table by a four-inch cabinet. Isolantite sockets are used throughout.

To the right of the oscillator stage is the grid coil, spaced 6 inches between centers from the oscillator coil. The grid coil tuning condenser, a standard 100 uufd. midget variable, is directly in front of the grid coil.

R-F portion of 47-210 transmitter

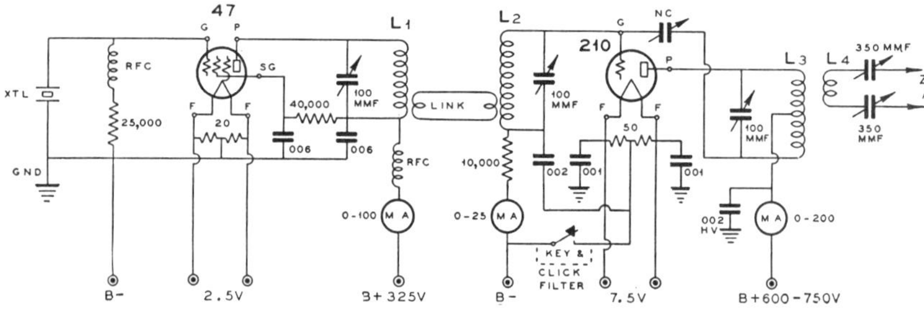

RF Portion of 47-210 Transmitter Circuit.

Link coupling is used between the oscillator plate coils and grid coil. The coupling link consists of a piece of No. 20 (or larger) insulated wire with a loop around both the plate and grid coils. The ends of the loops are connected to two of the prongs of the coil form, with the link line running under the baseboard. Variable coupling is not required for the coupling loop, the latter is simply slid over the coils and placed at a point about ¼ inch (or less) from the LOWER end of each coil. The high-potential (plate) lead of the oscillator **must** connect to the top turn of the plate coil and also to the stator of the tuning condenser.

Center-tap resistors are shunted across the filaments of the 47 and 210 tubes. The resistor across the crystal may have any value from 10,000 to 50,000 ohms, although many crystals will not "start" unless 20,000 or more ohms is used.

The 47 screen dropping resistor has a value of 40,000 ohms with a 5 to 10 watt rating, although values from 40,000 to 60,000 ohms will suffice. The .006 ufd. screen by pass condenser and the blocking condenser are of the mica type. The choke (RFC) must be of a good manufacture; a small receiving type of Hammarlund choke is satisfactory.

In the 210 amplifier stage, the bias resistor has a value of 10,000 ohms, 2 watt rating. As in the oscillator stage, the bypass condensers are of the mica type. The negative B lead from the oscillator stage is connected to the negative B lead of the amplifier stage. The neutralizing condenser in the 210 stage is a 35 uufd. double-spaced midget. The key, for CW transmission, is in series with the center-tap of the filament resistor and the negative B terminal.

Connection to a Zepp antenna only requires a simple coupling coil consisting of from 6 to 13 turns of No. 16 DCC wound on a 1¼ inch bakelite coil form, which is placed directly on top of the plate coil. The antenna coil is tuned with a .00035 ufd. receiving type variable condenser in parallel with the Zepp feeders; for series tuning, a .00035 ufd. receiving type condenser is placed in series with each feeder and the respective ends of the antenna coil. Coupling may be varied by placing a hinge on the antenna coil, or it may remain fixed by merely holding the plate and antenna coils in place by means of a cardboard sleeve, slipped into both coil forms.

Tuning: The tuning process is simple—secure a 2½-volt dial light, or small neon glow-lamp. If the dial light is used, connect a two inch loop of wire to the terminals of the lamp, soldering one end of the loop to the base connection and the other end to the screw-thread—with a neon lamp the loop is not needed.

Technique: Light the filaments of the 47 and 210 tubes. Turn on the plate voltage for the 47 oscillator tube but not to the 210 stage. Hold the lamp near the top of the plate coil and rotate the oscillator plate condenser until a point is found where the lamp glows brightest. This oscillation "peak" will not hold when the next stage is tuned. Now slip the coupling link over both the oscillator plate coil and the grid coil, placing the link at the bottom of both coils. Retune the oscillator condenser for brightest lamp glow; then tune the grid coil condenser until the lamp glows brightest when held over the grid glow. If the lamp does not light when held over the grid coil, it is proof that the oscillator stage has stopped oscillating. Retune the oscillator. A certain setting of both oscillator condenser and the grid condenser will be found where the lamp glows about the same briliancy when held over each coil.

Now place the lamp over the 210 plate coil and, with no plate current connected to this stage but with the filament lighted, make sure that the crystal is oscillating. Now rotate the 210 plate coil condenser over its entire range; if the tuning lamp lights up, the stage is not neutralized. Still keeping the lamp over the plate coil of the 210 stage, slowly rotate the neutralizing condenser until a position is found where the tuning lamp will not glow over the entire swing of the 210 plate tuning condenser. While neutralizing this stage, it is well to frequently hold the tuning lamp over the oscillator and grid coils to make sure that they are still in resonance. If no indication is found in the lamp when going back over the oscillator, retune the oscillator and grid coil circuit, because different settings of the neutralizing condenser throws the oscillator out of oscillation and require the circuits to be retuned.

When a point is found on the neutralizing condenser where no glow is indicated in the tuning lamp, no matter where the plate condenser of the 210 stage is set, the stage can be considered as being neutralized. Now apply the high voltage to the plate of the 210, press the key, and tune the plate condenser of the 210 until the lamp shows MAXIMUM glow. Connect the antenna to the coupling coil, place this coil over the 210 plate coil, tune the antenna feed condenser, and retune the 210 plate condenser until maximum indication is had in the antenna RF meter. The transmitter is now ready for operation.

Summary of Tuning: The entire transmitter is tuned to resonance if the tuning lamp glows when held over the oscillator plate coil and when it is placed over the grid coil, but NO GLOW should appear when the tuning lamp is placed over the 210 plate coil with the high voltage disconnected from this stage, no matter where the 210 plate condenser is set.

Cautionary Measures: Watch the RF antenna meter. Slowly retune each condenser until maximum indication is had in the antenna when the key is pressed.

Care should be taken not to hold the tuning lamp too close to the coils, else it will burn out. The small neon glow-lamps, on the other hand, can be placed in very close proximity to the coils without danger of burning out when tuning low power stages.

The newcomer is advised to use the neon glow-lamp for tuning, in preference to the 2½-volt dial light, because it is a more sensitive indicator and will not burn out so readily. The cost of these lamps are approximately 40 cents, and may be purchased from most any well-stocked radio supply house.

If one experiences difficulty in properly tuning the transmitter, it is suggested that advice be sought from a local radio club. Fellow amateurs are always willing to be of service. However, if the instructions given herein are closely followed, no difficulty should be encountered.

In conclusion, it is of paramount importance that a good, standard crystal and holder be used in the design of the oscillator. The emitted note of this transmitter has a sharp, pure DC crystal note, which is in conformity with the rules and regulations of the Federal Communications Commission.

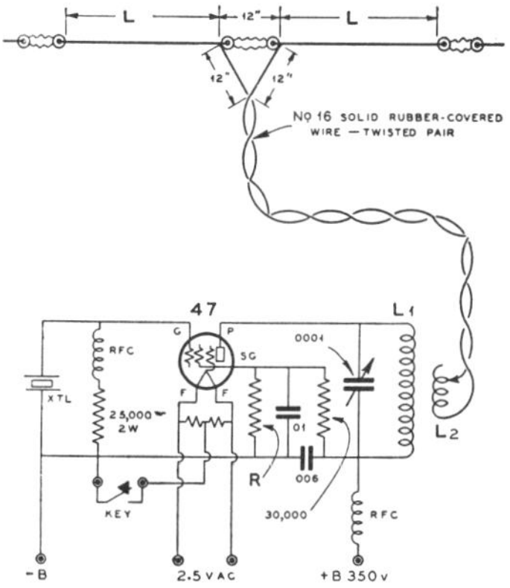

Experimental single-tube transmitter with 47 pentode. The plate coil (L1) is coupled to a twisted-pair feed line. Resistor R has a value of 10,000 ohms.

TRANSMITTER COIL WINDING DATA

For 47 Oscillator and 210 Amplifier Using Single Section Condensers

NOTE: L1—Oscillator Plate Coil. L2—Grid Coil. L3—210 Amp. Plate Coil.

		Tuning Condenser
	160 METERS	
Note: Coils L1 and L2 wound on 1½" Isolantite Forms.	**L1—70 Turns, No. 22 DSC, close wound**	**100 mmf.**
	L2—70 Turns, No. 22 DSC, close wound	**100 mmf.**
	L3—85 Turns, No. 20 DCC, close wound on 2" dia. form, 4" winding space.	**150 mmf.**
	80 METERS	
All 3 Coils wound on 1½" diameter Isolantite Forms	**L1—35 Turns, No. 22 DCC, close wound.**	**100 mmf.**
	L2—32 Turns, No. 22 DCC, close wound.	**100 mmf.**
	L3—45 Turns, No. 22 DCC, close wound. Center tapped.	**70 mmf.**
	40 METERS	
All Coils wound on 1½" dia. Isolantite Forms.	**L1—19 Turns, No. 20 DSC, space wound, one diameter spacing between turns.**	**100 mmf.**
	L2—Same as L1.	**100 mmf.**
	L3—24 Turns, No. 16 Enameled, 12 turns to inch. Center tapped.	**70 mmf.**
	ANTENNA COILS	
	160 METERS	
For 160 meters a Marconi antenna is well suited. Use the Collins Impedance Matching System. Wind antenna coil with 30 turns No. 12 enameled, on 2½" dia. form, 5 turns to inch.		**350 mmf.**
	80 METERS	
For Zepp Antenna, use coupling coil with 13 turns No. 16 DCC on 1¼" dia. form. If Collins System is used, 80 meter coil is same as 160 meter coil.		**350 mmf.**
	40 METERS	
For Zepp Antenna use antenna coupling coil same as for 80 meters. If Collins System is used, wind only 7 turns on antenna coil. This coil is also satisfactory for 20 meters.		**350 mmf.**

The Jones All-Band Exciter (Harmonic Oscillator), 1935 Model

Quantitative experiments with various types of quartz-crystal oscillator circuits proved the fact that a type 53 or 6A6 tube makes an excellent oscillator and harmonic generator which operates with the minimum heating of the crystal. The best results with a type 53 are obtained with one triode working as an oscillator and the other as a harmonic doubler or generator. The tube has a high mutual conductance and amplification constant; thus, it is well adapted for service in crystal oscillator circuits. in addition, both 53 and 6A6 tubes give a high output with low plate voltage. The harmonic content is higher with 300 to 400 volt plate supply than a Tritet oscillator having a 500 to 600 volt supply. This results in economy, since an ordinary BCL power supply is suitable.

up to the 8th harmonic. From the circuit shown in Figure 1, an output of 5 watts is secured on 80 meters from an 80-meter crystal, 5 watts on 40 meters, and 0.7 watts on 20 meters. These outputs are obtained from a 400-volt power supply with a total cathode current of from 50 to 90 milliamperes, which is about equally divided between the two triode plates. The crystal current is about 37 milliamperes.

If doubling is only desired, regeneration need not be incorporated in the general scheme. On the other hand, regeneration is required for the higher-order harmonics. The amount of feedback must be carefully controlled; otherwise the circuit may act as an oscillator at the frequency determined by the harmonic of the triode tuned circuit.

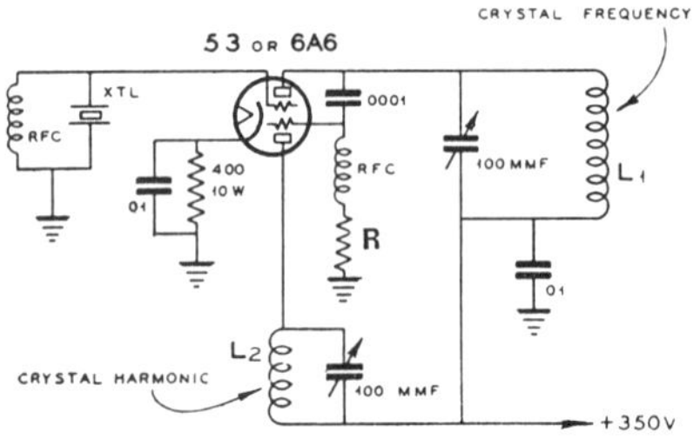

FIG. 1

Jones Exciter for two-band operation. R-50,000 ohms, 2 watts.

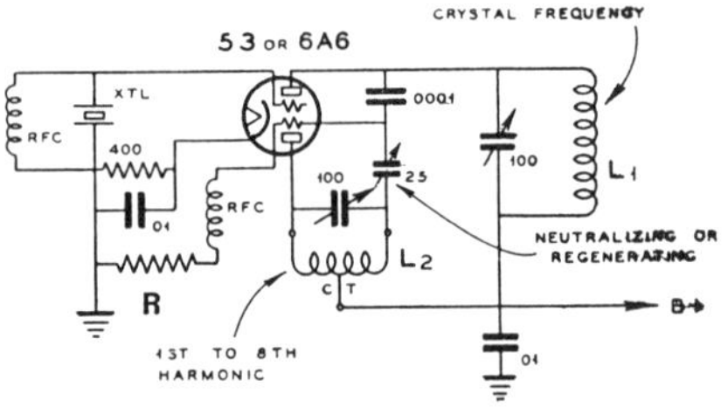

FIG. 2

Jones Regenerative Exciter for all-band operation. R-50,000 ohms.

The oscillation section works splendidly when heavily loaded and also with less crystal RF current, which results in minimum crystal temperature and frequency change. The other triode section is capacitively coupled to the oscillator, which acts as a very efficient harmonic generator.

In general, the simplest all-band exciter shown in Figure 1 is most desirable. It can be used to drive one or two 45 tubes as a buffer or low-power transmitter. The 45 tube will deliver as high as 20 watts output on 40 meters when excited by a type 53 tube and an 80-meter crystal.

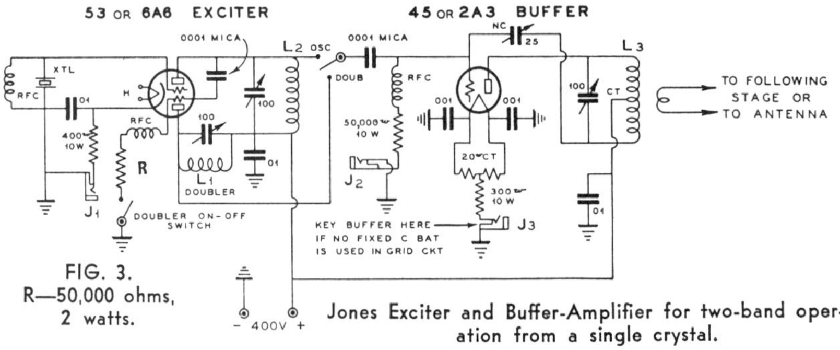

FIG. 3.
R—50,000 ohms, 2 watts.

Jones Exciter and Buffer-Amplifier for two-band operation from a single crystal.

Figure 1 shows the simplest oscillator-double circuit, while Figure 2 shows the same circuit with regeneration in the harmonic section to secure reasonable outputs

Over 15 watts output, see Figure 3, is obtainable on 20 meters from a 40-meter crystal in conjunction with a 53 or 45 tube supplied with 400 volts on the plate. In

Plan view of a complete Jones Exciter with 53 oscillator-doubler and 2A3 amplifier. Best results are secured when the components are arranged as shown above.

this case, the power source must be capable of supplying approximately 150 milliamperes.

The simplest test or adjustment of this circuit is to use a single turn of wire and a 6-volt pilot-light as an oscillation indicator. The oscillator stage is first adjusted and tuned to approximately the mid-point of its oscillating range and the doubler section is tuned to peak output. When the oscillator is not functioning, the cathode current will be approximately 20 ma.; but in a state of oscillation this current will rise to between 50 and 110 ma. If the doubler section is not employed, its grid-leak may be opened to remove the load from the oscillator section. The grid-leak increases the bias of the harmonic producing triode to a great many times cut-off bias. The 400-ohm cathode resistor provides a fixed bias to both the triode sections, and also stabilizes it for use on plate voltages of over 300 volts.

Note: Other than type 2A3 or 45 type tubes can be used in the buffer stage of a Jones Exciter. The following table is of inestimable value in determining the type of tube and method of coupling to give the best results.

Single 45 or 2A3	Capacitive coupling
Single 46	Link coupling
Single 841	Link coupling
Single 210	Link coupling
Single 801	Link coupling
Single 865	Capacitive or Link coupling
Single 53 Push-Push Doubler	Link coupling
Single 59 Regenerative Doubler	Capacitive coupling
Two 45s in Parallel or Push-Pull	Capacitive or Link coupling
Single 802	Capacitive coupling

Any of the above combinations will provide sufficient excitation for one or two type 211 tubes, or two 50Ts, or single 150T or HK354, for **grid modulation.** With CW transmitters having a very high power input to the final amplifier, an additional buffer stage should be used. Final amplifiers having two type 211 tubes are best driven by a pair of 210s or 801s operating with a plate voltage of at least 700 volts. A single 210 or 801 operating with 700 volts on the plate will give sufficient drive for CW when the final amplifier tubes are of types 150T or HK354.

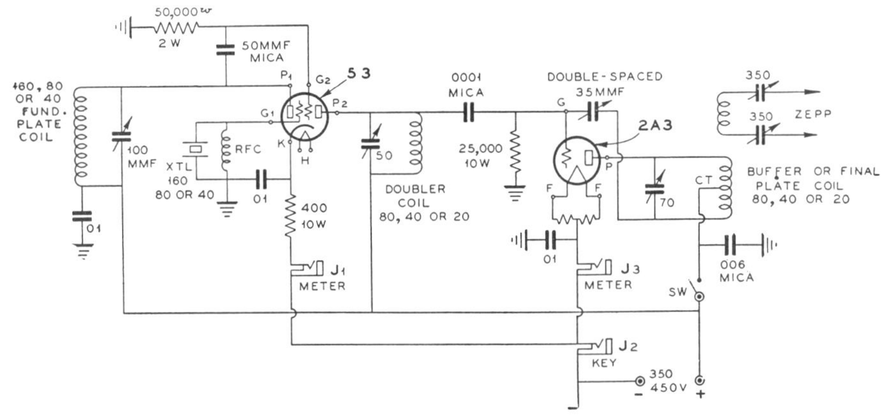

Jones Exciter for operation on the second harmonic of the crystal frequency. The output from the doubler drives the 2A3 buffer-amplifier stage. A 160 meter crystal is used for 80 meter operation, 80 meter crystal for 40, 40 meter crystal for 20.

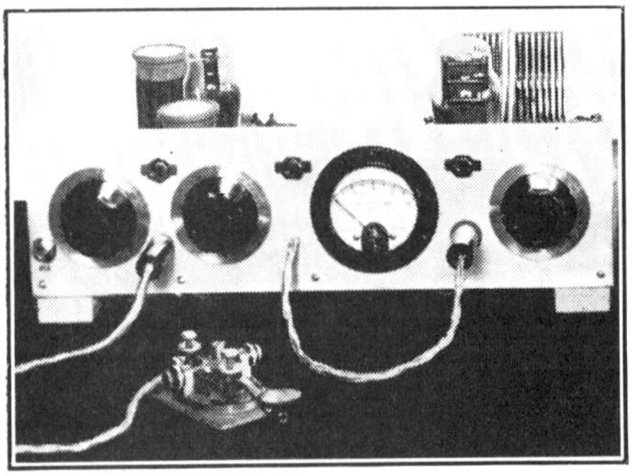

A business-like "bread-board" layout and control panel for the Jones Exciter.

Coil Winding Data

For 80 meter operation, using a 160 meter crystal in the oscillator.

Oscillator Coil—68 turns of No. 22 DSC wire, close wound, on a 1½-in. diameter standard 5-prong plug-in coil form.

Doubler Coil—27 turns of No. 22 DCS wire, slightly space wound, on a 1½-in. diameter 5-prong coil form.

Plate Coil for 2A3 Stage—Either a small coil, wound on 1½-in. plug-in coil form, with 25 turns of No. 20 DCS, slightly space-wound, or a large copper-wire-wound coil, 2¾-in. diameter, 28 turns of No. 14; "center-tap" must be taken at the mid-point of all plate coils in the 2A3 stage.

For 40 meter operation, using an 80 meter crystal in the oscillator.

Oscillator Coil—27 turns of No. 22 DSC wire, slightly space wound on a 1½-in. diameter 5-prong coil form.

Doubler Coil—13 turns of No. 16 enameled (or DCC) wire, space wound, on a 1½-in. diameter 5-prong coil form.

Plate Coil for 2A3 Stage—Either a 13 turn coil, No. 16 enameled (or DCC) wire, space wound on a 1½-in. diameter form, center-tapped, or a large copper-wire-wound coil 2¾-in. diameter (as shown in photo), with 16 turns of No. 14 enameled wire, spaced ⅛-inch between turns, and center-tapped.

It is important to mention here that a few turns must sometimes be added to, or removed from, the coils shown above if the circuits cannot be tuned to resonance. However, the coil data here given is **correct** for use with the tuning condensers specified in the circuit diagram.

For 20 meter operation, using a 40 meter crystal in the oscillator.

Oscillator Coil—16 turns of No. 16 enameled or DSC wire, spaced one diameter on a 1½-in. diameter coil form.

Doubler Coil—10 turns of No. 16, spaced one diameter on 1½-in. coil form.

Plate Coil for 2A3 Stage—11 turns No. 16, spaced one diameter on 1½-in. coil form and center-tapped. Or large diameter "air-wound" plate coil (2¾-in.) with 10 turns of No. 14 wire, space wound, can be used.

Jones Exciter and power supply on relay-rack.

C.W. Transmitter for One-Band Operation

A very low-cost C.W. transmitter having an output of 80 watts for operation on one-band is diagrammed herewith.

General Details: The oscillator plate circuit is tuned by a standard two-section, 35 uufd. per section double-spaced midget condenser; the two sections are connected in parallel, thus giving a total capacity of 70 uufd. The grids of the 210 tubes are driven with approximately 20 ma. from the oscillator.

The neutralizing condenser in the 210 stage is a 35 uufd. double-spaced midget variable. The final plate tuning condenser

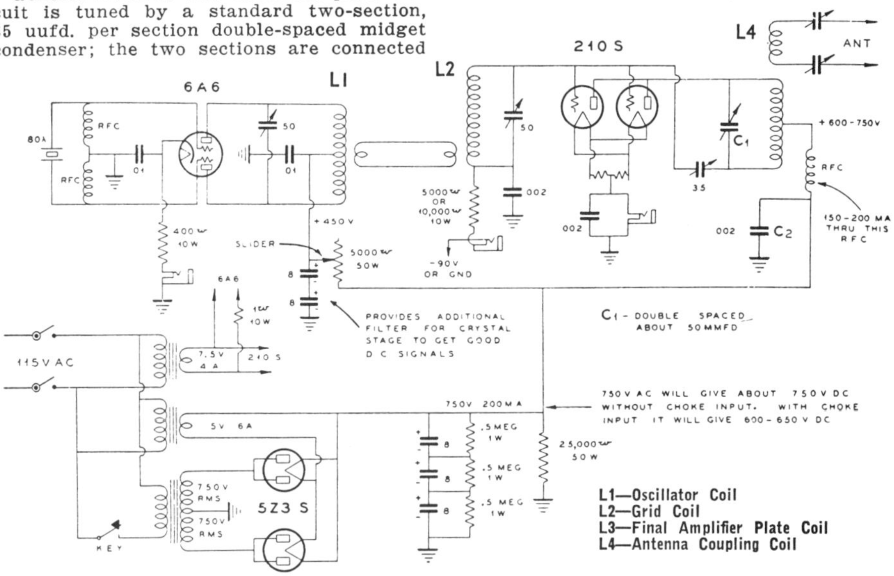

Here is the complete circuit diagram of the crystal oscillator, link coupling system, 210 final amplifier, Zepp antenna feeder coupling and tuning system, and the simplified power supply unit which operates the entire transmitter. The telegraph key is in the primary circuit of the high voltage transformer.

This illustration clearly shows how the various components are arranged on the base-board.

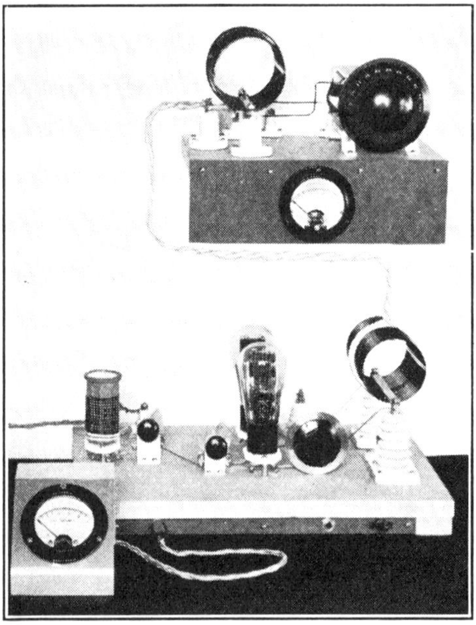

The 210 amplifier and its link coupled antenna coil with feeder tuning condenser.

(C1 in the circuit diagram) is the same as the condenser in the oscillator plate circuit. The 210 amplifier can be keyed by plugging into the jack in the filament center-tap circuit.

Condenser C2 has a 2,500-volt rating and is of the mica type. All other fixed condensers have 1,000-volt rating and are also of the mica variety.

Coils: For 40-meter operation, the final plate coil is wound with 16 turns of No. 14 enameled wire, space wound, 6 turns to the inch on a form 2¾ inches in diameter. The 80 meter final plate coil has 34 turns of No. 14 enameled wire, 9 turns to the inch, on a form 2¾ inches in diameter.

The oscillator plate coil, for 40 meter operation, has 20 turns of No. 18 enameled wire, slightly space wound, on a standard 1½-inch diameter plug-in coil form. The 80 meter coil has 33 turns of No. 18 DSC or DCC wire, close wound.

The grid coil, for 40 meter operation, has 22 turns of No. 18 enameled wire, slightly space wound, on a 1½-inch diameter plug-in coil form. For 80 meter operation, the grid coil has 33 turns of No. 18 DSC or DCC wire, close wound.

210s or 801s in parallel can be driven by a 53 or 6A6 push-pull oscillator, shown in the circuit below.

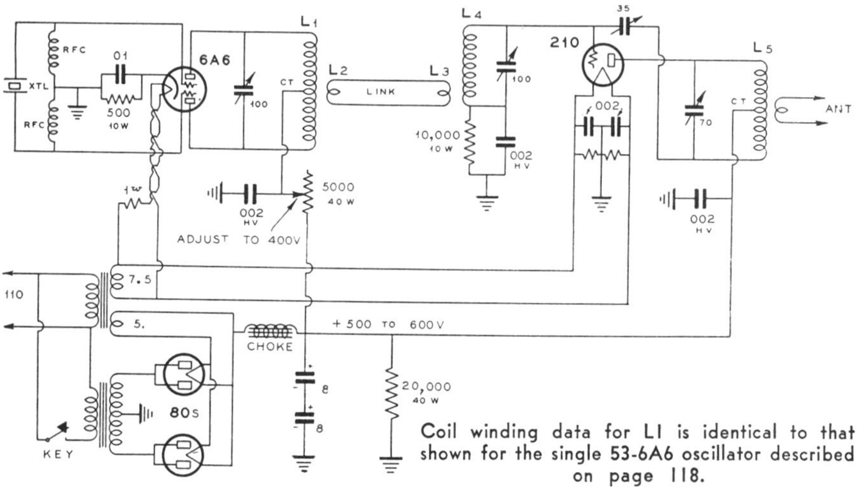

Coil winding data for L1 is identical to that shown for the single 53-6A6 oscillator described on page 118.

Modern all-band Jones Exciter, 2A3 doubler and 210 (or 801) final amplifier. 20 meter coils are shown in the picture. The final amplifier can be operated on 10 meters as a regenerative doubler by merely increasing the capacity of the neutralizing condenser and increasing the value of the final amplifier tube bias resistor to 25,000 ohms.

53-2A3-210 Multi-Band Transmitter: The illustration shows a modern, highly efficient and compact three tube r-f unit which incorporates all of the best features of present-day amateur design. Two-band operation can be had from a single crystal by the mere throw of the S-P-D-T switch which connects either the oscillator plate coil or the doubler plate coil to the grid circuit of the 2A3 buffer stage. This particular transmitter uses a 40 meter crystal oscillator, 20 meter doubler, 20 meter 2A3 grid coil and a 20 or 10 meter final amplifier plate coil. Link coupling is used between the 2A3 buffer and the grid of the final amplifier. The coupling loop on L3 has 2 turns, wound around the center of the coil. The other loop on L4 also has 2 turns, wound around the bottom of the final grid coil. 20 milliamperes of grid current is supplied to the 210. The final amplifier can be keyed by connecting the key in the filament center-tap circuit, or primary keying can be used if desired. All fixed condensers shown in the circuit diagram should be of the 1000 volt mica type, except for the cathode by-pass which can be a 600v paper condenser.

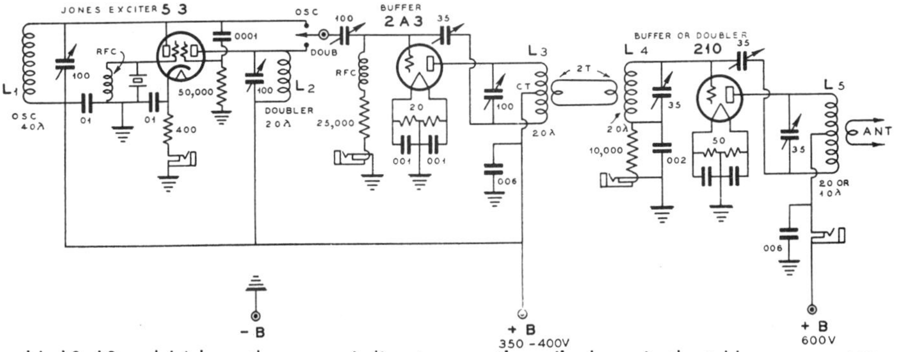

L1, L2, L3 and L4 have the same winding turns as the coils shown in the table on page 118.

90-Watt CW Transmitter: A splendid CW breadboard-type transmitter capable of developing a 90-watt RF output on 20-meters and approximately 100-watts on 40-meters is described in the subsequent text with reference to the wiring diagram of Figures 1 and 2.

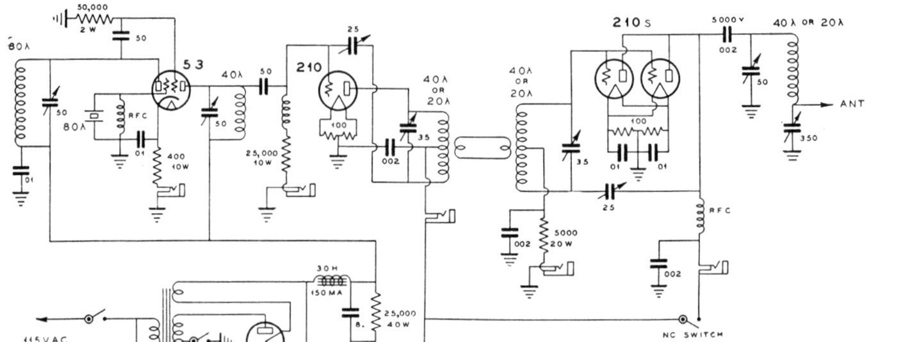

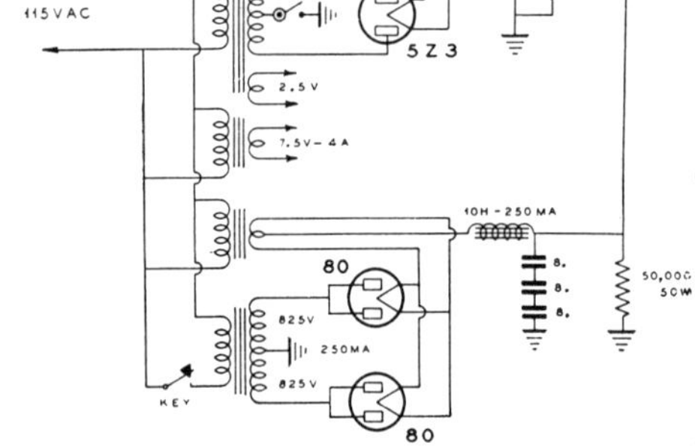

FIG. 1. Grid-neutralized final amplifier.

Technical Notes: A little study of the schematic will reveal that it is similar to other transmitting circuits previously described; therefore, to avoid repetition only the salient points will be touched upon.

condenser must be set for proper neutralization, but when the stage functions as a doubler, the capacity must be increased to almost twice that required for neutralization. In this case, the plate current drops from 110 MA to 75 MA, and more grid current is obtained into the final stage when the larger capacity is in the circuit. Too much capacity will cause oscillation, but the amount of capacity required will depend upon the value of the grid tuning condenser capacity and plate circuit loading.

All tuning condensers, except the antenna condenser and the two oscillator midget condensers, are double-spaced midgets. The antenna condenser is an old BCL tuning condenser of about 350 uufd. If a Zepp antenna is used this condenser should have a maximum capacity of approximately 500 uufd.

Antenna tuning: The final amplifier plate circuit is tuned and the antenna is matched in the same manner, as described on page 124 for the "portable relay-rack transmitter and receiver." In addition, a twisted-pair

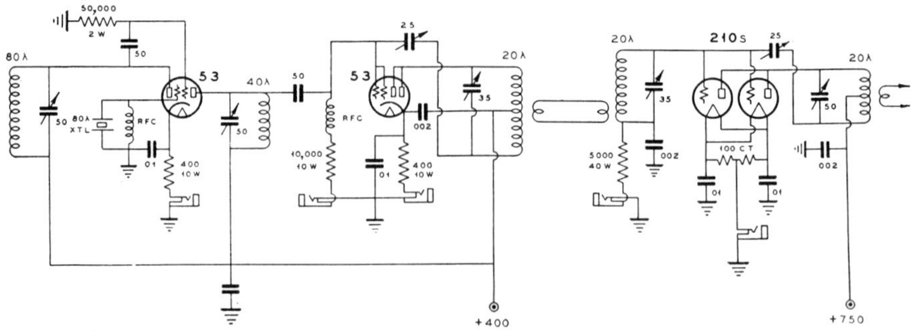

FIG. 2. Plate-neutralized circuit for 20-meter transmitter. The fixed condenser connected from the low voltage B to ground should have a capacity of 0.1 mfd.

The buffer stage can be capacitively coupled with the .00005 mfd. grid coupling condenser as shown or by one having a value of .0001. When the buffer is used for a straight amplifier, the neutralizing feeder can be coupled to the plate coil with from 1 to 3 turns wound around the condenser end of the coil.

Power Supply: A separate 400-volt power supply is needed for the crystal oscillator,

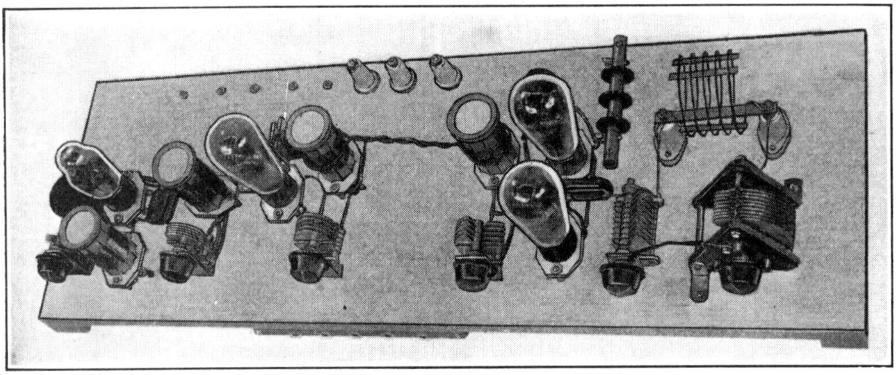

Breadboard layout for 20-meter transmitter.

and a 900 or 950-volt center-tapped power transformer with choke input will suffice for this purpose. The high voltage supply consists of a 1600-volt 250-MA center-tapped power transformer, two type 80 tubes and a choke input filter with 3 8-mfd. electrolytic condensers connected in series at the output.

Tests: If the constants of the circuit have been rigidly adhered to, the following readings will be obtained: With a 410-volt plate supply the cathode current of the oscillator-doubler will be 60 MA. The buffer grid current about 8 MA, and the final amplifier at approximately 30 MA, depending upon the antenna load. The 10-doubler plate current will be 75 MA and the final amplifier plate current 180 MA on 20-meters, with antenna load providing the plate voltage on these two stages is 700 volts.

The coil data for the circuit of Fig. 1 is as follows:

80 meter oscillator 53—32 t No. 18 E, close-wound, 1½-in. diameter.

40 meter doubler 53—19 t No. 18 E, spaced to cover 2 in. on 1½-in. diameter form.

40 meter amplifier 10—28 t No. 18 c, close-wound, 1½-in. diam. C.T.

20 meter doubler 10—14 t No. 18 c, spaced to cover 1¾-in. C.T.

40 meter final grid—22 t No. 18 c, close-wound, 1½-in. C.T.

20 meter final grid—10 t No. 18, close-wound, 1-in. C.T.

40 meter final plate—11 t No. 14 E, 2¾-in. diam., 2½-in. long.

20 meter final plate—5 t No. 14 E, 2⅝-in. diam., 2-in. long.

(E denotes Enameled wire, c denotes cotton-covered wire.)

Portable Relay-Rack Transmitter and Receiver: A simple transmitter and receiver that can be carried and set up in practically any locale is prized by many amateurs. Here is a 25 to 30-watt transmitter of modern design and a receiver of the Super-Gainer type. Most of the details and design specifications have been previously discussed; however, the following pointers will be of aid to the constructor, and reference should be made to the accompanying diagram.

Technical Notes: The 53 oscillator plate coil is split so that it can be used as a neutralizing system for the 2A3 by connecting the neutralizing condenser from the 2A3 plate back to the coil end opposite the grid connection. The antenna may be coupled inductively to the 2A3 plate coil if desired. This coil, L_5, is rather critical for the number of turns, depending upon the ratio of the two tuning capacities and antenna impedance. A satisfactory arrangement is to take taps about every 4 or 5 turns on L_5 and short-circuit the unused turns with a short lead. The coil can be of the plug-in type because the shorting wire would be between taps.

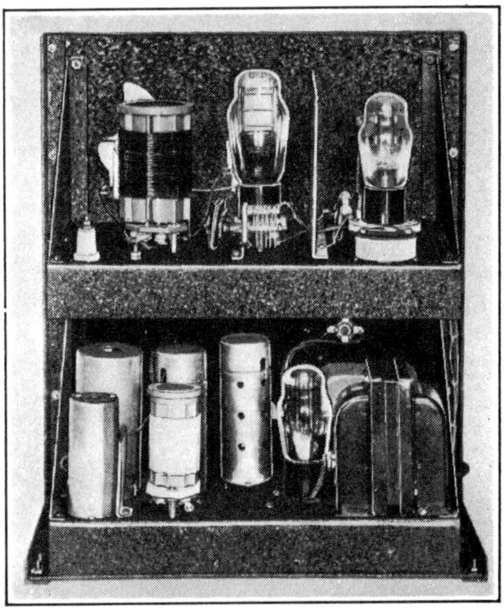

Portable relay-rack transmitter with Super-Gainer receiver on lower deck.

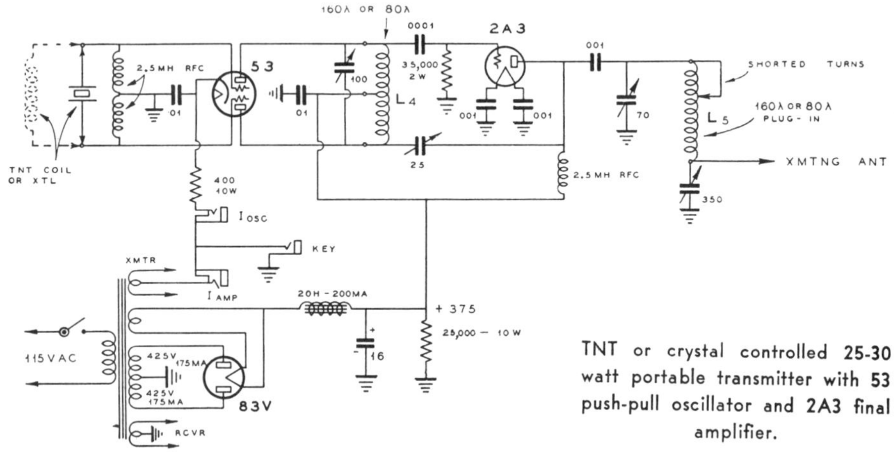

TNT or crystal controlled 25-30 watt portable transmitter with 53 push-pull oscillator and 2A3 final amplifier.

Antenna tuning: (1) The antenna 350 uufd. tuning condenser is set between the maximum setting and that which would correspond to 150 uufd., (2), the plate tuning condenser is adjusted for resonance. In order for an antenna of some given length to properly load the 2A3 amplifier, it is required that the two tuning capacities be adjusted to the proper ratio. With certain antennas, the coil turns in L_5 must be changed to obtain the necessary effect. This antenna matching system works best with a single wire feeder to a Hertz half or full-wave antenna.

The transmitter develops from 25 to 30-watts RF when the plate supply is approximately 400-volts, 100 MA to the 2A3. The 53 oscillator draws from 50 to 75 MA of cathode current and sufficiently excites the 2A3 for good class C operation.

Notes On Operating 20-Meter Transmitters

Tubes: Transmitters which employ type 50T tubes in the final amplifier are more difficult to drive than a 211D for the same output at medium wavelengths. A 50T has a maximum plate dissipation of 75 watts as against 100 watts for a 211D; therefore, a 50T must operate at higher efficiency in order to obtain the same output. Any tube which operates into a high load impedance with high plate voltage requires more grid excitation, unless its mutual conductance is extremely high. A tube operating into a moderate low impedance with moderate plate voltage requires less grid excitation. At 20 meters, the required grid excitation for a 50T or 211D are about equal for phone operation, but the low-C tube (50T)

Another method of construction for a 210 transmitter with 6A6 push-pull oscillator.

is easier to neutralize, which is a very important feature for phone operated transmitters. This applies especially to final amplifiers working with a single tube on 20 meters.

Antenna Loading: For phone, the antenna load does not have to be reduced as much from CW operation as when a 211D tube is used. The plate impedance under normal operating conditions for either a 50T or 211D is quite different, being several times as high with the low-C tube. This means that less C is needed in a tank circuit in order to maintain a "Q" of 10, for example, in the final tank circuit. Apparently the shift from CW to phone is more nearly an identical antenna load because the C of the tuned circuit was the same for both tubes. The 211D can be loaded more heavily for CW operation, although with less efficiency.

The complete 150 watt cw 20-meter transmitter and antenna coupler. The circuit diagram is shown below. A 40-meter crystal is used in the Jones Exciter Unit, capacitively coupled to an 801 buffer which drives the 50T in the final amplifier.

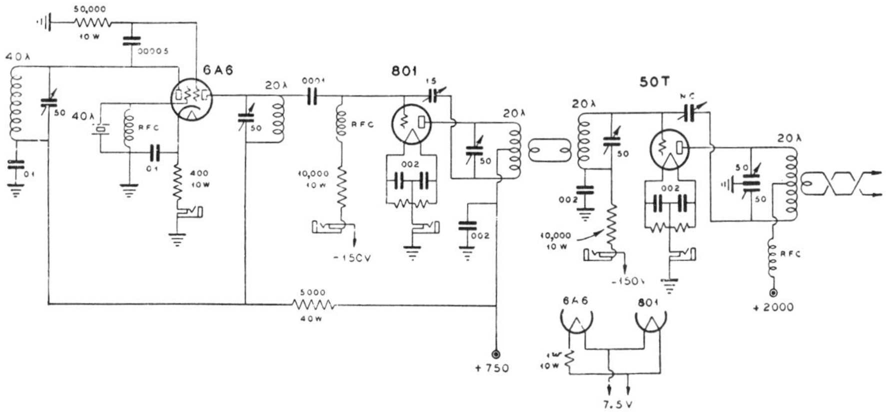

FIG. 1

COIL DATA FOR 20 METER OPERATION

Final Plate Tank	Final Grid Coil	Buffer Plate	Doubler	Osc.
10 turns No. 10 E. 3½-in. dia. 3-in. long. C.T.	11 turns No. 16 E. 1½-in. dia. 1⅜-in. long.	11 turns No. 16 E. 1½-in. dia. 1⅜-in. long. C.T.	9 turns No. 16 E. 1½-in. dia. 1⅜-in. long.	20 turns No. 18 E. 1½-in. dia. 1⅝-in. long.

Unity Coupled Jones Exciter, 1936 Model: Late technical improvements on the Jones 53-Exciter which, in the new design, incorporates three type-53 tubes gives more output per tube with less crystal heating than any of the other developments heretofore described. Some of the salient points of the new circuit are: The RF output is obtained on any three consecutive bands without coil changes or switching. Series link-coupling feeds the output to the external buffer tuned-grid coil. The grid tuning determines which frequency is absorbed from the exciter unit by the buffer stage. And lastly, the output is sufficient to drive an 801, 210, RK18 or 28, or 803 on any waveband from 10 to 160-meters.

will show that the two frequency-doubler stages have the grids in push-pull and the plates in parallel. This gives the plate tank an RF power surge each cycle instead of every other cycle as in most doubler circuits (for further detail see subtopic "Doubler Circuit Consideration"). The efficiency of a push-push doubler circuit is higher than in other circuits with the result that excellent output is obtained even on 10-meters from a 40-meter crystal. In the exciter circuit shown, two crystals and a total of seven coils will cover five amateur bands.

Design Specifications: The unit is built on a 10 x 17 x 2½-inch metal chassis mounted behind a 19 x 8¾-inch relay-rack

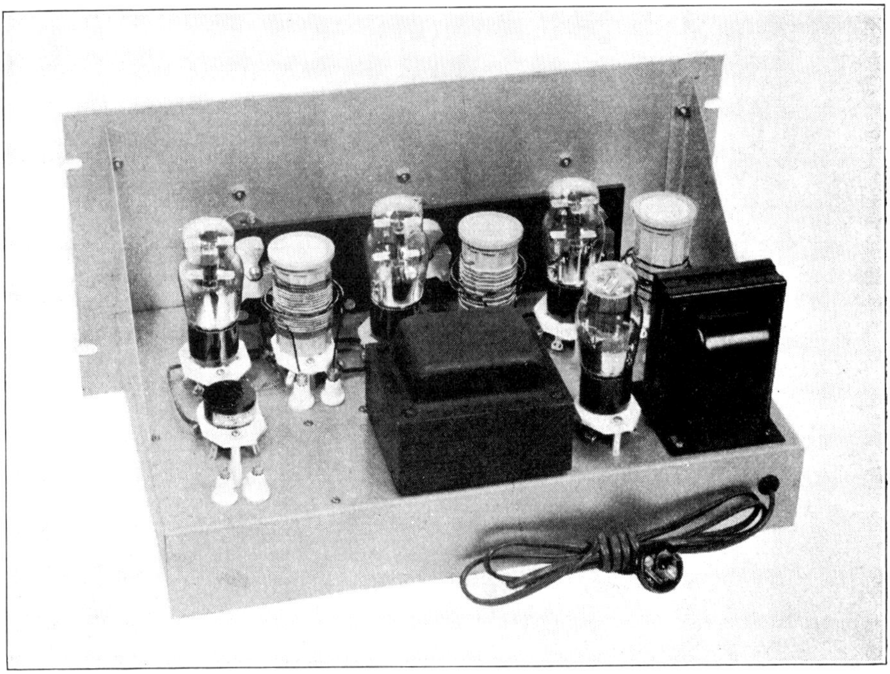

"3-53" 1936 Jones Exciter, 160 meters to 10 meters. The power supply, mounted on the Exciter chassis, makes this a complete and independent driving unit for a buffer-amplifier stage.

Technical Details: Three type-53 or 6A6 tubes act as push-push doublers and a push-pull crystal oscillator; the latter, with cathode resistor grid-bias and the push-pull or series connection places very slight mechanical strain on the crystal. The crystal RF current is not much higher than that in a pentode circuit but the plate output power is two to four times greater than from a pentode tube operating at the same plate voltage. The push-pull crystal oscillator delivers nearly twice as much output on the fundamental frequency of the crystal than can be obtained from a 53 as an oscillator-doubler in its usual form.

Reference to the accompanying circuit

panel. A built-in power supply completes this unit and renders the assembly adaptable to further additions of power stages without altering the exciter unit. A 12 x 3 x $\frac{3}{16}$-inch bakelite or "masonite" sub-panel supports the three midget tuning condensers which are near the coil and tube sockets. These condensers are connected through insulated couplings to the front panel tuning dials. The arrangement of the tube and coil sockets permits very short grid and plate leads. Since each stage operates on an octave higher frequency, no shielding is required. The power units are remotely mounted from the RF units so as to preclude the possibility of encountering hum problems. Only one filter-choke is shown, hence, it must be well designed and efficient.

The link-coupling circuits are wired in series, each has two turns around the center of each plug-in coil. These two-turn coils are large enough so that the plug-in coil-forms fit through them into the isolantite coil sockets. The links are supported on stand-off insulators and the output terminates on another pair of through-type insulators at the rear of the chassis.

The doubler grid-circuits are connected to interwound grid coils which are closely coupled to the plate tuned-circuits. As can be seen from the coil tables, only enough grid turns are wound on each coil to drive the push-pull grids at the proper frequency. Too many turns will cause excessive regeneration at the doubler plate-circuit frequency with self-oscillation. This is due to inductive reactance in the grid circuits and, when not excessive, improves the doubler efficiency and output.

Cathode resistors and grid-leaks combine to give high grid-bias to the doubler grid circuits. Cathode bias only on the crystal stage has the effect of keeping the crystal RF current low for relatively high power output.

An open-circuit plug in any cathode circuit, automatically cuts that and the succeeding stages out of the circuit. This allows about 10 to 15 per cent

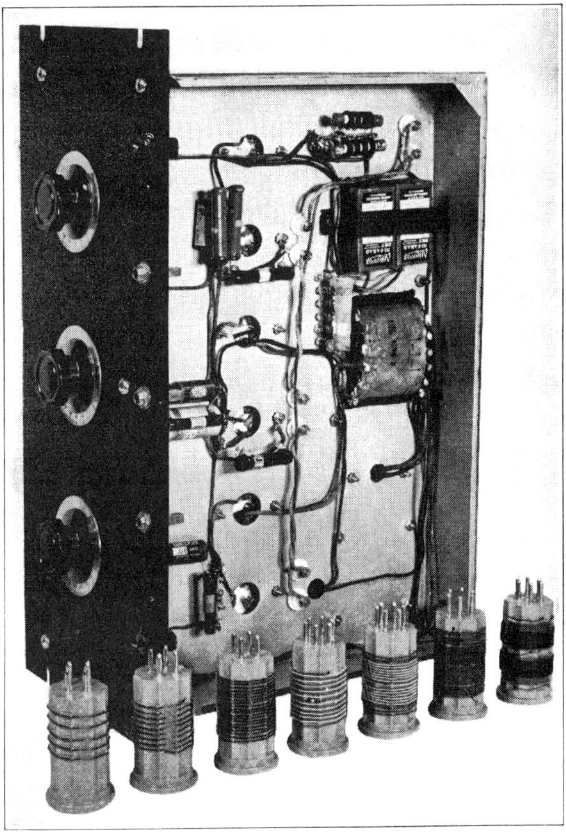

Only 7 coils are needed to cover the 160-to-10 meter bands.

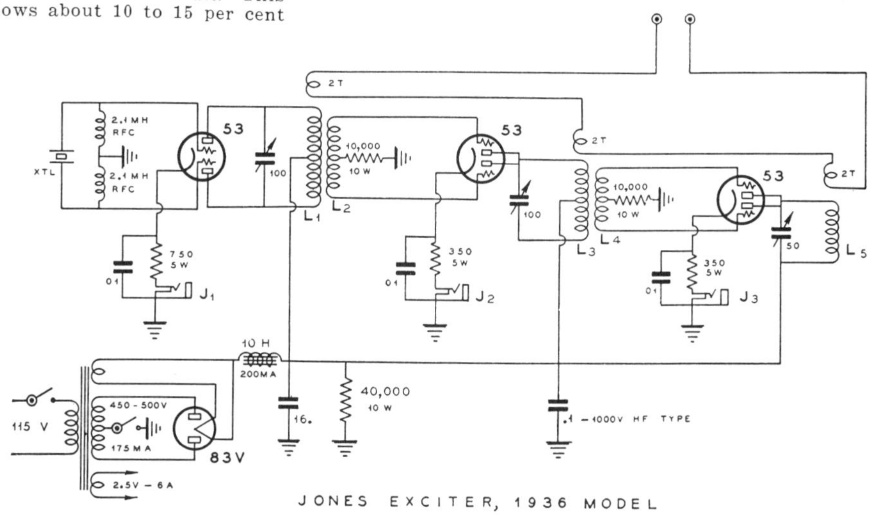

JONES EXCITER, 1936 MODEL

160 METERS TO 10 METERS

Coil winding chart is on page 128.

greater output from the stage which is actually driving the external buffer. The cathode current varies from 50 to 75 MA per stage depending upon plate voltage and RF load. Resonance in each doubler stage is indicated by a sharp decrease in cathode current of from 20 to 50 MA, providing the preceding stage is in resonance. (Note: a single turn of wire connected in series with a flashlight lamp will function as a resonator indicating device when coupled to the stage under test.)

The power supply has two switches, one for the mains control and the other to apply plate voltage after the 53 or 6A6 tube heaters have attained their normal operating temperatures. The DC plate potential should be between 375 and 425 volts; higher voltages will give greater output providing the cathode current per tube does not exceed 75 or 80 MA.

The greatest output from the oscillator stage occurs at the maximum condenser setting just before the stage drops out of oscillation, or near that setting. Generally, for stability, in keying the crystal stage, this oscillator condenser has to be set back a dial degree or two towards less capacity and higher cathode current. The cathode current in any stage should not be over 20 to 30 MA when the crystal is **not** oscillating, and between 50 to 75 MA when the stage is oscillating.

Tests made with an 801 or 10 buffer stage gave a grid DC current of 16 MA on 10-meters; 18 to 20 MA on 20-meters; 22 to 24 MA on 40-meters; and 22 MA on 80-meters. This was through a 10,000-ohm grid leak and without plate voltage on the buffer stage. A two-turn closely coupled link-coil was necessary on the buffer tuned-grid coil. These readings were obtained with 400 volts plate supply on the exciter.

COIL WINDING CHART FOR JONES 3-53 (1936) EXCITER

All Coils Are Wound on Standard 1½" Diameter Coil Forms

For 160 Meter Crystal:	Oscillator Coil has **60** turns of No. **24** DSC, with center-tap. First doubler grid coil has **50** turns of No. **24** DSC, with center-tap, wound over oscillator coil with a layer of celluloid to separate the coils. First doubler plate coil (**80** meters) has **40** turns of No. **24** DSC, center-tapped and space wound. Second doubler grid coil has **30** turns of No. **24** DSC, with center-tap, interwound with first doubler plate coil. Interwindings begin at center-taps and wind outward toward the ends of the coils. Second doubler plate coil (**40** meters) has **20** turns of No. **18** Enameled, wound to cover **1¾"** of space.
For 80 Meter Crystal:	Oscillator uses the **80** meter doubler coil described above. First doubler plate coil (**40** meters) has **20** turns of No. **20** DSC, center-tapped and wound to cover **1⅝"** of space. Second doubler grid coil has **14** turns with center-tap, wound with No. **20** DSC wire. This winding is interwound with the plate coil, starting the center-taps together and then winding outward towards the ends of the coil. Second doubler plate coil (**20** meters) has **9** turns of No. **18** Enameled, wound to cover a space of **1¼**-in.
For 40 Meter Crystal:	Oscillator uses same plate coil as described for first doubler coil when using **80** meter crystal. First doubler plate coil (**20** meters) has **10** turns of No. **20** DSC with center-tap. Second doubler grid coil has **10** turns with center-tap, interwound with the plate coil. The total coil length is **1⅜**-in. Second doubler plate coil (**10** meters) has **5** turns of No. **18** Enameled, wound to cover a space of **1¼**-in.

For **80** meter operation, either a **160** or **80** meter crystal can be used. For **40** meter operation, a **160, 80** or **40** meter crystal can be used. For **20** meters, an **80** or **40** meter crystal can be used. A **40** meter crystal is needed for **10** meter operation.

500 Watt C-W Transmitter: This transmitter is straightforward in circuit design and is laid out in breadboard fashion. It will put out a 500 watt CW signal in the 40 meter band when used with a suitable power supply. Breadboard construction simplifies the work of building such a set and tends to give higher efficiency than a set built into a metal frame or cabinet.

The oscillator uses a 6A6 twin triode tube with one section oscillating on 80 meters and the other section doubling to 40 meters. With 400 volts on the 6A6, sufficient output is obtained on the second harmonic to drive a pair of type 10 tubes to approximately class C operation as a buffer stage. The latter provides more than ample output to drive the pair of 211 tubes in the final amplifier.

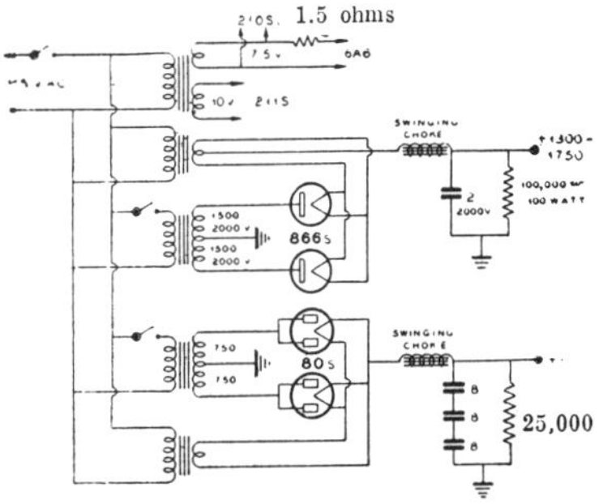

Power supply for $\frac{1}{2}$ k-w transmitter.

COIL TABLE

Final Amp. Plate

Meters
40—16 turns No. 10 E. 3½" d., 5¼" long. CT
20— 8 turns No. 10 E. 3¼" d., 5½" long. CT

Final Amp. Grid and Buffer Amp. Plate

Meters
40—15 turns No. 16 E. 2¼" d., 1¾" long. CT
20—9 1/3 turns 16 E. 1½" d., 1½" long. CT

Buffer Amp. Grid and Doubler Plate

Meters
40—22 turns No. 18 E. 1½" d., 1¾" long. CT
20—10 turns No. 18 E. 1½" d., 1¾" long. CT

Osc. Coil for 40 & 20 Meter Operation

Meters
80—31 turns No. 18 E. closewound, 1½" d.
40—22 turns No. 18 E. 1½" d., 1¾" long.
(E denotes enameled wire. d denotes diameter of coil.)

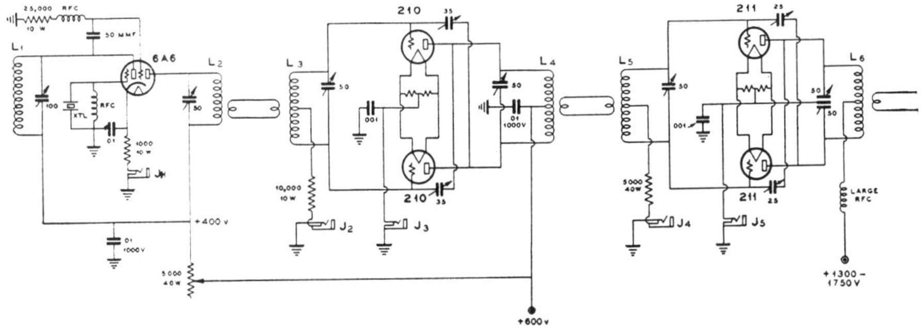

The r-f portion is straightforward and correctly designed.

Breadboard construction of the $\frac{1}{2}$ k-w transmitter with a pair of 211 tubes in the final.

High Power Transmitter Construction: Those who do not desire to use relay-rack construction will find some suggestions in the wood cabinet styles here pictured. The illustration to the right shows the HF-200 or HF-300 triode in a high-power amplifier, the entire unit mounted on a framework 18-in. long, 11-in. wide and 4-in. high. The grid coil and tuning condenser are at the extreme left, shielded from the plate circuit with a sheet of heavy aluminum. The final tank coil is mounted high on stand-offs, fitted with plug-in jacks. A low capacity, wide-spaced neutralizing condenser is also shown. The antenna coupler is link coupled to the final plate tank coil with a two or three turn loop around the center of tank coil.

The illustration below (center) shows a somewhat unconventional design for a complete r-f amplifier with Jones Exciter, push-push doubler and a 211 in the final stage. The panels are of white bakelite, the wood framework painted black. Each stage is widely spaced from the other so that no interaction between coils can take place. Push-through insulators support the final tank coil. The design for this r-f amplifier was copied from one of the early BCL receivers.

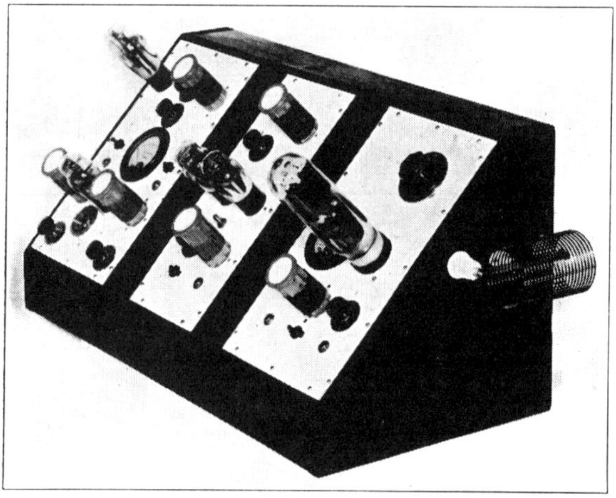

The illustrations at the foot of the page show two views of a buffer stage or final amplifier for a standard 211 or 203A tube. The mounting arrangement for the final tank condenser, neutralizing condenser and tank coil is ideal. Short, direct leads can be made to the vital parts of the circuit and the arrangement lends itself admirably to the use of tubes which have plate lead at the top.

The grid coil and condenser is well separated from the plate circuit. The coupling link is connected to standoff insulators directly behind the grid coil. The mounting cabinet is 18 in. long, 11-in. wide and 4-in. high. Small resistors and condensers are mounted under the baseboard. A 0-50 MA d-c milliammeter reads grid current, and a 0-300 MA d-c milliammeter reads plate current.

500 Watt Plate Neutralized Amplifier with Amperex HF-300 Tube.

H-F300 Amplifier: Here is a high power amplifier that puts out a carrier of over 600 watts on 20 meters, using a single Amperex HF300 tube. This amplifier can be driven by the 20 meter transmitter described on page 121.

The HF300 is designed to operate with plate currents up to 275 MA at voltages between 2000 and 3000. The filament takes 4 amperes at between 11 and 12 volts and the plate dissipation is rated at 200 watts. Its mu is 23, and its mutual conductance is rated at 5600. The plate-to-grid capacity is 6.5 uufd., the plate-to-filament capacity 1.4, and the grid-to-filament capacity is 6 uufd.

From this data it can be seen that twice cut-off bias at 2700 volt plate supply would be 235 volts. Bias is easily obtained with a 10,000 ohm leak, and fairly easy to obtain with a 5000 ohm grid leak using a 50T as a driver with 140 watts input. The DC grid current under load runs between 35 and 60 MA for grid-leak values of from 10,000 to 5000 ohms with various degrees of antenna loading.

For CW operation the power gain of this stage runs between 7 and 12, depending upon the allowable output and plate dissipation. A power gain of 9 is about as high as can be figured for an output of around 600 watts on 20 meters. For phone operation the grid excitation must be higher, but the plate load is lower, consequently a power gain of around 4 to 6 can still be figured on. These figures seem to be fairly high for operation at this frequency.

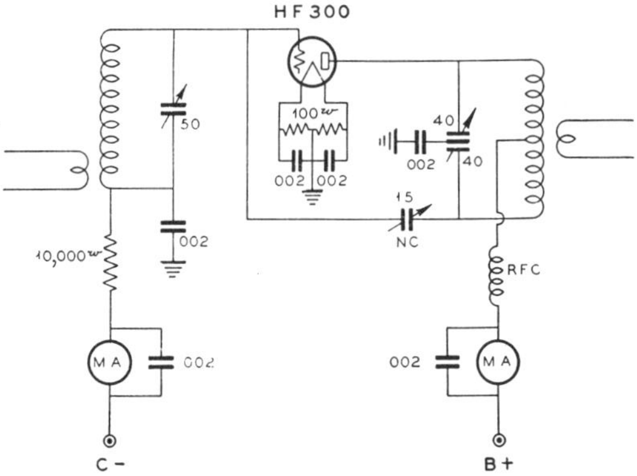

HF-300 Amplifier Circuit.

For CW operation a 53-45 exciter driving a pair of 10s or 801s will provide sufficient grid excitation to the HF300. The buffer stage should be operated with approximately 700 volts on the plate. For phone operation a 50T or 211D or H tube is recommended. The modulator should supply about 400 watts of audio power.

Interlock Crystal Control

By interlocking a crystal-controlled oscillator to an oscillating final amplifier it synchronizes the frequency of the latter and imparts to it a frequency stability depending upon the operating parameters of the crystal control. The proper condition of interlock is noted by monitoring the signal tone—pureness of note indicating correct interlock.

Details: By using the second harmonic of the crystal oscillator, shown in the circuit diagram, sufficient interlock is obtained without fracturing the crystal. Operation on the fundamental requires careful adjustment of coupling between the oscillator in order to prevent too much feed-back into the grid of the 47 tube, resulting in a fractured crystal. Doubling has the further advantage of allowing operation on 20 meters with a 40-meter crystal.

The 150T circuit is tuned up as an oscillator with enough feed-back through 25 uufd. condenser to allow moderate-to-weak self-oscillation. The larger oscillator plate tank-condenser is tuned to the point at which its frequency falls into step with that of the crystal oscillator's second harmonic. The condition of synchronism will cause the plate current of control oscillator and grid current of the power tubes to increase.

Interlock control is not advised for general use due to difficulties arising in the maintenance of synchronism. The system will, however, improve the CW note of a modern or relatively high-powered oscillator.

Interlock Crystal Control Transmitter. The home-built high-power final tank condenser is at the right. The constructional details for building this condenser are told elsewhere in this chapter.

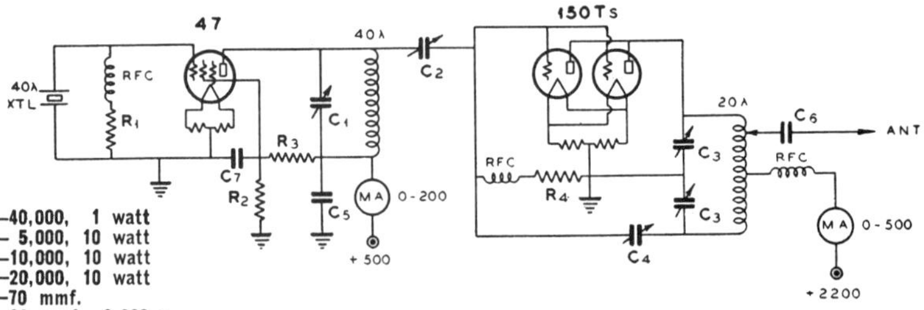

R1—40,000, 1 watt
R2— 5,000, 10 watt
R3—10,000, 10 watt
R4—20,000, 10 watt
C1—70 mmf.
C2—30 mmf. 3,000 v.
C3—45 mmf. 10,000 v.
C4—25 mmf. 10,000 v.
C5—.002, 2,500 v.
C6—.002, 5,000 v.
C7—.01 mica.

Complete circuit diagram of Interlock unit.

The giant EIMAC 300T triode in a super-power r-f amplifier. Grid and plate coils for all-band operation are pictured. The coils in the amplifier proper are for 10-meter operation. The circuit diagram is on the facing page.

High Power CW Amplifier: More than 800-watts output can be obtained from this amplifier which has ample capacity of handling inputs as high as 1 KW in the 10 to 80-meter CW amateur wavebands. In general, the capacity of the final tank tuning condenser is insufficient for phone operation on wavelengths above 20-meters, and would probably flash-over on plate modulation peaks. However, with grid modulation the amplifier functions splendidly on 10 and 20-meters.

Technical Notes: The filament transformer for the 300T Eimac tube is mounted near the tube socket in order to minimize voltage drop in the filament leads.

Plug-in coils are satisfactory since low-C circuits produce low values of circulating current. The plate tuning condenser is rated at 9000-volts per section. The latter has its rotor by-passed to ground if in the event of an RF arc is formed, the choke will not collapse due to a short circuit on the DC power supply. The capacity of the two-section 50uufd. split-stator condenser must be increased by the addition of two aluminum plates connected to the stators so that a standard heavy wire-wound coil can be used. This added capacity consists of two plate spaced ¾ths-inch apart and overlapped by 1½ x 3-inches. The tube capacity is of such a minimum value (excellent for 10-meter operation) that its shunt capacitive effect is insufficient to allow the regular coils to cover the amateur bands.

The plate coil stand-off insulators are mounted on small right-angle brackets above the tuning condenser. Front and rear supports on the condenser assure the rigidity of the assembly. The fixed plate of the neutralizing condenser is 2½ x 4½-inches and is fastened directly to the stator of the tuning condenser. The rotor plate is 3 x 4½-inches, separated about ¾ths-inch at neutralization. It is mounted on a standoff insulator with a large size coil-jack and plug for a rotor bearing. The baseboard is 11 x 16 x ¾ths-inches and the front panel is a piece of "masonite" 12 x 16 x 3/16th-inches, painted black (larger dimensions are discretional).

Miscellaneous Tests: Tests indicate that a 10,000-ohm grid-leak produces sufficient grid-bias when 25 to 30 MA is flowing through it at a plate potential of 1600-volts. At 2900-volts the grid current should be about 40 MA. And at 3500-volts

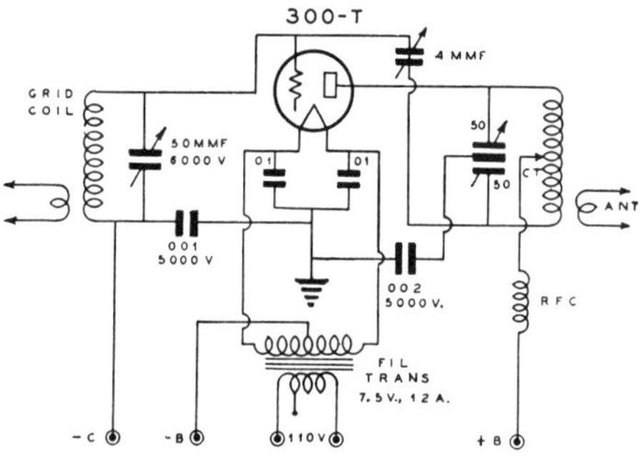

1-K.W. AMPLIFIER

it will be approximately 50 MA. These values are correct for CW operation at high efficiency on any of the bands. A bank of 100-watt lamps was used as a dummy antenna and grid currents below the aforementioned values show a decrease of power output. Higher values gave no indication of increased power output. As a regenerative frequency doubler, 700-watts was obtained on 20-meters when the grid current attained the value of 35 MA through a 25,000-ohm grid-leak at a plate potential of 2900 volts. The plate current, during this test, increased to about 350 MA showing an efficiency of 70%. In the original set-up, the neutralizing condenser had plate with twice as much area and so gave regeneration and high efficiency to the doubler whose coil center-tap was connected to the rotor.

Tests with grid modulation indicated a carrier output of between 150 and 200-watts with a 3000-volt power supply. A fixed cut-off bias of 175 volts and a cathode biasing resistor of 1000-ohms formed the grid-bias for high-efficiency grid modulation. The plate slightly colored (cherry-red—which is normal at 300-watts dissipation) and the DC grid current had a value from 0 to 2 MA. A type-10 doubler has sufficient output on 10-meters to excite this amplifier for grid modulation with part of the input power dissipated in a 40-watt mazda lamp, the latter acting as a grid stabilizing resistor. The grid coil was link-coupled to the lamp as well as to the buffer stage.

For CW operation a pair of 801's connected in push-pull will deliver the necessary grid excitation. Larger driver tubes will give a certain amount of excess excitation. However, 800-watts output may be obtained with 1000-watts input, using a pair of 10's or 801's as drivers with less than 700-volts plate supply.

Copper-Wire Tank Coils:

Tank coils for low power stages may be wound on receiving-type plug-in coil forms, providing that they are made of ceramic or Isolantite insulating materials. For high-powered stages, research with various types of tank coils definitely indicates that a tank coil wound with copper wire (size No. 10 to No. 14, depending upon the power) and supported by three or four narrow strips of thick celluloid fixed to the coil turns with collodion, or its equivalent, has the lowest losses. When properly built, this coil is quite rigid mechanically. Furthermore, it has the absolute minimum of di-electric in the field of the coil. The length of the coil should be about one and one-half times its diameter, and turns should be spaced more than the diameter of the wire. The proper mounting for the coil is about one diameter away from its associated tuning condenser.

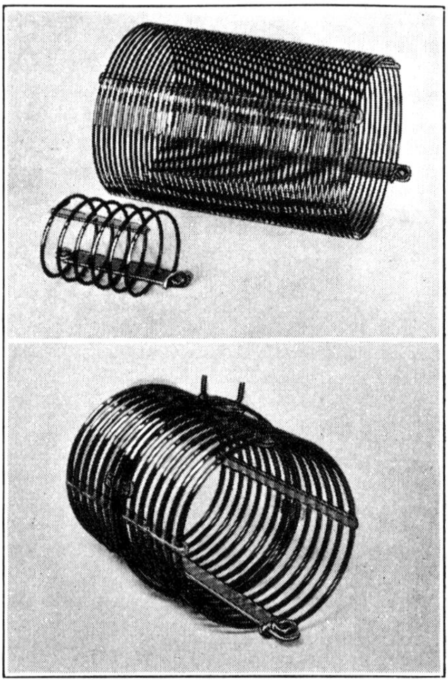

Typical copper-wire coils for final amplifiers.

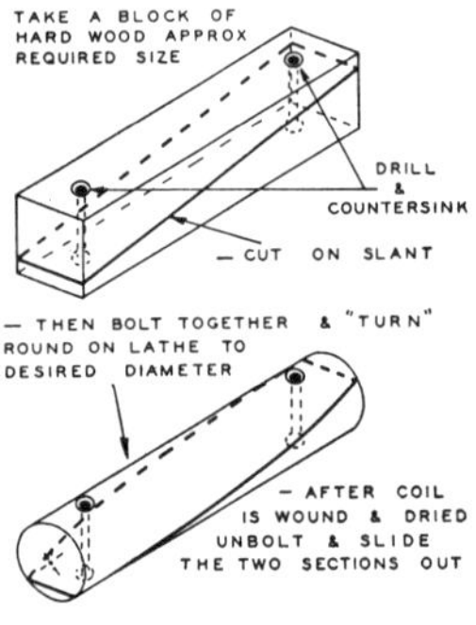

The home builder can wind his own tank coils by making a winding form as shown in the accompanying illustration. First, a layer of string is wound on the round wood form. Then a few layers of wax paper are wrapped over the string winding. Celluloid strips are then placed at four equidistant places. The copper wire is wound tightly on the form, space-wound by inter-winding a piece of heavy twine with the wire. The spacing twine is then removed. The wire is secured to the celluloid strips by means of collodion, airplane "dope," or Duco household cement. The job is then laid aside to dry. Unravel the layer of twine between the wood form and the wax paper and the finished coil can be easily removed from the form.

Neutralizing Condensers: The interlectrode electrode capacity of a vacuum tube determines the value of condenser capacity needed to neutralize a transmitter circuit. Tube interelectrode capacities vary from 2 to 15. The higher this capacity, the greater will be the value of the capacitor required to neutralize the circuit.

The capacity of the neutralizing condenser can be determined from the formula:

$$C = \frac{.225A}{d}$$

where A, is the area in square inches of the interleaved plates of the neutralizing condenser, multiplied by the number of air-spaces; d, is the distance in inches between adjacent plates; C, the capacity in microfarads.

For example, a condenser suitable for neutralizing a single 211 tube will have two plates, each 4 inches square, and spaced about ¼ in. apart. These plates can be secured to stand-off insulators in such a manner that the movable plate can be rotated through a 45-degree arc. Transmitting coil plug-in pins and jack make a good bearing arrangement for the rotor plate.

For the low-C tube, such as the 300T, 150T, 354, etc., the size of the neutralizing condenser plates would be approximately 3 inches square, spaced ½ in. apart. The corners and edges of the plates should be rounded and polished. The plate area sizes given here are those which actually overlap.

For voltages up to 1500, space the plates ¼ in.; up to 2500 volts, ½ in.; up to 5000 volts, 1 in.

Home-built high-power plate condenser.

Economical High-Power Transmitting Condensers: Applying the above formula, condensers can be built that will give entire satisfaction in high-power transmitters.

Referring to the accompanying photographs, it will be seen that the moving plates of the condenser are raised or lowered by means of a piece of dial silk-oil string, winding over a ¼ inch brass rod which is rotated by means of a knob. The brass rod is held in place by means of a pair of aluminum brackets. A long machine screw is run through these brackets, parallel to the rod. The latter provides an adjustable tension of the two end mounting plates against the end bearings on the brass rod. These end bearings can be a pair of washers between the two knobs, one on each end of the brass rod.

The rotor plates are mounted with flat-head machine screws to a back plate which is hinged to the main vertical mounting panel. The back plate on the split-stator condenser is a piece of No. 12 gauge aluminum, 7½ in. x 4 in., with a pair of hinges at one end. The vertical mounting strip is a piece of 4½ in. x 12 in. x 3/16 in. tempered "masonite" of "celotex" board. Bakelite or ½ in. wood will give even better rigidity.

The smaller condenser is made similar to its counterpart, the dimensions, of course, being proportionally made.

The large condenser plates are made by bending them on a sheet-metal brake. The plates are pieces of No. 14 gauge aluminum, 5 in. x 12 in., and then "U" shaped into pieces 5½ in. x 5 in. with 1 in. end section. These pieces are first cut to size and ¼ in. holes are then drilled in the center for stand-off insulator mountings. All edges and corners are rounded off on an emery wheel and then polished on a buffing wheel. The rotor plates are mounted so as to clear the ends of the stator plates when fully enmeshed by about ⅝th in. in order to give ½ in. clearance to the nuts holding the stator plates to the stand-off insulator machine screws.

The smaller condenser has a stator section made from a piece of No. 14 gauge aluminum, 7 in. x 3 in., to form sides 3 in. x 3 in. with a 1 in. end portion. The rotor is made of a piece of 8 in. x 2½ in. aluminum to give plates 3½ in. x 2½ in. with a 1 in. end portion. The sections overlap an area 3 in. x 2½ in. If more capacity is desirable larger plates can be used.

Cutting Aluminum: Short-cut methods simplify the working of aluminum, hence, when cutting a piece of sheet aluminum, first, lay a T-square, or other flat rule on the metal. With the aid of an awl, cut a deep groove into the material. Continue to gouge this groove (on both sides of the metal) until a small channel having a depth about half the thickness of the metal is obtained. Next, place the prepared sheet on a flat-top table with the channel at the edge of the table. Bend the over-lapping end back-and-forth until the sheet separates. Or, the sheet may be placed into a vice and the uppermost part given a series of rapid jerks until the metal breaks apart.

Making "U" Bends: Place two pieces of wide angle-iron into a vice. Place the metal sheet between the angle pieces. With the aid of a sturdy block of wood, laid against the metal where it protrudes from the iron angle pieces, press down on the wood block and the metal will bend easily. A wood or hard-rubber mallet can also be used to gently shape the metal to the desired angle.

Finishing Aluminum: Hold the aluminum sheet against a motor driven wirewheel brush and a beautiful satin-like finish will be secured. Less effective results are attained by rubbing the metal with steel-wool. Rub in one direction only.

Aluminum can also be given a dull, satin-like finish by immersing it in a hot solution of caustic soda and water. The metal is left in the solution for a few minutes, and is then removed with the aid of a pair of pliers. Rinse quickly in water. Stand upright while drying so that the water will run off without streaking the metal.

A "laboratory" finish can be applied to aluminum panels by holding the metal against a motor-driven cork. An ordinary cork is held in the chuck of a drill press or lathe and the spinning of the cork, when pressed against the metal will "grind" concentric circles into the panel. The circles can be made practically any size. Some beautiful effects can be secured by overlapping these concentric circles, or a border design of small circles can be run close to the edges of the panel.

High-power 150-T final amplifier with home-built plate condenser. The neutralizing plate (rear) is supported on a standoff insulator, fitted with jack and plug.

Eimac 150T push-pull amplifier.

Chassis Ground Connections: Steel chassis should not be depended upon for r-f ground connections. All r-f grounds in each stage should be returned to a common point of connection, if possible. These common ground points should be connected together by means of heavy copper wire or strips. Transmitter chassis should be connected to an external earth ground in order to prevent the r-f from reaching ground through the a-c power line.

HK-254, largest air-cooled tube made.

Coil Winding Charts for Copper Tubing Tank Coils

THE values given are a close approximation to your particular requirements in each case, but exact accuracy depends on the circuit arrangement and the length of the leads in the plate circuit of the tube to be used. The two factors mentioned become more important as the frequency increases. Long leads necessitate fewer turns on the coil, but the leads should be long enough to keep the tank condenser separated from the coil by at least the coil diameter.

All the values in the table are for the tubes specified when used as single-ended amplifiers with the neutralization tap near the center of the coil. If placed in the center of the coil, this tap will automatically give fixed neutralization on all bands. For push-pull amplifiers, decrease the number of turns by 25% for any given tube. The reason for this decrease will be apparent upon close comparison of single-ended and push-pull circuits. Just twice as much tube capacity is shunted across the tank in push-pull circuits as when single-ended circuits are used.

In low-C tanks, such as these, the voltage rating of the condenser should be equal to four times the plate voltage on the tube for single-section types, and twice the plate voltage (each section) for split-stator models.

CHART NO 1. For Coils Tuned With Split-Stator Condenser and Used in Circuits Employing Low-C Tubes, such as 150T, 50T, 354, 852, 800, 825, RK18.

BAND	2" Dia. Coil	3" Dia. Coil	4" Dia. Coil	5" Dia. Coil	6" Dia. Coil	Size of Tuning Condenser
160	N.S.	N.S.	N.S.	N.S.	80 Turns 36" Long 3/8" Tubing	250 Mmf. Each Section for Full Band Coverage.
80	N.S.	N.S.	60 Turns 20" Long 1/4" Tubing	50 Turns 18" Long 1/4" Tubing	40 Turns 18" Long 3/8" Tubing	100 Mmf. Each Section for Full Band Coverage.
40	N.S.	46 Turns 16" Long 1/4" Tubing	34 Turns 12" Long 1/4" Tubing	28 Turns 12" Long 1/4" Tubing	22 Turns 12" Long 1/4" Tubing	35 Mmf. Each Section.
20	32 Turns 15" Long 1/4" Tubing	20 Turns 12" Long 1/4" Tubing	16 Turns 12" Long 1/4" Tubing	14 Turns 12" Long 1/4" Tubing	10 Turns 12" Long 1/4" Tubing	35 Mmf. Each Section.
10	8 Turns 4" Long 1/4" Tubing	6 Turns 4" Long 1/4" Tubing	4 Turns 4" Long 1/4" Tubing	4 Turns 4" Long 1/4" Tubing	3 Turns 4" Long 1/4" Tubing	35 Mmf. Each Section. N.S. Indicates: NOT SATISFACTORY.

CHART NO 2. For Coils Tuned With Single-Section Condenser and Used in Circuits Employing Low-C Tubes, such as 150T, 50T, 354, 852, 800, 825, RK18.

BAND	2" Dia. Coil	3" Dia. Coil	4" Dia. Coil	5" Dia. Coil	6" Dia. Coil	Size of Tuning Condenser
160	N.S.	N.S.	N.S.	N.S.	60 Turns 36" Long 3/8" Tubing	100 Mmf.
80	N.S.	N.S.	50 Turns 20" Long 1/4" Tubing	40 Turns 18" Long 1/4" Tubing	30 Turns 18" Long 3/8" Tubing	100 Mmf. For Full Band Coverage.
40	N.S.	36 Turns 14" Long 1/4" Tubing	24 Turns 12" Long 1/4" Tubing	20 Turns 12" Long 1/4" Tubing	16 Turns 12" Long 1/4" Tubing	35 Mmf.
20	22 Turns 12" Long 1/4" Tubing	16 Turns 12" Long 1/4" Tubing	12 Turns 12" Long 1/4" Tubing	10 Turns 12" Long 1/4" Tubing	8 Turns 12" Long 1/4" Tubing	35 Mmf.
10	6 Turns 5" Long 1/4" Tubing	4 Turns 5" Long 1/4" Tubing	4 Turns 5" Long 1/4" Tubing	4 Turns 5" Long 1/4" Tubing	2 Turns 5" Long 1/4" Tubing	35 Mmf.

CHART NO. 3. For Coils Tuned With Split-Stator Condenser and Used in Circuits Employing High-C Tubes, Such as 50 Watters, 210, 204A, 849, 212D, 830, 46, RK20.

BAND	2" Dia. Coil	3" Dia. Coil	4" Dia. Coil	5" Dia. Coil	6" Dia. Coil	Size of Tuning Condenser
160	N.S.	N.S.	N.S.	N.S.	72 Turns 36" Long 3/8" Tubing	250 Mmf. Each Section for Full Band Coverage.
80	N.S.	N.S.	54 Turns 16" Long 1/4" Tubing	46 Turns 18" Long 1/4" Tubing	36 Turns 18" Long 3/8" Tubing	100 Mmf. Each Section for Full Band Coverage.
40	N.S.	36 Turns 14" Long 1/4" Tubing	24 Turns 10" Long 1/4" Tubing	20 Turns 10" Long 1/4" Tubing	16 Turns 10" Long 1/4" Tubing	35 Mmf. Each Section.
20	24 Turns 10" Long 1/4" Tubing	16 Turns 10" Long 1/4" Tubing	12 Turns 10" Long 1/4" Tubing	10 Turns 10" Long 1/4" Tubing	8 Turns 10" Long 1/4" Tubing	35 Mmf. Each Section.
10	8 Turns 5" Long 1/4" Tubing	6 Turns 5" Long 1/4" Tubing	4 Turns 5" Long 1/4" Tubing	4 Turns 5" Long 1/4" Tubing	3 Turns 5" Long 1/4" Tubing	35 Mmf. Each Section.

CHART NO. 4. For Coils Tuned With Single-Section Condenser and Used in Circuits Employing High-C Tubes, Such as 50 Watters, 210, 204A, 849, 212D, 830, 46, RK20.

BAND	2" Dia. Coil	3" Dia. Coil	4" Dia. Coil	5" Dia. Coil	6" Dia. Coil	Size of Tuning Condenser
160	N.S.	N.S.	N.S.	N.S.	60 Turns 36" Long 3/8" Tubing	100 Mmf.
80	N.S.	N.S.	50 Turns 20" Long 1/4" Tubing	40 Turns 18" Long 1/4" Tubing	30 Turns 18" Long 3/8" Tubing	100 Mmf. For Full Band Coverage.
40	N.S.	32 Turns 14" Long 1/4" Tubing	22 Turns 12" Long 1/4" Tubing	18 Turns 12" Long 1/4" Tubing	14 Turns 12" Long 1/4" Tubing	35 Mmf.
20	18 Turns 10" Long 1/4" Tubing	14 Turns 10" Long 1/4" Tubing	10 Turns 10" Long 1/4" Tubing	8 Turns 10" Long 1/4" Tubing	6 Turns 10" Long 1/4" Tubing	35 Mmf.
10	4 Turns 5" Long 1/4" Tubing	4 Turns 5" Long 1/4" Tubing	4 Turns 5" Long 1/4" Tubing	4 Turns 5" Long 1/4" Tubing	2 Turns 5" Long 1/4" Tubing	35 Mmf.

Coil Chart for 1½-in. and 2½-in. Dia. Coil Forms.

BAND	1½" Dia. Coil Form	Size of Tuning Condenser	BAND	2½" dia. Coil Form	Size of Tuning Condenser	REMARKS
160	Not Satisfactory		160	46 Turns No. 16 DCC. Close wound	100 MMF. or larger	The winding data shown here is for coils that are tuned with single-section variable condensers. See Chart below for coil winding data when split-stator variable condensers are used.
80	35 Turns No. 22 DCC. Close wound	100 MMF.	80	23 Turns No. 16 DCC. Spaced one dia.	100 MMF.	
40	19 to 21 Turns No. 16 DCC. Spaced one dia.	100 MMF.	40	16 Turns No. 16 DCC. Spaced one dia.	25-35 MMF.	
20	11 to 13 Turns No. 16 DCC. Spaced one dia.	25-35 MMF.	20	8 to 10 Turns No. 16 DCC. Spaced one dia.	25-35 MMF.	
10	5 to 6 Turns No. 16 DCC. Spaced one dia.	25-35 MMF.	10	5 Turns No. 16 DCC. Spaced one dia.	25-35 MMF.	

Coil Winding Chart for 1½-in. and 2½-in. Dia. Coil Forms and Split-Stator V.C.

BAND	1½" Dia. Coil Form	Size of Tuning Condenser	BAND	2½" dia. Coil Form	Size of Tuning Condenser	
160	Not Satisfactory		160	59 Turns No. 16 Enameled Close wound. Tap at center.	250 MMF. Each Section (smaller condenser can be used).	The standard Hammarlund 35 mmf. Each section split-stator double-spaced midget variable condensers are satisfactory. The Cardwell Trim-Air 100 mmf midgets can also be used by merely removing alternate plates from rotor and stator sections and ganging two of these condensers together. The capacity will then be 25 mmf. per section.
80	Not Satisfactory		80	55 to 57 Turns No. 16 DCC. Close wound. Tap at center.	35 MMF. Each Section	
40	35 Turns No. 16 DCC. Close wound. Tap at center.	35 MMF. Each Section	40	29 Turns No. 14 Enameled Space wound. To cover 3 inches.	35 MMF. Each Section	
20	19 Turns No. 16 DCC. Spaced one dia. Tap at center.	35 MMF. Each Section	20	15 Turns No. 14 Enameled Spaced one dia. Tap at center.	35 MMF. Each Section	

"LES-TET" COIL DATA

All forms 1⅝ inches outside diameter.

(See circuit diagram on page 101).

	20 meters	40 meters	80 meters
L1	same as 40 m. coil, no tap.	15 turns, #18 DCC, spaced 1/16". Tap, 5 turns up from bottom.	24 turns, #18 DCC, close wound. Tap, 8 turns up from bottom.
L2	7 turns #18 DCC, ⅛" spacing. Link coil, 4 turns #22 DCC, close wound, ¼" from cold end.	15 turns, #18 DCC, spaced 1/16". Link coil, same as for 20 m.	24 turns, #18 DCC, close wound. Link coil, same as for 20 m.

RADIOTELEPHONY

Definition: What can be defined as the average transmitter is that which consists of the following components: (1) a portion that generates and amplifies the radio-frequency carrier wave; (2) a portion that converts the sound waves into electrical waves; (3) a portion that takes amplified audio-frequency currents and mixes them (by a process known as modulation) with the radio-frequency carrier in such a manner that the power output of the transmitter varies in exact accordance with the variation in sound pressure applied to the microphone.

Modulation Fundamentals: In general, all communication systems utilize audio-frequency waveforms. These may be either pure tones and square-topped waveforms for use in code transmission, or the waveforms may be quite complex for conveying telephonic speech directly, without translating the intelligence conveyed into dots and dashes of the telegraphic or radio codes. The range of audio-frequencies required to transmit the intelligence varies from a few cycles to 10,000 cycles per second, depending on whether telegraphic or high-definition amplification is used. For amateur purposes, an audio-frequency range of from 200 to 2800 cycles per second will provide intelligibility, although fully natural and pleasing reproduction of the transmitted speech requires a range of from at least 100 cycles to 4000 cycles. For high-definition, frequencies between 80 and 8000 cycles must be faithfully reproduced at the receiving point, such fidelity is seldom secured in amateur practice, but should be attained whenever conditions permit.

In the transmission of telegraphic signals over a radio circuit, the carrier is radiated only during the "mark" period. The "space" is obtained and defined by an absence of the carrier. On the other hand, when telephonic communication is used on a radio channel, the carrier remains on between syllables and words. The audio signal periodically increases and reduces the amplitude of the carrier, while the average amplitude of the carrier remains constant. Certain commercial telephone circuits use a type of modulation, termed "Suppressed Carrier Single Sideband," but it is not very widely used because of the difficulty of obtaining satisfactory speech quality. The principal reason for the difficulties involved in this system or similar systems lies in the inability to maintain the oscillator in the receiver in exact synchronism with the oscillator in the transmitter.

When a modulated carrier is analyzed, it is found that the original carrier is present, plus two groups of the sum and difference frequencies, which have been named the upper and lower sidebands. These sidebands are generated in the transmitter by the familiar heterodyne process. Thus, one sideband consists of the waves whose frequencies equal that of the carrier **plus** that of all the individual audio components, and the other sideband consists of the waves whose frequencies equal that of the carrier **minus** all the audio components. In other words, the carrier and the audio signal were **heterodyned** together into a group of **beat-frequencies** by the action of the modulated amplifier.

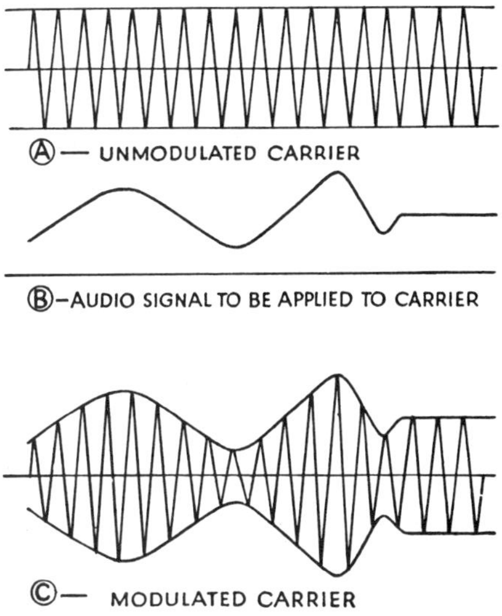

Curve (A) indicates the pure c-w wave applied to the grid of the modulated amplifier. Curve (B) shows the audio frequency output of the modulator. Curve (C) shows the combination of the two after being mixed in the modulated amplifier. Note that the average value of the modulated wave is constant.

The carrier takes up a relatively small position in the frequency spectrum, but, since each sideband contains all the audio signal components, the modulated signal will require a frequency band twice as wide as the highest audio-modulating signal. For example: If the transmitter responds to frequencies between 100 and 4000 cycles per second, then the bandwidth must extend 4000 cycles above and below the carrier. This 8000 cycle band will cause some interference to any other station whose sidebands extend into this particular portion of whatever amateur band the transmitter is working in. Almost 85 per cent of the power radiated in the sidebands consists of the frequencies in the register below 1500 cycles the remaining 15 per cent consist of frequencies of the upper register, which determine the quality of speech reproduction. The high audio-frequencies contain the greater portion of the harmonic content of sound which, if muted, depletes the fidelity and timbre of natural speech.

Power Distribution in a Modulated Wave: The amplitudes of the sidebands depend on the percentage modulation; the higher the degree of modulation, the greater the sideband amplitude. It takes **power** to modulate a wave which is expended in altering its amplitude. When a carrier is 100 per cent modulated by a pure audio tone, the power in each of the two sidebands equals one-quarter of the unmodulated carrier power output. Thus the power in both sidebands equals one-half the carrier wave and, therefore, complete modulation increases the average power output of the phone transmitter 50 per cent. If a class C radio-frequency amplifier is plate modulated, the plate power input must therefore be increased 50 per cent in order to get a 50 per cent increase in output, because the plate efficiency remains constant during modulation. This 50 per cent increase in plate input is obtained from the modulator tubes in the form of AC. It is superimposed on the DC plate input in such a manner that the instantaneous plate voltage (and current) is alternately raised to twice the unmodulated value, and then reduced to zero. In order to swing the plate voltage of the class C amplifier from zero to twice normal, the modulators must alternately supply and absorb power. This involves energy storage during the time the plate voltage is below normal. This energy is stored in the Heising choke, or in the modulation coupling transformer, depending upon whether capacitative or inductive coupling is used between the modulators and the modulated amplifier.

One hundred per cent modulation is approached only on the extreme voice peaks. Ordinarily these peaks should seldom be allowed to modulate a phone transmitter more than about 80 per cent, and the average modulation during the time that the operator is actually speaking should approximately average 40 per cent. However, the capability to modulate at 100 per cent is essential to minimize heterodyne interference between or with other stations.

All plate modulated RF amplifiers operate as class C amplifiers which require that the grids of the tubes be heavily excited by a buffer amplifier so that the power output of the stage will rise as the square of the plate voltage without any "dropping off" tendency as the instantaneous plate voltage approaches twice the normal value under modulation. HINT: In practice, choose tubes with as high a mutual conductance as possible to economize on driving power.

The plate input to a class C modulated amplifier increases during modulation, while the plate efficiency remains constant. On the other hand, the plate dissipation will increase when audio modulation is applied, necessitating that the tube operate below its maximum rating in order to allow some reserve dissipation for the heat radiated from the plate during complete modulation; incidentally, the heat increases upwards to 50 per cent during 100 per cent modulation.

Another reason for operating modulated amplifier tubes below their maximum rating is that the peak plate voltage and the peak plate current are doubled during complete modulation.

Shielding RF Portions of Phone Transmitters: Additional shielding or isolation of the RF portion of the transmitter will be required in order that all RF be kept out of the speech amplifier, such precautions will prevent the amplifier from overloading and "singing"; in some cases it will be even necessary to shield the entire speech amplifying equipment.

Phone Transmitter Components

The three principal parts of the phone transmitter are: (1) the radio frequency channel; (2) the audio-frequency channel; and (3) the power supplies. In the subsequent treatment, an analysis is given to the major components comprising each of the aforementioned parts.

The RF Channel: The principal function of this channel is to generate and amplify radio-frequency alternating-current oscillations which are ultimately modulated by the voice impulses and then radiated from the antenna.

The radio-frequency generator consists of a low-power AC oscillator tube whose frequency is held within very close limits in order that the period of oscillation does not appreciably drift. In practically all modern amateur transmitters the frequency is maintained at a near-constant value by a quartz plate oscillator.

While the crystal has a tendency to resist changes in frequency caused by changes in the plate voltage, or by other circuit characteristics, it cannot entirely compensate itself from factors tending to alter the generated frequency. For these reasons, the crystal oscillator cannot be modulated directly without some undesirable frequency modulation. These wide changes in plate voltage will have some effect on the circuit parameters of the most stable of crystal oscillators. Amplifiers which adjoin the crystal oscillator must not be modulated lest some reaction be reflected back into the oscillator which may have (it generally does) some effect on the frequency stability. A crystal oscillator must be isolated by at least one buffer stage between it and the succeeding radio-frequency amplifier which is modulated by the voice impulses.

The Modulated Amplifier: Power modulation, sometimes called "Heising Modulation," "Plate Modulation," or "Power Supply Modulation," is used in most amateur stations. In a previous explanation it was stated that all forms of plate-modulated amplifiers operated "class C," wherein the negative grid bias is greater than two times that value of bias which would reduce the plate current to zero if the RF grid drive is removed.

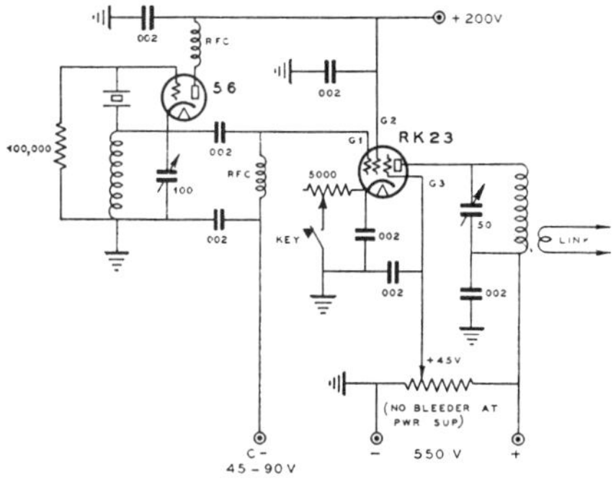

Crystal oscillator and low power buffer stage similar to "Les-Tet" Exciter.

The process of plate modulation occurs whenever the plate voltage is varied up and down over its normal value at an audio-frequency rate, the variations being exactly in accord with the voice impulses which strike the diaphragm of the microphone. If the class C RF amplifier is properly biased and driven, the radio frequency AC voltage measured across the plate tank coil will, at all times, be exactly proportional to the instantaneous DC plate voltage. By instantaneous DC plate voltage is meant the sum of the constant DC plate voltage, plus the instantaneous AC voltage which is superimposed on it, and which comes from the modulator tube or tubes. This variation of radio-frequency voltage across the tank coil obviously causes a variation in the power output of that amplifier stage, and if the antenna is coupled to the modulated amplifier the RF energy is modulated in accordance with the variation of sound applied to the microphone. The RF signal in the detector circuit of a distant receiver, when the carrier is unmodulated, is inaudible—unless a beat-frequency oscillator supplies a heterodyning signal. However, as soon as the amplitude (or voltage) of the carrier signal is varied and is present in the distant receiver, the variations are changed by electro-acoustic conversion in the reproducer and are heard as sound.

In order that the amplitude of the RF output shall be an exact replica of the voice impulses, it is essential that there be no regeneration in the class C amplifier. This means that the RF amplifier must be perfectly neutralized. It takes an appreciable amount of regeneration to make an amplifier break into self-oscillation; however, because an amplifier does not oscillate is not an indication that it is perfectly neutralized. There may not be enough regeneration to allow self-oscillation, but even a small amount of regeneration can seriously disturb the linearity of modulation and thereby cause distortion. The modulation must not only be linear, it must be perfectly symmetrical as well. In other words, the positive and negative peaks of modulation must be equal. This necessitates that the carrier output must swing up just as much as it swings down on the immediately succeeding half cycle. Non-symmetrical modulation causes a change in the average amplitude of the modulated wave, which results in carrier shift and serious interference, as well as introducing audio distortion. Interference due to non-symmetrical modulation is very much of the type as that resulting from over-modulation and is sometimes called "sideband splatter."

Non-symmetrical modulation is sometimes caused by having a very low C in the plate tank circuit of the modulated amplifier. If there is an excess of inductance and a deficiency of capacitance in the circuit, the proper amount of circulating current will not flow in the tank circuit to provide the necessary "fly wheel" effect.

See L to C Ratio Chart on page 103 for correct capacity to use at various frequencies.

Frequency Modulation: The oscillator frequency will vary during modulation unless one or more buffer stages are used between the modulated stage and the oscillator. This variation of frequency is called "frequency modulation."

Linear Amplifiers: To avoid distortion, any amplifier which amplifies a wave previously modulated in some preceding stage must produce output wave shapes which are exactly similar, except for size, to the input wave shapes which excite the grid. This type of amplifier is called a "Linear Amplifier" because its output is a linear function of its input. The most common type of linear amplifier is usually biased exactly to cut-off and it is called a "class B Linear Amplifier."

The reasons why these amplifiers are not more widely used in amateur stations are because they are quite difficult to adjust and they require a rather expensive amount of tube capacity for their carrier output.

It is desirable to operate a linear amplifier at as high a plate voltage as possible to obtain the maximum possible unmodulated plate efficiency. Because the grid current varies with the percentage modulation, the grid bias of a linear amplifier must be supplied from a separate source, such as batteries of a low-resistance power supply, to avoid distortion.

Linear amplifiers operate as efficiency modulated devices. The plate efficiency is controlled by the RF excitation voltage applied to the grid. The maximum theoretical unmodulated plate efficiency for a class B linear amplifier is 39 per cent, and 50 per cent for a class BC linear amplifier. In practice, the unmodulated plate efficiency of a class B linear amplifier seldom exceeds 30 per cent, and for the class BC linear amplifier the upper limit is about 40 per cent.

The modulated output is obtained by varying the instantaneous plate efficiency of the linear amplifier between the limits of zero efficiency and twice the normal unmodulated efficiency. Thus a class B linear amplifier might be 30 per cent efficient during periods of no modulation, and during periods of 100 per cent modulation the instantaneous efficiency would be varying at

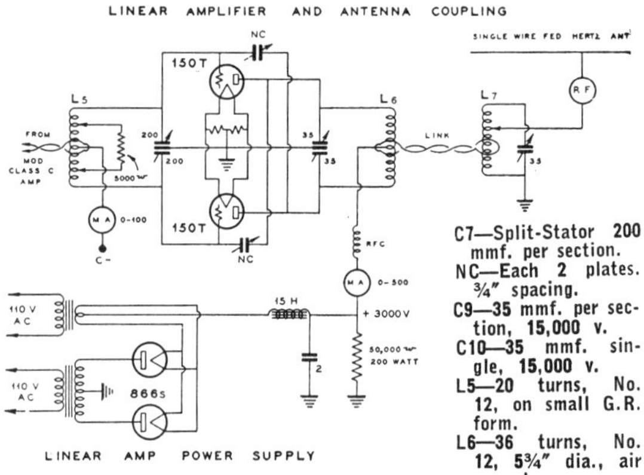

The coil data is for 75 meter operation.

C7—Split-Stator 200 mmf. per section.
NC—Each 2 plates. 3/4" spacing.
C9—35 mmf. per section, 15,000 v.
C10—35 mmf. single, 15,000 v.
L5—20 turns, No. 12, on small G.R. form.
L6—36 turns, No. 12, 5 3/4" dia., air spaced.
L7—22 turns, 3/16" copper tubing, 4 1/2" dia., air spaced.

an audio-frequency rate between zero and 60 per cent. During the period of 100 per cent modulation the **average** plate efficiency increases 50 per cent. Because the average plate efficiency is lowest when unmodulated, the plate loss is highest at that point, and it is therefore evident that all linear amplifiers cool off during modulation. Exactly the opposite occurs in a class C plate-modulated amplifier because its plate loss increases 50 per cent during periods of 100 per cent modulation and also because the plate efficiency of a class C amplifier remains approximately constant during modulation, although the average DC plate input is increased 50 per cent.

The Audio Channel: The fidelity and faithfulness by which the voice frequencies are amplified depends wholly upon the individual characteristics of the parts comprising the audio-frequency amplifying equipment. To satisfactorily pattern any group of instruments into a well-designed speech amplifying system requires that each part be better than just "ordinary" or "cheap."

Microphones: The function of a microphone is to convert sound energy into electrical energy. In a perfect microphone the electrical output would be an exact replica of the sound input caused by the successive compressions and rarefactions of the air in front of the mouth of the person who speaks into the microphone.

The various types of microphones in use today are:

(1) The carbon microphone (with one or two buttons).
(2) The condenser microphone (air and nitrogen filled).
(3) The crystal microphone (Piezo-electric type).
(4) The inductive microphone (moving ribbon) (velocity type).
(5) The dynamic microphone (moving coil type).
(6) The non-directional dynamic microphone.

An explanation embodying the principles and function of the above microphones are:

The Carbon Microphone: This type of microphone is the most common in use today. The electrical output results from the fact that the resistance of a group of carbon granules varies with the mechanical pressure exerted on them. The pressure is varied by a metal diaphragm whose oscillatory movement is conveyed directly to the pile of carbon particles which varies the microphone battery current flowing through the microphone. In the case of a double-button carbon microphone there are two groups of carbon particles located in metal buttons, one on each side of the diaphragm. This vibrating member is usually stretched so as to remove the mechanical resonant point of the diaphragm out of the most important part of the audio-frequency range. These two-button microphones are connected to a center-tapped primary winding on the microphone coupling transformer so that the buttons are effectively in push-pull. This tends to minimize the even harmonic distortion which is inherent in all carbon microphones.

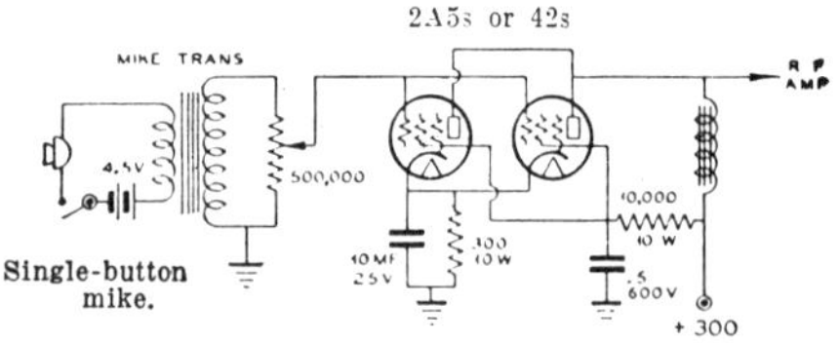

10-watt amplifier for use with a single-button microphone.

Unfortunately, all resistive types of acousto-electric converters have a rather high background hiss, due to the button current; in addition, are incapable of responding to wide frequency range, and generate more than a good portion of harmonic distortion. Fortunately, carbon microphones being low-impedance devices (200-400 ohms) require little or no shielding. Another fea-

T1 — Double Button Mike-to-Grid Trans.
T2 — 3:1 Interstage Trans.
Choke: 20 henry, 200 M.A.

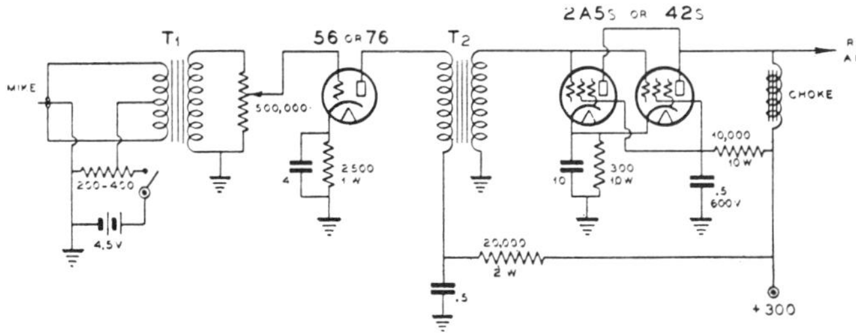

10-watt amplifier for double-button microphone.

ture is that the ruggedness and mechanical construction of the device allows it to be handled without the consideration that would necessarily be required with a more delicate instrument. The output of a highly damped two button type is rated at —45 D.B.

The Condenser Microphone: This type of microphone is capable of giving all the fidelity an amateur can use. It operates on the principle that any variation in the dielectric spacing of a small condenser generates a small AC voltage across the condenser terminals. A rather high polarizing voltage must be applied between the condenser plates to secure sufficient output; voltages from 90 to 180 volts are common. The condenser microphone has a heavy back plate in front of which is located a thin "dural" diaphragm, tightly stretched to eliminate resonance. The space between the diaphragm and back plate is sometimes filled with nitrogen to improve the over-all response characteristics. The diaphragm is usually a thousandth of an inch thick, with that amount of spacing between the vibrating member and the back plate. When sound waves are impressed on the diaphragm its oscillatory movement varies the electro-static capacity between the two condenser plates, which impresses an AC voltage on the DC polarizing voltage. This AC voltage is then transferred to the grid of the first audio amplifier through a small blocking condenser which isolates the polarizing voltage from the grid of the first audio tube. The output is about —95 D.B. as compared with an ordinary single button telephone microphone with unstretched diaphragm.

The condenser microphone has the advantage that its extremely light diaphragm gives somewhat better high-frequency response than a carbon microphone. In addition, there is no carbon hiss. The disadvantage lies in the fact that the audio output is so low that a "pre-amplifier" is necessary to augment the minute currents up to a level equal to that of a standard 2-button carbon microphone (—50 D.B.). Because the condenser microphone is a very high impedance device, it must be isolated from both RF and AC fields. Weather conditions, such as humidity and barometric pressure, affect the response characteristics of practically all electro-static devices. Cavity resonance, structural and resonance peaks tend to alter the fidelity of many types of condenser microphones, especially those of inferior manufacture. On account of these factors, prospective purchasers should carefully weigh both the electrical and mechanical features of a group of microphones before finally making their acquisition.

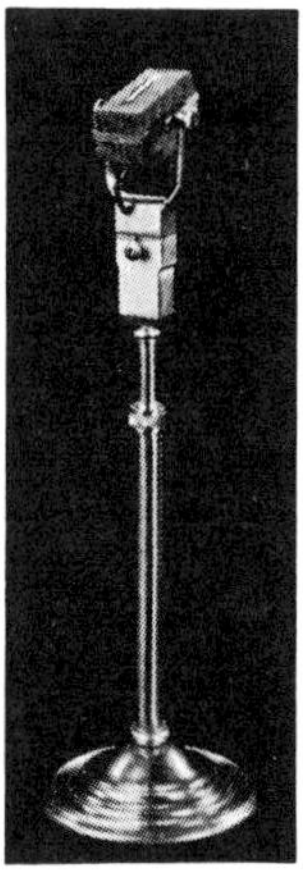

Brush sound-cell microphone.

The Crystal Microphone: There are two types of crystal microphones—the diaphragm type and the grill type. Both operate on the piezo-electric principles as defined by Curie. When a dielectric material in a condenser changes its mechanical dimensions or density, a change in capacity occurs. This change in capacity generates a small AC voltage. All crystal microphones use Rochelle salt crystals, which act like small condensers. If these crystals are subjected to a deformation by bending strains caused by an acoustic pressure, a small audio-frequency will be generated across the two small pieces of metal foil which are glued to opposite faces of the crystal. The voltage produced by the crystal is then fed into a pre-amplifier for the necessary amplification.

The **diaphragm type** is the most inexpensive of crystal microphones. While it is capable of somewhat better fidelity than the more common types of condenser microphones, its quality is not comparable to that secured from the better types of electro-static instruments. No polarizing voltage or magnetic field is required, and the audio output is approximately equal to that obtained from the highly-damped types of two-button carbon microphones. There is no background noise and the fidelity depends upon the care with which the diaphragm has been installed.

The **grill type** of crystal microphone is capable of almost perfect fidelity. It consists of a series of crystals (or sound cells) connected in series-parallel to produce a high output. The energy developed by this type of microphone is equal to that of the lower-level moving coil types, and is somewhat higher than that of the moving coil variety of microphones. Although the device is a high impedance source of audio voltage, its peculiar condenser characteristics allow the use of a shielded lead which can be 100 feet long between the microphone and its associated pre-amplifier. One important advantage of the grill type crystal microphone is that it is less directional than other types of microphones. The output level varies between —65 and —74 D.B., depending upon the construction.

The Inductive Microphone. These microphones operate on the principle that the movement of a conductor in a magnetic field induces a voltage in the conductor. The ribbon microphone utilizes a thin corrugated metal ribbon, or tape, a few thousandths of an inch thick, loosely supported between the poles of a form of horseshoe magnet. When actuated by sound waves, the diaphragm or ribbon oscillates in the magnetic field, which induces a very small current in it. The two ends of the ribbon are connected to the primary of a coupling transformer which steps-up the voltage output and applies it to the grid of a pre-amplifier tube. The ribbon microphone is very rugged and is of rather simple construction; in addition, is capable of high-definition in response with regard to the direction of the sound approach. Being actuated by velocity, rather than by pressure, the high frequency doubling is avoided which in other types of instruments impairs the fidelity. The microphone has an extremely low impedance (less than 1 ohm) and is therefore not affected by radio-frequency fields; on account of this feature the device may be placed close to the transmitter. Unfortunately, the microphone is sensitive to 60-cycle or power line fields if in the proximity of these areas. Low frequencies are unduly emphasized when speaking close to the ribbon. Because the audio output is approximately the lowest of all acousto-electric devices, a high-gain pre-amplifier is required to bring the output up to a usable level. The output is about —100 D.B.

The Dynamic Microphone: This type of microphone operates on the same principle as the ribbon type. However, it uses a small coil of wire attached to a diaphragm to generate the audio voltage. A permanent magnet supplies the magnetic field in most cases, and the audio output and fidelity are similar to those of the condenser microphone. The moving coil microphone is a low impedance device and thus can be remotely located from its associated pre-amplifier. The usual impedance of the moving coils is about 30 ohms and the rated output level is —85 D.B.

The moving coil microphone is rapidly gaining popularity amongst the amateur fraternity. It is quite rugged and has the outstanding advantage that its characteristics do not readily change with age or atmospheric conditions; once the device is equalized, its fidelity remains constant.

The Non-Directional Microphone: This is a type of microphone that will respond uniformly to all sound pressure and is built on the moving-coil principle. It differs radically from previous microphones in appearance, consisting of a two and one-half inch spherical housing with a two and one-half inch acoustic screen held a fraction of an inch off the surface. In this type of microphone the directional effect is so slight as to be imperceptible; this effect is largely a function of the size of the microphone relative to the wavelength of sound.

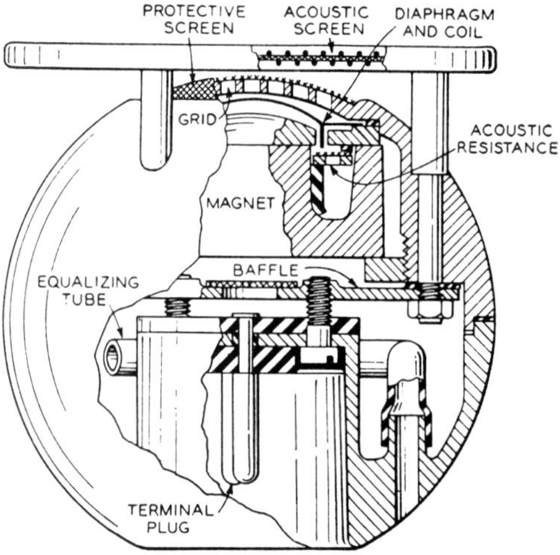

Simplified cross-sectional view of Western Electric non-directional microphone.

With a spherical microphone mounted with the diaphragm horizontal (see the accompanying figure) there would be a tendency for the response to be too high for high-frequency sounds coming down from above; that is, directly toward the diaphragm, and too low for similar frequencies coming from angles very much below the horizontal. The effects are completely avoided, and an essentially uniform response is obtained from sound coming from all directions, by mounting an acoustic screen in front of the diaphragm. This screen produces a loss in sound passing through it, but reflects back into the diaphragm all sounds coming from behind the microphone. It thus compensates for the unequal diffractive effects and makes the instrument non-directional in its response.

The microphone has a uniform characteristic from 40 to 10,000 cycles; it is also free from electrical interference and has such features as: high signal-to-noise ratio, ruggedness, dependability and freedom from temperature, barometric and humidity effects. Another characteristic is the low electrical impedance which allows its use several hundred feet from the amplifying equipment.

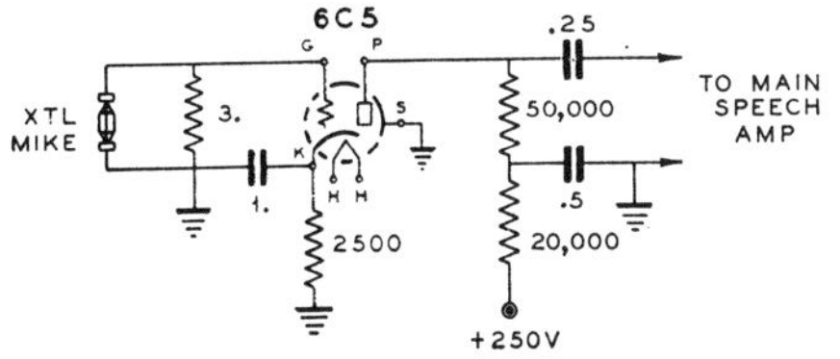

Crystal microphone pre-amplifier. Gain approximately 20 D.B.

Pre-Amplifiers: Practically all types of high-fidelity microphones have a very low audio-output and require an intermediate device between the microphone and main voltage amplifier to build-up the weak electrical output; this device is called the "pre-amplifier." It consists of two stages of resistance or transformer coupled triodes connected in cascade. The overall gain of most pre-amplifiers ranges from 35 to 55 D.B., depending upon the particular type of input microphone used; the decibel rating given here represents a voltage amplification of about 250 to 1000 times. (NOTE: It is almost prerequisite that all amateurs acquaint themselves with the D.B. unit; reference should therefore be

hum pick-up requires good shielding as well as RF chokes in the grid leads, the latter being required if the amplifier is to be operated close to a transmitter. The power supply leads energizing the equipment must be well shielded and not run closer than three feet to any choke or transformer which has AC flowing through it.

The heaters of the tubes can be operated from 6.3 volts AC if care is taken to completely by-pass and shield the filament leads. Otherwise a small storage battery may be necessary if high gain is desired.

Gain controls are seldom incorporated in a pre-amplifier; this function is best left to a voltage divider or attenuator in the main voltage amplifier. (NOTE: Design information on pre-amplifiers may be found in the section "Electrical and Radio Measurements.")

Pre-Amplifier With New Metal Tubes: The performance of a condenser microphone pre-amplifier cannot be improved on to any great extent, but the new developments in parts and tubes allow a more compact and better mechanical design.

From the photographs on page 146, it will be seen that two of the new metal-envelope type 6F5 and 6C5 tubes form major parts complement of the pre-amplifier. Electrically, metal tubes have little advantage over others except for the lower hum level obtained and, in addition, to a slight increase in the gain. These tubes, however, take up less space and greatly simplify the problem of shielding. The compactness of the pre-amplifier is evident from comparison with the microphone head, which is of standard size. The entire two-stage preamplifier fits into a case that formerly housed a "Stromberg-Carlson" audio transformer. The interior partitions are constructed from galvanized iron and then given a coat of lacquer to match the ap-

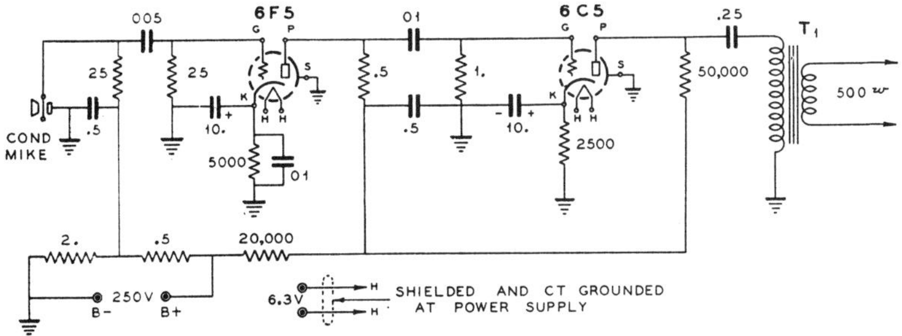

Metal tube pre-amplifier for condenser microphone. Gain approximately 50 D.B. T1—tube-to-line output transformer.

made to the discussion appearing in the section "Electrical and Radio Measurements.")

Since it is the function of pre-amplifiers to be associated with minute audio-frequency voltages, emphasis must be placed upon protecting the amplifier from all hum and background noises. To minimize the

pearance of the case. The mechanical construction is such that all parts in the amplifier are accessible.

A "Mallory" bias cell furnishes the bias voltage on the 6F5. A potentiometer arrangement (see diagram on page 146) reduces the polarizing voltage from 250 to 175 volts.

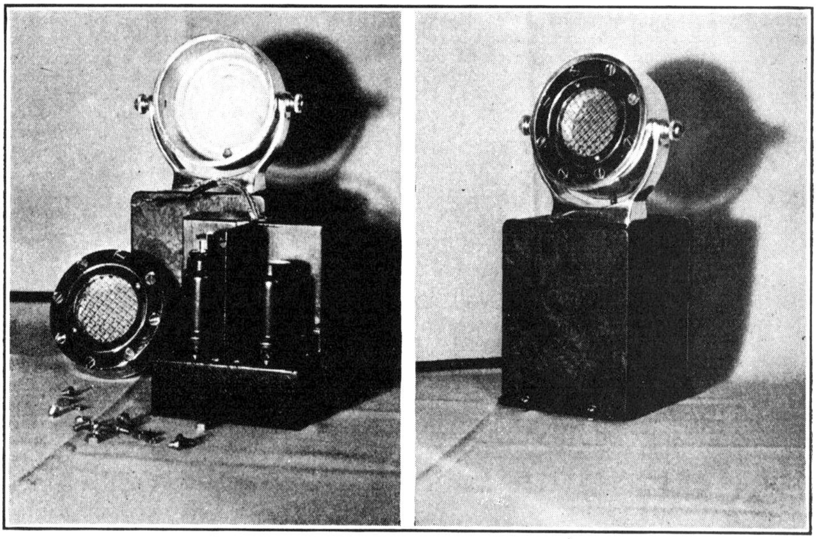

Metal tube pre-amplifier and condenser microphone.

With some types of condenser heads, it is not possible to apply more than 100 volts as a polarizing voltage; the differences in potential can, however, be changed by making the proper adjustments on the potentiometer.

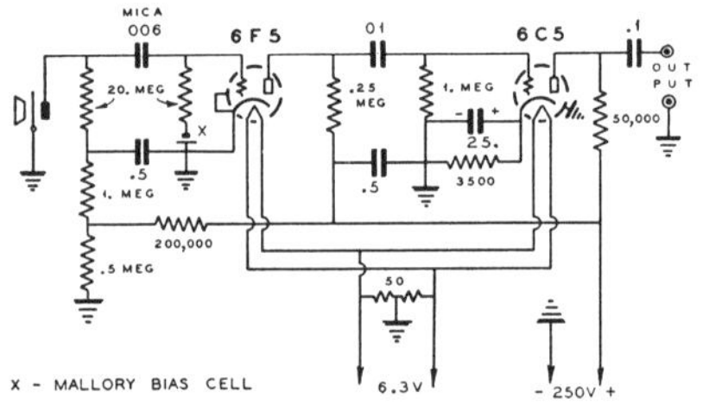

Circuit diagram for condenser microphone amplifier illustrated above. A small dry cell (1.5 volts) can be used in place of the Mallory Bias Cell.

It is recommended that well-designed "noiseless type" resistors be used. In the second stage the amplification level is of such a value that any good grade of carbon resistors will prove satisfactory. The shielded lead from the positive high-potential plate of the microphone head which energizes the grid circuit of the 6F5 must be well insulated between the shielding and the external wire. Considerable noise will be developed in this circuit if this insulation is faulty.

The output impedance of the pre-amplifier is low enough that no line coupling transformers will be necessary for distances up to 100 feet. The output is fed to the grid of the first tube in the main amplifier either through a regular interstage audio transformer or a 0.1 mfd. coupling condenser. The transformer should be located at the input of the main amplifier if transformer coupling is used. If resistor coupling is used, the 0.1 mfd. condenser should be located at the input of the main amplifier even though there is already one blocking condenser in the output lead at the pre-amplifier.

By rigidly following the mechanical arrangement as shown in the photographs, no trouble will be encountered. The filament leads must be shielded and the grid leads short as possible. It is important that the heaters of the tubes be grounded either at the center-tap winding on the filament transformer or by a center-tap resistor scheme.

The power supply must be well-filtered and provided with at least three filter sections with a total of more than 30 mfd. capacity. Three small 50 hy. (or those having higher values) 10 MA. chokes shunted at each terminus with 8 mfd. electrolytic condensers will provide a hum-free source comparable to battery supply.

The Main Voltage Amplifier: The main voltage amplifier is not clearly defined in most amateur stations, but is often combined with the driver stage for the high-powered modulator. Briefly, that part of the audio channel which starts at a point roughly corresponding to the output level of a damped high-quality two-button microphone which is approximately —50 D.B. below the zero level is termed the "main voltage amplifier." Throughout this HANDBOOK, a zero level equal to 6 milliwatts (.006 watts) of power will be used as an

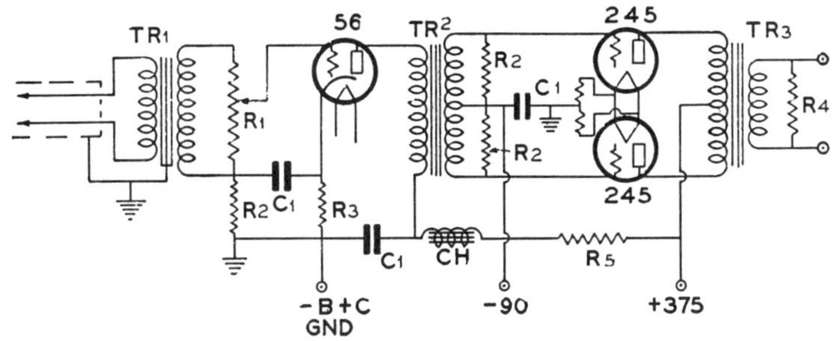

VOLTAGE AMPLIFIER AND DRIVER.

R1—500,000 ohms. R2—250,000 ohms. R3—3,000 ohms wire-wound. R4—1,200 ohms wire-wound. R5—30,000 ohms wire-wound. CH—50 henry, 15 MA. TR1—Plate to Grid Transformer. TR2—P.P. Input Transformer. TR3—Modulation Transformer. 3-1 stepdown.

arbitrary reference level. Thus —50 D.B. corresponds to one-one-hundred thousandth of 6 milliwatts. A good voltage amplifier must be capable of amplifying an input of —50 D.B. up to full output, that is, to the zero level. Such an amplifier can consist of two stages of type 6C6 triodes, transformer coupled.

To control the input to a main voltage amplifier requires the use of some type of attenuating device; this can consist of a potentiometer of about 250,000 ohms whose sliding contact connects to the grid of the first stage.

The main voltage amplifier drives either the "driver power amplifier" or, in certain cases, it directly drives the "low powered modulator." This amplifier operates at a considerably higher audio level than a pre-amplifier, and therefore does not require exceptional filtering or shielding to minimize background noises. However, as a precautionary measure, the first stage should be well-filtered and shielded because of the hum which might occur due to faulty construction or design.

The best inter-stage coupling in amplifiers in which the tube components are either pentodes or screen-grid types is that obtained by resistance coupling. Coupling high-mu tubes with transformers reduces their operating efficiencies to a very low value; and audio chokes, unless specially designed for this service are also taboo. On the other hand, choke or transformer coupling will give splendid fidelity when functioning in conjunction with medium- or low-mu triodes. Resistance-coupling is ideally suitable to the following screen-grid, pentodes or high-mu triodes; these are the 75, 2A6, 40, 6C6 or 57.

The Modulator and Its Associated Driver: A modulator which operates class A does not require that its driver supply power, but instead, it necessitates that its input be supplied with a certain voltage; this is due to the fact that the control grid of a class A amplifier, or modulator, is never driven positive, and so never draws grid current. The driver, therefore, will function splendidly with the following tubes as voltage amplifiers: the 6C6, 56, 76 and

Most Widely Used Main Voltage Amplifier and Audio Driver

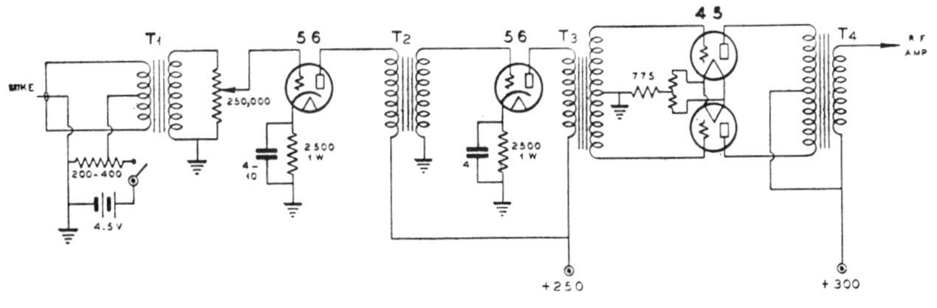

5-Watt Amplifier for use with highly-damped type of carbon microphones.

T1—Mike to grid transformer.
T2—Interstage audio transformer.
T3—Push-pull input transformer.
T4—Push-pull output transformer.

others. With the 845, 849 or 212D operating in a class A modulator, the driver should be chosen from among the small tubes, such as the 45, 42 (triode), 210, etc. The larger tubes operating class A require somewhat more grid voltage swing than can be supplied by the smaller voltage amplifier triodes.

When modulator tubes are employed in push-pull class B or class AB stages, the driver is frequently required to supply some grid power; this amount varies widely depending on the plate voltage and power output conditions. Class B modulators require slightly more grid driving power than class AB modulators, but class B modulators are often somewhat more economical to build and operate.

For example, class B 46s and class AB 50s have the same maximum undistorted power output of approximately 23 watts for two tubes at rated plate voltage. The grids of the 46s require almost two watts of power, whereas the two 50s require only about .4 watt for the same output. Thus another 46 operating as a low-mu triode must adjoin the driver for the class B 46s, while push-pull 76s can easily supply the small grid driving power required by the class AB 50s. However, the 46s are more modestly priced and operate with less plate voltage than the 50s. Offsetting this economy is the fact that the input and output transformers for the class B 46s are more expensive than the input and output transformers for the class AB stage with the type 50 tubes.

The choice of a modulator tube depends on the DC plate input power drawn by the class C RF amplifier which is to be plate modulated. The maximum undistorted audio power output of the modulator stage must be 50 per cent of the DC power input to the class C amplifier.

Reference to the tube tables will indicate the audio output to be expected from the more common modulator tube combinations, at commonly-found plate voltages. The same table will also suggest satisfactory tubes serviceable as drivers.

Low power modulators (up to 200 watts of audio power) often operate from either a single-ended or push-pull driver chosen from the following list of the most popular low power drivers: 45, 46 (low-mu triode), 59 (low-mu triode), 2A3, 71, 42 (triode), 2B6 and 50.

Power Modulation: Power modulation includes all forms of plate modulation because it involves the modulation of the source of power which is converted into RF carrier power by a vacuum tube amplifier. A radio-frequency class C amplifier normally operates under conditions such that the power output changes with the square of the plate voltage; thus the RF voltage output changes in exact accordance with the variation in the plate voltage. Ordinarily, all modulated class C amplifiers operate at a practically constant plate efficiency, but with a peak plate input varying above and below the normal unmodulated value in accordance with the audio-frequency AC supplied by the modulator. The plate efficiency of a plate-modulated class C amplifier can be made quite high; 92 per cent has been reached in laboratory amplifiers, although 65 per cent to 85 per cent is more common in amateur stations.

A study of the power distribution in a completely modulated wave shows that two-thirds of the total power consists of the carrier, and the other one-third is divided equally between the two sidebands. Thus the average RF power output must be increased 50 per cent for complete 100 per cent modulation, and proportionately less for lower percentages.

The plate efficiency remains approximately constant during plate or power modulation, and so the RF power output can be increased only by increasing the plate input power during modulation. In order to derive a 50 per cent increase in average power output during complete modulation, the plate power input must also be increased by 50 per cent. Because the audio-frequency modulator, or modulators, are the sole source of this increase in power, it is seen that the maximum undistorted power output of the modulators must be equal to 50 per cent of the constant DC plate input supplied to the unmodulated class C RF amplifier. The modulator, or modulators, must be coupled in the circuit between the source of DC plate power and the class C amplifier so that the peak AC voltage output and the peak AC current output of the modulators just equals the unmodulated DC plate voltage and plate current drawn by the class C stage. Under complete modulation, therefore, the constant DC plate input is alternately doubled and neutralized as the audio-frequency AC wave goes through its maximum positive and negative values. This shows that the impedances of the load represented by the class C plate circuit and the impedance of the AC power source, which is the modulator tube, or tubes, must be matched to each other if the AC voltages and currents are to exactly double and then neutralize the constant DC voltage and current, which represents the unmodulated plate current input power to the class C amplifier.

Efficiency Modulation: The average plate efficiency must increase 50 per cent during complete modulation of an efficiency modulated RF amplifier, and the plate peak efficiency can never exceed 100 per cent; hence, the unmodulated plate efficiency must be less than 50 per cent.

Efficiency modulated amplifiers include practically all forms of grid modulated amplifiers, whether they are modulated by variable excitation, in which case they are usually termed "linear amplifiers," or whether they are modulated by variable grid bias, in which case they are called "grid bias modulated amplifiers."

Grid Bias Modulation: When the axis of the AC grid excitation voltage is shifted by the audio-frequency modulating voltage, it is termed grid bias modulation. If the control grid of the modulated tube draws any DC grid current, then enough audio must be supplied from the modulator tube to modulate this DC grid current. Frequently this current is quite small in comparison to the DC plate current and a real economy of audio power can be effected by grid bias modulation instead of plate-power modulation. Under certain conditions, the vacuum tube amplifier can be operated so that the control grid draws no DC current, even when most positive, so that the modulator tube need not supply any power to effect deep modulation, as the effective grid impedance is, in that case, very high. It is poor economy to operate a RF amplifier control grid wholly on the negative side of zero bias because the efficiency of the plate power conversion is then low, unless high plate voltages are used together with a tube of exceptionally high mutual conductance. Most grid-bias modulated amplifiers operate so that some DC grid current is drawn, at least on the peaks of modulation.

Screen Grid Modulation: Practically all screen-grid tetrodes and pentodes built at the present time are incapable of complete and linear 100 per cent modulation when the AC modulating voltage is applied to the DC screen voltage.

It is theoretically possible to design a screen-grid pentode which will allow perfect and complete modulation to be effected by cascade screen voltage modulation, but such a tube has not been built to date, and even if such tubes were available, the use of two cascaded efficiency modulated stages would not be economical.

Suppressor-Grid Modulation: Suppressor-grid modulation is used quite extensively among amateurs in the United States. If some means can be found to increase the unmodulated plate efficiency around 40 per cent, suppressor-grid modulation should become universally acceptable, because it is probably the least critical modulation method of any in regards to adjustment.

Summary of Efficiency Modulation: In all known efficiency-modulation systems, the plate power input must remain constant, if linear modulation is desired. The unmodulated plate efficiency could be about doubled if it were possible to make some form of grid-modulated amplifier release its own additional plate input from the DC plate supply source during modulation.

In general, efficiency modulation is characterized by the fact that it is rather difficult to adjust without the aid of an oscilloscope; there is also some question as to whether it is more economical to employ a large tube operating at 35 to 40 per cent efficiency and a minimum of audio equipment, or to use a small high-efficiency class C amplifier stage together with extensive modulator and power supply apparatus.

Class AB Audio Considerations: The best load impedance for class AB tubes is difficult to calculate accurately. As in class B, for a limited grid voltage the output power will be greatest when a plate load is chosen such that the product of plate voltage swing and plate current swing is a maximum. For maximum power with minimum distortion, the load resistance will decrease as the driving power is increased. In other words, with greater driving power the plate current swing on the output tube can be increased and greater power output will consequently be developed across a lower load. This again is governed by the peak current which the plate supply can deliver. It is not good practice to place a low load resistance in the circuit if the plate supply regulation is poor. This and the foregoing factors are of more or less importance, depending upon the magnitude of values in the particular design. However, a general method of determining load impedance for push-pull tubes where the grids are not driven very positive is shown in Figure 1, applied to 845

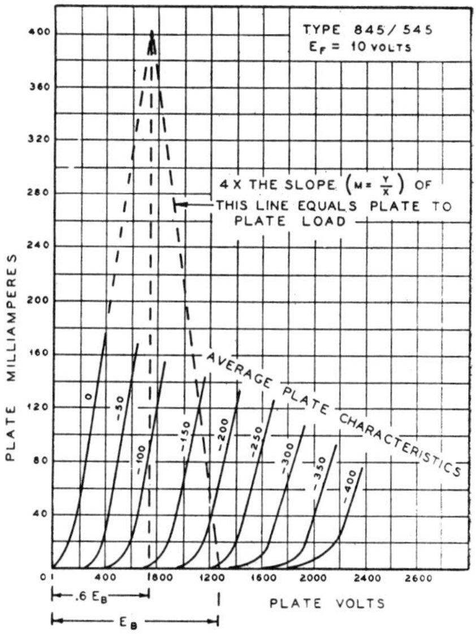

FIG. 1

Method of graphically computing proper load impedance (see text).

tubes. The published plate characteristic must be obtained and an operating voltage E_b selected. A vertical line is erected at $.6E_b$ and the $E_c = 0$ line is extended to meet it. A line is then drawn from the intersection to E_b. The slope of this line multiplied by 4 is the proper plate-to-plate

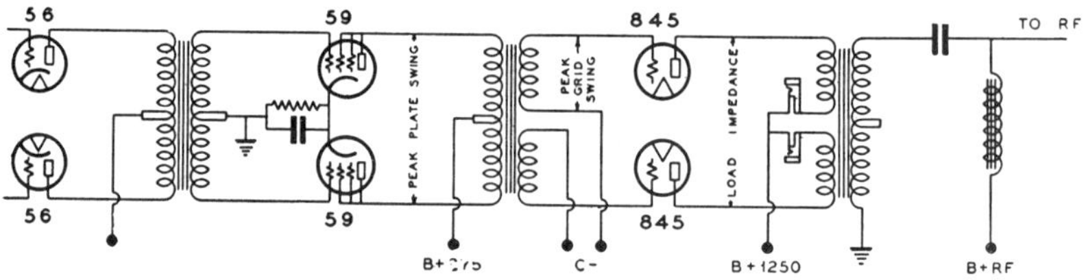

FIG. 2—A-prime 845s used for high level plate modulation.

load. In the example drawn this load is equal to

$$\frac{1250 - 750}{400} \times 4 = 5000 \text{ ohms.}$$

If the grids are driven sufficiently positive to make the normal output about four times that of a single class A amplifier with the same tube, this value of load impedance will have to be reduced about 20 per cent. If the plate supply regulation is better than 10 per cent, this load impedance can be reduced another 5 per cent. In the case shown, this would mean an effective plate load of 3750 ohms. The recommended RCA value is approximately this value.

The calculation of maximum power output is given herewith:

The power is equal to:

$$P = \frac{\text{Max. plate current} \times \text{plate voltage}}{5}$$

As shown in Figure 3, this gives

$$\frac{.40 \times 1250}{5} = 100 \text{ watts}$$

From a summary of the above notes it will be seen that AB amplification is a system lying between class A and class B; high biased near cut-off. Not all tubes are suitable for this class of service. Those most applicable are the 42, 245, 2A3, 250, 845, WE, 283A, 212D and E, and 849.

Class B Audio Considerations: For obtaining high quality amplification from a class B amplifying system requires the consideration of the following precautions:

(1) The driver stage must be able to supply about two or three times the actual power required to drive the grids of the class B stage. This reserve power is necessary so that the driving voltage shall have good regulation under the variations in the load represented by the class B grids. In general, the driver output should be from 5 to 15 per cent of the class B stage.

(2) The class B input transformer must have sufficient step-down so that the driver load impedance never goes below the plate impedance of the driver tube, when the class B grids are most positive. It follows that less step-down is necessary when the

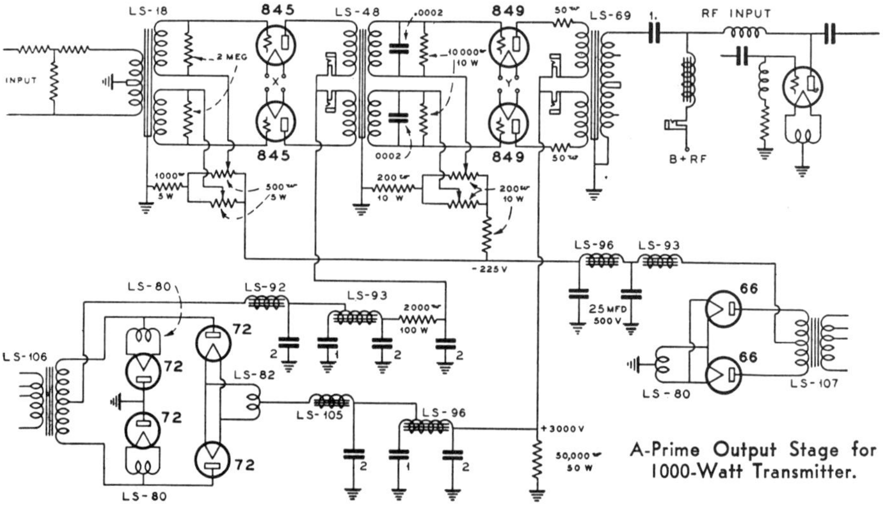

FIG. 3—The 849 bias should be adjusted so that the no-signal plate current is 40 M.A. per tube for class A-prime operation, or 10 M.A. per tube for class B operation. Other constants are not altered.

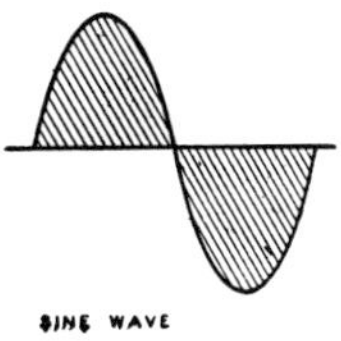

FIG. 4

FIG. 5

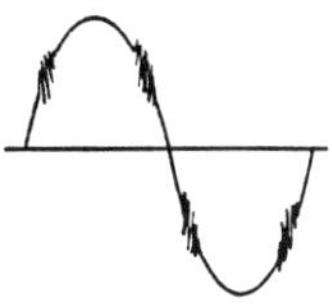

FIG. 6

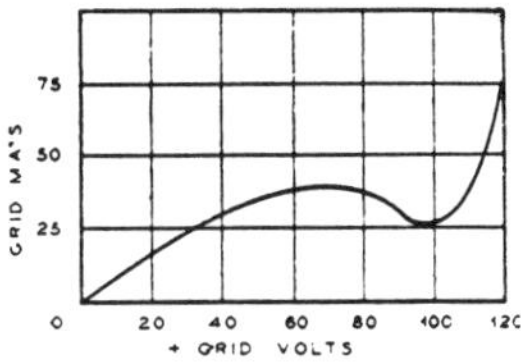

FIG. 7

class B tubes have a high grid impedance. By the same token, the choice of the driver tube with a low plate impedance, such as the 45, 50, 2A3, 2B6, and 42 triode, is necessary for minimum step-down ratios.

(3) The load impedance into which the class B stage works must be fairly high in comparison with the plate impedance of the class B tubes. The actual value of load impedance is not especially critical, and for practically all common tubes it can be between 5,000 to 20,000 ohms plate-to-plate.

When the plate load impedance of a class B stage is varied, the following action occurs: As long as the load impedance exceeds the static plate resistance of the tube, an increase in load impedance will improve the quality by reducing the harmonic distortion; in addition, the power output will be reduced for a given grid excitation, and therefore more energy will be required for the same power output, with higher loads. The plate efficiency increases as the load impedance is increased, so that more output can be delivered for a fixed plate loss by merely augmenting the grid drive. However, as the load impedance and the grid drive are increased, it is necessary to raise the plate voltage to prevent the maximum grid voltage from exceeding the minimum plate voltage, at the peaks of the grid drive.

(4) The two halves of the circuit must be accurately matched. Because each class B tube works for only half the cycle, it is essential that they receive exactly the same driving voltage and that they each draw the same plate current in the resting condition. No two tubes will maintain their characteristics for any length of time and it is essential that individual bias adjustments be provided so that the stage can be balanced. This precaution is only applicable with other than zero grid-bias tubes, which include the 46, 59, 19, 49, 89, 53, 79, NCB, RK31 and 838.

(5) The plate power supply must have good voltage regulation because the plate current varies quite widely with the grid drive. Any variation in plate voltage with changes in plate current will cause amplitude (harmonic) distortion, and is to be avoided. Low resistance windings on the power transformer and filter chokes are essential. The use of a saturated, or swinging, input choke helps to keep the output voltage constant with variation in current. Mercury-vapor rectifiers have an inherent voltage drop that is independent, to a great extent, of the load current, and thus cause no sacrifice of regulation, as is the case with thermionic rectifiers.

(6) With certain tubes, notably some makes of 210s, 203As, 211s, 800s, 204As and 849s, it is essential to take precautions against dynatronic distortion (see Figure 6). This type of distortion occurs when a stage starts to oscillate on the peaks, which gives rise to a rasping effect which greatly impairs the quality. This tendency toward oscillation is caused by a "dynatronic kink" in the grid characteristic of the tube; it can be "swamped-out" by placing 50-ohm parasitic resistors in each grid lead, combined with 5000 to 20,000 ohms shunted across each half of the input transformer secondary. Sometimes it is even necessary to shunt each side of the input transformer with .0001 ufd. condensers.

It is common practice to consider the average or effective audio power necessary to modulate 100 per cent, but the peak audio power is the correct accounting factor. When the peak audio voltage and power reaches a value equal to the DC input voltage and power on the modulated amplifier, 100 per cent modulation is attained. The average audio power at this point is of a value that is not known unless the wave form of the audio is known. The wave forms of voice or music are very complex and the effective power in them is much less than in a sine wave of equal peak voltage, although the peak voltage and peak power are the same. Because direct-current meters read average values, it is difficult to determine when the peak current has reached the correct value for 100 per cent modulation, the average values for voice and music being lower than for a constant sine wave input. The average audio power with a sine wave of constant amplitude necessary to modulate a carrier 100 per cent is 50 per cent of the DC input to the class C amplifier. But, with voice or music, the average power necessary is considerably less.

The shaded areas of Figures 4 and 5 show the average power in two different wave forms of equal peak voltage and power. Figure 4 shows the power in a pure sine wave with no harmonics. Figure 5 shows the power in a wave of the same fundamental frequency with a strong second harmonic. The aggregate or combined peak power of the wave is equal to that of Figure 4, but the average power over the entire cycle is much less.

The illustrations indicate that under certain conditions the output from a given tube or tubes is greater with a normal voice input than one having a constant tone.

Because the "saturation plate current" is the value flowing on peak audio swings

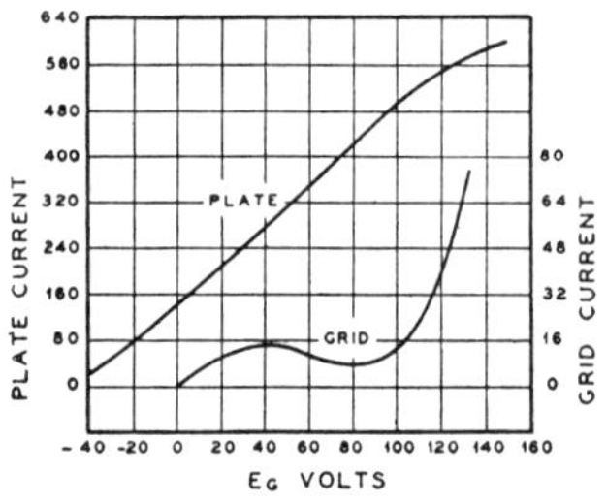

FIG. 8—203A operating characteristics.

during maximum output, further excitation will cause distortion. However, by increasing the plate voltage and load impedance, the power output can be raised up to such factors as that which limits the insulation of the tube terminals and the stem seal. Tubes with the plate-lead coming out of the top are ideal, as the plate voltage can be increased to a high value, resulting in higher audio outputs.

The grid voltage and grid current characteristics of a tube are the most important insofar as the quality or fidelity of the class B amplifier is concerned. When the grid goes positive, grid current flows, and if this curve is a straight line, little trouble will be encountered; however, such is seldom the case. As the grid becomes more positive, the grid current rises more rapidly until finally the grid current curve becomes almost vertical. Some tubes have a negative grid current slope such as shown in Figure 7. This gives rise to transient oscillation of the dynatronic variety and occurs only during a portion of the audio cycle. These parasitic oscillations cause a sort of "fuzz" to appear on the output. They can be analyzed only with the aid of a cathode-ray oscillograph; the transients are too fast to be recorded on galvanometric mirror type. In practice, the type 203A tube usually produces the spurious oscillatory effect; it can be reduced by placing a small capacitor (0.0001 ufd.) from grid to ground of both tubes, or by incorporating some scheme of neutralization.

Apparently it would seem that a tube of high amplification factor would be the most appropriate tube for class B service, due to the lower value of excitation voltage necessary, but actually more power is required to excite a pair of 203As to 200 watts output than a pair of 211s. The grid current rises to a higher value and there is a greater grid loss in the 203A type than in the 211. Of course, the C bias supply for the lower-mu type of tubes must be given consideration. Owing to the much lower grid current surges, the C bias supply can be of a type of small power supply employing an 83 rectifier, whereas if the C bias supply were to be used on 203As, it would have to maintain a constant of 30 volts at current changes as high as 75 milliamperes. Practically the same power output can be obtained with any of the 100 watt type tubes, such as the 203A, 211, 845, provided the proper excitation is applied. The high-mu types require lower excitation voltage, but better voltage regulation of the driver output is needed. The low-mu type tubes require more excitation voltage, but because of lower grid current the source does not need such good regulation. The tubes of medium-mu are usually the best, all points considered.

The Transformers: Many types of audio distortion can be produced in a class B amplifier if the transformers have been improperly designed. The input transformer must deliver perfect quality to the class B grids, even though the grids are drawing current from zero to maximum during any one audio cycle. The grids of the class B tubes offer a load that fluctuates from infinity down to several hundred ohms. It is therefore requisite that the input transformer supply a perfect reproduction of the signal wave-form without distortion, even though the load is of a varying character.

The driver must be capable of delivering sufficient power to maintain the grid voltage swing with the current of the class B tubes flowing through the secondary of the input transformer; furthermore, the secondaries must have a very low DC resistance so that the bias on the class B tubes does not vary appreciably with the grid current. This fault is common with most input transformers. All these points must be maintained with a fair degree of constancy over the entire frequency range.

The coils must be designed so that the primary has identical relationship with both halves of the secondary. The capacity and the leakage reactance must be the same for the primary and each secondary. If these precautions are not taken, the wave form of the voltage supplied to the class B grids is not the same for each grid and distortion of the wave form occurs, giving rise to harmonic distortion.

The input transformer must have a step-down ratio of such a value that the signal voltage applied to the class B grids is just sufficient to give the required output. This improves the regulation of the driver output voltage.

Most output transformers are designed to carry the current of the modulated stage, but this practice is not advisable if the best quality is to be had.

When the secondary is made to handle high currents, a large air gap in the core is necessary to prevent saturation. This, in turn, necessitates increasing the number of turns on the coils, which increases the DC resistance, the leakage inductance, and the distributed capacity to a point where the frequency response is impaired over a large portion of the frequency spectrum. When the secondary carries no DC, the core can be assembled without an air-gap, resulting in much better quality and greater output.

It is very important, however, when this

practice is followed, that the tubes have like characteristics and are adjusted to identical static plate currents. The output of a single tube working class B consists essentially of a fundamental and a series of even harmonics, chiefly the second harmonic. If two tubes are properly balanced in a push-pull circuit, the output will be free from even harmonic distortion. A correctly designed output transformer has a core of such dimensions that the flux density at peak plate current will be close to the upper bend of the B & H curve, in other words, close to saturation. Unless this is done, the incremental permeability will fall to a very low value for low percentage modulation, with a resulting loss of low frequencies. The unbalanced plate current will swing the iron through different ranges of flux density on alternate half cycles, producing high amplitude harmonics. These harmonics can produce severe overmodulation and cause the carrier to "splatter" over a wide frequency band, even though the fundamental frequency is modulating less than 100 per cent.

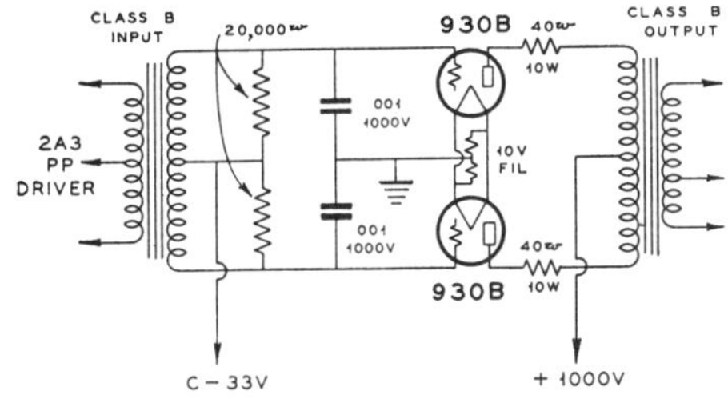

150-200 watt class B modulator with 930-B tubes in push-pull

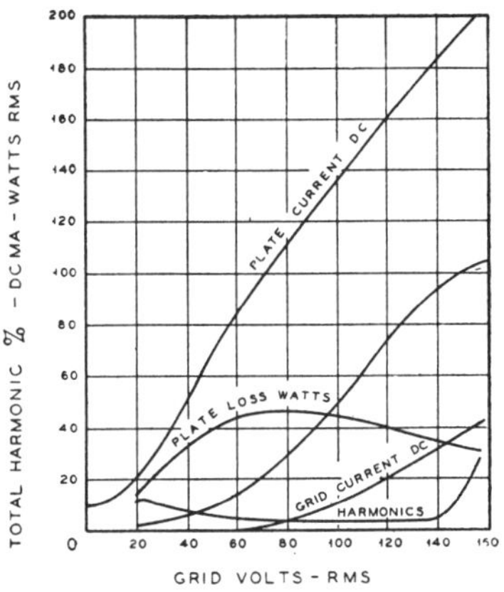

FIG. 10—Characteristics of a class B amplifier using push-pull 801 tubes.

In a class B circuit only one tube conducts at a time, so it is assumed that one tube is supplying all the power during one-half cycle and the other tube is supplying no power. The rated safe plate loss averaged over any audio-frequency cycle is 20 watts, which is a conservative value. Considering that either tube is supplying no power one-half of the time, the plate loss can be increased to double the rated value during the half cycle it is working. This would mean an average loss during the half cycle of 40 watts. The maximum loss does not occur at maximum plate current but usually near zero grid voltage, so that the plate voltage, plate current curves of the tube must be consulted to determine the average plate loss. The average plate loss of the 801 is computed from the curves

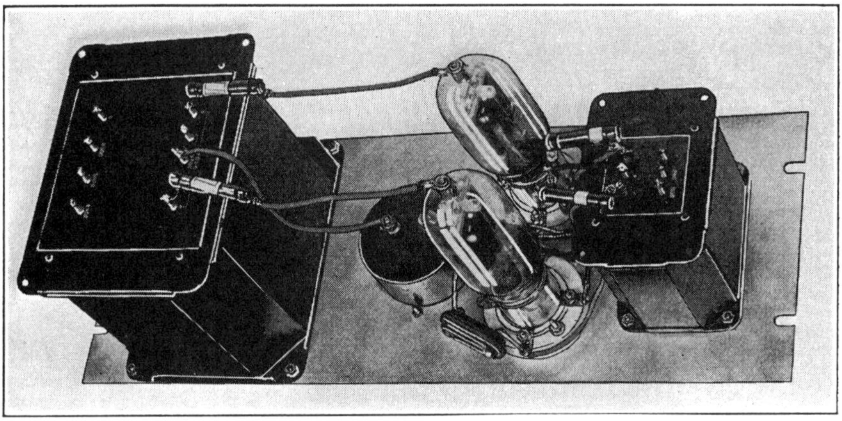

930-B (or 830-B) modulator shown in diagram above.

supplied by the manufacturer. The plate supply voltage is 750 volts, and the peak or maximum plate current is 250 MA. The plate losses for the different grid voltages are given below:

Average loss 36.2 over one-half cycle or 18.1 watts loss over the whole cycle. This leaves some margin of safety below the rated 20 watt dissipation. Using the equation.

$$I_p\ MAX^2\ R_L = PO_{max},$$

where R_L represents the load impedance, and I_p MAX, the peak plate current.

With the peak plate current of .250 ampere and a load of 2500 ohms, the result is 156.2 peak watts, or an effective power of 78.1 watts, which will modulate a DC input into a class C amplifier of 156.2 watts.

This output is possible with a pure sine wave of constant amplitude and with the tubes operating below rated plate loss, consequently, with normal voice or music input, the average or effective power being less than with constant sine wave input, the peak power can be increased to a value where the average plate loss is equal to the rated value. It must be understood that when operating tubes under these conditions the input to the amplifier should be normal speech or music, not sustained notes of appreciable duration and high amplitude, which would cause an average plate loss above the rated value.

Class C Amplifier Load Calculations: The correct terminal or output impedance of a class C amplifier is important for plate-modulated phone transmitters. A class B modulator must be matched correctly to the stage being modulated and the output transformer secondary to primary turns ratio is the method generally used. For example, if the class C stage is operating at 400-volts plate supply and drawing 110 MA. under load, the impedance to the modulator is 400/.110, or about 3600-ohms. The class B transformer then, must have a turns ratio such that this value of impedance will be correctly transformed for the class B tubes. This impedance ratio varies as the square of the turns ratio, hence, if the class B tubes work into a 5800-ohm load (class B 46 tubes), the output transformer must step-up the 3600-ohm load to 5800 ohms. The impedance ratio is 5800/3600, or 1.6. The turns ratio would be the square-root of 1.6 or 1.26, and the output transformer would require a step-down turns ratio of 1.26 to 1 into the load.

Quite often the class C load is higher in value so a step-up ratio is needed in the output or modulation transformer. For example, a class C 50T tube operating at 2000-volts plate supply at 100 MA. The impedance is 2000/.100, or 20,000-ohms. The class B audio power required is ½ × 2,000 × .1 or 100 watts. This power can be obtained from a pair of RK31 tubes at 1000 volts plate supply. These tubes require a load of 3400-ohms per tube or a total of 6800-ohms from plate to plate: 20000/6800 = 2.94 impedance ratio, with a turns ratio of $\sqrt{2.94}$, or approximately 1.7 to 1 step-up turns ratio. The secondary in this case would have 1.7 times as many turns as the whole primary winding.

25-Watt 160 Meter Phone Transmitter: Two typical 160 meter phone transmitters for newcomers are shown on these pages. In the illustration below, a 53 or 6A6 push-pull oscillator drives a pair of 45s in the final amplifier. The circuit diagram is shown in Fig. 12. Approximately 25 watts output is delivered by the final amplifier. These transmitters should be used only on the 160 meter band. If they are to be operated on the higher frequencies, a buffer stage should be added. The coil winding data for the oscillator is the same as that shown for the single 53 push-pull c-w transmitter described in the c-w chapter, i. e., 60 to 70 turns of No. 22 DSC wire, close wound and center-tapped, on a 1½-in. dia. plug-in coil form. The larger plug-in forms, 2¼-in. diameter, can also be used, but fewer turns will be required. 55 turns of No. 22 DSC close wound will suffice. The oscillator coil should be tuned with a 100 mmf. or 150 mmf. condenser. A reasonable amount of "C" should always be in use. Remove turns from the coil and increase the tuning capacity, so as to give a certain amount of lee-way in adjusting the 160 meter circuit.

Link coupling is used between the oscillator and the amplifier. The coupling loops have two turns each. One of the 2-turn loops is coupled around the center of the oscillator coil, the other around the lower end of the amplifier grid coil. The circuit shown at the bottom of page 155 (Fig. 13) uses a 47 crystal oscillator, link coupled to a pair of 45s in the final. It is just as satisfactory as the circuit shown in Fig. 12. From 60 to 70 turns of No. 22 DSC, close wound on a 1½-in. dia. coil form is suitable for the oscillator coil winding. The final amplifier plate coil has 55 turns, tuned with a 100 mmf. or 150 mmf. condenser.

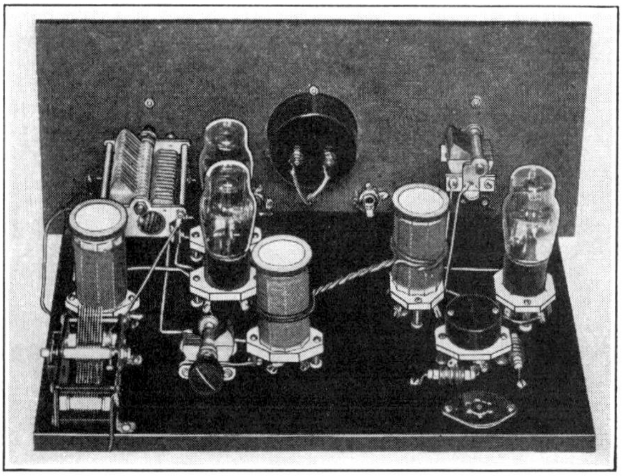

FIG. 11—160 meter phone, 53 or 6A6 and two 45s. See Fig. 12 for circuit.

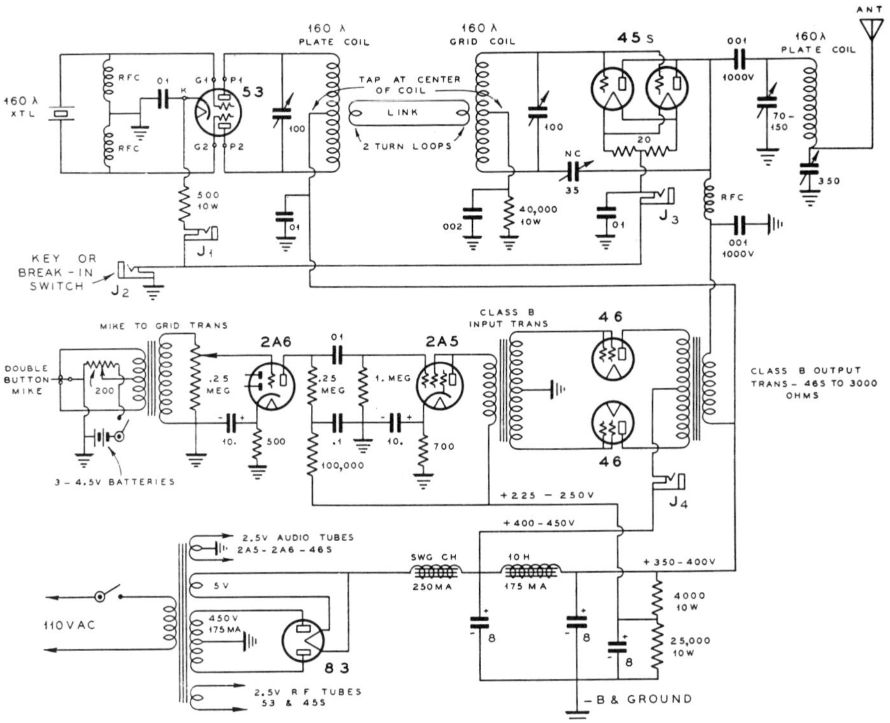

FIG. 12—R-F, speech and power supply for modern 160 meter phone, pictured in Fig. 11. All values are clearly shown.

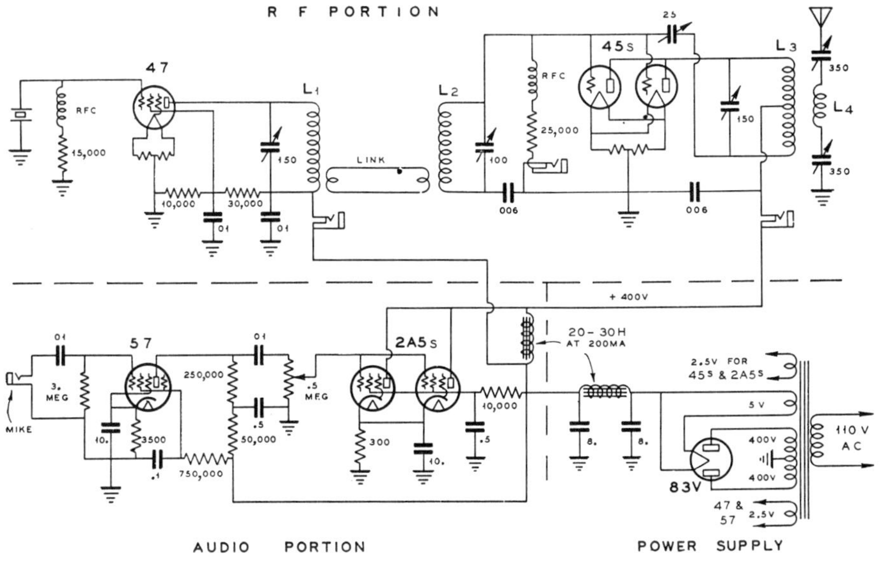

FIG. 13—Practical circuit for 25-watt phone with 47 oscillator link coupled to two 45s.

47-46-210 Phone—25 to 50 Watt Carrier

This is a standard low-power transmitter for operation on 160, 75 or 20 meters. It is extremely simple to construct. The C.W. portion of the transmitter should be constructed first. It is pre-requisite that the 46 and 210 stages be both perfectly neutralized. Next, the audio channel can be constructed and tested. The test simply consists of energizing the amplifier by either phonographic or radio program material; during the procedure the output ter-

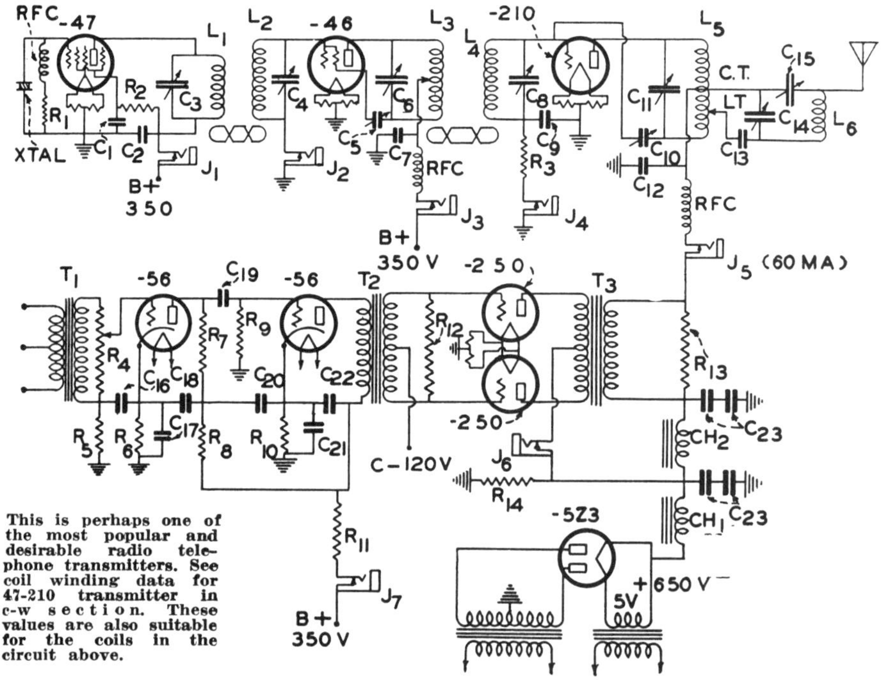

This is perhaps one of the most popular and desirable radio telephone transmitters. See coil winding data for 47-210 transmitter in c-w section. These values are also suitable for the coils in the circuit above.

Conventional phone transmitter of standard design.

minus R-13 should be shunted with a pair of headphones for aurally detecting the fidelity and over-all response. The audio modulation should not be applied to the plate supply of the class C 210 stage until both portions of the transmitter are working 100 per cent. The plate currents, read at the various plate current telephone-jacks, must remain constant during modulation, with the exception of the modulator plate current read at jack J6. The speech amplifier has ample gain to work out of the average double-button carbon microphone. If any of the low-level microphones are used, a pre-amplifier must be added and coupled to T1 in the conventional manner.

Legend for 47-46-210 Phone circuit on page 156.

R1—25,000 ohms. R2—30,000 ohms. R3—15,000 ohms. R4—200,000 ohms (tapered pot.). R5—100,000 ohms. R6—2,500 ohms. R7—100,000 ohms. R8—10,000 ohms. R9—500,000 ohms. R10—2,500 ohms. R11—5,000 ohms. R12—200,000 ohms. R13—100,000 ohms, 10 watts. R14—30,000 ohms, 100 watts. C1, C2, C7, C9—.001, C3, C4, C5, C6, C8—100 mmf. C10—35 mmf. double spaced. C11—50-70 mmf. double spaced. C12, C13—.006 mica. C14, C15—350 mmf. single spaced. C16—½ mfd. C17 to C22—Any value from ½ mfd. to 2 mfd. C23—16 mfd. 450 v. (2-8 mfd. units in series). T1—Mike-to-grid transformer. T2—Plate to push-pull grids. T3—Class A-prime output, 1.25-to-1 step-down. CH1 15 henrys, 200 MA. CH2—30 henrys, 75 MA. Two power transformers are required, one 650 to 800 v., one 1200 to 1400 v., center-tapped.

Battery Operated Phone: Many amateurs who reside in localities where no electric power is available will find this transmitter of inestimable value. The construction is such that the transmitter will operate on both 80 and 160 meter bands. The carrier output is about 2½-watts with 135-volt plate supply, and distances over several hundred miles can be attained when there is a minimum of QRM.

Circuit Details: The circuit is quite conventional and the designer will not encounter any difficulties in wiring and assembling the various components providing the arrangement, as shown in the photograph, is followed.

For 160-meter operation the oscillator and grid coils are wound with 72 turns of No. 22 DSC wire, close wound on a 1½-inch diameter form. The plate coil is wound with 44 turns of No. 18 DSC wire on a 2¼-inch diameter plug-in coil form. The two smaller coils are center-tapped.

For 80-meter operation the two smaller coils are center-tapped and are wound with No. 22 DSC wire on a 1½-inch diameter form, 30 turns about 1½ inches long. The plate coils are wound with No. 16 enameled wire on a 2¼-inch form with 28 turns about 2½ inches long. The plate coils for either band must sometimes be either shortened or lengthened slightly for certain antenna characteristics.

The oscillator grid leaks are not critical in value; 30,000 ohm 1-watt resistors give about the same or slightly more output than those values indicated in the wiring diagram.

Because of Federal regulations, an overmodulation indicator monitors the modulation quality. The improper condition is denoted by the slightest deflection of the meter needle during modulation. Talking in a lower voice or at a greater distance from the microphone, will prevent over-modulation when trouble is experienced from this source.) The RF pick-up coil is loosely coupled to the antenna coil by means of flexible leads.

The transmitter is adjusted in the conventional manner or by following the instructions given in the sub-topic heading "Transmitter Adjustments."

The complete battery operated phone and modulation indicating device.

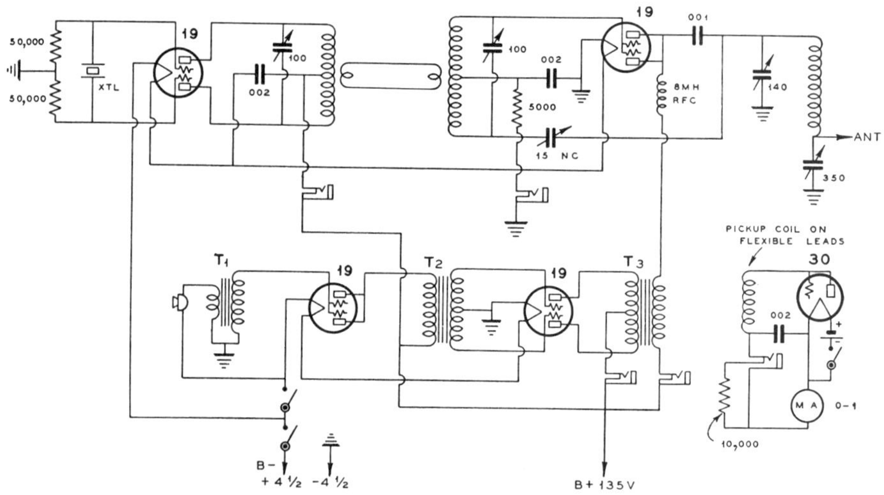

Circuit diagram of the 2½ watt battery operated phone. T1—Mike-to-grid transformer. T2—Class B input transformer, about 3-to-1 primary to each secondary. T3—Class B output transformer for 19 tube to 6,000 ohm load.

The antenna circuit is tuned by connecting any fairly long antenna wire and ground or counterpoise to the antenna connection and ground, and by varying the 350 and 140 uufd. tuning condensers. The 140 uufd. condenser is always tuned for minimum plate current; heavier antenna load increases the minimum reading.

Notes: The No. 14 gauge aluminum grounded shield, shown in the breadboard transmitter, prevents RF feedback into the grid or oscillator circuit and isolates the modulator apparatus. Some of the small parts are under the breadboard; the dimensions of the latter are 16 in. x 11 in. x 1 in. Isolantite sockets are used throughout; bakelite sockets may be substituted with a slight sacrifice in efficiency.

The daylight range of this phone transmitter is limited to a few miles, due to the low power output.

Coil Winding: For 160 meter operation the oscillator and grid coils are wound with 72 turns of No. 22 DSC wire, close-wound on a 1½-inch diameter form. The plate coil is wound with 44 turns of No. 18 DSC wire on a 2¼-inch diameter plug-in coil form. The two smaller coils are center-tapped.

For 80 meter operation the two smaller coils are center-tapped, and wound with No. 22 DSC wire on a 1½-inch diameter form, 30 turns about 1½ inches long. The plate coil is wound with No. 16E wire on a 2¼-inch diameter form, with 28 turns about 2½ inches long. The plate coils for either band must sometimes be "pruned" slightly for the correct number of turns for a given antenna.

The 14 gauge aluminum grounded shield, shown in the breadboard transmitter, prevents RF feedback into the grid or oscillator circuit and isolates the modulator apparatus. Some of the small parts are under the breadboard. The dimensions of the oak baseboard are 16 inches x 11 inches x 1 inch. Isolantite sockets are used throughout. Good bakelite wafer sockets can also be used, with only a slight sacrifice in efficiency.

Typical Readings for 80 and 160 meter operation

Meters	B. Bat.	I. Osc.	I. Grid.	I. Amp.	Carrier Output	Efficiency Final Amplifier
160	138	16	6	25	2.6	74%
80	138	15	5	23	2.4	74%
80	185	20	8	30	4.5	80%

Table Showing Audio Output of Modulator Tubes

	Audio Power	Type of Tube	Class	Driver Tube
Plate Volts 400	4 watts	1—2A5 or 42	A	56, 57, etc.
	4 W.	1—250	A	56, 27, 12A, etc.
	25 W.	2—46s or 59s or 53s. PP.	B	PP 45s
	40 W.	4—46s in push-pull Par.	B	PP 2A3s
Plate Volts 450	4.6 W.	1—250	A	56 or equiv.
	15 W.	2—250s in push-pull	A	56 or equiv.
	30 W.	2—46s, etc., push-pull	B	PP 45s
	50 W.	4—46s, etc., push-pull Par.	B	PP 2A3s
Plate Volts 500	5 W.	1—250	A	56 or equiv.
	7.5 W.	1—845 or WE284A	A	210 or 45
	35 W.	2—210s or 841s. PP.	B	PP 45s
	40 W.	2—46s. pp.	B	PP 45s
	80 W.	4—46s. ppp.	B	PP 2A3s
Plate Volts 600	4.5 W.	1—WE211D or E	A	210 or 45
	8 W.	1—845 or WE284A	A	210 or 45
	30 W.	2—250s. PP.	AB	45 or 250
	40 W.	2—210s or 841s. PP.	B	PP 45s
Plate Volts 750	5.5 W.	1—RK18 or WE211E	A	56 or equiv.
	6.25 W.	1—211 or WE242A	A	56 or equiv.
	10 W.	1—845	A	210 or 45
	15 W.	1—WE284A	A	210 or 45
	50 W.	2—210s or 841s. PP.	B	PP 2A3s
	70 W.	2—RK18s or 801s	B	PP 2A3s
	75 W.	2—WE211D or E, push-pull	B	PP 2A3s
	90 W.	2—50Ts, push-pull	B	PP 2A3s
Plate Volts 1000	8.5 W.	1—RK18	A	56 or equiv.
	10 W.	1—354 or 150T	A	210 or 45
	10 W.	1—211 or WE242A	A	210 or 45
	15 W.	1—845	A	210 or 45
	25 W.	1—WE284A	A	210 or 45
	40 W.	2—930s in push-pull	AB	PP 45s
	100 W.	2—800s, push-pull	B	PP 2A3s
	100 W.	2—WE211D or E, push-pull	B	PP 2A3s
	150 W.	2—50Ts	B	PP 2A3s
	175 W.	2—WE284As or 845s	AB	PP 2A3s
	200 W.	2—203As, 838s, or 211s	B	PP 2A3s
	250 W.	2—150Ts, HF200s or 534s	B	PP 2A3s
Plate Volts 1250	18 W.	1—211 or WE242A	A	210 or 250
	30 W.	1—845	A	210 or 250
	30 W.	1—WE284A	A	210 or 250
	106 W.	2—800s in push-pull	B	PP 2A3s
	125 W.	2—WE284As in PP.	AB	PP 2A3s
	175 W.	2—50Ts	B	PP 2A3s
	200 W.	2—211s or WE242As. PP.	B	PP 2A3s
	225 W.	2—203As or 838s	B	PP 250s
	225 W.	2—845s in push-pull	AB	PP 2A3s
	400 W.	4—203As, push-pull parallel	B	PP 2A3s
	275 W.	2—150Ts, HF200s or 354s	B	PP 250s
Plate Volts 1500	35 W.	1—WE212D	A	210 or 250
	75 W.	1—HK255	A	203A or 211
	165 W.	2—50Ts. PP.	B	PP 2A3s
	240 W.	2—845s, push-pull	AB	PP 2A3s
	350 W.	2—WE212Ds in push-pull	AB	PP 250s
	350 W.	2—150Ts, HF200s or 354s	B	PP 2A3s or 250s
Plate Volts 1750	87.5 W.	1—HK255	A	203A or 211
	400 W.	2—354s, HF200s or 150Ts	B	PP 250s
	400 W.	2—WE212Ds in push-pull	B	PP 845 or equal
Plate Volts 2000	35 W.	1—354, HF200 or 150Ts	A	210 or 45
	40 W.	1—204A	A	210 or 45
	60 W.	1—849	A	210 or 45
	100 W.	2—354s, HF200s or 150Ts	A	210 or 45
	100 W.	1—851	A	210 or 45
	100 W.	1—HK255	A	203A or 211
	500 W.	2—354s, HF200s or 150Ts, push-pull	B	PP 250s
	500 W.	2—204As in push-pull	B	PP 845 or equal
	600 W.	2—849s in push-pull	B	PP 845 or equal
Plate Volts 2500	600 W.	2—204As in push-pull	B	PP 845 or equal
	600 W.	2—849s in push-pull	B	PP 845 or equal
	600 W.	2—150Ts, 354s or HF200s in PP.	B	PP 250s or equal
	750 W.	2—WE270As, push-pull	B	PP 845 or equal

75-Watt Modulator: More than 75 watts of audio output power is supplied by the unit described herein. This power is sufficient to plate-modulate a 211 or 203A operating at 1000-volts with 150 to 160MA of plate current. The device will economically modulate an input of at least 150-watts, or will fully modulate a 100-watt carrier at 100 per cent.

The advantage of using four 46s, rather than a pair of 801 tubes in this class B audio circuit, is in the lower cost. No C-bias is required with these tubes, hence, only a very modestly-priced power supply is needed.

75-watt class B modulator for modulating any class C stage with inputs up to 150 watts.

watt resistors must be inserted in series with the plate leads to prevent parasitic oscillation.

If the microphone is of the diaphragm-crystal-type, the input transformer should be eliminated and the first 57-tube connected as a high-gain pentode with a 250,000 ohm plate-resistor, 2 megohm series screen resistor and a .1 ufd. by-pass, and a 3500 ohm cathode resistor. The input should be shunted with a 3 or 5 megohm resistor, as well as being fully shielded.

The cases of the transformers, the volume control mounting bracket, and either center or one side of the input transformer

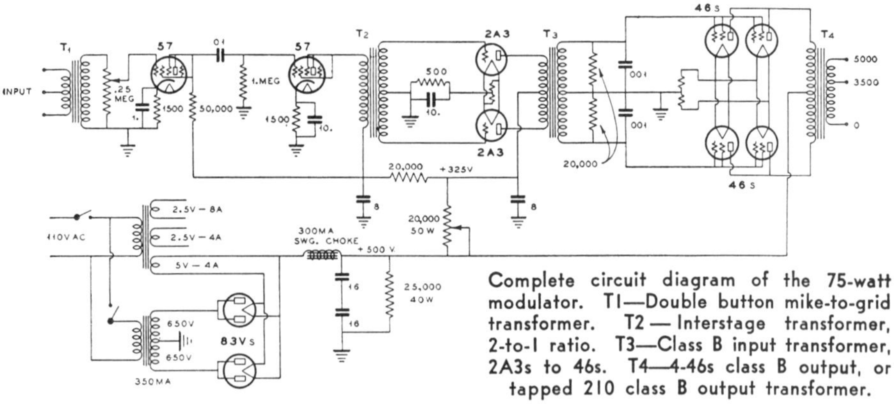

Complete circuit diagram of the 75-watt modulator. T1—Double button mike-to-grid transformer. T2 — Interstage transformer, 2-to-1 ratio. T3—Class B input transformer, 2A3s to 46s. T4—4-46s class B output, or tapped 210 class B output transformer.

Technical Details: The 20,000 ohm grid resistors across the input class B transformer provides a more constant load to the driver stage. The .001 1000-volt mica condensers across each grid to ground tend to prevent high-frequency oscillation under conditions of high input. In some cases, 40 ohm 10-

must be grounded. In cases of audio-frequency feedback, it is suggested that the transformer connections be reversed. Often the input transformer and tube must be double-shielded to prevent RF feedback or hum pick-up.

Other than the power supply filter, additional filtering is given to the 2A3 stage by means of a semi-variable 50-watt 2000-ohm resistor and another 600-volt 8 ufd. condenser. The voltage at this point is between 325 and 350 volts. Further filtering in the form of a 20,000 ohm 5-watt resistor and an 8 ufd. condenser is imposed on the two 57 stages. This in effect gives three sections of filter for the first two stages of hum-free audio amplification. Triode 57 stages give more gain than 56 stages.

The amplifier baseboard is 24 in. x 11 in. x 1 in. with a cleat at each end so that resistors, wiring and terminal strips can be mounted underneath. The power supply baseboard is 20 in. x 8 in. x 1 in.

The total plate current drain is about 150MA at 500 volts with no speech input, and on speech peaks the meter should not read more than 300MA. By exercising these precautions, the 46 tubes will never operate at more than about 160MA on voice input. On a steady sine wave tone input the plate current may increase to a total value of 350 to 400MA for the same peak power output.

Power supply for 75-watt modulator.

High-power modulator using RCA-838 tubes in final stage.

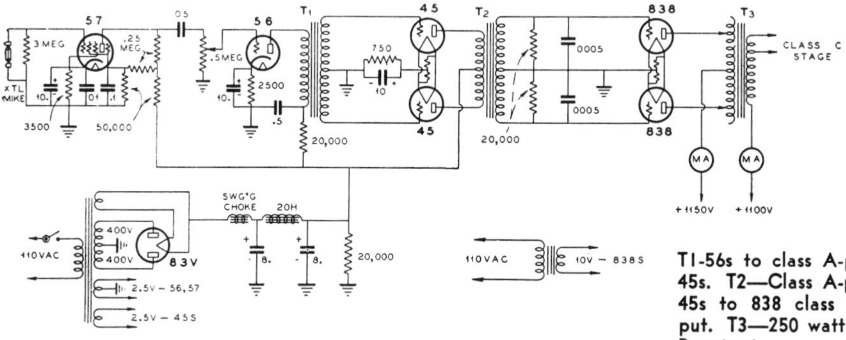

T1-56s to class A-prime 45s. T2—Class A-prime 45s to 838 class B input. T3—250 watt class B output.

Complete circuit diagram of the 838 modulator for crystal mike operation.

Practical Grid-Bias Modulation

The economical factors entering into the carrier powers obtained from either a bias or high-level plate-modulated transmitter are approximately equal. The reason may be attributed to the fact that certain tube combinations happen to work better for one or the other two systems of modulation.

The most outstanding feature of a bias-modulated transmitter (see the fundamental circuit shown in Figure 2) is that a minimum of audio equipment is required, in comparison to a plate modulated transmitter of the same power output. This is an invaluable feature for anyone who spends the greater part of the time operating on C.W. Instead of having over half the transmitter idle, as when working C.W. with a plate modulated phone, less than 10 per cent remains inoperative when the bias-modulated transmitter is used for this class of service.

Several different types of grid bias modulation systems have been brought forward in the past years, but only one bias system has proved itself capable of complete, linear and symmetrical modulation; this is the Hawkins class BC bias development which is characterized by a constant angle of plate current flow. See Figure 1. The most practical test for linearity of any bias modulated amplifier is to determine whether or not the average plate current remains absolutely constant for all percentages of modulation up to 100 per cent. The class BC amplifier is the only one, at the present time, which fulfills this requirement.

Amplitude or harmonic distortion occurring in other systems manifests itself because the average value of plate current is not proportional to the peak value of the same current during modulation. As the grid swings more positive (during modulation), plate current flows during a longer time interval, and as the grid goes more negative, plate current flows for a shorter time interval than when the bias is unmodulated. For distortionless modulation, the interval during which the plate current flows must remain constant, regardless of the percentage modulation. This condition is fulfilled only when the fixed bias is exactly equal to the cut-off value. However, while fixed cut-off bias is entirely workable for a class B linear amplifier, which is amplifying a wave which was modulated in some preceding stage, it cannot be used in a grid modulated amplifier because the operating bias must always exceed cut-off by an amount equal or greater than one-half the audio signal voltage, in order to keep the negative halves of the RF excitation cycles from crossing the cut-off point during modulation.

Fundamentals of Grid-Bias Modulation: Grid modulation is characterized by the fact that very little audio power is necessary to modulate the grid bias of an RF amplifier. However, the complexity of adjustment in the older systems has prevented about 99 per cent of those attempting it from obtaining satisfactory results. Usually terrific distortion and over-modulation followed most attempts to obtain the combination of 100 per cent modulation capability and high plate efficiency.

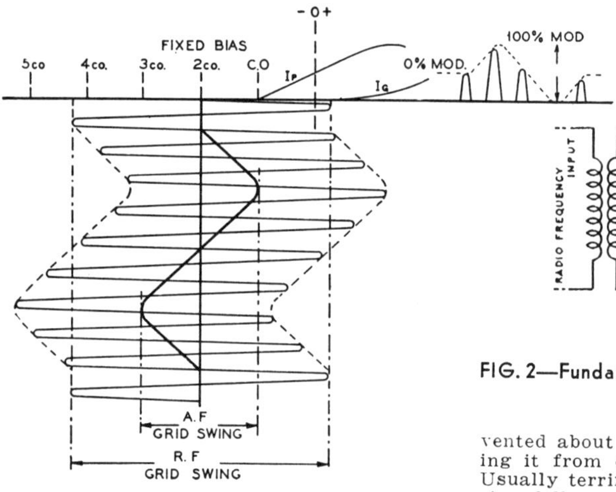

FIG. 1—Graphic representation of class BC bias modulation.

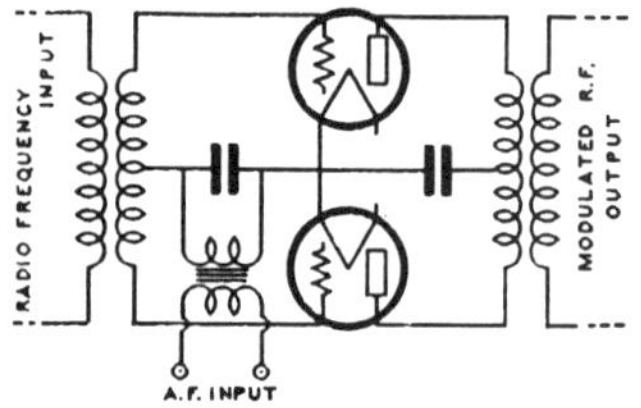

FIG. 2—Fundamental circuit of grid modulation.

When audio modulation is used the radio-frequency carrier has two sidebands which carry the transmitted intelligence. Mathematics shows that one-third of the power in a completely modulated signal is contained in the two sidebands, while the other two-thirds consists of the carrier. Thus the problem which faces the builder of a phone transmitter is to increase the power output of the transmitter, during modulation, up 50 per cent for complete modulation. This additional power must be released in exact accordance with the variations in sound pressure which the operator's voice impresses on the microphone.

The source of power that increases the carrier output when the biased-modulated amplifier is modulated is explained as follows: The fundamental nature of a vacuum tube amplifier might be defined as a device which converts DC (plate power) into AC (RF output) power; but, since the conversion process is never 100 per cent efficient, it must be concluded that the difference between the plate input and the power output energies are dissipated from the plate of the tube in the form of heat. The efficiency of a vacuum tube amplifier depends on numerous factors which vary widely for different types of amplifiers. If a given amplifier is, for example, 25 per cent efficient under a given set of conditions, it will have a certain power output. By maintaining a constant DC plate input to the amplifier and by changing some of its operating conditions, it is possible to increase the average plate efficiency 50 per cent by swinging the instantaneous plate efficiency between zero and twice the unmodulated value.

By measuring the power output, it will be found that when the average plate efficiency is increased, while the plate input remains constant, the power output increases one and one-half times more than its former value. Thus, if some means is found to cause the plate efficiency to increase 50 per cent, it will be possible to obtain a similar increase in average output necessary for complete modulation. Under certain conditions the grid-bias voltage affords this means of varying the plate efficiency of the amplifier.

Therefore **all grid modulation systems, whether they use the control-grid, screen-grid or suppressor-grid for the audio control, operate with constant plate input and variable efficiency.**

In class BC amplification (see Figure 1) the fixed bias is equal to cut-off bias. This bias should usually be supplied from batteries. Additional bias approximately equal to the fixed bias is obtained from a cathode bias resistor connected in the conventional manner. The extra bias supplied by the voltage drop through this resistor is proportional to the plate current and therefore to the grid voltage. When the ratio of grid voltage to this excess bias voltage is a constant, a condition arises where the plate current impulses all flow for the same time interval, regardless of their peak amplitude.

In the older bias modulation systems the distortion increases almost directly as the ratio of fixed bias to cut-off bias. This limits any attempt to increase the plate efficiencies by using higher values of bias and driving voltage. However, in the class BC the total bias may be as high as desired in search for a higher plate efficiency and the absolute value of the total bias, and therefore the driving power is dictated by the usual class C amplifier considerations. A limiting factor in class BC is the voltage drop across the cathode resistor which represents a waste of plate volts, as the bias is increased. There is no objection to driving the grid of the class BC amplifier to positive saturation, although extremely high values of grid current will cause some slight distortion because the grid current flows through the cathode resistor. Therefore the plate voltage should be as high as the tube insulation and gas content will allow, so that positive saturation will be as close to the zero bias line as possible.

The best tubes for class BC amplifier service are those of medium mu, such as the 210, 211, 800, RK18, 242A, 852, 50T, HK354, and 150T. The high-mu tubes, such as the 841, 203A, 830B, 46, 838 and the screen-grid tubes have an advantage in that a smaller cathode-resistor can be placed in the circuit because less bias is necessary to reach any given number of times cut-off. However, the high plate impedance of these tubes makes their use undesirable because it is difficult to secure a linear dynamic characteristic. This limits the undistorted power output.

The low-mu tubes (245, 2A3 or HK255) have the most linear characteristic, but the cathode-bias resistor must be so large in order to get enough bias for efficient operation that a large proportion of the plate voltage is lost. If there are no limitations to the plate voltage available, the low-mu tubes will give slightly better results than the medium-mu tubes. Perhaps the best single index of merit is the grid-plate transconductance, although this factor of tube merit is measured under such widely varying conditions that direct comparisons should be made with caution, except for tubes of the same general type.

Table of Data for Class BC Amplifier Operation

Tube Type	Input W	RF Unmodulated Carrier Power Output W	Plate Loss W	μ
210	25	10	15	8.3
801	33	13	20	8.5
800	60	25	35	15
50T	83	33	50	13
211-242A	166	66	100	12
852	166	66	100	12
354	250	100	150	13
150T ...	250	100	150	13
212D ...	333	133	200	16
204A ...	416	166	250	24
270A ...	500	200	300	16
849	583	233	350	19
851	1000	400	600	20
251A ...	1250	500	750	10.5

Designing the Bias-Modulated Amplifier: The relationships which exist in the class BC amplifier circuit are given below so that the designer can calculate the unknown factors from those already known.

Technical note: The unmodulated plate efficiency of a class BC amplifier is approximately 40 per cent, but rises up to 60 per cent during complete modulation. The limitation on the output of all bias modulated amplifiers is the available plate dissipation of the tube (or tubes) used in the amplifier.

The known factors in designing the bias-modulated amplifier are:

(1) E_b = DC plate supply voltage, in volts

(2) $W_{plate\ loss}$ = rated plate dissipation of the tube in watts.

(3) μ = amplification factor of the tube.

The above factors can be determined from tube tables and the plate supply voltage with a high voltage voltmeter. From this information, the designer must determine in advance all of the unknown factors, in order to allow the amplifier to operate properly. This is the only bias-modulation system which allows such data to be accurately determined in advance. The unknown factors which are to be determined from these shown above are:

(4) W_{input} = DC plate input power, in watts.
(5) W_{output} = RF unmodulated carrier output in watts.
(6) I_p = DC plate current, amperes.
(7) E_{cco} = DC battery bias equal to theoretical cut-off bias (one-half total bias).
(8) R_k = cathode bias resistance, in ohms.

The information given above simply describes the conditions under which the class BC amplifier wil operate when properly adjusted. E_{cco}, which equals the amount of DC bias equal to theoretical cut-off at the plate voltage used, is the battery bias which must be used, and is also equal to the voltage drop across the cathode bias resistor. The following formulas define the unknown factors in terms of those already known:

(9) $W_{input} = 1.66 W_{plate\ loss}$
(10) $W_{output} = .66 W_{plate\ loss}$

(11) $$I_p = \frac{1.66\ W_{plate\ loss}(1 + u)}{\mu E_b}$$

(12) $$E_{cco} = \frac{E_b}{1 + \mu}$$

(13) $$R_k = \frac{E_b^2 \mu}{1.66\ W_{plate\ loss}(1 + u)^2}$$

The above formulas are based on 40 per cent plate efficiency, which can be realized from any tube operated at, or above, its rated plate voltage. The class BC amplifier requires closer coupling to the antenna than is commonly used in CW transmitters.

Note: The class BC amplifier makes an exceptionaly good linear RF amplifier for amplifying a previously modulated wave. It is capable of somewhat better linearity and plate efficiency than the conventional class B linear amplifier. The class BC linear amplifier is a modulation-gaining device because it doubles the percentage modulation of the excitation wave. Thus, the amplifier which precedes the class BC linear amplifier (when twice cut-off bias is used on the class BC linear) must not be modulated more than 50 per cent if overmodulation of the output wave is to be avoided. When the exciting wave is modulated 50 per cent, the output wave delivered to the antenna is exactly 100 per cent modulated. In certain cases this permits a distinct economy of audio power to be realized in modulating the preceding stage. Fifty per cent modulation of a given plate-modulated class C stage only requires one-fourth of the audio power necessary for 100 per cent modulation.

The Smallest Economical Bias-Modulated Transmitter

In this transmitter the grid bias on the final stage is varied at an audio-frequency rate by means of a low-power modulator. Less than 150 volts of audio swing is required for complete modulation; four times as much power is required with plate circuit modulation for the same 50 watt carrier.

The oscillator and buffer stages are similar to those described under "The Jones Harmonic Oscillator." Either the fundamental or second harmonic of the 53 tube drives the 2A3 or 45 buffer stage. The latter gives from 14 to 18 watts for driving the grid of the final 211 stage, which is needed for CW operation at the plate voltage available. However, for phone operation this is too much power, and part of it is swamped out by means of a 100 watt lamp shunted across the link circuit and the remainder reduced by slightly detuning either the 211 grid circuit or the crystal oscillator circuit. The second harmonic is needed for operation in the 20 meter band. It also allows operation on 40 meters with an 80 meter crystal. The output from the 2A3 neutralized buffer stage is about the same for either fundamental or second harmonic operation of the crystal oscillator.

The final stage has a very effective method for coupling to either a single wire fed Hertz or to an end fed Fuchs antenna. The coupling is similar to the Collins antenna tuning system, but, unlike the Collins system, there are no losses introduced by the additional tuned circuit. In the circuit diagram condenser C1 provides a low impedance path for harmonics and the inductance acts as an RF choke; therefore there are practically no harmonics across the antenna coupling condenser C2, or C2 and C3. The fundamental frequency has its proper impedance drop across the antenna condenser for matching from 2000 ohms down to 50 ohms. The antenna is easily matched with this system and more power is delivered into the antenna than with other types of antenna coupling schemes.

A well-spaced receiving condenser, rated at 1000 volts, is sufficient for C2, but C1 should be rated at 2500 volts to prevent it from rupturing on RF peaks. With the power supply used (1800 volts), the carrier output power is approximately 60 watts for phone, and over 200 watts for C.W. operation. The transmitter is conservatively rated as a 50-watt phone and 150-watt C.W. transmitter. Keying is conveniently accomplished in the cathode circuit of the crystal oscillator because all the stages are biased beyond cut-off and the tuned circuits between the antenna and crystal stage tend to eliminate spurious sidebands caused by key-clicks.

Constructional Details: The transmitter can be built into a relay-rack mounting of the table type, as illustrated. The power supply is built into the lower deck on a 10-in. x 17-in. x 1½-in. chassis pan, behind an 8¾-in. x 19-in. front panel. The chassis rests directly on the bases, which prevents the heavy power equipment from putting a strain on the front panel. Most interconnections between panels are made by means of 5-conductor patch cords; tube sockets are attached to the rear of each chassis to accommodate plug-in receptacles affixed to the end of the patch cords.

The power deck holds the high voltage transformer, rated at 1500 volts RMS, each

The Jones Exciter and the buffer stage. The tuning condensers are mounted on a long bakelite strip. The condenser shafts are insulated from the front metal panel. At the rear of this unit, to the left, is the 83V rectifier tube and alongside of it is the power supply choke for the low-voltage supply. The filter condenser bank is at the extreme right.

side of center-tap, and capable of supplying at least 200 milliamperes of DC load out of the filter; the latter consists of two 2 ufd. 2000 volt condensers connected across a 15 to 20 henry choke of 250 MA capacity. This gives ample filtering for phone use, as the load current is only 100 MA for grid modulation.

The remaining decks are of the same size as the power deck. The second or middle chassis contains the low voltage rectifier, 83 with the filter system; associated on the same deck are the crystal harmonic oscillator and buffer stage. The tuning controls are brought out to the front panel through insulated couplings. The buffer stage is neutralized by means of a screw-driver adjustment through a hole in the front panel. An aluminum shield is placed between the crystal and buffer stages to minimize RF feedback. The tube, coil and crystal sockets are arranged for convenience around the midget tuning condensers. Since Isolantite sockets are used, ¾-in. holes are made under each socket for wiring, most of which is under each deck. Ordinary good radio hook-up wire is satisfactory for wiring this deck. All high voltage leads are made with small flexible wire, insulated for 10,000 volts breakdown.

The top deck mounts behind a 10½-in. x 19-in. front panel and holds the modulator system and final RF stage. The entire modulator system occupies a space about 2-in. x 6-in. The grid circuit of the 211 tube is link-coupled to the 2A3 buffer stage. An aluminum shield is placed between the grid and plate circuits of the 211 stage to insure ease of neutralization. Grid neutralization functions perfectly and is comparable to the simplicity of plate neutralization. The plate and antenna condensers are mounted on the front panel, with the rotors grounded. The grid and neutralizing condensers must be mounted on insulators. The three controls on the top panel are arranged to be symmetrical with those on the middle deck in order to give a pleasing appearance.

The current measuring jacks are mounted on the front panels without insulators; however, the two grid current measuring jacks are insulated from the panel because they must be connected in opposite manner to the cathode circuit jacks. It is desirable to use a 0-25 MA meter for grid current measurements. Either a 0-200 or 0-300 MA meter is suitable for the other measurements.

The speech amplifier has the input leads

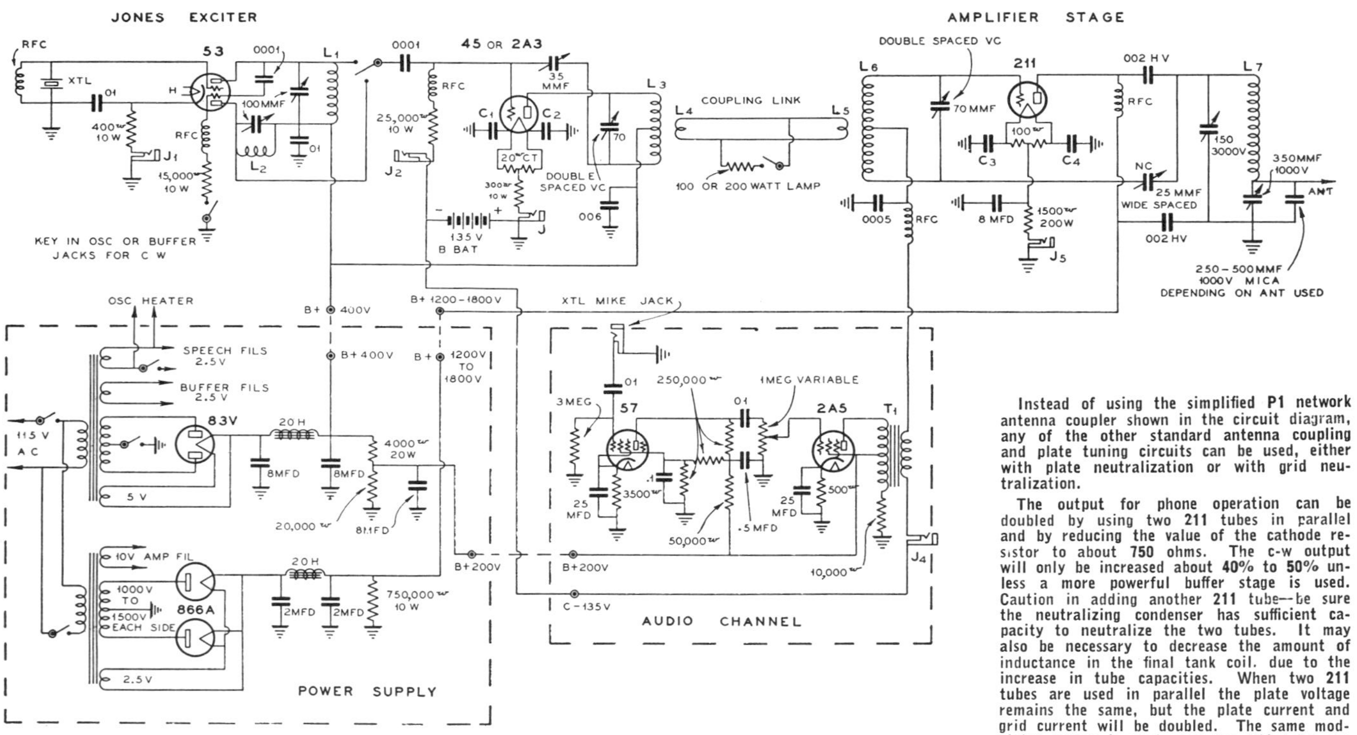

L3 and L4 each have two turns in the coupling loop. Condensers C1, C2, C3 and C4 are .001 or larger, fixed mica type, 1000 volt rating.

Instead of using the simplified **P1** network antenna coupler shown in the circuit diagram, any of the other standard antenna coupling and plate tuning circuits can be used, either with plate neutralization or with grid neutralization.

The output for phone operation can be doubled by using two **211** tubes in parallel and by reducing the value of the cathode resistor to about **750** ohms. The c-w output will only be increased about **40%** to **50%** unless a more powerful buffer stage is used. Caution in adding another **211** tube—be sure the neutralizing condenser has sufficient capacity to neutralize the two tubes. It may also be necessary to decrease the amount of inductance in the final tank coil, due to the increase in tube capacities. When two **211** tubes are used in parallel the plate voltage remains the same, but the plate current and grid current will be doubled. The same modulator system is used when two **211** tubes are used in parallel. High MU tubes, such as the **03A**, **838** and **830B** do not give good results when used as grid-modulated amplifiers.

The Final Amplifier and the simplified PI Antenna Network. To the extreme right is seen the 57 tube (shielded) and the 2A5 tube, these two tubes constituting the entire audio channel. The small modulation transformer is seen alongside the grid coil. The grid tuning condenser is mounted on a bakelite support and the condenser shaft is insulated from the front metal panel. The 100 (or 200) watt lamp bulb is in shunt with the coupling link, although this lamp may not always be required. The coupling link around the grid coil has two turns, 2 inches in diameter.

shielded to prevent audio and RF feedback. The resistors and condensers can be mounted beneath the upper deck by means of terminal strips. For a small modulation transformer the DC is balanced-out in the primary winding as shown in the drawing. This loading resistor also provides a needed load on the pentode modulator because the grid-impedance of the 211 is not constant. This transformer is a small universal type class B transformer, originally designed for either input or output circuits, with several taps. A number of manufacturers have this type of transformer on the market.

The total speech amplification is only great enough for close-speaking operation with a crystal microphone. One possible source of distortion or "downward" modulation swing is in the modulation transformer. If this transformer saturates trouble will occur. Usually a 2-to-1 impedance step-down ratio from the 2A5 plate to 211 grid is satisfactory in this transformer.

T1 is not very critical and practically any of the conventional class B transformers will give good results. It is necessary, however, that the transformer be capable of handling 2 watts of power, plus 25 MA of DC in the primary at a 2:1 step-down ratio.

Because the high-voltage supply has a condenser input filter 1800 volts should be delivered under no load. The low-voltage supply should deliver 400 volts under a load of approximately 160 MA.

The output of the crystal oscillator should light-up a 6-volt pilot lamp connected in series with a single loop of wire, when it is coupled to either the oscillator or doubler coils. The lamp also provides a convenient indicator for neutralizing the buffer and final amplifier circuits. Plate voltage can be removed from these circuits by merely plugging-in an open-circuit plug in the cathode circuits. It is advisable to shut off the high voltage supply when neutralizing the final amplifier as a precaution against a flash-over occurring at the open plug.

The link-coupling and preceding tuning adjustments should be made to obtain maximum grid current in the 211 tube. Without plate voltage, this grid current should be between 30 and 40 MA. Under load this current will be less, because the cathode resistor builds up additional bias due to the plate current.

Neutralizing the final amplifier can be simplified by using a tapped-coil instead of a split-stator tuning condenser in the grid circuit. Neutralizing on 20 meters is slightly more difficult than on other bands.

So that the antenna circuit will remain in resonance when tuning, it is desirable to make the necessary adjustments simultaneously with the plate condenser C1; the proper setting is indicated by a minimum plate current reading. The antenna or dummy antenna must never be disconnected

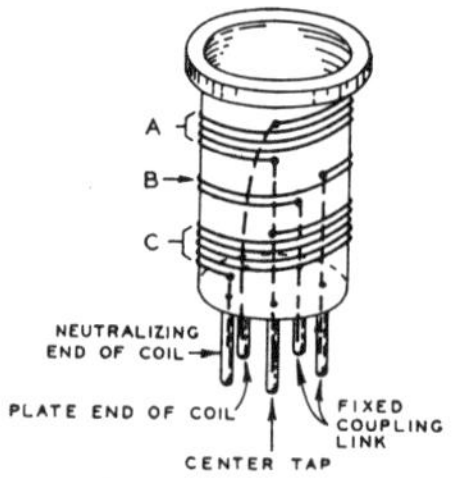

Showing how the buffer plate coil is wound. 5-prong Hammarlund forms are used. The link coupling loop portion of the winding has two turns, in the center of the form. A and C are the two halves of the plate coil winding. B is the two-turn coupling loop, L4 and L5.

The Buffer Plate Coil for any band is center-tapped. At the center of the buffer plate coil winding, this winding is spread apart so that a 2-turn coupling loop can be wound directly in the center of the coil. This is a fixed coupling loop and the connections are brought to two of the prongs on the coil form base. The link coupling loop of two turns on each of these buffer plate coils does not require a variable adjustment. The same size wire is used for winding the two-turn coupling loop as is used for winding the coils proper. On the 160-meter buffer plate coil this 2-turn loop is close wound with the other winding. On the 80-meter buffer coil this 2-turn loop is separated by 1/16-in. from both sides of the coil winding proper. On the 40-meter buffer plate coil the loop is separated by 1/8-in.; on the 20-meter buffer plate coil the separation is 1/4-in.

Coil-Winding Table for Grid-Modulated Phone

NOTE—5-Prong 1½-in. Dia. Coil Forms Used Throughout, Except for 211 Amplifier Plate Coil.

Band	Oscillator Coil L1	Doubler Plate Coil L2	Buffer Plate Coil L3	211 Grid Coil L6	211 Plate Coil L7
160 Meters	68 turns, No. 22 DSC, close wound on 1½-in. dia. form. Winding space occupies 2 1/16-in.	NONE	78 turns, No. 22 DSC, close wound on 1½-in. dia. form. Winding space occupies 2¼-in. tap to be taken at center of winding. Also 2-turn winding is to be wound in center of coil for coupling link. See drawing of coil form for data.	77 turns, No. 22 DSC, close wound on 1½-in. dia. form, center-tapped. Winding space occupies 2¼-in.	51 turns, No. 14 enameled wire, wound on 2½-in. dia. ceramic or bakelite form. Winding space occupies 3⅝-in.
80 Meters	27 turns, No. 22 DSC on 1½-in. dia. form. Space wound to cover 1½-inch winding space.	Same coil as 80 meter oscillator coil for doubling to 80 meters from 160 meter oscillator.	45 turns, No. 22 DSC, close wound on 1½-in. dia. form, center-tapped. Winding space occupies 1⅞-in. A 2-turn coupling link is wound in center, same as for 160 meter coil.	44 turns, No. 22 DSC, wound on 1½-in. dia. form, center-tapped. Winding space occupies 2-in.	26 turns, No. 12 bare wire, space wound, on 2½-in. dia. form to cover a winding space of 3½-in.
40 Meters	13 turns, No. 18 DCC on 1½-in. dia. form. Space wound to cover winding space of 1½-in.	Same coil as 40 meter oscillator coil for doubling to 40 meters from 80 meter oscillator.	22 turns, No. 18 DCC on 1½-in. dia. form, center-tapped. Winding space occupies 1¾-in. A 2-turn coupling link is wound in center, same as for above buffer coil.	22 turns, No. 18 DCC, wound on 1½-in. dia. form, center-tapped. Winding space occupies 1½-in.	12 turns, No. 12 bare wire, space wound on 2½-in. dia. form to cover a winding space of 1½-in.
20 Meters	Use 40 meter oscillator coil.	7 turns, No. 18 DCC on 1½-in. dia. form. Space wound to cover a winding space of 1⅜-in.	12 turns, No. 18 DCC on 1½-in. dia. form, center-tapped. Winding space occupies 1¾-in. A 2-turn coupling link is wound in center. Same as for above coil.	10 turns, No. 18 DCC on 1½-in. dia. form, center-tapped. Winding space occupies 1⅜-in.	5 turns, No. 12 bare wire, space wound on 2½-in. dia. form, to cover a winding space of ¾-in.

unless the plate voltage is reduced to not over 1000 or 1200 volts, as C2 will flash-over on RF peaks. For C.W. operation, use all of the available grid excitation and adjust the antenna condenser C2 until maximum output with normal plate current is obtained. C1 must always be set for resonance, or lowest plate current. For C.W. the plate current ought to be in the neighborhood of 200 MA.

For phone adjustment turn on the modulator filament switch—increase the antenna current to maximum, even though the plate-current is high—then reduce the grid RF excitation until only a few grid millamperes flow.

If the load is proper, the plate current will be about 100 MA with an 1800 volt supply. Since the cathode bias resistor has a value of 1500 to 1700 ohms, the actual plate voltage will be approximately 1600 volts; this gives 160 watts input. At 40 per cent unmodulated plate efficiency there is about 60 watts of carrier output, and 100 watts of plate loss. Note that the tube cools off during modulation and if completedly modulated with tone, the input will remain at 160 watts. The RF output will increase to 90 watts, and the plate dissipation will fall to 70 watts.

The flashlight lamp and single turn of wire, or an antenna RF meter, should be used to ascertain that upward modulation is being secured when the microphone is whistled into. The antenna current should rise 20 per cent on a steady tone. When talking, the antenna current must not rise more than 10 per cent as with any plate modulated phone when operating within the 100 per cent modulation limitation. If downward or insufficient modulation is obtained, it is indicative that there is too much RF excitation or not sufficient antenna load.

The "buffer plate coil" for any band is center-tapped. The center of the coil winding is spread apart so that a 2-turn coupling loop can be wound directly in the center of the coil. This is a fixed winding and the connections are brought out to two of the prongs on the coil form base. The two-turn coupling loops on each of the buffer plate coils do not require a variable adjustment. The same size of wire used for the two-turn coupling loop is also satisfactory for the other coils. On the 160-meter "buffer plate coil," the loop is close wound with the other winding. On the 80-meter "buffer coil," the loop is separated by 1/16 in. from both sides of the coil windings. On the 40-meter "buffer plate coil," the loop is separated by ⅛ in.; on the 20-meter "buffer plate coil," the separation is ¼ in.

Any other standard antenna coupling and plate tuning circuits can be used with either plate or grid neutralization, instead of the simplified PI network antenna coupling shown in the schematic.

The output for phone operation can be doubled with two 211 tubes in parallel and by reducing the value of the cathode resistor to about 750 ohms. The C.W. output will only be increased about 40 to 50 per cent unless a more powerful buffer stage is used. CAUTION: In adding another 211 tube, the neutralizing condenser must have sufficient capacity to neutralize the two tubes. It may also be necessary to decrease the amount of inductance in the final tank coil, due to increase in tube capacities. When two 211 tubes are paralleled, the plate voltage remains the same, but the plate current and grid current should be doubled. The same modulation system is used for paralleled operation.

The indicators on the grid and plate current meters should not move on modulation, except for a very slight upward motion on over-modulation peaks.

For phone operation, a 100 watt lamp is connected across the coupling link between the buffer and the final amplifier to stabilize the load on the buffer during modulation. This is required because the instantaneous grid current in the final amplifier varies with modulation. The lamp introdues no losses when phone operation is used because the buffer supplies considerably more excitation than is needed. In fact, the grid circuit of the final amplifier must be detuned to still further reduce the excitation. Normally a good C.W. antenna load will be sufficient for phone operation, and so only one adjustment is needed—**reduce the grid RF excitation until a good upward antenna current swing is obtained when the microphone is whistled into.**

For phone operation the C.W. key, or the low-voltage power supply center-tap switch can be used for push-to-talk control.

The external C battery of 135 volts is correct for a total applied plate voltage of 1800 volts. To determine the correct value divide the plate-to-cathode voltage under load by the amplification constant of the tube. This is about 12 for a 211 tube.

Technical Notes: A condenser input type filter is necessary with a 1500 volt power transformer in order to obtain at least 1700 volts DC for maximum output. For C.W. operation, the 2A3 tube must deliver sufficient excitation to the final stage so that the grid current will be about 20 milliamperes under load. The grid current can be raised by increasing the number of turns in the coupling link (within limits) or by increasing the plate supply voltage on the 2A3 up to but not higher than 500 volts. The grid blocking condenser in the 211 stage must have a rating of 2500 volts DC working voltage. The filament by-pass condensers on the 211 tube must be .001 ufd. or larger, 1000 volt rating, mica type. The condensers enable the filament of the tube to reach RF ground potential as well as being essential for complete neutralization.

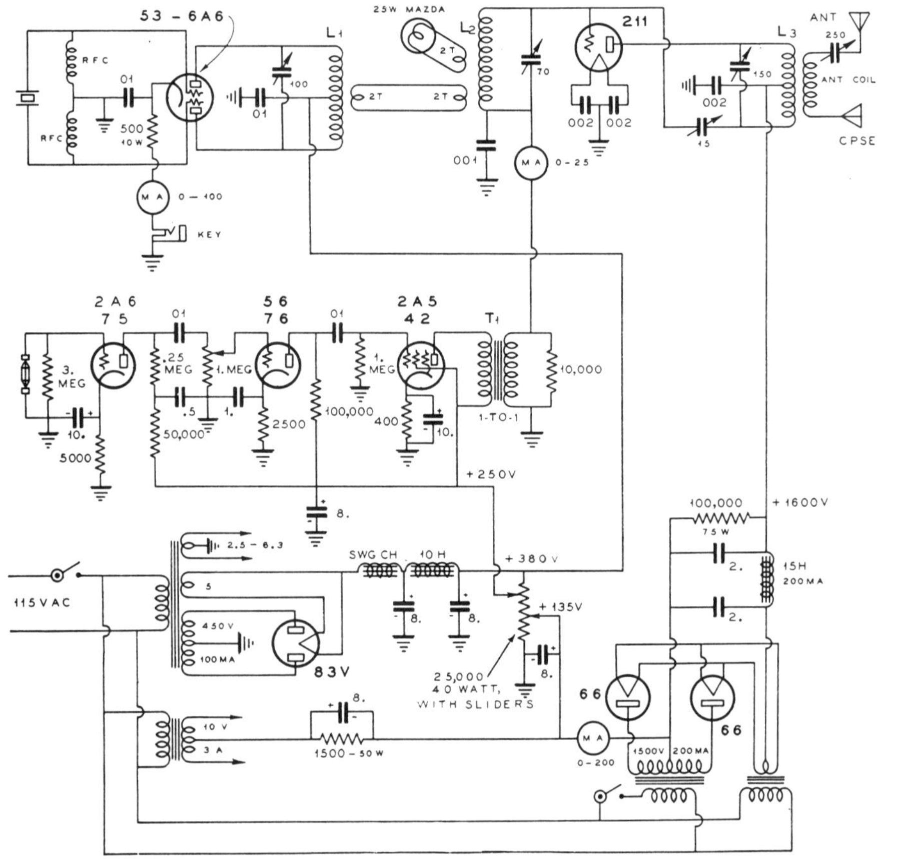

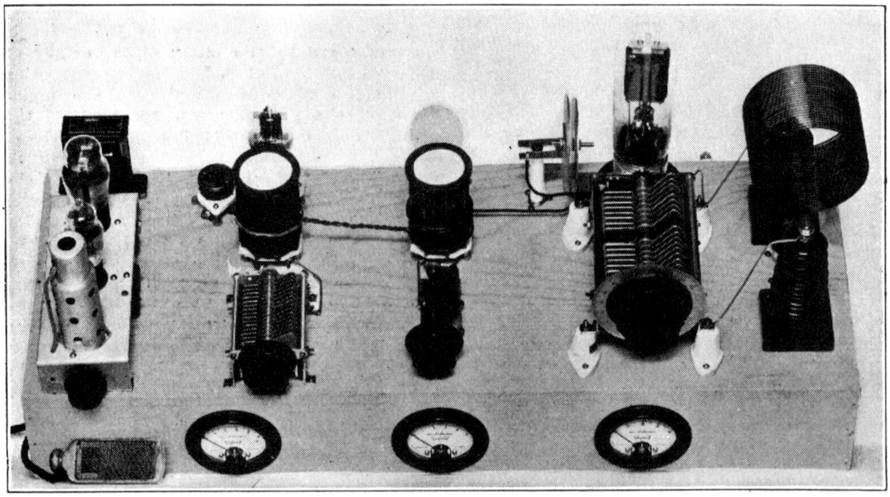

Circuit (above) and illustration of 160 meter grid bias modulated phone, 50 watts output. L1—53 turns, No. 16 DCC, close-wound on 2¼-in. dia. form, center tapped. L2—40 turns, No. 16DCC, close-wound on 2¼-in. dia. form. L3—37 turns, No. 14 Enameled, air-wound, 4-in. dia., 3½-in. long. A 2-turn loop is wound around the grid coil L2.

600-Watt C.W.—200-Watt Phone Transmitter

In this transmitter only 5 per cent of the equipment is idle on C.W. The design features "high-quality grid modulation"; "Jones all-band exciter;" "push-to-talk break-in"; crystal keying for full break-in on CW;" "grid neutralization;" "completely self-contained relay rack construction which eliminates the use of chassis;" "power gain of 20 through the final amplifier," and a "construction cost of less than $300." The carrier power is approximately 200 watts on phone and is slightly greater than 500 watts on CW in the 20 meter band. The quality of speech is excellent, due to the use of cathode resistor plus fixed bias equal to cut-off in the operation of the grid modulated stage.

General Construction: The entire transmitter is built into a standard relay rack, 36 units in height—that is, 36 1¾-in. units or 63 inches of space for mounting panels. The RF amplifier panel at the top is 14 in. x 19 in. of No. 14 gauge iron. The next panel, 5¼ in. x 19 in., contains the speech amplifier and modulator. The oscillator-doubler and buffer stages are mounted behind an 8¾-in. x 19-in.. panel with four dials for controls and three panel holes for regeneration and neutralizing adjustments to the sub-panel. The low-voltage power supply is mounted on a 10½-in. x 19-in. panel and the high voltage supply on a 24½-in. x 19-in. panel at the bottom of the rack. The entire unit makes a pleasing appearance and takes up but little space. A rear ventilated cover should be used to prevent personal contact with the 3,000 volt power supply.

Technical Details: The coils for this transmitter are wound for operation in the 20 meter band. If the set operates satisfactorily on 20 meters, it will invariably operate as well on the lower frequency bands. The 150Ts are easily driven on 20 meters and a pair of 45s or 2A3s will drive them to over 500 watts output on CW and approximately to 200 watts on phone with modulation capability of 100 per cent.

The oscillator, a 53 tube, functions as a crystal oscillator and frequency multiplier and provides sufficient excitation to drive the 150Ts to 200 watts for phone operation. Link coupling is used between the doubler plate coil and the buffer grid coil, as well between the buffer and final stages. No. 18 hook-up wire is satisfactory for this purpose, with one turn link coils wound over the center of each coil.

The neutralizing condenser in the final stage is made with three aluminum plates about three inches square with one-half inch spacing. The 50 uufd. tuning condenser may flash-over without an antenna load. No trouble will be encountered in neutralizing the higher power stages when using the grid current meter as an indicator.

The constants for grid modulation are calculated from the formula appearing under the subtopic heading, "Designing the Bias Modulated Amplifier." By calculating

Medium-power plate modulated phone transmitter wherein much of the equipment is idle when c-w is used. The grid-bias modulated transmitter on page 172 overcomes this difficulty, but with lower output.

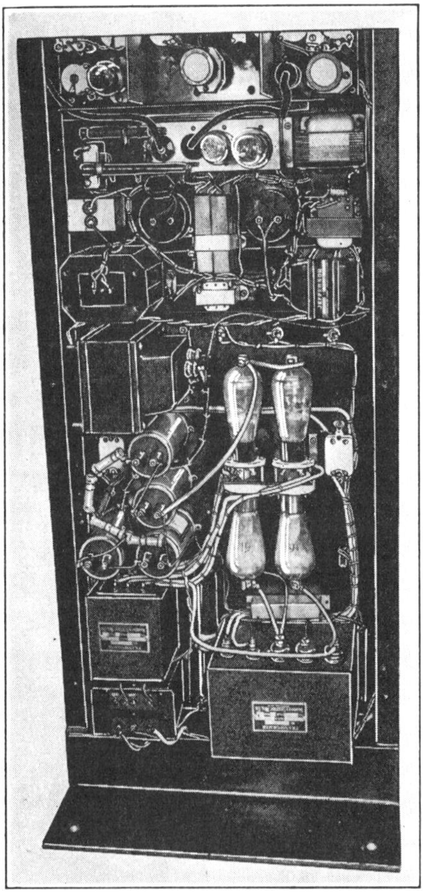

Power supply and portion of exciter unit.

the values before the transmitter is built, a close approximation to the actual values can be found.

On CW, about half of the cathode resistor is normally cut out by means of a small snap-switch on the lower panel. This increases the plate current up to about 300 ma. at 2,900 volts, with an output approximately between 500 and 600 watts. The plates show no color on CW and the efficiency with a twisted-pair feeder 20-meter antenna calculates to better than 70 per cent. The actual C bias is slightly less than class C, due to lack of grid excitation, but 500 or 600 watts output on 20 meters will produce remarkable results on distant transmission. The input to the 45s or 2A3s is between 40 and 50 watts with a power output of 25 to 35 watts. This gives a gain of about 20 in the final stage.

The low voltage supply should deliver about 250 ma. at nearly 400 volts. A key-click filter is desirable. It consists of a 1 or 2 henry choke in series with the key, plus a .1 ufd. or larger condenser in series with a 500 ohm resistor across the key contacts. Fixed bias is used on all stages following the oscillator-doubler and keying can therefore be accomplished in any of the current measuring jacks.

Coil Design: The oscillator coils are made on 1½-in. plug-in forms. The 80-meter coil consists of 27 turns of No. 22 DSC, 1⅜ inches long. The 40-meter oscillator coil consists of 16 turns, 1¼ inches long, of No. 18 DSC wire.

The 20-meter doubler and the 20-meter buffer grid coils are each 10 turns of No. 18 DSC on a 1½-in. form, 1 inch of winding space, with a center-tap. The 20-meter buffer plate coil consists of 9 turns on a 1¼-in. form, wound to cover 1½ in., with a center-tap. The final amplifier grid coil for 20 meters consists of 9 turns on a 2¼-in. diam., 2-in. long form. The latter can also be wound on a form 1½ in. diam., 1 in. or 1¼ in. long with 10 turns. These two coils should be wound with No. 16 wire because No. 18 shows some heating effects at this frequency.

The final tank coil on 20 meters is wound on a porcelain tube 2½ in. diameter, with 7 turns per inch of No. 10 wire. This coil is wound with 16 turns, 10 of which are shorted-out for 20-meter operation; the whole coil is used for 40-meter operation. By using a very low resistance shorting connection across the 10 turns, and very low C in the tank condenser, this coil will not heat excessively on 20 meters with 800 watts input to the final stage. For connec-

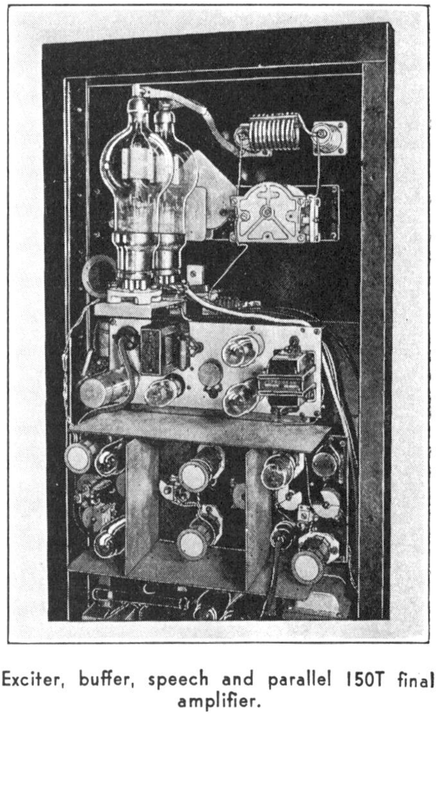

Exciter, buffer, speech and parallel 150T final amplifier.

tion to a Collins coupler, this coil should be tapped at about 3 turns for 20 meters and 8 turns for 40 meters. For use with a twisted-pair feeder, two turns of No. 14 rubber-covered wire wound over the +B end turn is correct for proper loading.

The RF chokes, with the exception of the one in the final amplifier plate circuit, are of the small 2.1 mh. type.

Transmitter Adjustments: The oscillator is tuned for maximum output at lowest cathode current. The doubler section must provide sufficient excitation swing to the buffer stage grids. The cathode current of the 53 tube is adjusted to between 60 and 70 ma., and the doubler output must drop to zero when the crystal is short-circuited. Too much regeneration will cause the doubler to go into self-oscillation. A neon bulb, an RF galvanometer and loop of wire, or a lamp and a turn of wire are good oscillation indicators. An open-plug should be inserted into the jack of the buffer stage while neutralizing; the plug is afterwards removed. In neutralizing, the two condensers are simultaneously adjusted until the plate tank has no RF.

The plate current of the buffer stage reaches between 110 and 160 ma. when loaded by the grids of the 150Ts. For CW operation with half of the cathode resistor cut out and about 200 volts fixed bias, the final grid current will attain a value between 25 and 45 ma. when the buffer stage is operating properly. The high value allows better efficiency and a little more output. The plate current in the final amplifier should be from 250 to 350 ma. on CW.

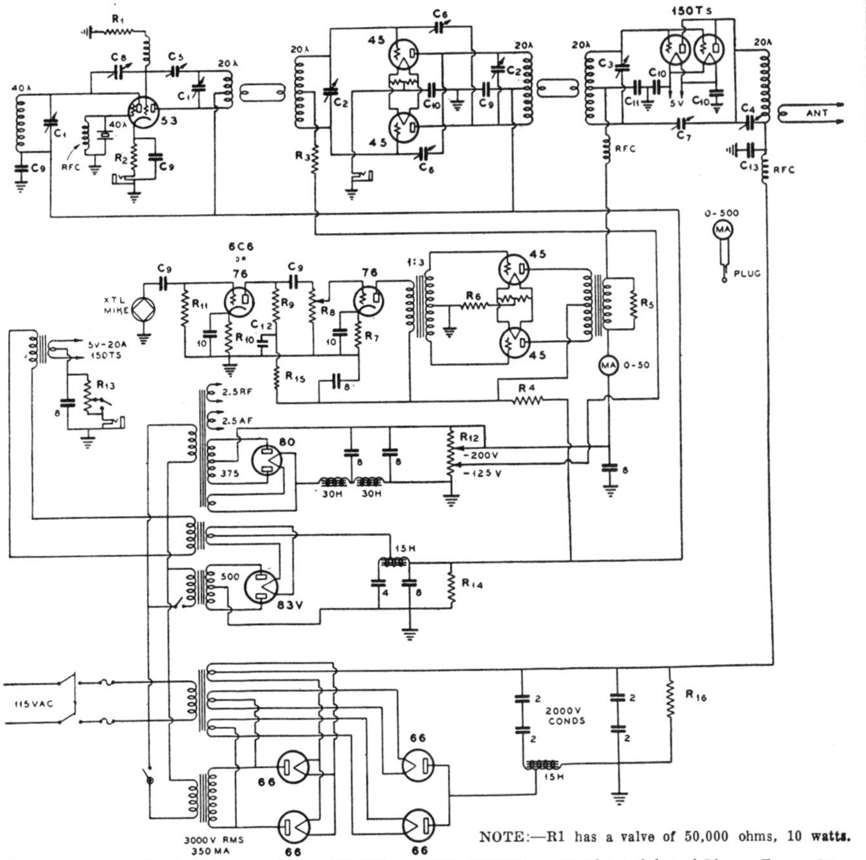

NOTE:—R1 has a valve of 50,000 ohms, 10 watts.

The Complete Circuit Diagram of the 600-Watt C.W., 200-Watt Grid Modulated Phone Transmitter

LEGEND: C1—100 mmf., 500 volts. C2—50 mmf., double spaced. C3—50 mmf., double spaced. C4—50 mmf., 6500 volts. C5—15 mmf., 500 volts. C6—25 mmf., double spaced. C7—6 mmf., 10,000 volts. C8—3 to 30 mmf. mica trimmer. C9—.01, 1000 volts. C10—.001, 1000 volts. C11—.0005, 5000 volts. C12—½ mfd., 600 volts. C13—.001, 5000 volts. R2—500 ohms, 10 watt. R3—10,000 ohms, 10 watt. R4—3500 ohms, 20 watt. R5—5000 ohms, 10 watt. R6—750 ohms, 10 watt. R7—2500 ohms, 1 watt. R8—½ megohm, 1 watt. R9—¼ megohm, 1 watt. R10—3500 ohms, 1 watt. R11—3 megohms, 1 watt. R12—4,000 ohms, 40 watt. R13—1000 ohms, 200 watt. R14—25,000 ohms, 40 watt. R15—50,000 ohms, 1 watt. R16—Four 100,000 ohms, 20 watt resistors in series.

400-watt grid modulated amplifier with four 150Ts in push-pull-parallel.

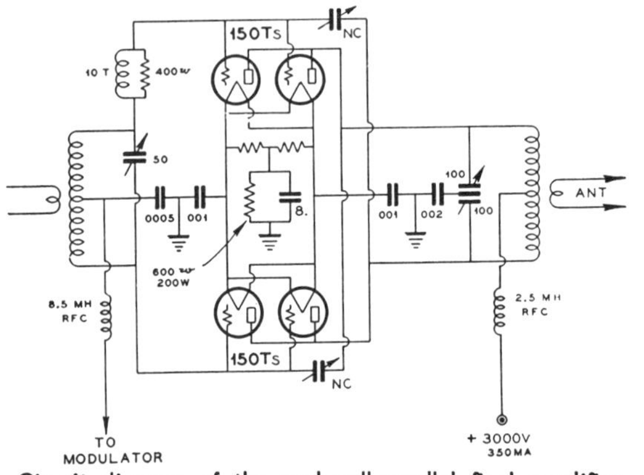

Circuit diagram of the push-pull-parallel final amplifier. Only one grid choke is required.

On phone operation the grid circuit of the buffer stage is detuned to drop the grid excitation on the final amplifier to about one to three ma. under load. The plate current then has a value between 180 and 200 ma. at a plate supply voltage slightly under 3,000 volts. The full 1,000 ohm cathode resistor must be in the circuit. Less plate current is often indicative of insufficient antenna load.

To change from CW to phone, the microphone is plugged into its jack on the speech amplifier panel, the buffer-grid dial is detuned until only a fraction of grid current flows; the cathode-resistor switch is thrown to the "off" or phone position.

A steady tone will increase the antenna current from 20 to 22 per cent. On normal speech the gain control should be reduced until only a very small RF current is indicated. Too much RF grid excitation (too much grid current) will prevent modulation and zero grid current reduces the plate efficiency and drops the power output. About one or two milliamperes is the optimum value.

NOTE: A high-pitched AC hum is sometimes introduced across the electrolytic by-pass condenser which is connected across the first speech amplifier tube. The hum can be removed by connecting a .00025 mica condenser from cathode to ground.

Increasing Carrier-Power Up to 400 Watts: To increase the carrier power of the transmitter just described up to 400 watts requires the use of four type 150T tubes in a push-pull parallel connection; such a scheme is shown in the accompanying figure where the RF amplifier is driven by a pair of 2A3s and modulated by a pair of 45 tubes.

Technical Notes: A 4-meter parasitic oscillation may develop in the final amplifier described above and may be eliminated by placing a small RF choke in series with one of the grid leaks. This choke consists of 10 turns of No. 14 wire wound on a half-inch diameter form, shunted by a 400 ohm resistor.

Cut-off bias is supplied by means of a C-bias power-pack, and cathode automatic bias furnished by mean of a 600 ohm 200 watt resistor in the center-tap lead to the filaments.

For phone operation, the above RF unit need only be supplied with about 15 to 25 watts of grid power.

50-Watt Suppressor Modulated Phone: The RK28 and 803 RF pentode tubes make possible the construction of a very simple phone transmitter which requires no neutralizing and has relatively few parts. Band switching works quite satisfactorily. The transmitter can be operated on phone or CW on 80-, 40- or 20-meters with only a few seconds' delay in switching and tuning for resonance. While the large pentode is more expensive than other types of 100-watt tubes, the saving in parts and elimination of a buffer stage appreciably offsets the tube cost.

In the circuit shown, an Ohmite band-changing switch places any one of three different crystals into the circuit, one for 80 and 40-meter CW, one for 75-meter phone and a 40-meter crystal for 40-meter CW or 20-meter phone. The 80-meter CW crystal will also serve on 20-meter CW, quadrupling in the 53. The drive is sufficient for 150-watts of CW output on 20-meters from the pentode when the 80-meter crystal is in service, but is insufficient for 20-meter phone. The 40-meter crystal will drive the final to nearly 200-watts output on 20-meter CW and 50-watts of phone carrier. On the other bands the output is slightly higher.

One of the triodes in the 53 tube functions as the crystal oscillator. This circuit is comprised of a 24-turn coil of No. 18 enameled wire on a 1½ in. diameter coil 1¼ in. long, tuned with a 100 uufd. midget condenser. When this circuit is tuned to 40 meters for the 40-meter crystal, half of the coil is shorted-out by means of an ordinary snap-switch mounted just beneath the tuning dial associated with this circuit. All controls are on the front panel.

The frequency doubler or quadrupler section of the 53 employs regeneration to obtain good efficiency, especially when quadrupling. A small 3-30 mmfd. semi-adjustable trimmer condenser is mounted along side the coil socket and only enough capacity inserted to provided regeneration below the point of self oscillation. This is easily checked by placing the final amplifier into operation and setting the crystal switch between contact points; no output for any dial setting should result.

Capacity coupling from either section of the 53 provides grid coupling to the pentode. A 100 uufd. midget variable condenser acts as a control of grid excitation. The condenser is fitted with an extension shaft and knob on the lower edge of the

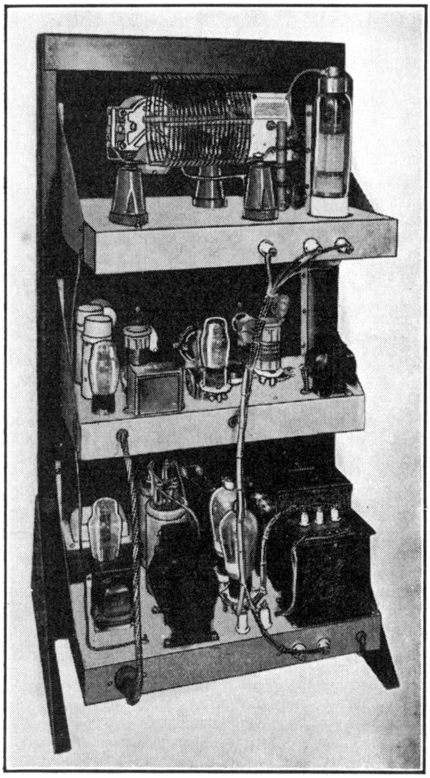

Rear view of 50-watt phone.

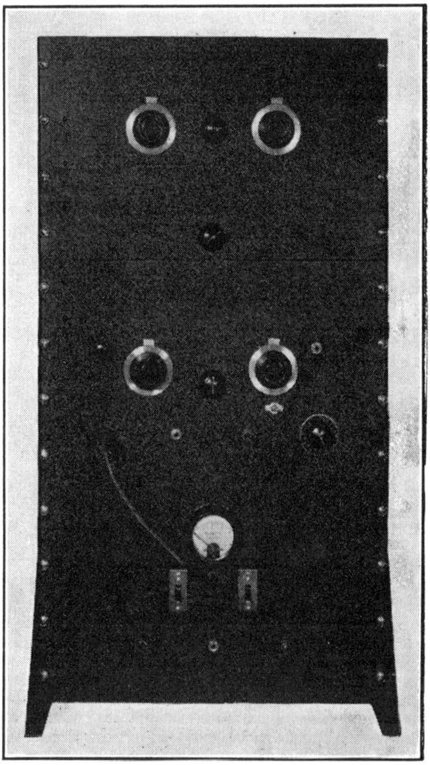

Front panel view, 3-ft. rack.

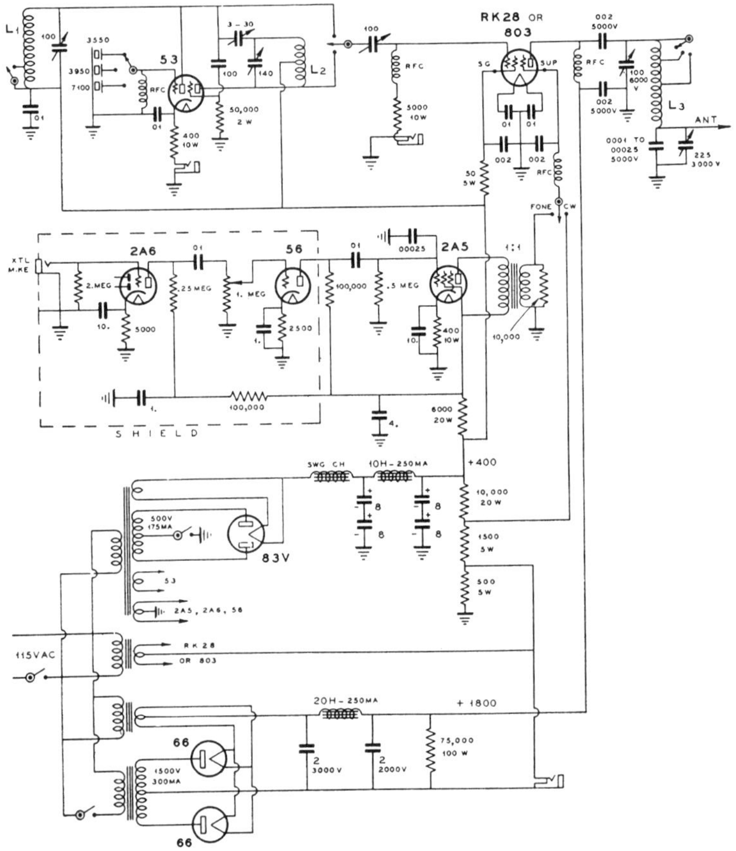

Circuit diagram of the entire transmitter.

top panel. Either section of the 53 can be brought into service depending upon whether straight-through amplification of the crystal fundamental is used, or if one of the harmonics are desired. Switching over to the crystal section requires less tuning capacity due to the added capacity of the RK28 grid circuit. Generally, the second section of the 53 is tuned to resonance on the second harmonic, as indicated by a sharp decrease in the cathode current of this tube, in order to keep this total current in the neighborhood of 50 to 60 MA.

The doubler section has a 140 uufd. midget variable condenser for tuning in order to cover both the 20 and 40-meter bands without coil changing. This means the low C is used on 20 meters and high C on 40 meters, but since there is more than ample output on 40 meters when doubling, this high C tuning arrangement is satisfactory. Regeneration is obtained by connecting the "plus" B tap 3 turns from the grid end of the coil. The coil has 10 turns, 1¼ in. long, of No. 16 enameled wire on a 1½ in. diameter form. The pentode must feed from the plate end in order to obtain the best impedance match and consequently greatest grid drive.

The pentode plate circuit consists of a simplified Pi network for reducing harmonic radiation. Any length of antenna is satisfactory but infinitely better results are secured with a 132-foot aerial wire connected to an off-center single-wire feeder. This length is a compromise, being a trifle long for 75 meter phone, but is the correct length for 80-meter CW, 40- and 20-meter CW and phone.

The plate circuit is tuned by a 100 uufd., 6000-volt condenser. The coil and antenna condenser complete this circuit, resonance always being obtained with the 100 uufd. condenser. The other condenser properly loads the antenna. A low impedance antenna requires more capacity than given by the 225 uufd. 3000-volt condenser and a .0001 or .00025 ufd. mica 5000-volt condenser must be shunted across it at the condensers terminals. This condenser is not always necessary with a single-wire feeder, but with any odd length antenna, the fixed condenser may be required. The coil for the final stage consists of No. 10 wire on celluloid strips, 4 turns per inch, with 19 turns total on a 4½-in. diameter. The 20-meter tap is at three turns, and the 40-meter connection at eight turns. The switch short circuits all of the other unused turns so that at 20 meters only three turns are left in the circuit. For this reason the Ohmite switch is mounted very close to the coil for short connections. An extension shaft and knob provides front panel control.

All by-pass condensers on the filaments and grids are connected together at the socket and to a common ground point on the chassis.

Two power supplies provide all of the necessary voltages. A 400-volt supply furnishes 50 to 60 MA to the 53 tube, and from 60 to 70 MA to the pentode screen. It also supplies 250 volts to the speech amplifier and modulator through a 6000-ohm 20-watt resistor. The main voltage divider across the 400-volt supply provides various bias voltage for the pentode tube grids. The negative B lead from the 1800-volt supply and 10-volt filament center-tap connects to a point about 50 volts above ground potential on the voltage divider. The latter is made up of individual resistors mounted underneath the middle chassis. This gives an additional 50 volts to the control grid so that it operates with a negative excitation potential of 100 volts. It also provides a 50-volt negative potential to the suppressor when the switch is thrown to the "phone" position. On CW the suppressor-grid must be approximately 50 volts positive with respect to the tube filament, hence, another tap higher up on the voltage divider gives this voltage. This tap is 100 volts above ground but only 50 volts with respect to the RK28. The plate current does not flow through any part of this voltage divider.

The modulator unit is conventional; however, a carrier-shift indicator consisting of any diode tube, RF pick-up coil or lead, and a 0-1 milliammeter should be available at all times to check over-modulation. Tests have indicated that the RK28 would modulate properly when 9 to 12 MA of DC grid current was flowing. The 53 tube supplied 12 MA on 20 meters, 13 to 15 on 40 meters and 14 to 18 on 80 meters.

The transmitter is mounted on a 3-ft. relay-rack. Three 1-ft. No. 14 gauge panels hold the three chassis and controls. The chassis are of No. 20 gauge steel with corner end-supports for bracing to the front panels. The top and bottom decks are 12 x 17 x 2 inches, and the middle deck is 9 x 17 x 2 inches. The lower deck holds the high voltage power supply, the low voltage transformer, rectifier and first filter choke. The second deck holds the remainder of the low voltage filter on the end opposite to the audio amplifier. The 53 crystal-oscillator and doubler circuits mount in the middle of this deck. The top decks holds the final amplifier.

Close-up of the RK-28 (803) final amplifier.

The grid circuit of the final stage is shielded from the output circuit by the chassis. The five-prong socket is mounted on standoff insulators below the chassis which has a hole cut in it for the tube.

Note: The new low temperature-coefficient AT or V-cut crystals are recommended because the 53 operates at fairly high power, which may cause a frequency drift with ordinary X-cut types.

Controlled Carrier Modulation

With controlled-carrier modulation a high percentage modulation is obtained at all levels; in addition, the average carrier varies in magnitude with the audio output. The advantages of this type of modulation are enumerated below:

(1) Increased DX possibilities for the amateur, or greater blanket coverage for the broadcast station. This improvement is attributed to the high-percentage modulation at low levels, which is very important, as the integrated audio output of a transmitter is seldom exceeded by 10 per cent of its rated maximum output. The theoretical side of this phase of controlled-carrier modulation indicates that it is possible for a small station to have the same coverage as a station many times larger, and with less interference.

(2) Increased tube economy. Due to the fact that the class C input is low for the major part of the time, which is similar in effect to class B, permits much higher output from a given tube arrangement.

(3) Reduction in interference. This is of extreme importance to both the amateur and broadcast station. Because the carrier level is low for the major portion of the time, interference and so-called "chatter" is reduced.

(4) Reduction in power consumption. This is due to the low power taken by the final amplifier over a large portion of the operating time. If the tubes are not operated above their normal ratings, this power reduction will tend to prolong tube life.

(5) Higher definition is possible from the broadcast station from the standpoint of volume range. One of the factors con-

trolling this range occurs when the audio level drops to a very low level, which decreases the percentage modulation with the square of the audio drop. This results in poor and possibly no reception at the receiving end. A higher percentage modulation at low levels would tend to overcome this difficulty and permit the transmission of wider audio volume ranges.

(6) If high power output is required, the controlled class C can be fed into a class B linear amplifier. The power rating of output tubes operating in this manner is at least doubled; in addition, the efficiency is increased due to the class B operation. As the audio level rises and falls, the class C input is varied; these transitions are applied to a clas B linear stage whose power output is caused to vary over a wide range. One of the feature points of this modulation system is that modulation need not take place at a very high level. A 500 watt station will require only 50 watts, or less, of audio output; this modulates a corresponding class C tube which drives the 500 watt class B linear amplifier. For higher powers, the class B linear amplifier need only be made to excite another higher powered class B linear stage.

Class B Series Control Using Same Tubes for Audio and Syllabic Modulation: Numerous systems of controlling the average input and carrier output of a class C plate-modulated amplifier have been developed and successfully applied to high-frequency phone transmitters.

In one of the earliest systems the variations in the average plate-to-cathode resistance of a class B audio stage series modulated a class C amplifier. The audio output of the class B amplifier was fed through a conventional class B output transformer into the plate circuit of the controlled and plate-modulated class C amplifier.

This system is quite simple, but it has two disadvantages. These are:

(1) The resistance of the plate-to-cathode path of the class C amplifier is in series with the B-plus high-voltage lead to the modulators, which is the equivalent of placing a 5000 ohm resistance in the B-plus lead. This materially affects the voltage regulation of the DC plate-voltage supplied to the modulators, with some consequent audio distortion. This variation of the plate-voltage applied to the class B modulators also causes the "cut-off" point to drift as the audio signal varies. Without a zero-biased modulator tube, the modulators would be operating class C part of the time, which is not conducive to high-definition.

(2) During the non-operative condition, the plate-voltage across the modulators must be about twice the operating value, which means that lower than normal operating plate voltages must be used because there are no zero-bias modulator tubes available which will stand plate voltages of twice the normal operating voltage. Consequently, the maximum power output must be reduced below that which the same tube capacity could deliver in the conventional system of constant carrier modulation.

Thyratron Control: Another system of controlled-carrier modulation of a class C amplifier employed grid-controlled rectifiers in the plate power supply. These grid-controlled mercury-vapor rectifiers allow a simple means of controlling the DC plate voltage at syllabic frequencies.

For high-power operation this system is practically unexcelled. This is largely due to the elimination of lag when the system is applied to a three-phase power supply; the latter also requires little hum filtering.

Variactor Control: Another simple and effective modulation system has been developed by I. A. Mitchel, in which the variation in DC plate current drawn by a class B audio amplifier is used to control the saturation of a voltage-dropping reactor in series with the primary of the class C amplifier plate power supply. As the DC drawn by the modulators increases, the reactor core becomes more saturated, which reduces the AC resistance and allows more current to flow in the primary circuit of the class C power supply. This increased current, which corresponds to raising the primary voltage, increases the DC plate voltage and thus causes the input and average carrier output of the class C stage to increase with modulation.

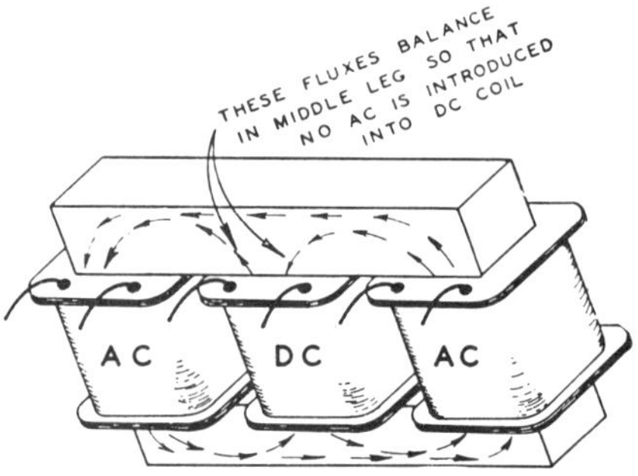

FIG. 1—Appearance of saturable reactor.

The success of the Mitchel modulation system depends completely on the fact that the modulator plate current in a class B amplifier varies practically linearly with the power output. This plate current saturates a control reactor which in turn controls the plate supply of the class C final amplifier. With a class A modulator, other means of obtaining this control current are possible. The general nature of a saturable-core reactor is illustrated in Figure 1. A shell type of laminated core of somewhat different proportions than in an ordinary transformer forms the magnetic circuit. Three coils are placed on the respective legs of this core, the outer two being connected in series with the AC line and so related in polarity that their respective magnetic circuits are in accordance with the arrows shown. It is seen that the MMFs of the two magnetic circuits are of opposite direction to the middle leg and so tend to neutralize each other. If the coils and magnetic circuit are perfectly bal-

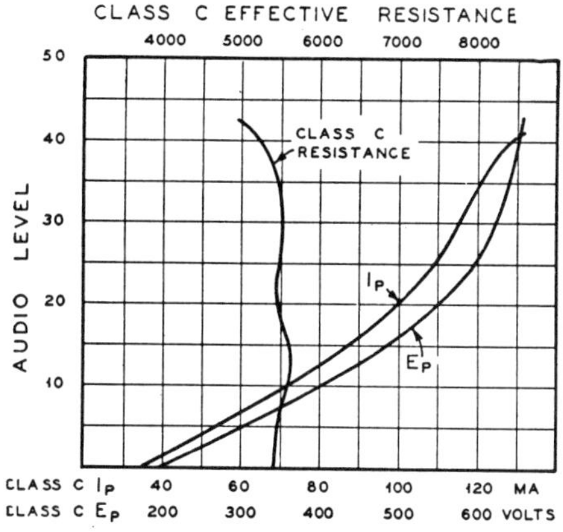

Relation of class C operating characteristics to modulator level in controlled carrier transmitter.

anced, these fluxes will cancel out and no AC will traverse the central magnetic circuit. The control coil is placed on the middle leg of the core and the plate current of the class B modulator is passed through it. The reactor functions by reason that as the DC current is increased in the middle winding the inductance decreases, which augments the strength of the DC magnetic field and retards the flow of the AC magnetic flux coupling the two associated magnetic paths. By properly designing the reactor, a fairly linear relation and a wide range of inductance can be obtained. The relation of the saturating DC to AC impedance, used in the transmitter previously referred to, is illustrated by a curve drawn in Figure 2; the linearity of this curve can be increased still further.

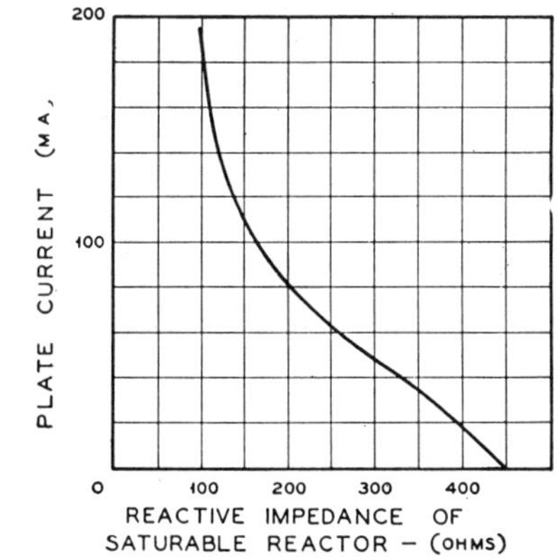

FIG. 2—This curve shows the change in reactance of a-c coils in a saturable reactor as the d-c is increased.

The saturable core reactor is placed in series with the primary of the final plate transformer. It is seen from Figure 2 that with no audio signal (minimum DC) the reactance is quite high (450 ohms). This effects a great voltage drop to the primary of the plate transformer, as the effective impedance of this primary is quite low. However, as the saturating DC is increased, the reactance as well as the voltage drop is decreased. The primary voltage rises in accordance with this, and with proper design reaches almost maximum at normal maximum-audio output. Even with the reactor practically saturated, a small reactance and consequent voltage drop exists. To compensate for this, an auto-transformer is placed on the line-side of the reactor, which increases the total impressed voltage. The auto-transformer is not needed if the plate transformer primary-winding is wound or tapped for reduced voltage obtained after the reactor drop. In either case, this voltage drop does not represent a power or efficiency loss, because the drop is almost entirely reactive and results primarily in a change of power factor; that is, only the ratio of VA/watts increases.

Notes on Adjustments of Controlled-Carrier Transmitters: The first step in adjusting a controlled-carrier transmitter is to properly modulate the class C stage with the carrier control disconnected. In other words, the plate voltage on the class C stage is first fixed at the highest value obtainable when the carrier control is turned on. For conventional modulation, the neutralization, RF excitation and grid bias must be correct; in addition, there must be no detectable movement of the pointer in the plate milliammeter during audio modulation. After the class C stage and the audio modulators are functioning correctly, the carrier-control can be turned on. Some means must be provided to vary the carrier-control, or syllabic modulation, in relation to the audio modulation.

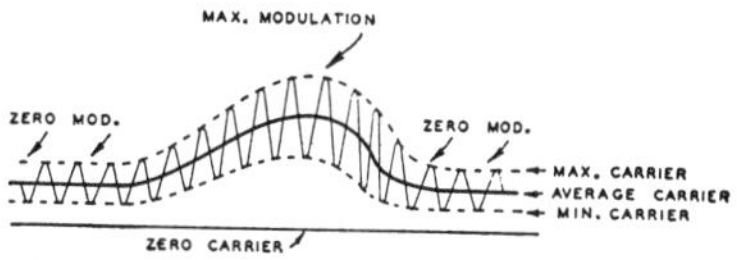

Ⓐ TOO MUCH CARRIER CONTROL: INEFFICIENT AND NOISY.

Too much carrier-control is indicated by the curve drawn in Figure A. The undulatory diagram depicts that the DC plate voltage on the class C stage swings up much farther than required by the amount of audio modulation present. The only disadvantage involved here is that unecessary heterodyne interference is created, and a class B linear amplifier following the class C stage will operate inefficiently because the average percentage of audio modulation of the carrier is not high enough. (Note: class B linear amplifiers are only efficient at high percentages of modulation.)

Where there is not enough carrier-control, as indicated in Figure B, the average carrier power becomes less than twice the maximum side-band power, which repre-

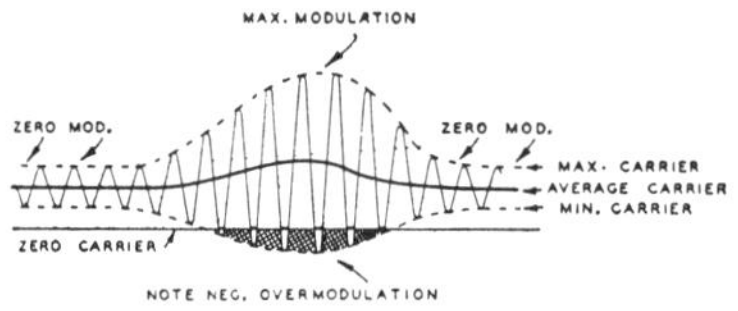

(B) NOT ENOUGH CARRIER CONTROL: OVERMODULATION.

sents over-modulation. The shaded area under the zero line represents an absolute absence of RF output which causes bad side-band "splatter" and undesirable interference. This negative over-modulation can only be detected by means of the overmodulation indicator as shown in Figure 10. The remedy is to increase the carrier-control, or else reduce the audio modulation.

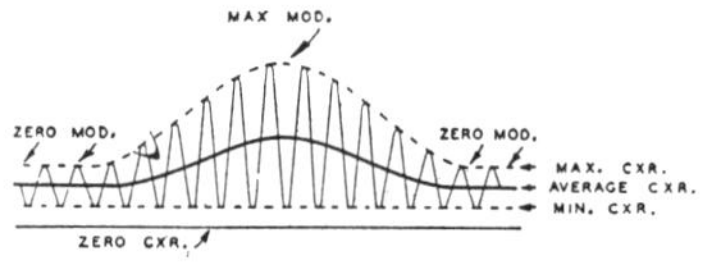

(C) PERFECT CARRIER CONTROL. NOTE PRACTICALLY CONSTANT MIN. CARRIER.

Perfect carrier-control is shown in Figure C. Here, the minimum carrier level remains constant and the carrier-control has just enough effect to keep the average carrier power a little over twice that of the audio side-band power. While the wave shown in Figure C has been idealized for the purposes of illustration, the smoothness of the undulations will not be encountered in practice. However, it is desirable to keep the minimum carrier value as low and as constant as possible without letting the negative peaks cross the zero line (overmodulation).

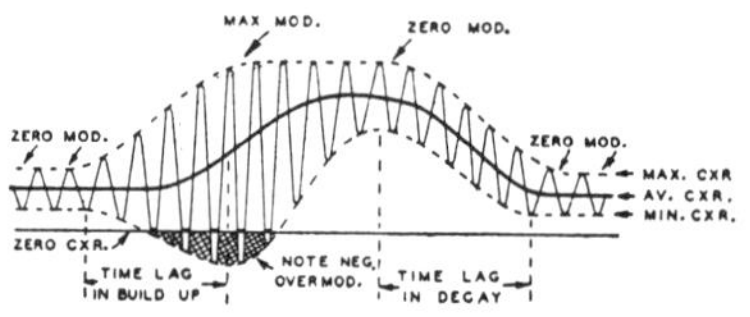

(D) SATISFACTORY DEGREE OF CARRIER CONTROL BUT BAD TIME LAG IN CONTROL: OVERMODULATION AND LOW EFFICIENCY.

A satisfactory degree of carrier-control is portrayed in Figure D. In general, the slow time-delay (or lag) in the filter circuit, which separates the audio component from the syllabic component, delayed the rise in the carrier until considerably after the audio modulation has become effective. Thus the carrier is over-modulated at the start of the syllable, and under-modulated at the end of the syllable. This trouble can be traced to too high capacitance in the syllabic filter.

¼-KW Variactor Controlled Carrier Phone: This transmitter employs the variactor system for controlled carrier phone operation. A linear RF stage is used to efficiently amplify the controlled carrier output of the class C stage. The efficiency of a linear stage under these conditions is much higher than for a normal straight linear stage, consequently an effective carrier of about 250 watts is developed. The output is almost twice that obtainable with the same tubes in an uncontrolled-carrier transmitter at the same plate voltage.

The complete circuit wiring diagram of the transmitter is shown. The transmitter is constructed on five panels about 19 in. x 12 in., made from No. 14 gauge iron, and mounted on a standard relay rack. The high voltage power supply is built so that all of the transformers and chokes are mounted on the front panel, with only the 866 rectifier tubes, 2 ufd. 2000-volt condensers, bleeder resistor and terminal strip mounted on the chassis. The latter is 17 in. x 9 in. x 1½ in. and is made of No. 20 gauge lead-plated steel.

The low voltage power supply has a chassis 17 in. x 11 in. x 2 in. on which are mounted the three power transformers, variactor and auto-transformer AV-1 and chokes. The filter condensers and bleeder resistors are mounted underneath the chassis. The filament windings on the transformer supplying heating current to the class C stage must be on a separate transformer on account of the primary voltage variations which are in accord with the syllabic modulation.

The chassis for the modulator deck is 17 in. x 8 in. x 2 in. The two meters, a 0-50 and a 0-500 or 0-300 milliammeters are also mounted on this panel. The current measuring jacks are arranged in the circuits to open the center-tap leads in order to eliminate sparking or flash-over. These jacks may have to be mounted in insulating washers since the jack-sleeve is ungrounded in some circuits. A telephone plug and flexible cord permits either meter to be inserted via the plug into any circuit jack.

The low power RF deck is made with a vertical bakelite subdeck, 16 in. x 8 in. x $\frac{3}{16}$th in., mounted on 1¼-inch studs from the front panel. This 1¼-inch space allows space for wiring, resistors, condensers and insulated flexible shaft couplings for the variable condensers. Two 8 in. x 6 in. aluminum shields are placed around the buffer 45-stage to prevent reaction between RF stages.

The top deck is made with a chassis 17 ins. x 11 ins. x ½-in. of No. 12 gauge aluminum, with a vertical partition between the plate tank circuit and the remainder of the final amplifier parts. This vertical partition strengthens the chassis, fastens to the front panel, and provides a mounting surface for the two neutralizing condensers on stand-off insulators. All the parts are arranged symmetrically so that all leads on each side of the push-pull stage are equal in length.

Circuit Details: On 75 meters, a 7000-volt single-section tank condenser having an ap-

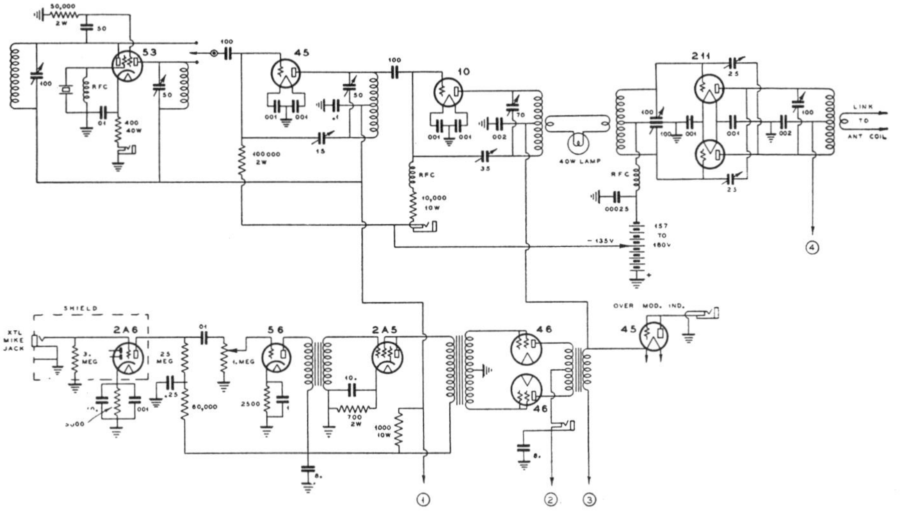

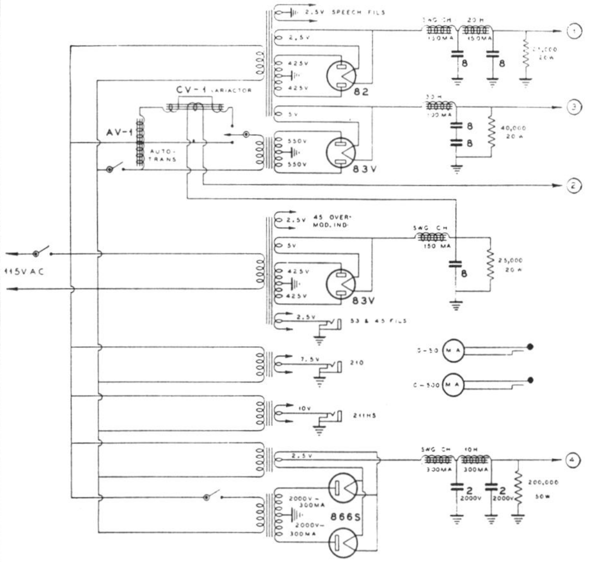

Complete circuit diagram of the r-f, speech, Variactor control and power units for the 1/4 KW phone transmitter. Although the circuit diagram shows type 211 tubes in the final r-f amplifier, the tubes actually used in this transmitter were Amperex type 211D. Standard 211 types can also be used without change in the circuit. This transmitter has been given a thorough try-out on the air, and reports from amateurs 1000 miles away state that the voice quality is good, the signal strength R9 plus.

As can be seen from the circuit, the Jones Exciter is used to drive the 45 stage which, in turn, is capacitively coupled to the single 210 driver for the 211s in the final amplifier.

proximate capacity of 100 uufds. is required to properly load the antenna for linear amplification. The coil is center-tapped and by-passed to ground with a 5000-volt .002 ufd. mica condenser. No plate RF choke is needed.

If the transmitter is to be operated on 20 meter operation, a split-stator plate condenser is necessary and no grid r.f. choke should be used.

A 25 or 40-watt Mazda lamp functions as a plate load for the class C stage. This resistor provides a more constant load than would be provided by the grid circuit of the linear stage. More than half of the output from the class C stage is dissipated in this resistor. Placing the lamp in series with the link coupling requires that 5 turns be placed around the center of the 75-meter plate coil and 2 turns around the center of the 211-grid coil. The loops and links are made with ordinary No. 20 hook-up wire.

When the transmitter is operated on CW, a 5000-ohm 50-watt grid-leak must be plugged into the final grid current measur-

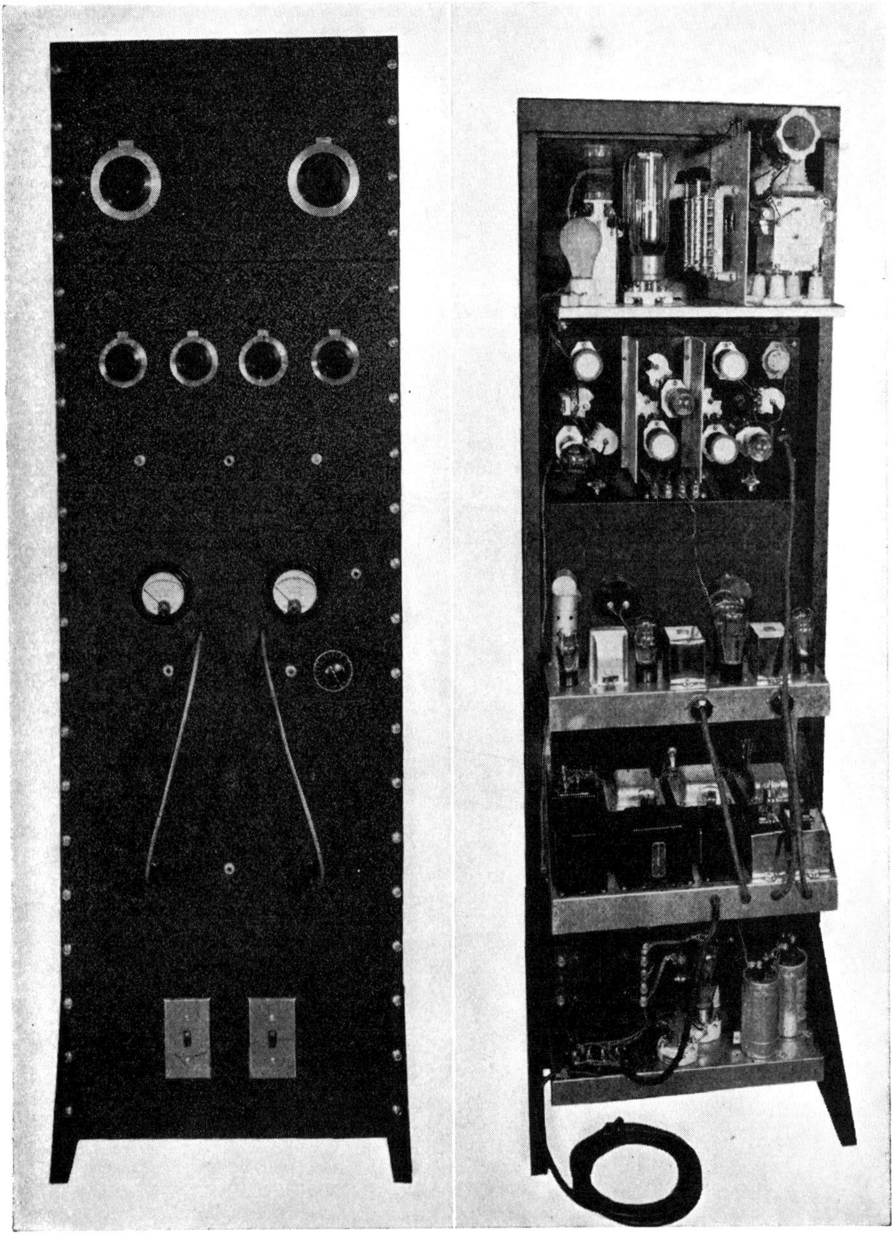

Front and rear views of the 1/4 k-w Variactor controlled carrier phone transmitter.

ing jack, or in series with the C-battery lead. The link coupling is then made with one turn at each end and without the series lamp. Keying may be done in the 53-cathode current jack as fixed-bias and grid-leak bias is placed on all the stages. External B batteries are normally employed for furnishing C-battery potentials; the total charging current is low, between 10 and 15 MA for phone operation.

The final tank coil is link coupled by 2 or 3 turns to an external antenna coil circuit. This plate coil for 75 meters is made of No. 12 wire on a 2¾-inch form, seven turns per inch, with a total of 22 turns, center-tapped. All coils, except those in the oscillator are center-tapped. The grid coil is made of 40 turns of No. 18 DSC wire, close wound on a 1½-inch diameter form. The 10-stage plate coil is made of No. 18 DSC wire on a 1½-inch diameter plug-in coil form, with 32 turns close wound. The buffer coil has 38 turns of a similar winding. The doubler is the same as the 10-stage coil, except that no center-tap is required. The 160-meter oscillator coil is close wound with 70 turns of No. 22 DSC wire, also on a 1½-inch form.

Each panel must be provided with either a terminal strip or with power sockets for plug and cable cross connections. No. 12 gauge flexible wire will amply carry the filament load to each tube socket without a drop in voltage. All filaments are by-passed each side to ground with .001 ufd. condensers placed at the tube sockets.

Miscellaneous Adjustments: An ordinary 2½-volt lamp and single turn of wire may be used in making the circuit alignment. When the oscillator and doubler is properly tuned, the cathode current must not be over 60MA. The buffer stage current is approximately 20MA in this case, and the 10-stage late current, on peaks, will be between 50 and 70MA. It is not necessary to increase the 46 class B current to over 90 or 100MA for full modulation and output. The Mazda lamp will increase in brilliancy when modulating with or without a controlled carrier; with a controlled carrier, the lamp will increase from minimum to moderate brilliancy.

The final stage is first adjusted without carrier control, as is the practice with any linear stage, preferably at low plate and grid voltage. Without carrier control, the grid current will be zero and the plate current about 150 to 200 MA (at 1800 volts), with no modulation, and not over one or two milliamperes of grid current when modulated with a steady tone.

With carrier control, the grid current will deflect upwards to between 5 and 10 MA and the plate current on the final stage will indicate up to about 300MA. If this value is not attained, with good quality on speech as received on a monitor, it is possible that the antenna is insufficiently loaded or the C-bias voltage is not slightly less than the cut-off value.

Variactor Controlled-Carrier Unit For Small Transmitters

This unit can be quickly incorporated in nearly any moderate or low powered phone transmitter in order to provide controlled carrier transmission. The unit described here will control power of from 25 to 50 watts input to the class C stage, the operation being effected without disturbing the type of modulation used.

Circuit: In general, the circuit follows the principles outlined in the foregoing paragraphs; however, it is required that the power supply for the 2A3 have poor regulation, such as that shown, in order that linear operation can be secured. It is desirable that the 2A3 plate current through CV-1 vary directly as the ratio of input AF voltage from the modulator.

The input from the modulator choke or driver stage consists of two of the new midget ½-ufd. oil-filled 1,000-volt condensers in series with a 25,000-ohm, 40-watt voltage divider. This provides a simple means of setting the actual AF voltage to the desired value which is to be supplied to the grid.

C-bias should be of a value that will bring the plate current of the 2A3 to between zero and ten milliamperes with no input.

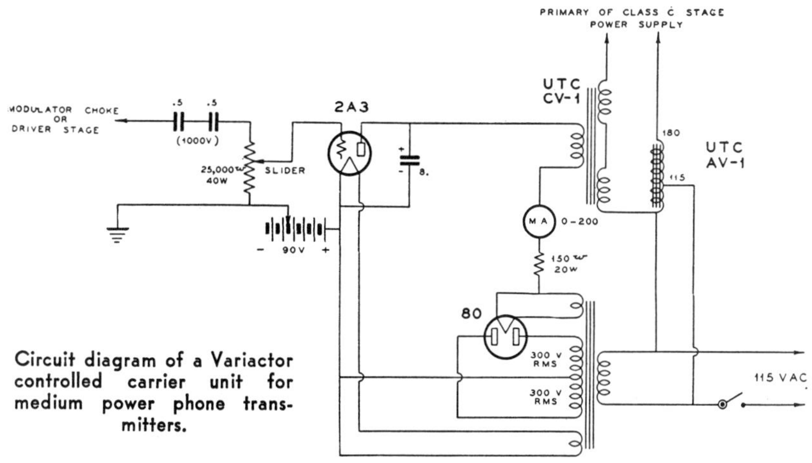

Circuit diagram of a Variactor controlled carrier unit for medium power phone transmitters.

Controlled Carrier Grid-Bias Modulation

Excellent results can be obtained with grid bias modulation when used in conjunction with controlled carrier. The final modulated amplifier stage operates more efficiently, with the result that more output for a given amount of plate dissipation, or the same equivalent output with much less tube heating is obtained.

In the transmitter illustrated and shown in the circuit diagram, the grid bias modulation system is exactly the same whether controlled carrier is switched in or out. The controlled carrier system uses a type 46 high mu tube in series with the buffer or doubler stage. This 46 tube acts as a cathode bias resistor, reducing the plate current of the buffer r.f. tube to nearly zero when there is no speech input. When the microphone is spoken into, the 46 tube acts as a V.T. voltmeter, with an increase of its plate current just as in a class B audio amplifier. This current increase also means an increase of buffer tube current and r.f. output.

The 46 tube requires a little audio power to drive its grid circuit, and thereby to lower its plate impedance. In the circuit shown the 2A5 modulator supplies this power, as well as the power needed for the modulation of the two type 50T tubes. The 46 grids should be tapped across about one-half of the load resistor so as to not cause overload trouble in the 2A5 pentode output circuit. A pair of 45 tubes of lower impedance would be more suitable for a modulator.

Since only the DC change of plate current is desired in the control of grid bias of the buffer stage, an audio filter choke and a pair of ½ mfd. condensers prevent voice modulation in this stage. The DC plate current follows the syllabic variations of the voice, and therefore carrier power is supplied to the grids of the final amplifier in proportion to the voice input.

The 50T tubes operate cooler because their plate current when idling is about ⅓rd to ½ of that when fully modulating. If greater control is desired, choke input filter from the power supply would give

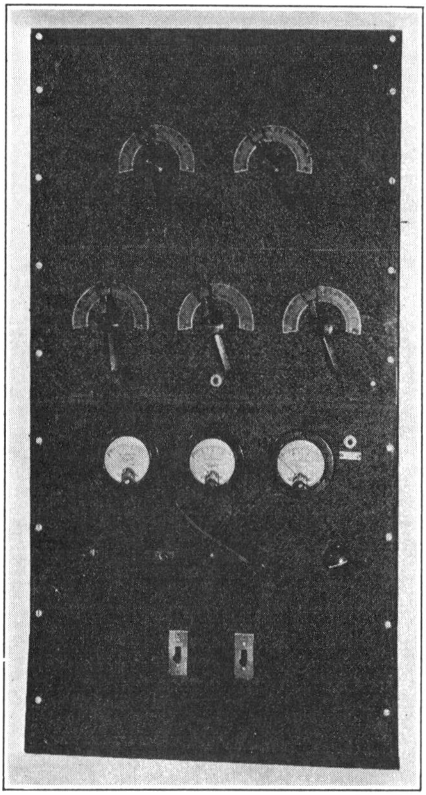

Front and rear view of Jones controlled carrier grid-bias modulated phone transmitter.

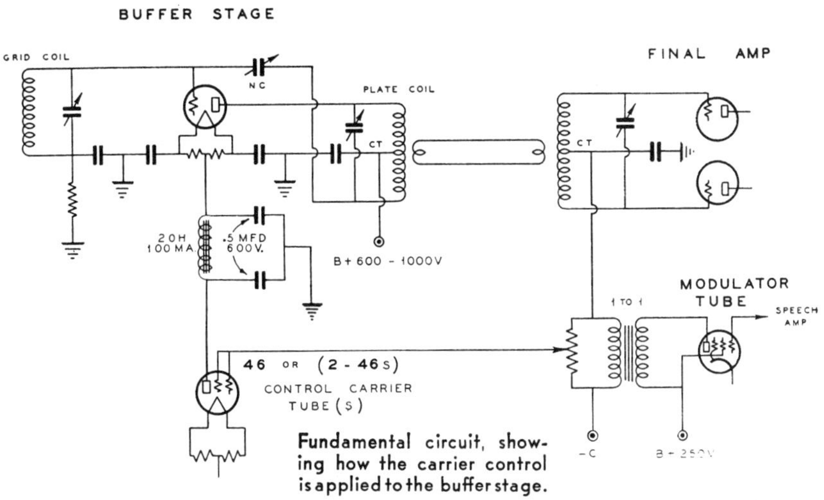

Fundamental circuit, showing how the carrier control is applied to the buffer stage.

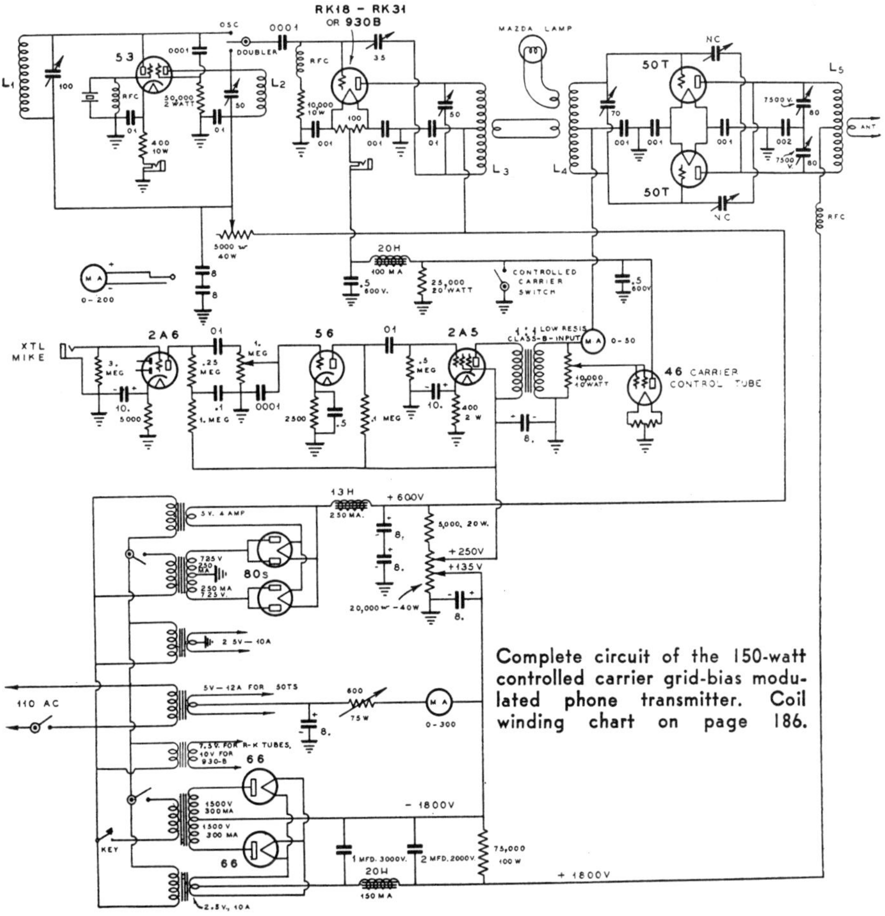

Complete circuit of the 150-watt controlled carrier grid-bias modulated phone transmitter. Coil winding chart on page 186.

COIL WINDING CHART
For Grid Bias Controlled Carrier Phone

Wavelength in Meters	L_1 Oscillator	L_2 Doubler	L_3 Buffer	L_4 Final Grid	L_5 Final Plate
160	56 turns No. 24 DSC. Close wound. 1½" diam.				
80	29 turns No. 20 DSC. 1½" long. 1½" diam.	29 turns No. 20 DSC. 1½" long. 1½" diam.	32 turns No. 20 DSC. Center tapped. 1⅜" long. 1½" diam. 1 turn link.	34 turns No. 20 DSC. Center tapped. 1½" long. 1½" diam. 2 turn link to lamp.	44 turns No. 16 "air wound" Center tapped. 2⅛" diam. 2½" long.
40	16 turns No. 20 DSC. 1½" long. 1½" diam.	16 turns No. 20 DSC. 1½" long. 1½" diam.	20 turns No. 20 DSC. Center tapped. 1½" diam. 1½" long. 1 turn link.	20 turns No. 20 DSC. Center tapped. 1½" diam. 1½" long.	22 turns No. 12 "air-wound" Center tapped. 2⅛" diam. 3" long.
20	7 turns No. 20 DSC. 1½" long. 1½" diam.	7 turns No. 20 DSC. 1½" long. 1½" diam.	11 turns No. 20 DSC. Center tapped. 1½" long. 1½" diam. 1 turn link.	10 turns No. 20 DSC. Center tapped. 1½" long. 1½" diam. 1 turn link to lamp.	7 turns No. 10 "air-wound" Center tapped. 3⅝" diam. 2" long.
10			5¼ turns No. 18. Center tapped. 1¼" long. 1½" diam. 1 turn link.	5¼ turns No. 18 Enam. Center tapped. 1¼" long. 1½" diam. 1 turn link to lamp.	4 turns No. 10 "air-wound" Center tapped. 2⅝" diam. 2" long.

better voltage regulation in this stage. Oscillograph pictures of steady tone or voice input look exactly the same for carrier output with either controlled or regular output, except in amplitude.

This system is easily applied to any grid modulated phone transmitter, providing the buffer stage has sufficient power output for normal cw operation. With controlled carrier, this buffer output is reduced even on peaks to about ⅕th or, at the most, ⅓rd of the normal output. This peak output should be about the same as the steady output needed for normal grid modulation in driving the final stage. A tube such as a 2A3, with fixed bias set to near cut-off, would allow more output to be obtained from the buffer stage on voice peaks than that which is secured with the 46 tube shown in the diagram. The latter requires no fixed bias, but requires grid driving power, and does not reach as low an output impedance value as when a 2A3 or 45 tube is used.

Speech channel for high-power transmitter shown on facing page.

High-Power Controlled Carrier Grid-Bias Modulated Phone: One or two 2A3 tubes, biased to approximately cut-off, serve as a control for carrier output in the grid-modulated phone transmitter illustrated on page 187. A picture of the carrier control and speech channel is shown.

Two 2A3 tubes in parallel will enable the RK-31 buffer stage to operate at a lower value of plate voltage than that shown. The 2A3 tube serves as a cathode grid bias resistor which varies in accordance with the syllabic modulation of the voice. The audio components are filtered by means of a 20 henry filter choke and two ½ mfd. condensers. These condensers should have a rating of 600 volts. The grid bias on the 2A3 tube should be no higher than that required to reduce the RK-31 plate current to about 10 or 20 MA., and the link coupling

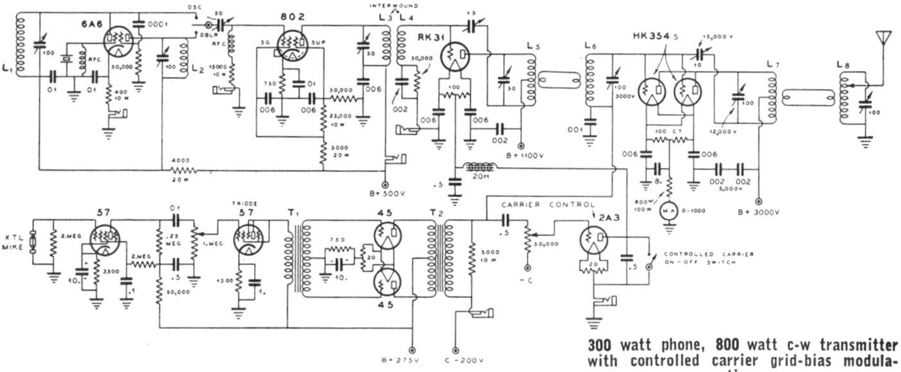

300 watt phone, 800 watt c-w transmitter with controlled carrier grid-bias modulation.

should be adjusted so that the final amplifier will deliver from 25 to 50 watts carrier. The buffer and final amplifier plate currents increase when modulation is applied, and thus 300 watts of carrier can be obtained.

With a plate supply of 3,000 volts, the actual plate-to-filament voltage will run approximately 2,750 volts because there should be a drop of more than 200 volts across the cath-

Right: Jones Exciter, 802 buffer-doubler and RK-31 buffer for high-power final amplifier shown below. Two HK-354 Gammatrons are used.

ode or filament center-tap resistor. On voice peaks the plate current should run at about 320 watts, since the side-bands use up the remainder of the power, being 50% of the carrier at 100% modulation.

Although the carrier output efficiency will not run over 40%, the fact remains that the actual output efficiency during modulation runs about 60%, which explains why 50% more power output can be obtained when using carrier control with less plate heating than when normal grid-bias modulation is used. When no voice input is applied, the final amplifier is not putting out more than a small fraction of its normal output, which means that the plate dissipation is below normal, even if the efficiency during such times was as low as 20%.

This transmitter uses a 6A6 oscillator-doubler with either an 802 or RK-25 tube in the buffer or additional doubler stage. The screen grid tube requires no neutralizing when used as a buffer, and as a doubler the output is ample for driving the RK-31 tube which is rated at 40 watts plate dissipation. For c-w operation this tube has sufficient output to drive the final amplifier to more than 1 k-w input at high efficiency on any band from 20 to 160 meters. If 10 meter operation is desired, the final amplifier can be operated as a doubler. It would be better, however, to use push-pull connection for the HK-354 tubes because the circuit shown on page 187 has too much regeneration at 10 or 20 meters to allow satisfactory doubling in the final amplifier.

A pair of 45 tubes act as modulators when phone operation is used, and they also drive the carrier control tube. Primary keying prevents key clicks when operating with c-w. The full output of the RK-31 buffer is needed when the set is operated as a c-w transmitter.

COIL WINDING DATA FOR HIGH-POWER CONTROLLED CARRIER GRID-BIAS MODULATED PHONE

Band	6A6 Osc. or Doubler	802 Plate Coil	RK-31 Plate Coil	Final Grid Coil	Final Plate Coil
160	56 turns, No. 24 DSC, 1½-in. dia. Close-wound.	56 turns, No. 24 DSC, 2-in. long, 1½-in. dia. 50 turns, No. 24 DSC, wound over plate coil for grid winding.	65 turns, No. 16 SCC, 2½-in. dia. Close-wound, center-tapped.	40 turns, No. 16 Enam., 2¾-inch dia. 3-in. long.	40 turns, No. 12 Enam., 5-in. dia. 7-in. long. Center-tapped.
80	30 turns, No. 20 DSC, 1½-in. dia. 1½-in. long.	36 turns, No. 22 DSC, 1¾-in. long, 1½-in. dia. 28 turns, No. 24 DSC, close-wound.	44 turns, No. 16 Enam., 2-in. dia. 3¾-in. long. Center-tapped.	20 turns, No. 14 Enam., 2¾-in. dia. 2¾-in. long.	24 turns, No. 10 Enam., 4½-in. dia. 6½-in. long. Center-tapped.
40	20 turns, No. 20 DSC, 1½-in. dia. 1¾-in. long.	16 turns, No. 22 DSC, 1½-in. dia. 1¾-in. long. 16 turns, No. 24 DSC, Interwound.	22 turns, No. 12 Enam., 2-in. dia. 3-in. long. Center-tapped.	11 turns, No. 14 Enam., 2¾-in. dia. 2½-in. long.	18 turns, No. 10 Enam., 3½-in. dia. 6½-in. long. Center-tapped.
20	10 turns, No. 20 DSC, 1½-in. dia. 1¾-in. long.	7 turns, No. 20 DSC, 1½-in. dia. 1½-in. long. 7 turns, No. 20 DSC, Interwound.	10 turns, No. 12 Enam., 2-in. dia. 3-in. long. Center-tapped.	6 turns, No. 14 Enam., 2½-in. dia., 2¾-in. long.	9 turns, No. 10 Enam., 3¼-in. dia., 6½-in. long. Center-tapped.

10-Meter Equipment

Amateur stations are licensed to operate in the bands from 28,000 to 30,000 KC, and voice communication between 28,000 and 29,000 KC. The 10-meter band, during certain sun-spot cycles, lends itself admirably adaptable for communication over both moderate and extremely long distances with relatively low power. No special license is required for telephonic communication in this band. Licensed radio amateurs can, therefore, add to their already existing pleasures the attainment of 10-meter DX possibilities.

The 10-meter band is subject to extreme skip-distance effects and the advantages of this band are best realized during daylight hours. A well-designed crystal-controlled 25-watt transmitter, or one with less output, will practically enable world-wide communication under favorable conditions.

For local or medium distances, up to 1000 miles, a vertical half-wave antenna, approximately 16-feet long, has proven most satisfactory. Either a twisted-pair feeder or two-wire spaced feeder is quite effective for coupling the transmitter or receiver to the antenna. A horizontal half-wave antenna, or directional array, is ideal for long distance work; such a system has the minimum parasitic pick-up for receiving and is more efficient than a vertical antenna.

Diamond antennas are excellent where transmission or reception is to be confined roughly to one or two directions.

Transmitter Considerations: Crystal-controlled signals are always desirable, and 40 or 20-meter crystals can be used in oscillators having relatively high outputs. This adaptation simplifies the problem of driving frequency doublers without auxiliary intermediate buffer stages. Frequency doublers should be of the regenerative push-push or regenerative single-tube types so that reasonable output is secured without excessive tube heating. For moderate or low power, the output from the frequency doubler is sufficient to drive a final amplifier to 40 or 50-watts output on 10 meters. Grid modulation is advantageous as it requires less excitation than that of plate modulation.

Tubes for 10-meter operation should have fairly low inter-electrode capacities, such as types 53 or 6A6, 10 or 801, 50T, 150T, HF200, 838, HK354, 300T, HF300, 800, RK18, RK25, RK20. The plate efficiency in operating at very high frequencies is generally lower than can be obtained on other amateur bands, and the designer must make an allowance for this fact. All RF leads must be very short and direct. In general, neutralized amplifiers should have split-stator tuning condensers in order to minimize regeneration. Low-C tubes will neutralize more readily than the medium or high-C types.

The type 802 tube is not satisfactory for 10-meter service; however, it performs quite effectively as a doubler but has twice as much output capacity, and less power output than the RK25.

Link coupling is most practical between stages, except where space is at premium. In such cases, interwound, untuned grid coils which have unity coupling to the tuned plate coil work quite effectively. Capacity coupling is not desirable due to power loss difficulties of neutralization. The circuits should be tuned with low-C, since the frequency is high enough to provide ample flywheel effect. Low-loss dielectric sockets are a necessity and small coil forms are desirable as less coil heating and higher output efficiency can be obtained.

Any of the conventional methods of antenna coupling are suitable. Link coupling to a tuned antenna circuit, or directly to the twisted-pair feeders is most practical for the small coils recommended for use in a final amplifier.

A field intensity meter is almost indispensable for tuning the antenna circuit to maximum output (such an instrument is described in this section).

10-Meter Receivers: It is important that all the RF leads be kept as short as possible, and fairly low-C form the circuit complement specifications. The special receivers described in the subsequent sections were developed primarily for 10-meter operation and the constructor is advised to closely adhere to all details. Most all-wave receivers, due to inherent design difficulties, produce a high noise level with very poor sensitivity for actual signal reception; however, a remedial expedient is to inductively couple the receiver to the feeders from a resonant antenna; this will always give the optimum signal-to-noise ratio.

10-Meter Antennas: A practical 10-meter antenna is described on page 262. A 10-meter field strength meter is shown on page 263.

10-Meter Portable Receiver: This receiver, the circuit of which appears in Figure 1, has been especially designed for 10 and 20-meter operation.

In general, a 19 tube serves as a regenerative detector and audio amplifier. Large coils, wound on plug-in coil forms, make the L over C ratio exceptionally high, resulting in good sensitivity.

Regeneration is controlled by a tapered 250,000-ohm variable resistance of the audio volume-control type. The filament switch is mounted on the shaft of this control of which the tapered side connects to the audio transformer. A .5 mfd. condenser shunted across this resistor minimizes circuit noises and by-passes the AF currents. A 100,000-ohm or ¼th megohm resistor connected across the secondary terminals of the AF transformer prevents "fringe howl." The audio plate of the 19-tube is by-passed on account of its closeness to the detector plate which, if not included, would tend to make the phone leads act as an antenna.

Two dry-cells and a small 45-volt B bat-

I-tube 10 meter portable receiver. Suitable also for 20 meter operation. Both coils are shown.

The 10-meter coil is wound as shown on a standard 1¼-inch Isolantite plug-in coil form. The windings should be spaced as shown in the drawing. The 20-meter coil winding data is shown in the text.

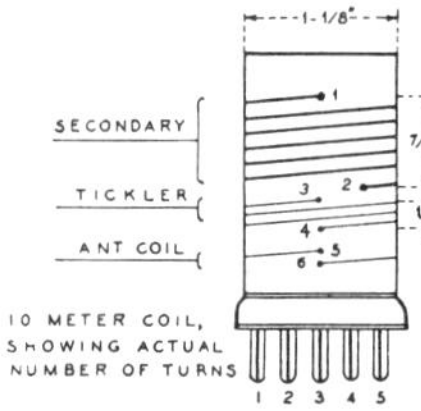

tery provide the necessary power supply. These batteries fit into the metal cabinet behind the receiver assembly. If the receiver is to be placed in service as a monitor on other bands, larger coils with bandsetting trimmer condensers soldered across the coil secondaries will be required.

Because of the complete shielding of the receiver, including that of the power supply, it will be found that the set has a surprisingly low noise level. The antenna is very loosely coupled to the grid coil; such inductive coupling in conjunction with a doublet receiving antenna will insure the minimum interference from auto ignition systems and other forms of man-made parasitics.

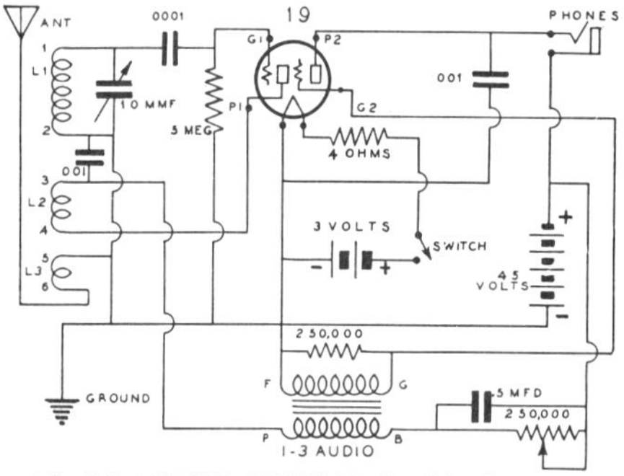

10-20 METER PORTABLE RCVR

FIG. 2

Coils: The coils are wound on isolantite 5-prong coil forms 1⅛-inches in diameter. The secondaries on both 10 and 20-meter coils are wound over a space of ⅞-inch with the grid end-lead towards the top. The tickler windings are wound in the same direction as that of the secondary, the last turn farthest from the grid end connects to the plate; there is a ⅛-inch space between windings and the coil being space wound to cover ¼-inch. The antenna coils are wound ¼-inch below the tickler, near the bottom of the coil form. No. 20 DSC wire is used on the secondaries and No. 30 enameled wire on the other windings. The 10-meter coil has a 5¾-turn secondary, 3-turn tickler, and a single-turn primary. The 20-meter coil has a 12¾-turn secondary, 5-turn tickler and 3-turns on the antenna coil. The coil turns can be cemented in place at a few points after the receiver is tuned to cover both bands. The coil turns may be expanded or compressed during this adjustment.

Chassis: The chassis is 4¼-inches wide, 4-inches long and 2-inches deep. The metal cabinet has a hinged top to allow coil changes. The case dimensions are: 9⅝-inches long, 6½-inches high, 5¼-inches wide and is made of No. 20 gauge steel. The front panel is 6½-inches x 5¼-inches and is of No. 12 gauge aluminum. The phone jack must be insulated from the front panel.

Note: Other receivers of the multi-band type covering the 10-meter range are described in the section devoted to ultra-high frequency communication equipment.

10 Meter Superheterodyne Converter: More than ordinary precautions should be taken in the design of a receiver for 10 meters. It is often advisable to use a separate tuning unit, or converter, primarily designed for optimum efficiency on the "daylight" bands. Such a unit is here illustrated. It is a 10 meter converter, consisting of a regenerative r-f stage, detector and high-frequency oscillator. In turn, this converter connects to the intermediate amplifier of any superheterodyne receiver. Obviously this converter will give good results on 20 meter with proper coils.

The converter here described gives excellent results on 10 meters. Three metal tubes are used, a 6K7 in the r-f amplifier stage, 6L7 mixer and 6K7 high-frequency oscillator. These types of metal tubes have extremely low capacities and this feature, coupled with the small physical size of the tubes, makes it a simple matter to run short, direct connecting leads, none over 1¼-inches long. The 954 acorn tube can be used in the r-f stage, if desired.

Mechanical construction of the converter chassis is somewhat unconventional. Each stage is well shielded and there is ample spacing between coils and shield plates. The heater leads from all tubes must be run through shielded braid and they must be well spaced from other circuit wiring. The alignment of the circuit is similar to that of any other superheterodyne. The oscillator is set to the higher difference frequency and the other stages are then brought to resonance by adjusting the trimmer condensers. This adjustment should be made

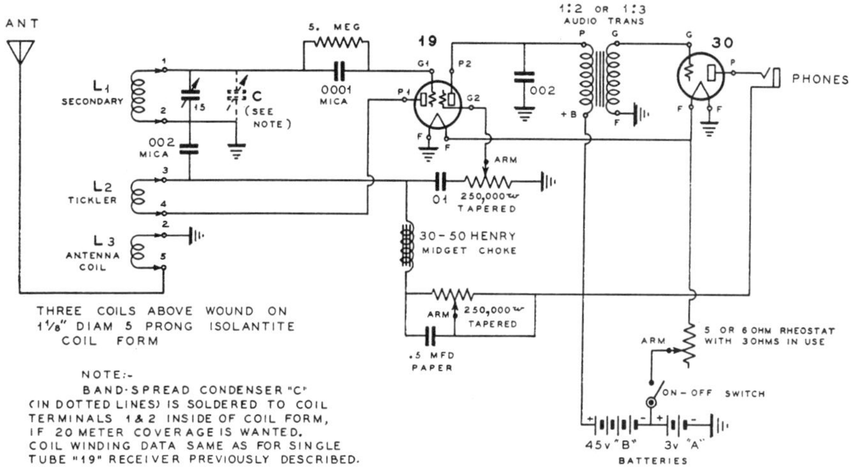

2 VOLT - TUBE BATTERY OPERATED 10 - 20 METER RECEIVER

FIG. 2A—Another audio stage can be added to the single '19 tube receiver as the circuit above shows.

with the main tuning condensers set to minimum capacity. If signal strength falls off at the lower frequency setting of the tuning condensers, the rotary plates should be bent out slightly so that perfect tracking is had over the entire band.

The regenerative r-f stage gives worthwhile gain on 10 meters; its use is of paramount importance. Spacing between cathode coil and grid coil in the r-f stage must be adjusted to the characteristics of the particular antenna in use, the tighter the antenna coupling the closer the cathode coil must be coupled to the grid coil in order to enable the r-f stage to regenerate properly. Furthermore, the spacing of the cathode coil from the grid coil should be adjusted so that smooth regeneration is had with approximately 100 volts on the screen of the r-f tube, lower screen voltages will result in a decided loss in sensitivity.

The main tuning condensers, three in number, each have a capacity of 10 mmf. (3 plates). These condensers are ganged by means of flexible couplings. Approximately 60 degrees dial spread is secured on 10 meters, if the coils are wound as shown in the table. All coils are 1½-inch diameter, 6 prong plug-in-type. The "stubby" type of form can be used. Coil data is on page 193.

10 meter superheterodyne converter assembly.

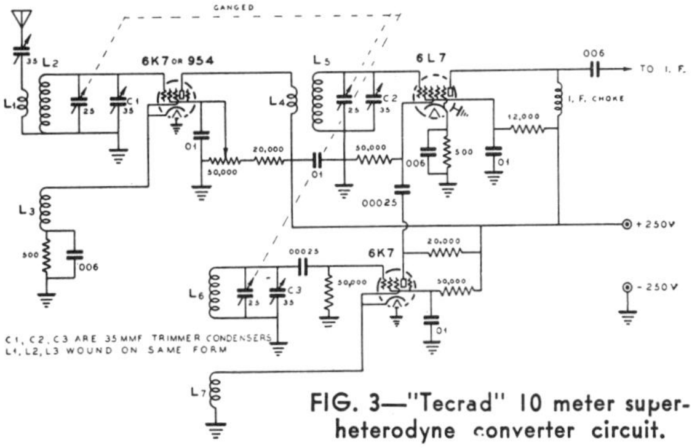

FIG. 3—"Tecrad" 10 meter superheterodyne converter circuit.

10-Meter "Super Gainer" by Frank C. Jones. An Iron-Core I.F. Transformer is used. The air-wound coils are soldered directly to the tuning condenser terminals. A double-shielded partition isolates the two stages. A shield can is placed over the 6A6 tube, but the 57 or 6C6 tube does not require shielding.

Coil Data for 10-Meter Converter
R-F Coil.

(L1) Antenna Winding:—4 turns No. 22 DCC, close wound.
(L2) Grid Winding:—4 turns No. 22 DCC, spaced 2 diameters.
(L3) Cathode Winding:—2 Turns No. 22 DCC, close wound.
Spacing Between L1 and L2, 1/4 inch.
Spacing Between L2 and L3, 1/4 inch.

Detector Coil—

(L4) Grid Winding:—4 turns No. 22 DCC, spaced 2 diameters.
(L5) Plate Winding:—2 turns No. 22 DCC, interwound with bottom two turns of L4.
6 Prong coil form used for above coil, but only 4 of the 6 prongs are connected.

H-F Oscillator

(L6) Grid Winding:—3 turns No. 22 DCC, spaced two diameters.
(L7) Cathode Winding:—3 turns No. 22 DCC, close wound, spaced 1/16-inch from L6.

10-Meter Super-Gainer: This receiver is specially designed for operation on 10-meters only, since the coils, being soldered to the tuning condensers, cannot be very well interchanged. The set design embodies an autodyne detector which greatly simplifies construction and lowers the overall cost. Image interference is troublesome on any ultra-short-wave receiver, hence no attempt has been made to prevent this phenomena from developing in the detector circuit. However, the regenerative tuned RF stage serves to prevent image interference and at the same time greatly enhances the sensitivity of reception on 10 meters. This receiver is more sensitive to weak signals than most multitube all-wave superheterodynes which are capable of covering the 10-meter band, in addition, the sensitivity is higher than that of a set with a good regenerative detector and one stage audio amplifier.

Design Specifications: The regenerative RF stage has the cathode tap up one turn on the grid coil. Inductive coupling gives an ideal signal-to-noise ratio when a twisted-pair feeder to doublet antenna is available for receiving. For short antennas, capacity coupling by means of a twisted insulated wire around the grid lead will give excellent results. Capacity coupling to the first detector simplifies construction but requires an ultra-short-wave RF choke for shunt DC plate supply. This choke consists of approximately 75 turns of No. 34 DSC wire wound on a 3/8-inch dowel rod. Two or three different types of 5 and 10-meter RF chokes will prove satisfactory. (Note: a 2.1 mh. pie-wound midget RF choke causes the RF stage to improperly function). The coupling condenser is mounted on the detector tuning condenser and adjusted to about 10 or 15 uufds.

FIG. 4—Shield cabinet for 10-meter "Super Gainer."

The detector tuning condenser has two plates, and three plates for the RF condenser. This combination gives a good band-spread on 10 meters for the detector tuning and enough capacity in the RF stage to allow the use of either inductive or capacitative antenna coupling.

The first detector also has a cathode tap on the grid coil, and screen-grid voltage variation for control of intensity of oscillation. This tube must oscillate at a frequency ±200 KC from the incoming signal in order to heterodyne it into the second detector through the 200 KC I. F. transformer. Since 200 KC off tune is less than 1% at 28 MC, practically no signal intensity is lost. By allowing the first detector to oscillate instead of having a separate H. F. oscillator, eliminates an additional

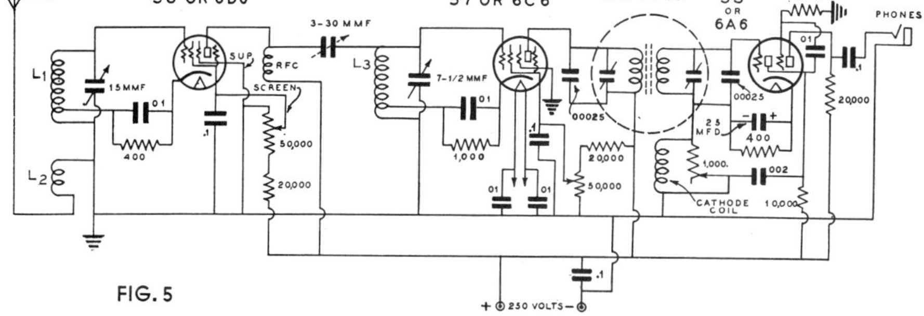

FIG. 5

10 METER -SUPER-GAINER

tuning circuit and also eliminates about 75% of the design difficulties encountered in 10-meter superheterodynes. The receiver will perform almost immediately after it has been properly built and, will outperform larger and more complicated sets.

Technical Notes: The I. F. transformer can be used at 456 KC or it may be substituted with a 175 KC type. The 456 KC iron-core transformer, shown in the circuit diagram, has a pair of .00025 mica condensers shunted across the tuned circuits to lower the I. F. frequency. The coupling between the I. F. coils is increased from that of the original factory adjustment by turning the lock-nut on the side of the I. F. shield can. The higher value of C to L improves the frequency stability of the second detector when it is oscillating for CW reception.

Oscillation or regeneration in the second detector is obtained by a cathode coil wound with 200 turns of No. 30 enameled wire on a ¾-inch dowel rod (jumble wound). Control is had by means of a 1000-ohm variable resistor shunted across the coil. The inductance of this coil is greater than that used in other Super-Gainer receivers, because of the lower I. F. frequency.

A 6A6 or 53 tube acts as a second detector and audio amplifier. It was found that a pair of .01 condensers shunted across the heater leads to ground greatly reduced the residual hum in the output of the receiver, when no center-tap resistor was connected into this circuit. This would probably apply to all Super-Gainers.

The metal chassis is 1¾ in. x 6¼ in. x 10 in. with a double partition 6 in. wide and 5 in. high between the RF and detector stages. The metal cabinet is 11 in. x 6½ in. x 7 in. and an external power pack supplies the 250-volt plate and 6.3-volt (or 2½-volt) filament potentials; connections being made through a plug and cable at the rear of the chassis and cabinet. The metal cabinet is a necessity because of its shielding properties.

The RF and detector coils consist of 15 turns of No. 14 wire, wound on a ¾-inch diameter. A five turn winding near the grounded end of the RF coil acts as an antenna coupling coil. The RF cathode tap is at slightly less than one turn and the detector cathode tap at three turns from the grounded ends of the coils.

Super-Selective Phone Receiver: A remarkable improvement in operating performance is accomplished in this receiver by the use of duplex I.F. circuits. In addition to super-selectivity as a result of the special band-pass I.F. circuits, it has high sensitivity on both 10- and 20-meter bands.

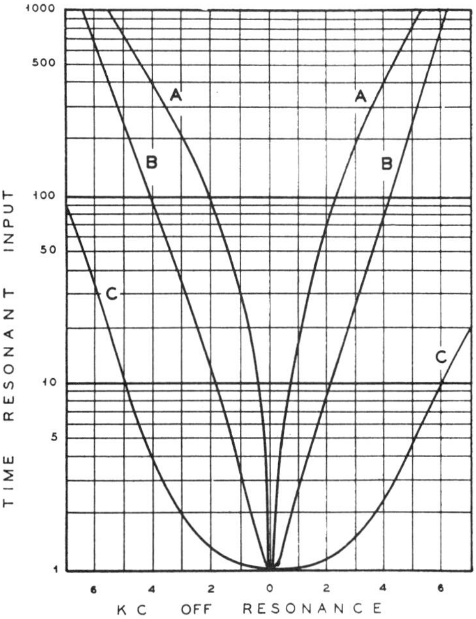

FIG. 6

The receiver must be lined up by means of an all-wave test oscillator. The I. F. may be aligned by connecting the test oscillator to the first detector grid. The first detector must oscillate in actual operation and the RF stage should tune sharply when the regeneration control is well advanced. The second detector should go into oscillation smoothly. The same receiver can be used on 5-meters providing the transmitters are crystal-controlled. The set selectivity is very high due to the regeneration in the second detector.

Technical Details: Besides the conven-

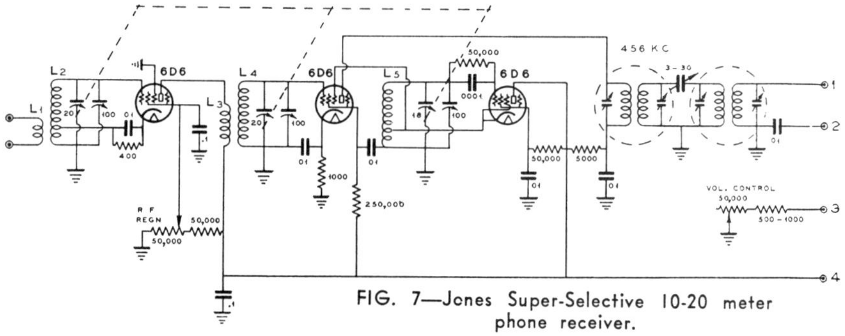

FIG. 7—Jones Super-Selective 10-20 meter phone receiver.

The Jones super-selective receiver.

Top view showing internal construction.

tionality of the circuit, emphasis is placed on the following details: The I.F. transformers are connected so as to have four tuned circuits between each stage, intercoupling being accomplished by means of three 30 mmfd. trimmer condensers set to about 5 mmfds. Air dielectric tuning condensers would be desirable in the I.F. units.

I. F. oscillation will take place with two stages of iron-core transformers if the 6D6 tubes are operated at normal voltages. Since full gain of the two stages is not needed, a large fixed cathode resistor holds the maximum obtainable I.F. gain to a value only a little higher than can be secured with a single stage. The two stages merely serve as a coupling convenience and to compensate for the "band pass" circuit losses.

The set is built on a 14 x 9 x 2-inch plated steel chassis with an 8 x 15-inch 10-gauge aluminum front panel. The three RF shield cans are 5-inches high, 6½-inches long, and 4-inches wide in their outside dimensions, and are made of No. 12-gauge aluminum. A drum dial drives three midget condensers for band tuning. Shunt 100 mmfd. band-setting condensers are individually controlled from the front panel. The R.F. and detector tuning condensers are rated at 20 mmfd. maximum capacity and the oscillator at 15. By resorting to bending the condenser plates and by expanding or compressing the coil-turns, good tracking can be obtained over the narrow amateur bands. The antenna coupling has to be capable of slight variation in order to obtain regeneration with different antennas.

A separate midget condenser with knob control out of the rear of the chassis is shunted across the second detector input. This allows oscillation frequency control for single signal effect. Regeneration or oscillation is accomplished by means of a small coil shunted by a tapered variable resistor in series with the cathode circuit.

Under-chassis view of the super-selective receiver.

This coil consists of about 100 turns of small wire wound on a ¾th-inch diameter

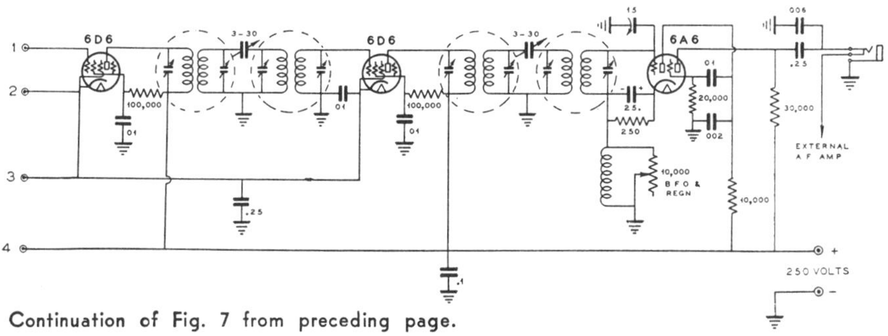

Continuation of Fig. 7 from preceding page.

Coil Data for Super-Selective Phone Receiver.

Wavelength	L_1	L_2	L_3	L_4	L_5
160 Meters	12 turns No. 32 DSC.	1¾" of No. 24 e. tapped at 1¼ turn closewound	25 turns No. 34 DSC. closewound over lower end of L_4	Same as L_2 no tap	1¼" of No. 24 e. closewound Tap at 2/5 of turns.
80 Meters	8 turns No. 32 DSC.	38 turns No. 22 DSC. 1¾" long. Tap at ¾ turn.	15 turns No. 34 DSC. interwound	Same as L_2 no tap	32 turns No. 22 DSC. 1¾" long. Tap at 10 turns.
40 Meters	6 turns No. 32 DSC.	12 turns No. 22 DSC. 1½" long. Tap at ½ turn.	8 turns No. 32 DSC. interwound	Same as L_2 no tap	11 turns No. 22 DSC. Tap at 3 turns. 1¼" long.
20 Meters	2 turns No. 22 DSC.	6 turns No. 20 DSC. 1" long. Tap at ¼ turn.	4 turns No. 32 DSC. interwound	Same as L_2 no tap	6 turns No. 20 DSC. 1" long. Tap at 1½ turns.
10 Meters	2 turns No. 22 DSC.	3½ turns No. 20 DSC. 1" long. Tap at ¼ turn.	3 turns No. 32 DSC. interwound	Same as L_2 no tap	3½ turns No. 20 DSC. 1" long. Tap at 1 turn.

over a winding length of approximately one inch. Too many turns will cause strong oscillation, whereas the effect should function smoothly in a manner similar to an autodyne.

Circuit Alignment: The I.F. system alignment is quite difficult as each circuit must be accurately peaked to about 465 KC. A test oscillator should be coupled into the individual transformers starting with the one feeding into the second detector. As each transformer is aligned, the oscillator can be capacitively coupled to the next preceding transformer until the whole system is completely lined up to the desired. I.F. frequency. The R.F. circuits may be lined up with the same oscillator, providing it is of the all-wave type. When proper alignment is obtained, phone signals that would ordinarily be unreadable on a conventional "sharp" super-heterodyne can be copied with ease, though the fidelity may be impaired very slightly due to attenuation of the higher frequency sidebands.

Other Receivers for 10-meter Operation: Other Super Gainers and general receivers suitable for 10-meter operation are described in the chapter devoted to "Receivers."

10-Meter M. O. P. A. CW Transmitter: This is a 10-meter CW transmitter of striking simplicity. The circuit consists of a 6A6 and type 10 (801) tube working as an oscillator-doubler and final amplifier. The oscillator-doubler differs slightly from the conventional type in that a resonant grid-coil is used instead of a quartz-crystal and RF choke. Oscillation is obtained on either 20 or 30 meters in one triode section of the 6A6, and the other section of the same tube is utilized as a frequency doubler, or tripler.

Circuit Details: The oscillator, TNT circuit, has a resonant grid coil or choke and a fairly high-C tank coil. A 20-meter tank coil gives excellent output on six and two-thirds meters in the 6A6 output tank. The second harmonic on ten meters even gives greater output. The condenser coupling the oscillator tank circuit to the grid of the frequency multiplier has a value between 3-35 uufd. and is adjusted to give optimum output from one stage to the other.

All circuits except the oscillator tank circuit are made low-C. One turn link coupling between the doubler plate-coil and the 801 grid-coil is sufficient. The circuit is easily neutralized by any of the methods described in the section "Transmitter Adjustments."

The output circuit is simply adjusted for any antenna and is quite efficient for impedance matching. The 35 uufd. double-spaced tuning condenser is the normal tank-tuning condenser, and the BCL type 350 uufd. condenser serves as an impedance matching system for antenna coupling.

A combination grid-leak and C-battery bias on the final stage allows keying the oscillator-cathode circuit. A small 1½-henry choke-coil placed in series with a 500-ohm resistor is shunted across the key contacts to minimize key-clicks. There is less frequency creep when keying the oscillator tube than when keying the final amplifier. Typical current readings are: 70 MA, at 400 volts on the 6A6 tube; 120 MA at 600 volts on the 801 or 210, and approximately 5 MA grid current.

Coils: The oscillator-grid coil consists of 40 turns of No. 16 DCC wire, wound on a piece of three-eighths inch dowel rod. The

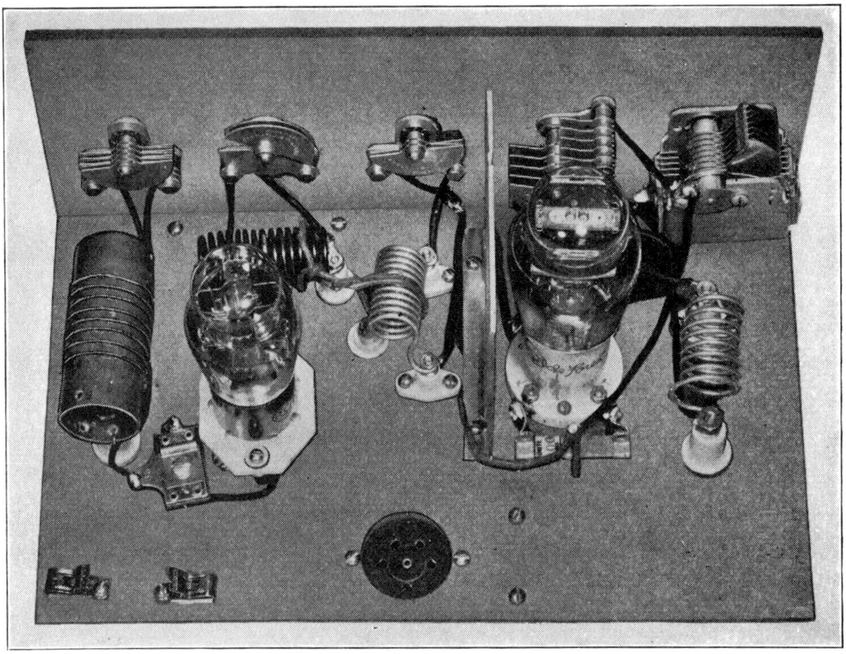

The 10-meter c-w transmitter is mounted on a Masonite or Celotex baseboard and front panel.

oscillator plate-coil has 9 turns, on a 1¼-inch diameter form, 1¼-inches long. The doubler-plate and final grid-coils are of 12 turns of No. 14 or No. 12 wire, wound on a form ¾-inch in diameter and 1¾-inches long. The final amplifier plate-coil has 8 turns, No. 12 wire, wound on a form seven-eighths inch in diameter and 1½-inches long. The RF chokes are ordinary 2.1 mh. section-wound midget RF chokes.

Notes: The transmitter may be adjusted by using a 25-watt lamp for a dummy antenna. All by-pass condensers are of the 1000-volt mica type. Double-spaced condensers are only used for neutralizing the final amplifier tank circuit.

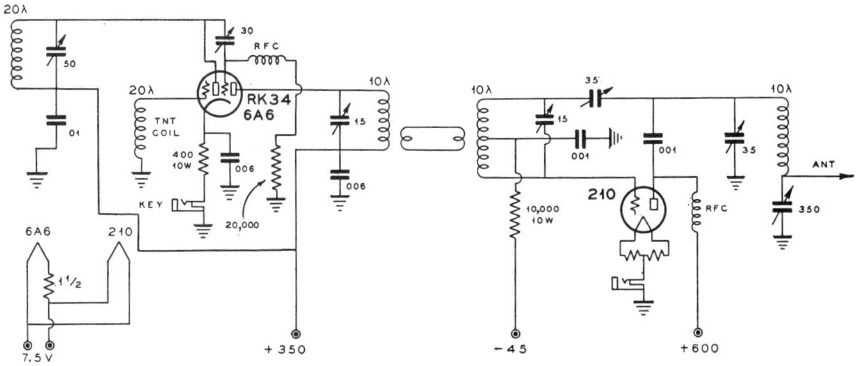

Complete circuit of the 10-meter M.O.P.A. c-w transmitter. Either the RK-34 or type 6A6 tube can be used in the oscillator.

53 Jones Exciter, 210 doubler and push-pull 210 final amplifier, breadboard mounted.

10-Meter Phone-CW Set: This transmitter is economical to construct, crystal-controlled and is capable of developing 15 to 20-watts of carrier power on "phone" and from 40 to 60-watts of RF power for CW on 10 meters. The power for telephonic communication is ample for international coverage on the 10-meter band during such times the conditions for transmission are favorable. Economical grid-bias modulation is one of the complementary features of this transmitter. A good quality single-button microphone working into a 2A5 pentode modulator tube has sufficient output to grid modulate a pair of type 801 or 210 tubes in the final stage.

Technical Details: From the circuit diagram shown on page 199 it will be seen that the exciter, with a 53 tube, functions as a 40-meter crystal-oscillator 20-meter frequency doubler. Unity coupling energizes a type 10 tube either as a neutralized buffer for 20-meter band operation, or as a regenerative doubler to 10 meters. For the latter purpose the neutralizing condenser is set from 2 to 3 times higher capacity than that for normal neutralization. The 10-meter output of this type 10 tube under the specified conditions is between 15 and 20-watts and is, therefore, sufficient to drive the final stage on CW. On "phone," the excitation power has to be reduced as only about 1 to 2 MA of grid current should flow when the stage is not being speech-modulated. This final grid excitation is adjusted by means of sliding taps on the plate voltage dropping resistor in the supply lead to the 53 tube. On CW, as high as 450-volts plate potential may be applied (measured from plate to ground) if a special cut 40-meter crystal is used. On "phone" this voltage is reduced until the output of the 10 doubler is just sufficient for proper modulation. A small toggle-switch will serve to change from CW to "phone" operating conditions—the variable 10,000-ohm resistor is located under the baseboard.

The transmitter is mounted on a baseboard 17 in. x 9 in. x 1 in., cleats are placed on each end to make room underneath for all resistors, current measuring jacks and by-pass condensers. One grounded aluminum-baffle shield separates the 53 exciter unit from the output stages. As shown, it isolates the final tank coil from the buffer input coil to permit 20-meter operation. It is not required on 10 meters. The final stage and its driver are crowded into a small space so as to minimize the length of the plate leads; the latter cross over each other to opposite stator sections to preserve the symmetry and balance of the neutralizing circuit. Type 10 or 801 tubes may be used, preferably the latter, as the plate dissipation and 10-meter efficiency is higher. With type 10 tubes, it is necessary to cut slots in the tube base with a hacksaw, cutting between the prongs improves the efficiency and simplifies circuit neutralization under operating conditions; in addition, prevents spurious RF circulatory currents in the base from generating heat which would cause blisters.

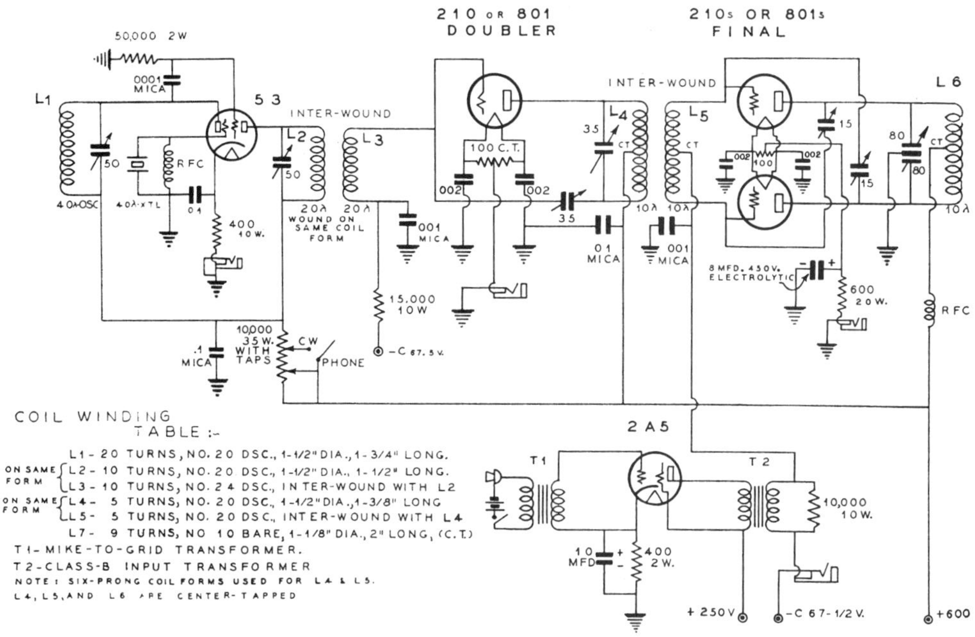

ECONOMICAL 10-METER PHONE-CW SET

An 83 rectifier system or two type 80 tubes with plates connected in parallel as half-wave rectifiers may be used for the power supply. On "phone," the total plate current flowing is approximately 160 MA, while on CW, the value will increase up to as high as 250 MA. About 600 volts plate supply is the maximum that can be safely applied on the 801 tubes, and between 500 and 600 volts on the type 210 tubes. The final and buffer stages must have separate 7.5-volt filament windings.

The 10-meter coil is mounted as shown above. Note crowding of final amplifier components to shorten all connector leads.

RK-20 All Band Suppressor Modulated Phone: This transmitter can be used on other bands as well as 10 meters by merely changing the coils. On 10 meters the carrier output is from 30 to 35 watts on phone, and from 130 to 150 watts on cw. Crystal control maintains a constant carrier frequency.

RK 20 r.f. pentode tubes are used in push-pull in order to minimize the tube input and output capacities for 10 meter operation. These tubes have suppressor grids which can be used for voice modulation by means of efficiency variation. On cw, the suppressor grids are normally run at plus 45 volts; on phone at—45 volts. Audio frequency voltage at voice frequencies varies this suppressor voltage and consequently modulates the r.f. carrier wave. The more negative the suppressor grid with respect to filament, the less will be the output and (tube output to plate input) efficiency. 100% modulation can be obtained with less than normal output from a 41 pentode audio amplifier.

Relatively small grid drive or excitation is needed for r.f. pentodes and therefore the output from a good frequency doubler is sufficient to drive two RK 20 tubes on 10 meters. A push-push doubler (grids in push-pull and plates in parallel) gives relatively high output with only a 400 volt power supply. A twin triode tube, such as a 53 or 6A6, serves admirably for this purpose because it has two tubes in one glass envelope.

35-watt phone, 150-watt c-w transmitter for all-band operation.

Another 53 or 6A6 tube serves as a crystal oscillator and a frequency doubler, driving the second 6A6 or 53. Thus a 40 meter quartz crystal controls the carrier frequency on 10 meters, with a minimum number of tubes for a given output. The 53 or 6A6 tube is a good crystal oscillator as well as doubler because of its low grid-to-plate-and-ground capacities, and high amplification constant.

The grid circuit of the RK 20s can be tuned by means of the proper coil to any of the three frequencies in the exciter unit of 40, 20 or 10 meters. A series link coupling automatic system provides output on any one of these bands.

When phone operation is used, the cathode current in either 6A6 tube runs between 60 and 70 ma. The total DC grid current on the final runs at from 8 to 10 ma., while the screen current is about 60 ma. and the plate current 80 ma. For cw operation with positive suppressor grids, the plate current with antenna load should be about 150 ma. The antenna can be connected across a tuned circuit and the latter link coupled to the final tank coil. This coil and its double spaced split-stator condenser are mounted on small panels in order to provide very short leads to the RK 20 plates.

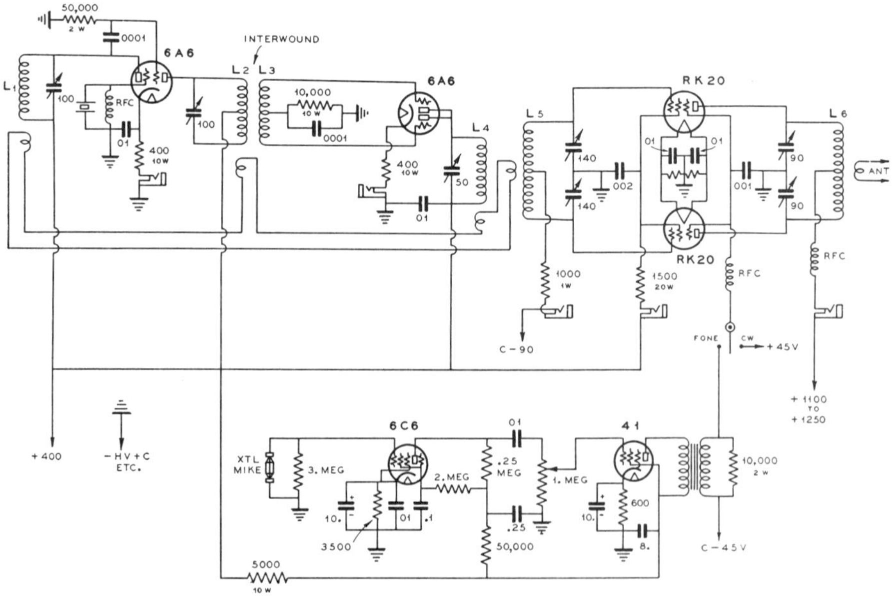

Circuit diagram for the transmitter pictured above. See page 201 for coil-winding chart.

COIL WINDING CHART

RK-20 All-Band Suppressor-Modulated Phone

Wavelength in Meters	L_1 Oscillator	L_2 Doubler	L_3 P. P. D. Grid	L_4 P. P. D. Plate	L_5 Final Grid	L_6 Final Plate
10	None	None	None	4½ turns No. 18 Enam. 1½" diam. ¾" long.	6 turns No. 18 DSC. Center tapped. 1½" diam. 1" long.	8 turns No. 12 "air-wound." Center tapped. 1½" diam. 2½" long.
20	None	10 turns No. 18 Enam. Center tapped 1½" diam. 1½" long.	10 turns No. 26 DCC. Center tapped. Interwound with L2.	9 turns No. 18 Enam. 1½" diam. 1¼" long.	10 turns No. 18 DSC. Center tapped. 1½" diam. 1" long.	11 turns No. 12 "air-wound." Center tapped. 2⅛" diam. 3" long.
40	18 turns No. 20 DSC. 1½" diam. 1½" long.	20 turns No. 20 DSC. Center tapped 1½" diam. 1¾" long.	20 turns No. 26 DSC. Center tapped. Interwound with L2.	20 turns No. 18 Enam. 1½" diam. 2" long.	33 turns No. 20 DSC. Center tapped. 1½" diam. 1½" long.	21 turns No. 14 "air-wound." Center tapped. 2⅛" diam. 3" long.
80	34 turns No. 20 DSC. 1½" diam. 1½" long.	40 turns No. 22 DSC. Center tapped 1½" diam. 2¼" long.	40 turns No. 26 DSC. Center tapped. Interwound with L2.	None	42 turns No. 20 DSC. Center tapped. 1½" diam. 1¾" long.	36 turns No. 16 "air-wound." Center tapped. 2⅛" diam. 3" long.
160	65 turns No. 22 DSC. 1½" diam. 2" long.	None	None	None	70 turns No. 22 DSC. Center tapped. 1½" diam. 2¼" long.	54 turns No. 16 DCC. Center tapped. 3" diam. 3¼" long. Close wound.

All by-pass condensers in the final stage are mounted right at the sockets and connect to a common point on the filament centertap ground lead. Link coupling is adjusted for maximum grid drive while tuning the various stages. An absorption type wavemeter, covering the range of from 50 to 5 meters, is a great aid in tuning the circuits to the proper harmonics for operation in the 10 meter band.

The modulator, consisting of a 41 and a 6C6 tube, gives ample gain for use with a crystal microphone for close talking purposes. Two power supplies, plus three 45 volt "C" batteries, are needed. One supply should provide 400 volts at 200 ma., while the other should provide from 1100 to 1250 volts at 150 ma., well filtered. The batteries provide various bias voltages for the control and suppressor grids in the RK 20 tubes.

RK-18 Phone—CW Transmitter: A splendid breadboard-type crystal-controlled transmitter which has a carrier output of 30-watts on phone and nearly 100-watts on CW in the 10-meter band can be constructed by following the simple details given here.

Technical Notes: The construction of the transmitter follows along the same lines as those previously discussed except for these notations: The variable condensers have sufficient capacity for operation in the 80-, 40-, 20-, or 10-meter bands by changing coils and crystals. Control of excitation to the grid of the RK25 is done by the 50 mmfd. variable midget condenser. Regeneration allows the operation of the 6A6 as a quadrupler to 10-meters from a 40-meter crystal, but somewhat less tube heating is experienced when the RK25 is used as a doubler to obtain 10-meter RF output.

For CW operation, a switch changes the suppressor-grid from the negative 45-volt bias to a positive bias of approximately the same voltage. This increases the RF output, with the result that more grid drive can be given to the final stage.

For phone, the final amplifier is operated as a class B linear stage, amplifying the modulated RF wave up to about a 25 or 30-watt carrier. For CW operation, a 5000-ohm grid-leak is cut into the grid return for additional bias at approximately cut-off.

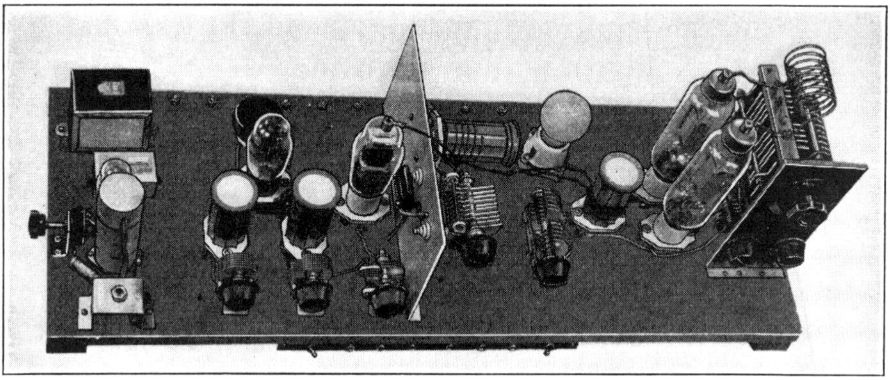

Speech and r-f channel, breadboard mounted. The arrangement of parts should be as shown.

The Mazda "ballast" lamp is unscrewed from its socket and the RK25 output increased several times by switching over the suppressor-grid. These three operations allow class C operation in the final amplifier with an output from 90 to 100-watts at much higher plate efficiency. For phone operation it is necessary to operate the final amplifier at cut-off grid-bias in order to obtain linear amplification, the latter taking place at rather low plate efficiency.

A one-turn coupling link is satisfactory between stages, and the grid load Mazda lamp is connected by a coupled link to the grid coil.

Tests: For phone, at a carrier output of approximately 30-watts with a 1000-volt plate supply on the final amplifier and 385-volts on the RK25 and crystal stage, the following readings can be obtained: final amplifier grid current 1 MA., plate current at from 100 to 110 MA. For these readings the antenna load must be tightly coupled to the linear stage in order to obtain linear amplification and upward swing on antenna current with modulation. The 6A6 cathode current will be about 60 MA, the RK25 plate current 25 MA, and the screen current at 20 MA.

On CW operation the readings from the crystal oscillator will be the same as above; however, the RK25 plate current will be 50 MA; grid current on the final amplifier 10 to 15 MA through a 5000-ohm grid-leak and 45-volt C-battery; plate current on the final amplifier 175 MA. In this case the antenna coupling need not be as great as for phone operation.

Note: The complete set, less power supplies, is built on a 11-in. x 30-in. x ¾-in. baseboard, mounted on end cleats to facilitate under-board wiring as well as allowing space for connection terminal and testing strip.

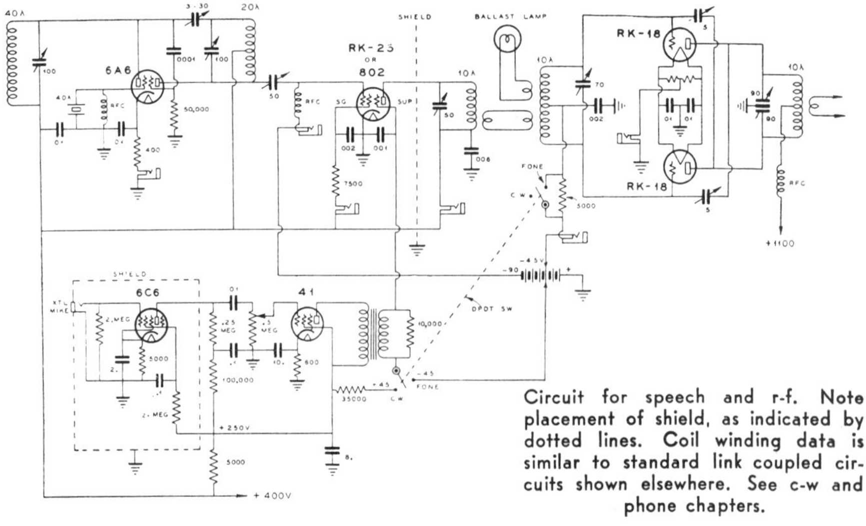

Circuit for speech and r-f. Note placement of shield, as indicated by dotted lines. Coil winding data is similar to standard link coupled circuits shown elsewhere. See c-w and phone chapters.

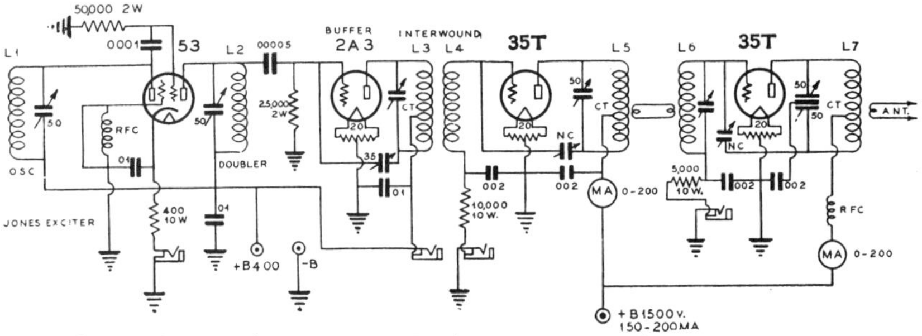

Circuit diagram showing new 35T tube transmitter with 125 watts output.

The New Eimac 35T Low C Triode: This new tube is ideally suited for use in doubler circuits because it has a high amplification constant. Its low inter-electrode capacities make operation on 5-meters practicable. Two of these new Tantalum-plate tubes are shown in the transmitter pictured below, the circuit diagram for which is at the top of this page. This transmitter will put out 125 watts, even on 10 meters, without exceeding the 35-watt plate dissipation of the tube. A pair of 35Ts in the final amplifier will deliver more than ¼ k-w output.

A Jones exciter drives a 2A3 buffer stage which, in turn, excites the first 35T either as a doubler or buffer. Unity coupling provides ample grid excitation from the 2A3 to the first 35T. Optimum grid excitation with respect to the 2A3 plate load can be adjusted by varying the number of turns in the grid coil L4. The value of grid leak shown in the circuit is suitable for doubling. However, it will also function satisfactorily for buffer operation where the output efficiency is higher, less input plate power therefore being required. The grid leak in the final should have a value of from 1,000 to 5,000 ohms. The 35T has a 5-volt filament, 4.0 amps. Maximum plate voltage 1,500. Maximum plate current 100 MA. Mu 27. Grid-to-plate capacity approximately 2 mmfd. Base—4-prong isolantite. Plate at top. Grid connection to socket. The 35T has no internal insulators. It will function as a zero-bias operated tube in class B audio at 750 volts.

Coil Data for the Transmitter here described are on page 206.

The new 35T low-C triode in a modern c-w transmitter. This new tube is the answer to the doubler problem.

FIG. 1—M.O.P.A. parallel plate oscillator and 801 amplifier for 10 meter operation.

10-Meter M. O. P. A. Phone Transmitter: Figure 1 shows the RF portion of this transmitter which, from the mechanical standpoint, presents some interesting features. Note the measures taken to support the various coils and link-coupling loops so as to avoid the slightest mechanical vibration. Even the condensers are double-spaced to lessen the likelihood of change in capacity due to vibration. The speech amplifier and modulator are conventional.

Technical Notes: The coils, as shown in the photographs, are permanently fastened to standoff insulators. In the case of the oscillator coil, it would probably be advisable to mount a hard rubber strip across the top to lessen the tendency for this coil to start vibrating. The 10-meter coils have so few turns that no trouble will be experienced from this source.

The transmitter is tuned in the conventional manner, the only precaution being that the relative plate spacing of the oscillator condensers must be adjusted for best stability and output. One turn of link coupling provides ample transfer of R-F to the final amplifier.

The three jacks shown in the circuit diagram are, respectively, +B lead of the oscillator, C bias lead of the final grid circuit, and high voltage lead of the final amplifier. The meter can be plugged into the C-bias lead to determine the correct adjustment of the excitation from the oscillator, and the grid meter can be further used to neutralize as in ordinary technique.

Note: Other transmitters suitable for 10-meter operation are described in the transmitter chapter.,

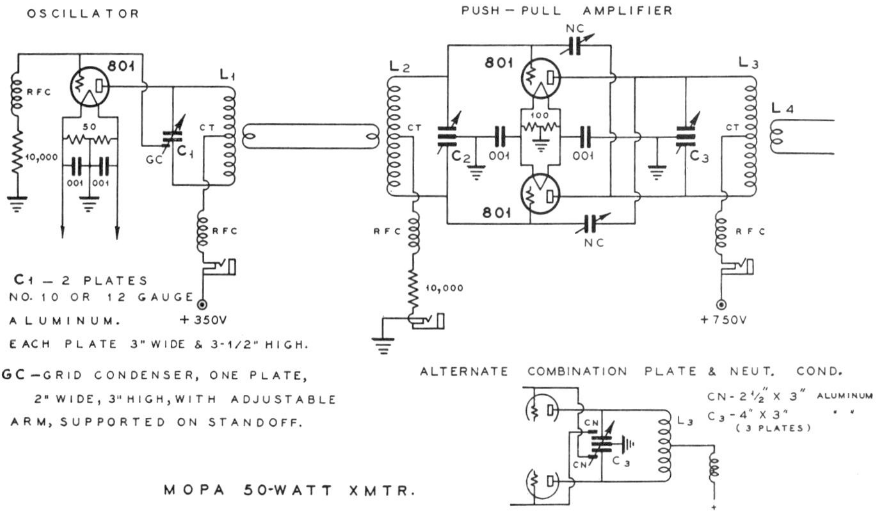

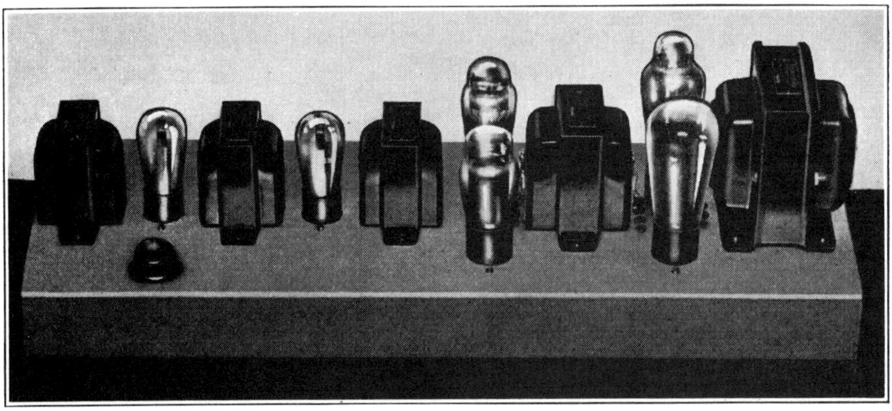

60-watt modulator system for M.O.P.A. phone transmitter described on preceding page. The circuit diagram for this modulator is shown below.

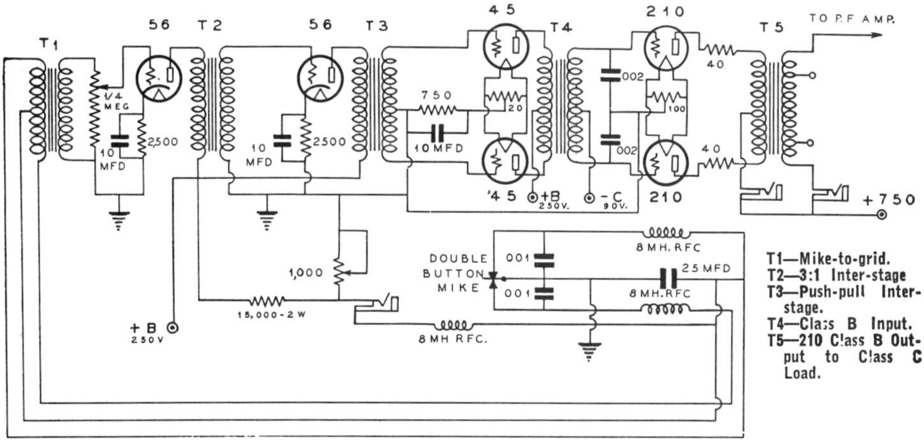

T1—Mike-to-grid.
T2—3:1 Inter-stage
T3—Push-pull Inter-stage.
T4—Class B Input.
T5—210 Class B Output to Class C Load.

HK-354 Gammatron Doubler: Many amateurs are using the final amplifier of their transmitters for the purpose of doubling from 20 to 10 meters. The low-C tubes are particularly suitable for this service. The HK-354 can be used in the final amplifier with from 1,500 to 3,000 volts on the plate for 10 meter operation. A simple medium power 10 meter transmitter could consist of a Jones exciter with 40 meter crystal and 20 meter doubler coil, capacity coupled to a 2A3 buffer operating on 20 meters. This buffer is then link-coupled to a 210 buffer, also on 20 meters, which drives the HK-354 final by link coupling. The final amplifier tank coil is wound for 10 meters and a home-built 2-plate variable condenser, shown in the photo, will tune the coil. The 10 meter coil has 10 turns of No. 12 enameled wire, 1¾″ diameter, 2½″ long. The home-built variable condenser has a stator plate, 4″ x 3″ and a rotor plate to correspond. The spacing between plates is ¼″. A 50,000 ohm, 50 watt grid leak is required for doubling in this final amplifier. The filaments should be by-passed at the tube socket and the grid and plate by-pass condensers should be grouped closely around the tube base.

HK-354 Gammatron in a high power doubler for 10 meter operation.

Coil Winding Table for 35T Transmitter Described on Page 203

L1	53 Osc. 40 Meters.	16 turns, No. 20 DSC on 1½-in. dia. form. Space wound to cover winding space of 1½-in.
L2	53 Doubler. 20 Meters.	8 turns, No. 20 DSC on 1½-in. dia. form. Space wound to cover winding space of 1⅜-in.
L3	2A3 Plate. 20 Meters.	10 turns, No. 20 DSC on 1½-in. dia. form. Space wound to cover winding space of 1⅜-in. This winding must be center-tapped.
L4	35T Grid, 20 Meters.	8 turns, No. 22 DSC on 1½-in. dia. form. This winding is interwound with L3.
L5	20 Meters.	12 turns, No. 12 Enameled, 2-in. dia., 2-in. long. This winding must be center-tapped.
	10 Meters.	10 turns, No. 8 wire, 1½-in. dia., 3-in. long.
L6	20 Meters.	10 turns, No. 12 Enameled, 2-in. dia., 2½-in. long.
	10 Meters.	8 turns, No. 8 wire, 1½-in. dia., 3-in. long.
L7	20 Meters.	12 turns, No. 12 Enameled, 2-in. dia., 2½-in. long. This winding must be center-tapped.
	10 Meters.	10 turns, No 8 wire, 1½-in. dia., 3-in. long. This winding must be center-tapped.

The Heintz & Kaufman Gammatron line, showing the HK-354 in the lower foreground, HK-352 (special tube for lifeboat transmitters) and a group of high-powered Gammatrons for use on shipboard and at commercial land stations. The intermediate size tubes, three of which are pictured, are the gridless Gammatrons used in Globe Wireless equipment on the Robert Dollar fleet and at Globe Wireless shore stations.

Ultra-High Frequency Telephony

Technique and Principles of Ultra-High Frequency Communication

The technique and principles involved in the transmission of radio energy below 7.5 meters differ greatly than other frequencies commonly employed in the higher amateur wavebands. The transmission characteristics of ultra-high frequencies are restricted to practically local communication. This is attributed to the fact that the radiated wave does not return to the earth by reflection from upper ionized layers of atmosphere however, for communication purposes, the radiations travelling along the surface of the ground are utilized and no fading or variation in signal strength occurs. The range of transmission may be governed by the elevation of the transmitting antenna and to some extent by the transmitter power. It is limited to a distance somewhat in excess of the radius of the horizon as seen from the transmitting antenna when the receiving points are at ground level and to a distance somewhat in excess of the combined horizon distances when both transmitter and receiver antenna are elevated.

5-meter radiotelephony plays a major role in aiding the bridge builder. The illustration shows the San Francisco Bay Bridge operator conversing with the construction crew.

Optical Phenomena: Ultra-high frequencies travel in optical paths like light rays and behave according to the physics to light-ray phenomena. The wavelengths of the radiowaves, however, are millions of times greater than that of light. In general, light waves are slightly bent in passing from one point to another, this effect becomes more and more pronounced as the wavelength increases. In the ultra-high frequency bands this effect has become so pronounced that the optical path is no longer approximately straight but is curved along a line which is the circumference of a circle about four times the radius of the earth. This is due to the refraction caused by the earth's atmosphere and its exact curvature is dependent upon the variations of the density of air with altitude. Since this changes from time to time, the range beyond the true horizon may vary. One result which has been observed is the gradual increase in range or in the signal between two points after the sun has gone down and darkness approaches. The reason is that the density of the atmosphere near the earth's surface increases as the temperature falls, this results in the optical path through the air becoming more curved so that it remains closer to the earth after passing the true horizon. In another case, it has been observed that where the two stations were beyond the "light horizon" no signal was obtained during the day. Soon after the sun went down the signal began to come through and by the time it was dark a very reliable signal was received.

The "light horizon" in miles from an elevated antenna location can be found by taking the square root of the height in feet of the antenna above sea level and multiplying it by 1.23. The "radio horizon" is greater and the multiplying factor is approximately 1.4. In other words, communication is reliable over sea or over land at sea level for a distance approximately 20 per cent greater than the "light horizon."

When one station is located in the shadow of a hill, other facts enter into the problem due to reflection and refraction which make individual problems in themselves, but these may be solved by changes in the antenna location, which may amount to only a few feet.

The power required for transmission is astoundingly small; a "transceiver" capable of energizing an antenna with .5-watt has a range of from 6 to 15 miles in a practically flat country or area. The relative efficiency of such a low-powered transmitter is due to the use of a resonant antenna. Where there are obstacles within the area to the horizon, an increase in power will produce a higher field intensity, hence, at points where the received signal is weak, more power will lay down a stronger signal.

In free space, from an airplane when line of sight exists, power of the order of .5-watt is often sufficient for ranges up to 100 miles although greater power is required for the return circuit to produce a strong signal intensity at the plane. This is necessary so as to overcome the exceedingly high surrounding noise level through which the signal must be intelligible. With a plane flying at 1500 feet the "radio line of sight" is between 50 and 55 miles. Using .5 watt antenna power the attenuation is such that reliable communication with a super-regenerative receiver is just possible to the ground. By the reciprocal law, .5-watt at the ground station will produce the same signal strength at the plane but this would not be reliable on account of the plane noise.

Receivers For Ultra-High Frequencies: The super-regenerative type of receiver is almost universally used for reception on wavelengths between 1 and 10 meters. Radio frequency amplification and present day super-heterodyne circuits are coming into prominence for 5-meter operation, but super-regeneration provides a practical method of receiving weak signals.

Super-regeneration is a highly developed form of detection which is many times more sensitive than other systems. The function of the detection, its advantages and disadvantages are outlined in the following explanation.

Super-regeneration carries an ordinary regenerative detector beyond the point of oscillation without distortion to the quality of the signal; this is done by allowing the tube to oscillate, then damping-out the oscillation a great many times per second by an auxiliary oscillating source functioning above audibility.

The damping or quenching effect can be accomplished in a number of different ways. Sometimes a regular oscillation circuit working in the range of from 20,000 to 200,000 cycles per second is used as a means of controlling the ultra-high frequency oscillations. The latter take place in the detector circuit so the other low frequency (called the interruption frequency) oscillator can feed a little energy into the grid or plate circuit. The most common method is to couple the two tube plate circuits together for a form of Heising or plate modulation. In this case, the interruption frequency varies the detector plate voltage enough so that this tube goes in and out of oscillation at a rate determined by the interruption frequency. The same detector tube can also be used as an interruption frequency oscillator by putting the tuned circuits for the latter into the detector circuit.

Another form of super-regeneration makes use of a blocking grid-leak condenser action so that no extra tube or low-frequency coils are necessary. Such a circuit functions as an ordinary oscillator in which the grid-leak is too high to allow the electrons on the grid to leak off at a rate to give constant value of grid-bias voltage. This causes a change in the average bias and stops the oscillation because the plate current is decreased and the mutual conductance of the tube drops. If the circuit constants are correct, including a fairly high decrement in the detector tuned circuit, the blocking action takes place at an inaudible rate and super-regeneration is accomplished. The decrement of even a low-loss five-meter circuit is sufficiently high to allow this circuit to function well.

Super-regeneration will amplify a weak signal many thousands of times, while a radio-frequency amplifier will only amplify it about five times at five meters.

Unlike ordinary regeneration, super-regeneration always broadens the tuning, or gives much less selectivity. Practically all ultra-short wave receivers tune to cover at least 100 KC at any point on the tuning condenser. Such action is comparable to a transmitting 5-meter oscillator—decreasing or increasing the DC plate voltage 60 per cent usually varies the oscillator frequency from 60 to 120 KC.

The hiss or rushing sound (which is somewhat similar to static) audible in a super-regenerative set is mostly due to the thermal and contact circuit noise. The detector is in an extremely sensitive operating condition when no signal is present, thus the noise is greatly amplified and made audible in the loudspeaker or headset. A carrier signal automatically reduces this amplification or sensitivity and decreases the background noise. A strong signal will completely eliminate the hiss or roar.

One particular disadvantage of the super-regenerative detector is due to radiation. When receiving, the detector oscillates intermittently and radiates a signal fully modulated by the quenching frequency. Another super-regenerative receiver operating within receiving range of the radiating receiver's carrier, picks it up and the beat notes between the quenching frequencies of the two receivers cause very serious interference. The radiating range may be greater than one mile.

It is possible to place a RF amplifier ahead of the detector to function as a blocking tube and, in addition, derive some gain from the added stage; however, such preventive measures on short-waves do not frequently remedy the condition. Even the best screen-grid-tubes at ultra-high frequencies allow considerable energy to be by-passed in the wrong direction. Then again the power cable to the set may be of such lengths as to act as a very efficient antenna. Choke coils in the individual leads do little good since the spurious capacities to the

The Jones Economy Transceiver. See Page 218 for details.

receiver are sufficient to allow considerable RF power to pass to the cable.

The chief advantage of the super-regenerative receiver, namely, its extreme sensitivity, should be an incentive to the experimenter and engineer alike in developing improvements to remove its disadvantages.

Transmitters for Ultra-High Frequencies: Almost any type of circuit will oscillate quite efficiently at frequencies down to 70 or 75 megacycles (mc), if a few simple precautionary measures are observed. A popular circuit has been the tuned-grid tuned-plate arrangement. At the highest frequencies this has a distinct advantage, since the tube capacities are in series across the tank circuit, but at frequencies up to 60 mc. there is little choice between it and the same circuit single-ended, other than the increased power resulting from two tubes. In designing any circuits for these frequencies, short and direct leads are very essential. For example, at high frequencies, a straight piece of wire a few inches in length has sufficient inductance to offer a reactance of 400 ohms at 60 mc. The design of ultra-high frequency equipment is as much mechanical as it is electrical, and the test "breadboard" set-up cannot be very well transformed to a different layout in the finished set with equal success.

Unfortunately, the greatest number of transmitters operating in the amateur bands are constructed of directly modulated oscillators which are unstable as far as frequency is concerned; in the country this circuit does not cause much interference, but in densely populated areas the interference problem assumes serious proportions. Directly modulated oscillators develop a frequency modulation when the modulating percentage is high, which causes a receiver to detect a weak signal spread over quite a wide band. In a super-regenerative receiver, the signal can be heard over a large portion of the silent region. On the other hand, if a well-designed M. O. P. A. transmitter is used, the voice is observed quite sharply in the center of the carrier, and since the sideband power is concentrated at one point, the signal is louder for the same modulation percentage, and consequently greater range can be expected, in addition, the amplifier may be modulated 100 per cent. In the M. O. P. A. transmitter, the oscillating circuit must have such operating parameters as to maintain the generated frequency during slight variations in the supply voltages. The oscillator must be sufficiently powerful so that the coupling between it and the amplifier may be reduced to a certain degree to prevent intercircuit reaction from the modulated amplifier. Tubes of the same type are satisfactory for both oscillator and power amplifier.

The power amplifier may be modulated by a class B audio amplifier and, in small transmitters, a single power supply will furnish the necessary power for both units providing care is taken to insure extremely good regulation. For larger units the class B modulator must have its own power supply to prevent any frequency fluctuations in the oscillator due to the voltage drop in the plate supply during modulation. The oscillator and power amplifier may have its own supply, on the other hand, separate supplies for each unit can be used.

A well-designed 5-watt transmitter will suffice for practically all amateur purposes. Increased power accomplishes little in extending the signal beyond the horizon and, except in those cases where the location is shadowed, will produce sufficient strength within the horizon limits.

Conventional 5-meter inductances and chokes.

Doubling the transmitter power produces little if any gain. The signal strength will, however, increase to 3DB which is hardly noticeable. For this reason, power increases should be made in multiples of 10, which gives a 10DB gain for each step.

Antenna and Transmission Lines: The antenna almost universally used for fixed stations is the vertical half-wave dipole. A horizontal dipole is directional at right angles to its axis and exhibits this characteristic very noticebly in free space and to a less degree where local reflections caused by buildings and hills change its pattern. A wave radiated from a horizontal dipole is polarized in such a way that it also must be received on a horizontal dipole.

In the 56 to 60 mc band a rod one-half inch in diameter is reasonant if cut to approximately 95% of the actual half-wavelength of the frequency used. Since such an antenna has a high radiation resistance (74 ohms), its resonance curve will be very broad and the physical length will not be critical. A rod cut for 58 mc will operate well anywhere in the 56 to 60 mc band. The antenna should be mounted as high and as free from all surroundings as possible; the supports should be near the middle rather than at the ends where voltage maxima exit, to reduce losses. If conditions permit, brackets should be used to support the antenna so as to hold it away from the mast by at least 4 inches.

The antenna may be supplied with power from the transmitter by a transmission line, either of the matched impedance type or of the resonant tuned type. In either case some method must be employed to determine when the antenna is in resonance. One of the most reliable methods is to use a field strength meter to be definitely assured that the system is radiating properly.

To transfer the maximum amount of power from one circuit to another requires that the impedances of both the source and the load be closely matched. A mismatch of 2 to 1 is not very serious, but one should always endeavor to match as closely and as correctly as possible.

For a dipole antenna, a transmission line can be made up of two No. 14 bare copper wires spaced 2 inches apart, such a line has an impedance of approximately 500 ohms; to match this line to the antenna requires that the feeders be attached at two points equidistant from the center where an impedance measurement would indicate 500 ohms. The points at which this measurement would occur would be equal to a spacing of 24% of a half wavelength, or 12% each side of center. For example, assume it is desired to set up a matched impedance antenna for 58 mc. This wavelength is 5.17 meters or 203.5 inches, one-half wavelength is 101.8 inches. The ½-inch diameter rod is cut 95% or 96 inches long. The tap-offs are made so that the points are 24% of the actual ½-wave length apart or 24% of 101.8 inches which equals 24.4 inches or 12.2 inches each side of center.

The line made up as indicated above is spread out from a point about 30 inches away and at right angles to the rod and attached. It may be run for any length to the transmitter. If the match were perfect there would be no standing waves on the wires and a neon light will show the same brilliancy when touched at any point of either wire. Such a perfect match is seldom obtained in practice and standing waves exist to some degree in some installations. Standing wave phenomena can be minimized by changing the line length a few inches at a time, noting for each length the transmitter setting for minimum standing waves. A position will finally be found for the transmitter setting and length which is best.

This type of line may be connected directly to the tank circuit through fixed blocking condensers and no series or parallel tuning of the line is needed.

There are two other methods of matching which are adaptable in certain locations. Both of these make use of a length of transmission line as a transformer. If a section of transmission line ¼-wave-

Raytheon ultra-high-frequency transmitter. Note compact arrangement of parts and leads.

length long resonating at the operating frequency is shorted by a jumper-wire at one end, and the antenna is attached to one wire at the other end, points can be found along these lines where the impedance is 500 ohms. To these points equidistant from the jumper-wire the line is attached. This quarter-wave transformer can be hung directly beneath the antenna rod and may be convenient to use in some cases, although no better results will be obtained than when the antenna itself is employed for matching.

The same principle can be employed by connecting a ½-wavelength of wire to the center of the antenna, shorting the far end. In this case two 500 ohm points can be found which are approximately the same distance from the antenna end and the shorted end. The line can be attached at either point and standing waves eliminated.

These ¼- and ½-wave line transformers are frequently used in setting up directional antenna arrays (see section: Antennas).

The antenna may be energized by means of concentric tube lines, but such a system is more difficult to construct and is very expensive; its advantage, however, is that the energy is all confined inside the outer tube, hence, the line cannot radiate. The outer tube may be grounded at any point along its length or may even be buried in the ground with no loss in efficiency.

Two-wire lines cannot be constructed to have a very low impedance. For 6-inch spacing the impedance is 628 ohms, for 4-inch spacing 578 ohms, for 2-inch spacing 495 ohms, for 1-inch spacing 413 ohms, and when the wires are spaced only .1-inch or practically in contact, the impedance is 137 ohms. This drastic change of 60 times in spacing has reduced the impedance to only 137 ohms from 628 or by a factor of 4.5. It can be seen from these figures that the wire spacing of a two-wire line is not very critical and that variations in spacing, unavoidable in construction, will have little effect. On the other hand, a concentric tube line can be constructed to have much lower impedances. If the ratio of the outer diameter of the inner conductor (which may be either solid or tubing) to the inner diameter of the outer tubing is 3.44, the line will have an impedance of 74 ohms independent of the size of the pipe and will form a matched impedance system into the center of the dipole, the outer tube to one side and the inner to the other. A line made of ¾-inch outside diameter tubing having a 1/32-inch wall for the outer sheath, and No. 4 B&S copper wire for the inside, meets these specifications very closely. Thin isolantite spacers can be used at intervals to hold the inner conductor in place.

Such types of line as applied to ultra-high frequency uses are more particularly adaptable to mobile and plane installations, since they can be bent to conform to the car body or plane fuselage much easier than an open-wire line.

For such mobile installations another type of antenna is often more convenient to install. A quarter-wave rod is extended upward through the car roof or through the fuselage in the rear of the plane. The metal framework of the car or plane acts as a counterpoise, extra foil, metal screen, or wires can be added around the base of the rod if necessary. This type of antenna is actually a ¼-wave Marconi radiator, and it shows an impedance between its base and the surrounding counterpoise of 37 ohms, half that of a dipole or ½-wave antenna. A concentric tube line can be made to feed this system, the ratio of diameters to make the line 37 ohms being 1.86. Using ¾-inch o.d. pipe with 1/32-inch wall, the inner conductor will be .367-inch o.d. The use of ⅜-inch outside diameter is satisfactory. Smaller tubing with a copper wire threaded through isolantite beads or spacers is more practical for most installations.

Five Meter Circuits: A typical 5-meter receiver circuit is shown in Figure 1. By

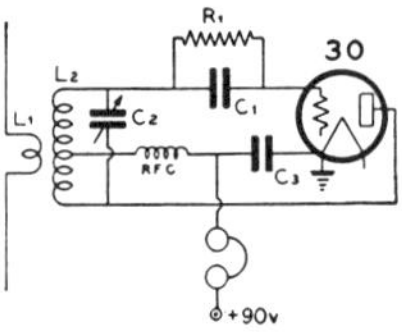

FIG 1

making R1, C1, C3 and the plate voltage supply of certain values, the super-regenerative effect will take place. R1 and C1 cause a blocking action which throws the detector in and out of oscillation at a high rate of frequency. R1 can be returned to the filament or B + as shown, depending upon its value, but for less overloading and distortion effect on strong 5-meter signals, the connection shown is highly desirable. C3 must be large enough to bypass the high super-regenerative surges back to the filament, but not large enough to short-circuit the audio frequencies in a modulated signal which must be impressed across the telephone receivers or audio amplifier. Common values for R1 are from ¼ to 2 megohms, C1 of .00025 to .0001 mfd., and .006 mfd. for C3.

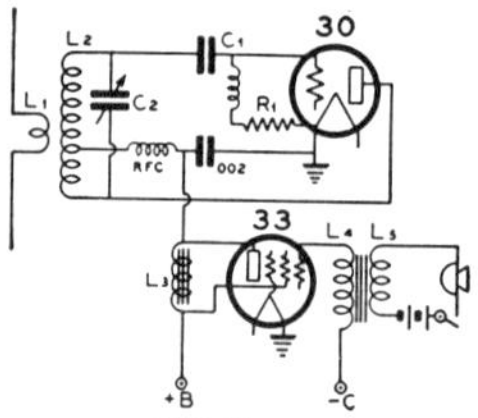

FIG 2

In Figure 2 is shown a 5-meter transmitter of very low power. This circuit is similar to that of Figure 1 except that the grid-leak component has a lower value. The lower value of R1 allows steady oscillations to take place, and energy can be fed to the antenna system through the coupling between L2 and L1. As an alternative, capacitive coupling will give the same results.

Technical Notes: In the design of ultra-short wave apparatus, it is of utmost importance that all leads be kept as short and as direct as it is possible to make them; this applies especially to grid and plate leads. All connections must be soldered by a rosin-flux solder. Tuning condensers must be kept away from metal panels, and control shafts must be extended to the control dial by bakelite coupling connections and extension rods made of the same material. The polarity of connections and batteries must be carefully observed, if the apparatus is expected to operate. The layout of "breadboard" receivers and transmitters must be closely followed to the line-drawing specifications. Equipment supplied with B batteries are required to have at least one 8 mfd. electrolytic condenser shunted across the power input terminals; this condenser prevents "fringe howl" when the batteries become old and have high internal resistance. The equipment should be completely shielded whenever possible or placed inside of a metal cabinet.

For either transmitting or receiving, the antenna must be erected as high above the ground as local conditions will permit. For 5-meter work, a half-wave vertical antenna made from a rod or wire about 8 feet long will suffice. The antenna may be coupled either capacitively or by means of a RF feeder; the latter consists of a line of any practical length run off for at least three or four feet at right angles to the antenna, the feeder is attached 12 inches below the physical center of the rod. Ordinarily, a simple broadcast receiving antenna will give fairly good results for short-wave reception on account of the harmonic effect.

In the design of ultra-high frequency equipment, it is necessary to quite accurately know the frequency upon which the apparatus is functioning. The technique for making these measurements appears below.

Wavelength and Frequency Determination: Some means of adjustment of the transmitters and receivers must be made in order to operate within the 5-meter band which extends from 56 to 60 megacycles. This band is over four times as wide as the whole American broadcast band, yet it covers only a third of a meter in this range. In localities where there is some 5-meter activity, a frequency check can be given by other amateurs who have calibrated frequency meters or receivers. Where it is impossible to acquire the use of such instruments either directly or indirectly, an alternative is to employ a parallel or Lecher Wire resonating system to measure the frequency, such a system is accurate to within 1 per cent.

Technique: Erect a parallel line of two clean bare-copper wires of No. 14 and 18 gauge, about four or five feet from the ground, spaced 3 inches apart by small drilled bakelite separators slipped over the wires and slid along into position. It is of major importance that the wires clear all objects by at least 3 feet, be straight and absolutely parallel. The length of the line should be between 35 and 40 feet, open at one end, and coupled by a single turn loop at the other end to the oscillator coil in either the transmitter or receiver. When these wires are coupled to the oscillator, standing waves are set up having current anti-nodes, that is, points of maximum current, at distances equal to a half wave-length. Resonance is attained by sliding a short-circuiting bar or copper-link along the wires starting from the **farthest** end of the parallel line (the link must be moved by an insulated rod or dry rolled newspaper that is at least 2 feet long—body capacity will give an erroneous reading). The indicating device can be either a milliammeter placed in the oscillator plate circuit, or by means of a variation of the RF current. The latter indicator consists of taking a small turn of wire and connecting it in series with a 6-volt radio dial light or RF thermogalvanometer; these indicating instruments are coupled to the plate side of the oscillator coil. A deflection in plate current or dimming of the lamp will be noticed when the shorting link is across some half wave point on the parallel wires. Sliding this link between the first and second points of indication, making note of the points, then measuring the distance with a scale or tape-measure will give the exact half-wavelength of the frequency of the oscillating circuit.

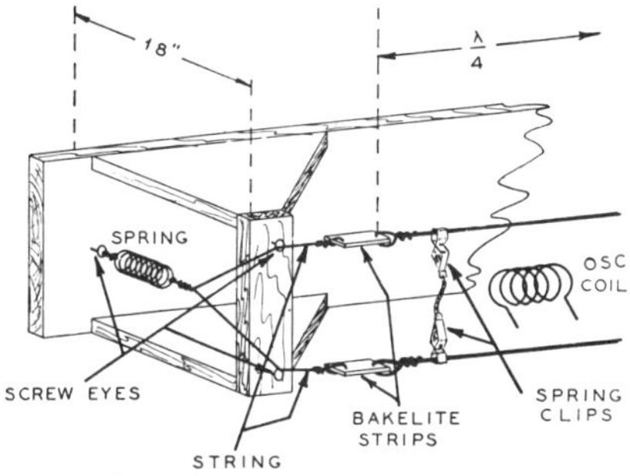

FIG. 3—Lecher Wire measuring system.

Another method of determining the frequency or wavelength which does not require the use of a milliammeter is to connect a 6-volt radio dial lamp or neon bulb in series with the short-circuiting link, as described above, then sliding the lamp bridge along the wires, starting from the farthest end. The lamp will light up at each half-wave current anti-node, the distance between any two current anti-nodes is approximately half the wavelength at which the circuit is oscillating. This method is not very accurate.

In either of the above methods the measurement must be reduced to meters (metric measure) to give the wavelength, then the results must be multiplied by 2.

Wavemeters (Absorption Type): While it is not always convenient to erect a Lecher Wire system each time it is necessary to measure the wavelength, an alternative is to construct a wavemeter, calibrated from the parallel line and oscillating receiver or transmitter. A handy form of absorption wavemeter, having a range between 4.7 to 7 meters, consists of a 25 mmfd. midget variable condenser paralleled with a 12 mmfd. mica fixed condenser and a coil wound with 5 turns of No. 10 wire, spaced in one inch.

Another form of absorption wavemeter having a range from 4 to 14 meters can be made from a 150 mmfd. variable condenser in parallel with a 2 turn coil of No. 10 wire, wound in a 2-inch diameter. The coil should be supported with bakelite spacers. A neon lamp may be shunted across the circuit for indicating resonance.

In the design of the above wavemeters, it is required that the whole circuit be rigidly constructed; any deformation in either the coil or condenser after calibration will give an erroneous reading. Hand capacity effects must also be eliminated, hence, the condenser must be manipulated by means of an extension handle preferably made of bakelite; the complete wavemeter assembly must also be held by a small bakelite rod about 6 inches long attached to the condenser supports.

Technique: The wavemeter is calibrated on any arbitrary scale attached to it by simply bringing the wavemeter within the proximity of the oscillating circuit of either a super-regenerative receiver, transmitter or Lecher Wires being energized by an oscillating instrument. In the case of Lecher Wires, tuning the wavemeter will absorb some of the RF energy from the loop end near the transmitter; this will cause the plate milliammeter to slightly deflect, indicating, of course, that the wavemeter is in resonance with the oscillating circuit. The wavemeter must be coupled very loosely to the circuit and the smallest deflection of the meter must be sought; the most accurate reading is obtained with an almost imperceptible deflection. By moving the short-circuit bar along on the wires, the wavemeter can be calibrated for harmonics by measuring the distances between more than one current anti-node.

After the wavemeter has been calibrated properly it always can be used for measuring the wavelength of oscillating receivers and transmitters. In a super-regenerative receiver, simply bring the wavemeter close to the plate coil, and by tuning the condenser will cause a diminution in the "hiss level."

Transceivers: A transmitter capable of operating as a receiver, or conversely, a receiver capable of operating as a transmitter by a simple switching of the circuit elements is called a transceiver. The number of circuits, tubes and combinations entering into the design of a transceiver are infinite; however, a technical view of some of the better circuit developments, commensurate with good amateur practice, is given herewith.

Combination One-Tube Transceivers: For the newcomer in the 5-meter band, the set shown in Figure 4 is about as simple as can possibly be built. The transmitter radiates modulated signals, and in the receiver position functions as a super-regenerative circuit.

The circuit, operating with a type 19 tube, functions as a push-pull oscillator and detector. As an oscillator or transmitter, the carrier is grid-modulated. The microphone, an ordinary single-button telephone transmitter, is in the negative B battery lead and the voltage drop and variation of voltage is used as a grid bias. There is a steady voltage drop across the resistance of the microphone, and when the latter is spoken into, the variation of resistance causes a variable grid-bias to be applied to the oscillator.

The receiver is frequency stabilized in both transmitter and receiver positions by unity coupling. Tuned-grid and tuned-plate

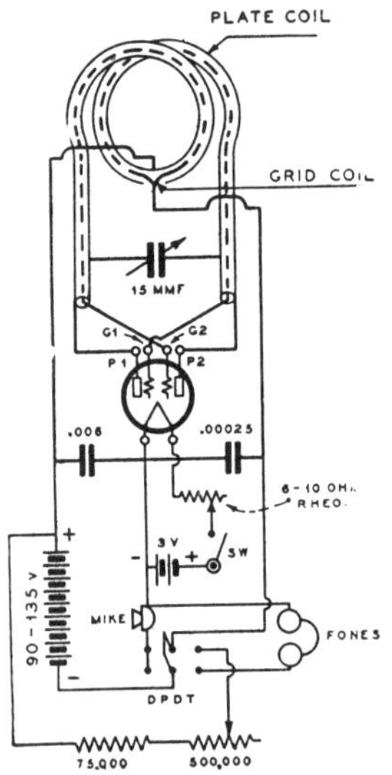

Revised Jones 5-meter circuit as used by Allied Radio Corporation in the "Knight" Transceiver.

(TNT) oscillator circuits require a compensator on one of the switch positions, which adds complication to the circuit. Unity coupling is obtained by placing the grid coil inside the plate coil. Two turns are used in order to conserve space and coil external field, and also to give short leads to the by-pass condensers C2 and C3.

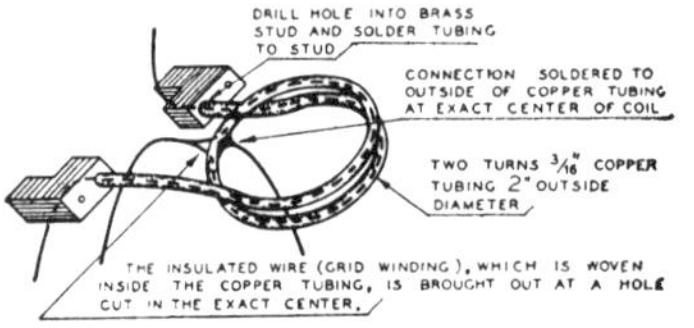

Showing how to make the plate coil. Grid winding is inside of plate coil.

Technical Details: The oscillator coil consists of a small coil 3/16-inch or ¼-inch soft copper tubing with a well-insulated piece of rubber- or cambric-covered wire threaded through it for the grid coil. The copper tubing coil consists of 1-⅝ turns, two inches outside diameter, with a center-tap on both coils. The grid coil center-tap can be made most easily by cutting a small slot (about ½-inch long) in the wall of the copper tubing, at the center of this plate coil. The grid-coil can be threaded through the tubing in two sections with the center connection soldered in a small "pig-tail" form about ¼-inch clear of the copper tube center opening. The ends of the plate-coil tubing can be fastened into small brass end-blocks or soldered directly to the two plate terminals on the 19-tube socket. The ends of this coil extend down about an inch, or slightly less, in order to keep the coil center-taps clear of the other socket terminals. The two inside, or grid leads cross over to the opposite socket grid terminals. The tuning condenser mounts near the tube socket and thus the leads to the condenser are only one inch long. A bakelite extension to the dial shaft is necessary to eliminate hand capacity effects.

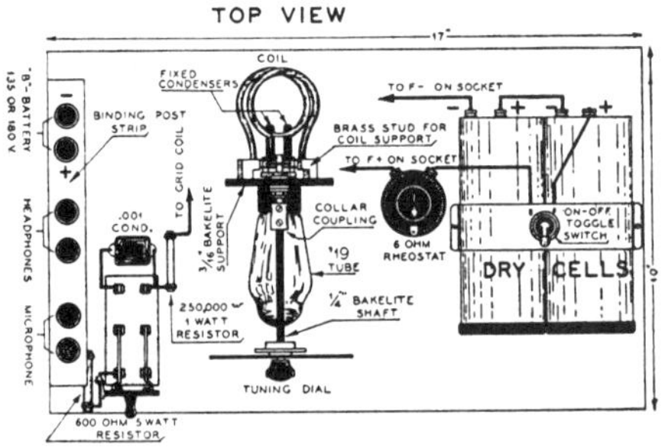

Looking down into the assembled transceiver.

For convenience, the battery leads, microphone and headset connections are brought out to six binding posts. Either 135 or 180 volts of B batteries or a small B eliminator may be used. The plate current is from 30 to 50 milliamperes on transmission, and about 5 on the receive position.

The transmitter will illuminate a 6-volt radio dial light, when the latter is coupled to the oscillator coil by means of a two-inch turn of wire soldered to the lamp terminals.

The receiver will give a hissing sound when it is functioning properly. The antenna can be most conveniently coupled to the set by means of a clip on the copper tube inductance. The clip is set near the center-tap, but as far away from it as possible to still get the super-regenerative hiss over the tuning dial range. Usually the clip will not be over an inch along the inductance from the center-tap. Any wire will act as an aerial, and wires up to several hundred feet in length will give good results. For most local work, a four-foot wire or rod connected to the oscillator by means of the aforementioned clip will be satisfactory. For better results, a wire 12 feet long is recommended; it gives a quarter-wave plus a half-wave antenna. The four-foot section acts as a quarter-wave antenna with the set and batteries acting as a ground or counterpoise.

Notes: To check the 19 tube, insert a milliammeter in series with the B battery leads, it should read from 50 to 60 ma. when transmitting, and when the set is not oscillating, the reading should drop to

10 or 15, such as when touching a plate or grid terminal with the antenna or one's finger-tip. For receiving, the plate current reads approximately 5 ma.

If it is possible to obtain a high-level single-button microphone of about 200 ohms resistance, the 600 ohm plate resistor can be eliminated and more power output obtained without excessive plate current. This resistor maintains the plate voltage to about 100 to 120 volts, as the microphone in the set originally designed had only a resistance of about 20 ohms.

2½ and 5-Meter Transceiver: For reception, the sensitivity of this transceiver is good due to super-regeneration, and for phone transmission the output is appreciably high. The component costs are so low and the construction is so simplified that 2½-meter operation should become more popular in many localities where now exists a dearth of such activity.

FIG. 7

Design Specifications: A 76 tube acts as a grid-modulated oscillator for transmitting and as a super-regenerative detector for receiving. Switching is done with a DPDT switch, the "on" position shorting out the headset and connecting in the grid-modulating transformer. The latter may be of any type of carbon microphone to grid transformer which has from 3000 to 5000-ohms secondary (grid winding) resistance. The primary is connected in series with a single-button microphone and a battery. If AC is used on the heater of the 76 (or 56), a 4½-volt C-battery may be connected in series with the microphone, switch and transformer primary without a connection to the tube heater.

B-voltage can be from 135 to 225-volts. A 6-volt storage battery or so-called "hot-shot" battery serves admirably for the filament supply for a 76 tube.

When receiving, the grid-leak is connected back to the plus-B for obtaining high audio output. The grid-leak current biases the grid as long as the tube is strongly oscillating in a super-regenerative condition. The plate current is less than 6MA. while receiving, hence, will not cause the headphones to burn out through excessive current. When transmitting, the plate current rises to 25 or 30 MA and the headphones are short-circuited by means of the send-receive switch.

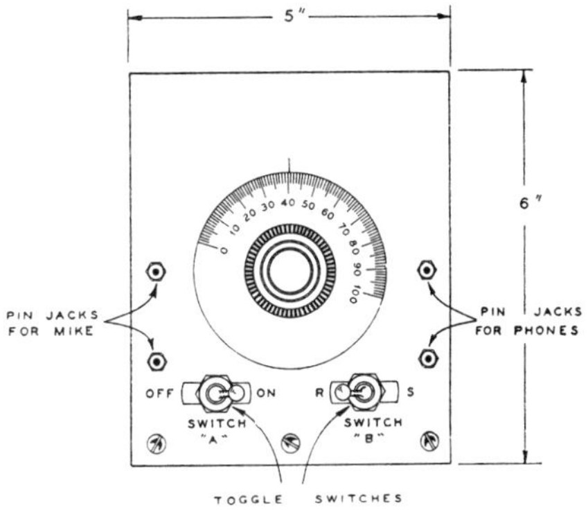

FIG. 8—Front panel layout.

The .01 mfd. condenser must be connected as shown because of the 400-ohm cathode resistor. The latter is not needed for receiving but is essential for good modulation when transmitting. The .01 mfd. condenser by-passes the interruption frequencies from the plate to cathode so a blocking grid-leak and condenser action can easily take place.

Placing the grid condenser in series with the grid coil rather than in the center of the tuning coil, gives better sensitivity on weak signals for a wide variation in battery voltages, antenna coupling and circuit constants.

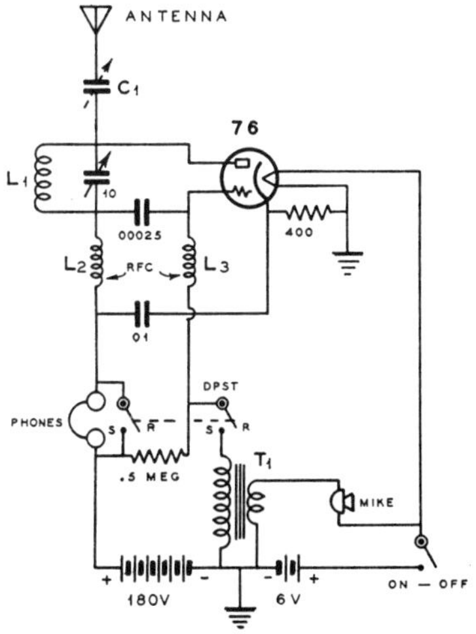

FIG. 9—Circuit diagram.
T1—Mike-to-grid transformer.
C1—Antenna lead twisted around plate lead.
L2-L3—(RFC)—Two r-f chokes, each 100 turns, No. 34 DSC, 3/8" diameter.

The RF chokes are not critical if operation in only one band is desired. If both 2½ and 5-meter operation is planned, the chokes must be wound with the correct number of turns so as not to resonate at any harmonics in the bands being worked. If this resonance occurs, the receiver will not super-regenerate at one or more spots on the tuning dial, even without an antenna

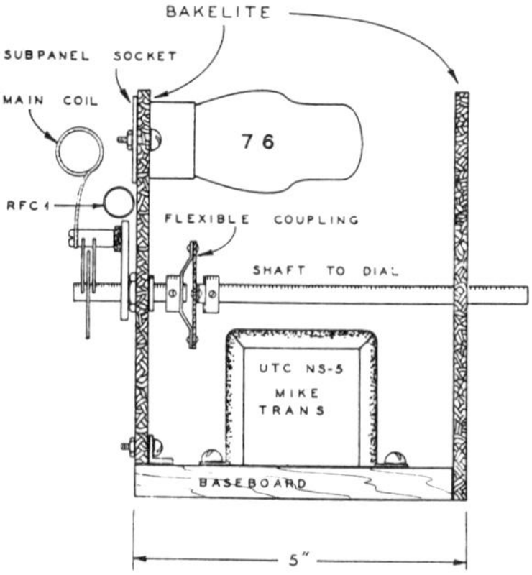

FIG. 10—Side view, showing coil support and shaft coupling.

coupled into the detector. The chokes wound with 100 turns of No. 34 DSC wire on a ⅜th-inch diameter form work satisfactorily, as the nearest "dead-spot" is then near 4½-meters, outside of either the 2½ or 5-meter bands.

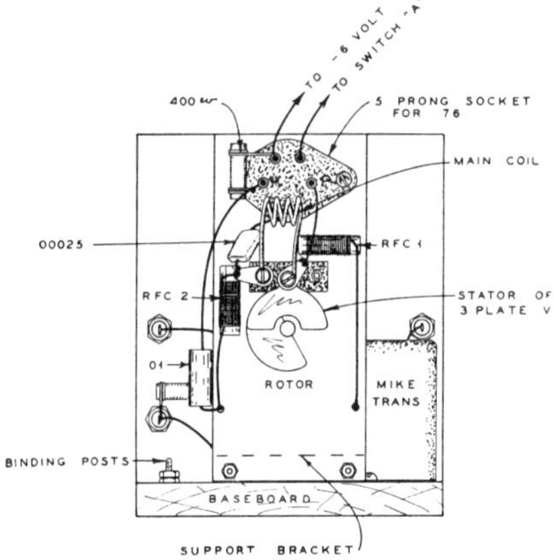

FIG. 11—Rear view.

A three plate old-type Cardwell midget tuning-condenser has machine-screw terminals which facilitate coil changes. The 2½-meter coil is made of 4 turns of No. 14 wire on a ⅜th-inch diameter spaced out to be about ⅝th-inch long. The 5-meter coil consists of 9 turns of No. 14 wire on a ½-inch diameter with a turn spacing equal to about the wire diameter.

Very short leads to the grid and plate terminals on the socket permit operation down to as low as 2-meters. The tuning condenser shaft is insulated from the front dial with a flexible shaft coupling and an extension shaft. This eliminates hand-capacity.

The antenna is coupled by twisting one to three turns of **insulated** hook-up wire around the plate lead. This capacity coupling should be as great as possible without causing a cessation of the super-regenerative hiss heard in the receivers when the control switch is in the receive position. For the reception of very weak signals, the antenna coupling must be adjusted to the optimum operating conditions. Tight antenna coupling allows a higher degree of modulation when transmitting.

With the antenna disconnected, a 6-volt pilot lamp connected in series with a small turn of wire can be used for a modulation indicator. The lamp should light up to about ⅔rds normal brilliancy when the turn of wire is coupled to the 2½ or 5-meter coil in the send position. Talking into the microphone will cause a downward variation in the light intensity.

Another receiver may serve as a monitor of voice quality and degree of modulation, or a monitor may be quickly assembled from a simple tuned circuit in which a diode and headset is connected in series; the receivers being by-passed with a .001 mfd. condenser. This transmitter will modulate easier than the other single type-tube 5-meter transceivers previously described.

Two-Tube Transceiver: This transceiver will operate in both 5 and 10 meter wavebands, its power output ranges from about one watt carrier at 160 volts plate supply to about 3 watts at 250 volts. These powers are suitable for use in cities or level forests of from two to six miles on five meters. These same sets will transmit and receive up to any visual distance (100 miles or more) between mountain sides. High sensitivity of the set is obtained by relatively tight coupling to a resonant antenna and operation of the super-regenerative detector at a moderate value of actual plate potential and grid bias, followed by a high-gain audio stage.

The circuit consists of two tubes which, in the receive position, one tube acts as a super-regenerator and the other as an audio amplifier. In the transmit position, the actual plate voltage on the former tube is increased greatly and a low value of grid leak makes it into a powerful oscillator. The audio amplifier becomes the modulator and the headset is cut out and the single-button microphone is cut in.

The transmitting oscillator draws relatively high plate current on these short wave-lengths and best results are obtained when the modulator has a step-down output transformer or choke for coupling. A center-tap output transformer or center-tapped 30 or 40 henry choke will give high percentage of modulation as compared to the usual Heising choke coupling to the oscillator. The choke carries the combined oscillator and modulator plate current, hence, it must have a suitable air-gap to preserve the quality of speech.

The microphone transformer can be any single-button grid-input type. The volume control for receiving has no effect on the transmitter except to act as a fixed load resistor across the microphone transformer secondary winding.

Coil Data: The plug-in coils have two pin-jacks mounted very close to the tuning condenser terminals, preferably on the same piece of bakelite or hard-rubber sub-panel. These coils must be at least ¾-inch away from any metal shields. The RF chokes are mounted beneath the sub-panel near the grid condenser and grid terminal of the oscillator socket. The chokes are made by winding about 50 turns of No. 30 DSC wire, on a ⅜-inch diameter bakelite rod, the winding must be confined to a space of approximately 1¼ inches. The chokes are mounted by means of very short 6/32 machine screws, the latter must not extend into the RF choke winding. The finished coils should be dipped in clear lacquer or coil "dope" and dried before using.

Technical Notes: The antenna coupling condenser consists of two right-angle brackets about ⅛-inch apart and ⅝-inch square. A slot is cut in the mounting screw hole of one of these brackets in order to have a slight variation in coupling in order to adjust it to a point where the receiver has a tendency to pull-out of super-regeneration with the regeneration control set at the half-way position. The 10 mfd. electrolytic by-pass condenser across the 2-watt 600-ohm cathode resistor can be of the 25-volt type. For coupling into a single-wire feeder, a condenser spacing of about 1/16- to ⅛-inch will suffice. A quarter-wave antenna will give good results with this transceiver.

Two-Tube Transceivers (2): A simple transceiver circuit for operation on 5 meters is shown in Figure 15. In general, this circuit eliminates the need of a 4-pole double-throw switch for changing from transmit to receive, and vice versa. The switch used is an ordinary single-pole snap-type with the moving arm grounded. When receiving, the switch short-circuits practically all of the modulator output and provides a low impedance path for the detector plate circuit. When transmitting, the grid leak is lowered in value so as to obtain a steady, strong carrier output and, in addition, the modulator tube is allowed to function. The modulation components and their adjustments are similar to the transceiver described above. The modulation capabilities are tested with the usual RF

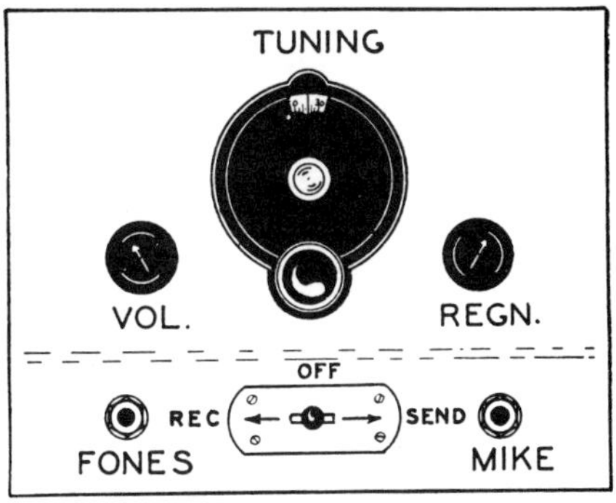

FIG. 12—Front view of transceiver.

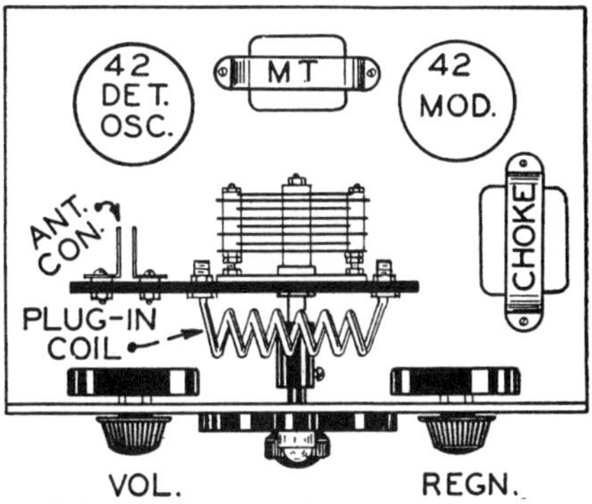

FIG. 13—Correct placement of parts.

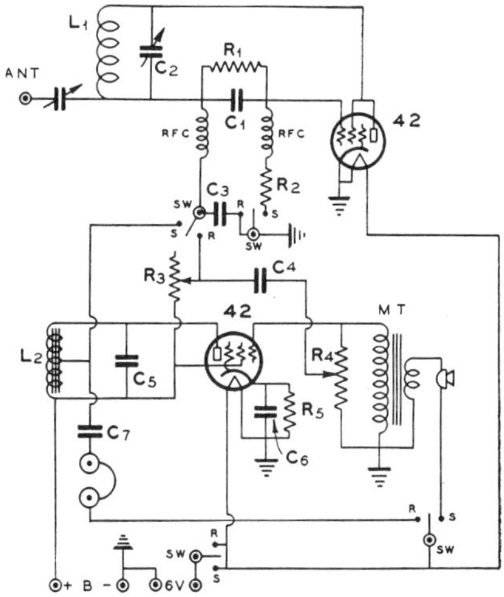

"42-42" Transceiver.

L1—6 turns, ⅝" dia., spaced ⅛" between turns.

L2—Center-tapped 20 henry choke.

MT—Microphone transformer.

R1—1 meg. R2—5,000 ohms. R3—50,000 ohms, R4—250,000 ohm. Pot. R5— 400 ohms. SW —4PDT switch. C1—.00035. C2—15 mmf. C3—.006. C4—0.1. C5—.006. C6—10 mfd. C7—0.5 mfd.

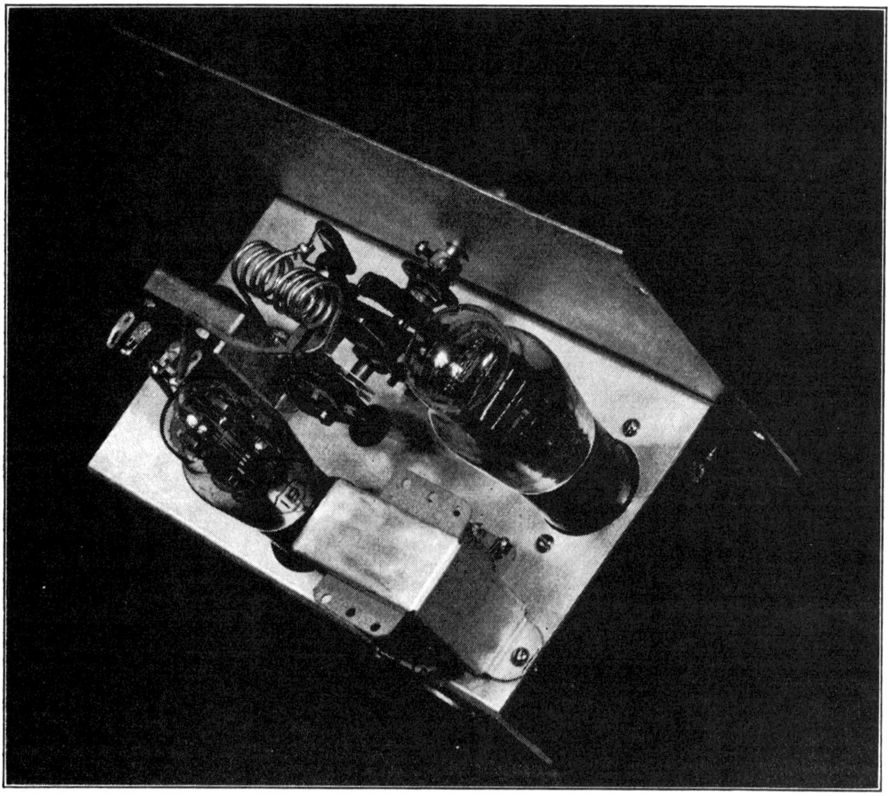

FIG. 15—Looking into the 2-tube transceiver. Parts should be placed as shown.

dial lamp indicator; in this transceiver, the brilliancy of the lamp will increase when the microphone is spoken into. Lack of sufficient modulation intensity can be traced to saturation of the modulation choke; this effect can be eliminated by inserting a piece of paper into the air gap along the butt joint of the choke so as to increase the air-gap length.

Coils and Chokes: The RF choke consists of about 50 turns of No. 22 DSC wire, wound on a ¼-inch dowel rod, and then slipped off the rod. The coil is supported from the oscillator coil to a small hole in the bakelite sub-panel which holds the tuning condenser. The 2500-ohm grid-leak is mounted in a similar manner on the other edge of the bakelite panel. The two oscillator coils consist of about 4 to 4½ turns, wound on a ½-inch diameter, are mounted on the tuning condenser, and the mica condenser (see schematic) is soldered across the inside ends of the two coils. The turns are spaced one diameter of the wire.

Antenna Coupling: The antenna coupling condenser should be a 3-30 mmfd. adjustable trimmer condenser. Normally, the con-

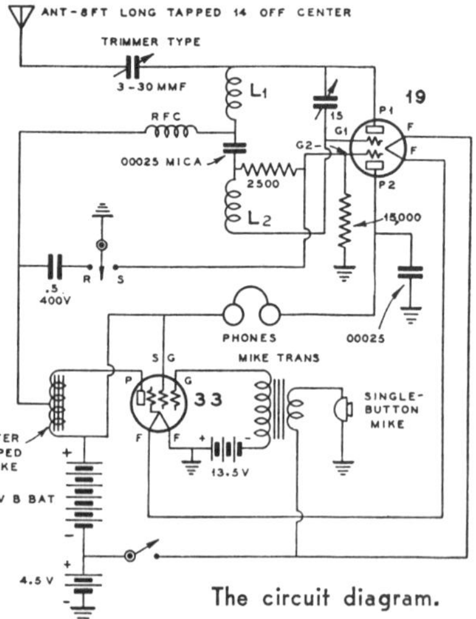

The circuit diagram.

L1, L2—Each 4 turns, No. 14 wire, ½-inch dia., air supported, spaced one diameter of wire.
L3—R-F choke, 50 turns No. 22 DSC, ¼-inch dia., air supported.

denser must be set so that the two plates are about 1/32- to 1/16-inch apart. This is much more coupling than can be used with a 30 tube oscillator.

Technical Notes: The set can be built on an aluminum panel and chassis, the latter being about 5-in. x 8 in.; the front panel dimensions are approximately 6½-in. x 8-in. It is convenient to use an extra tube socket for a plug and cable battery-connection system. Either bakelite wafer or isolantite sockets are suitable. Tip jacks or a plug and phone jack for headphone and microphone connections may be made through the front panel; it is necessary, however, to insulate both sides of the headphone connections from the front panel. Too much capacity in the antenna coupling condenser will prevent the receiver from properly functioning as a super-regenerator.

Two-Tube Commercial Transceiver Circuits: For those who wish to design two-tube transceivers after the circuits adopted by manufacturers will find the following schematic diagrams of practical interest. In general, constructors are advised to follow the details as close as possible to those given if good results are to be expected.

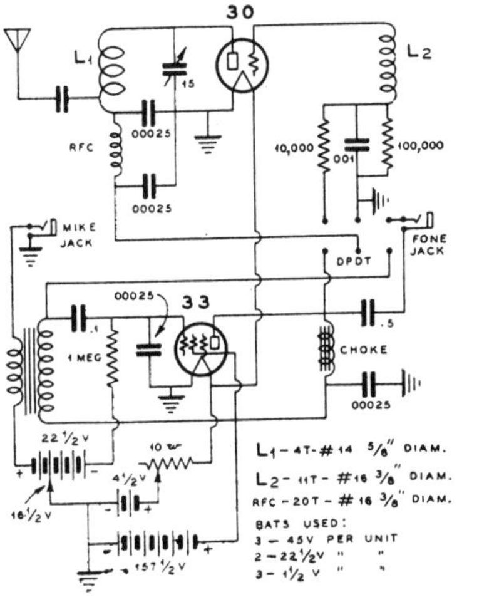

FIG. 17—Haigis Portophone model PF-1, showing resistors, condensers and coil constants.

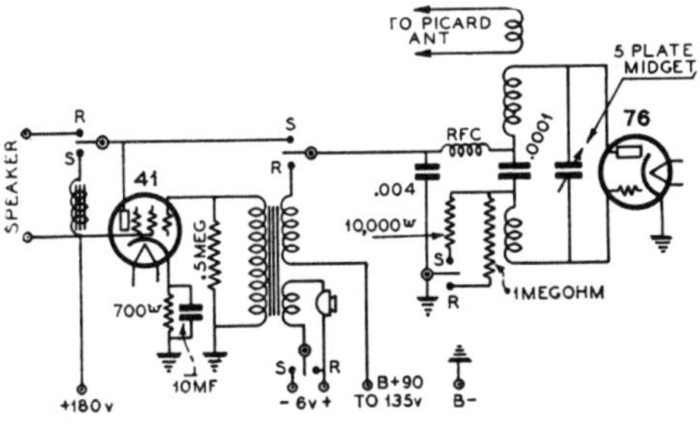

FIG. 18—Chauncey Wing's Sons Transceiver.

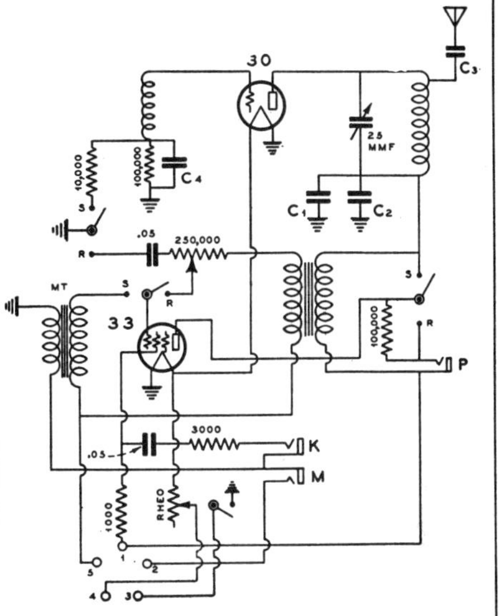

FIG. 19—Haigis "A-M" DeLuxe transceiver.

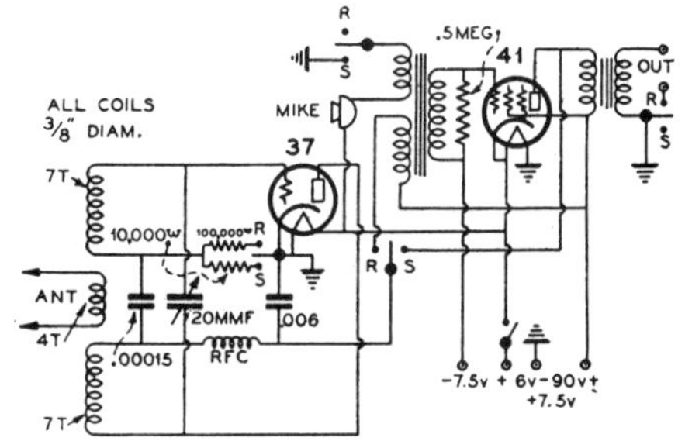

FIG. 20—Harvey Radio Laboratories 5 Meter Transceiver, 6 volt model.

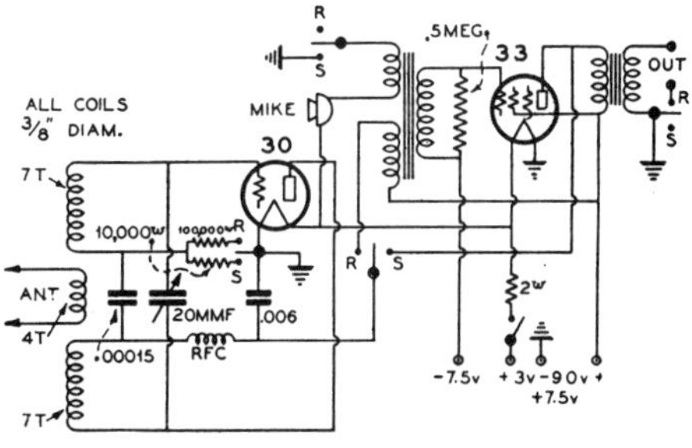

FIG. 21—Harvey Radio Laboratories 5 Meter Transceiver, 2 volt model.

Parts List for the Lafayette Transceiver

C1—15 mmf. midget.
C2—.002 mfd. mica.
C3—.002 mfd. mica.
C4—.00025 mfd. mica.
C5—.004 mfd. mica.
C6—.00005 mfd. mica.
R1—1 megohm.
R2—5000 ohms.
R3—200,000 ohms.
R4—1.5 ohms.
L1—Tank coil as described.
T1—Special Lafayette double primary transformer.
T2, T3—Class C AF transformers.
V1, V3—Type 19 tubes.
V2—Type 30 tube.

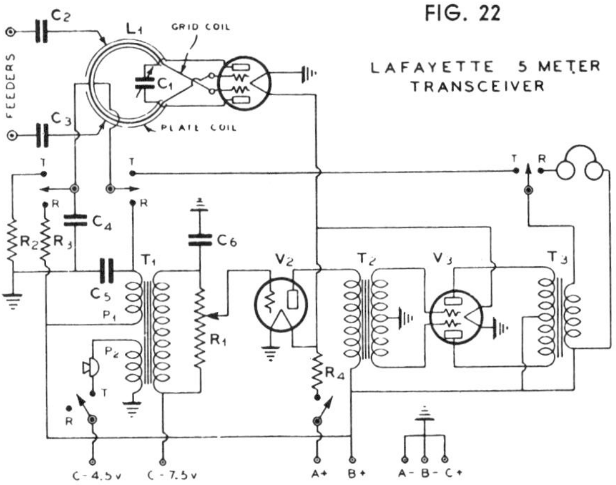

FIG. 22
LAFAYETTE 5 METER TRANSCEIVER

Three-Tube Transceivers: The push-pull oscillator, class B modulator transceiver herein described has a power output from 10 to 20 times that of the conventional transceiver employing type 30 and 19 tubes.

In operation, when the send-receive switch is thrown to the receive position, the RF circuit becomes a push-pull superregenerative detector feeding into a special primary winding on the microphone transformer. After being amplified by the class A driver and class B amplifier tubes, sufficient energy is developed to drive a loudspeaker. This transceiver circuit delivers about 2 watts of undistorted output, which is greater power than developed by many medium broadcast receivers. Throwing the switch to the transmit position changes the RF circuit into a high-powered oscillator and at the same time connects the microphone to its transformer. In general, the circuit design follows that of the aforementioned two-tube transceivers.

"Duplex" Transmitter-Receivers: The greatly increased popularity of the 5-meter amateur band has resulted in the use of transceivers which, unfortunately, have some disadvantages if very many of them are used in one locality at any one time. The receiver portion radiates strongly and the radiation can be heard nearly as far as the transmitter itself, in some cases. The transmitter is tuned to the same frequency as the receiver; it crowds-up all of the stations on one frequency. Some transceivers possess the annoying feature of not transmitting on the exact frequency of the receiver. Thus, two similar sets will sometimes drift in opposite directions throughout the entire waveband, and sometimes even beyond the band during a QSO. The power output is low because the antenna coupling must be very loose in order to prevent pulling the detector out of superregeneration.

As more 5-meter sets come into use, some means for overcoming these faults must be found. At the same time, the cost of construction must not increase appreciably. The circuit diagram shown in Figure ? has several advantages over the ordinary transceiver.

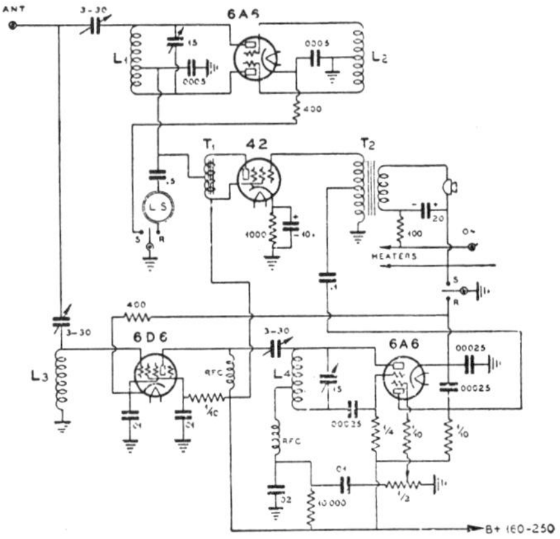

FIG. 23—Circuit Diagram.

L1—6 turns, No. 12 wire, ⅝-in. dia., ⅞-in. long.
L2—15 turns, No. 14 wire, ½-in. dia., 1½-in. long.
L3—18 turns, No. 22 DSC wire, ¾-in. long, ¼-in. dia.
L4—6 turns, No. 12 wire, ⅝-in. dia. 1¼-in. long.
T1—Center-tap 30 henry choke, 100 MA rating.
T2—Mike transformer with secondary center-tapped.
RFC—60 turns No. 30 DSC wire, ⅜-in. dia.
Send-Receive Switch is a D.P.D.T.

Circuit Details: The 42-tube modulator acts as a second stage of audio for loudspeaker reception when the send-receive switch is on the receive position. The second triode unit in the 6A6 acts as a high-mu resistance coupled audio stage. The microphone transformer gives a step-up ratio in the receive position for the audio amplifier. This additional gain is not

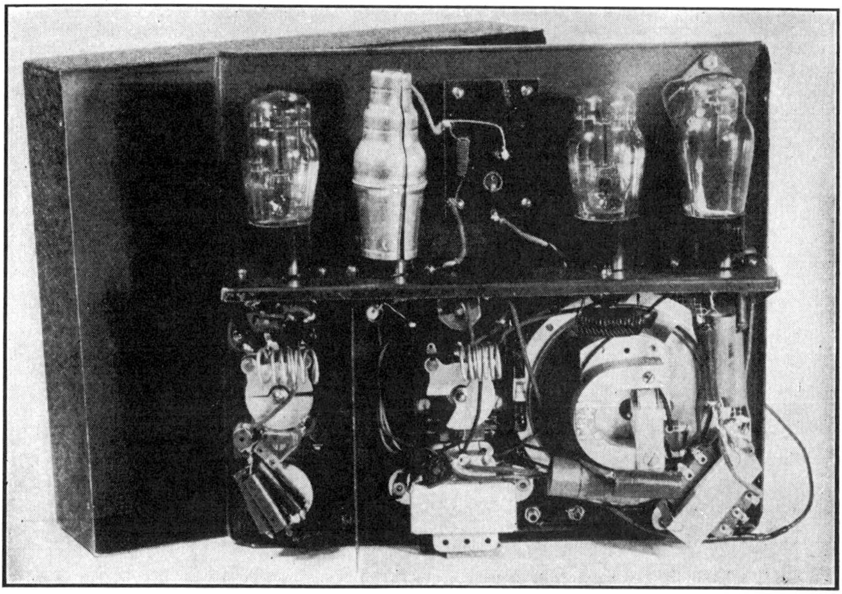

FIG. 24—Interior view of 5-meter auto set. Note shield partition between coils.

usually needed but the center-tap connection prevents the 6A6 plate circuit and the 100,000 ohm resistor from loading the modulator in the transmit position. If the 6A6 plate is connected across the entire secondary of the microphone transformer, it would be necessary to have an additional switch to cut this load off while transmitting.

The tuned RF circuit has a resonant grid coil which resonates with the tube and antenna coupling capacities to the low-frequency end of the amateur band, or preferably just outside of the band if the transmitter is to be used near this portion of the transmitting spectrum. This stage must be detuned 2 megacycles, from the transmitter, if no power is to be absorbed.

The grid coil is made by winding 18 turns of No. 22 DSC wire on a quarter-inch diameter rod to cover a length of ¾-inch. The coil is slipped off the rod and supported by its ends soldered to a pair of soldering lugs. Once its correct length is determined it can be coated along one side with Duco cement so that it will retain its proper inductance.

The semi-variable coupling condensers, marked 3-30 mmfd. in the circuit are of the small compression type with mica spacers. The one in the transmitter (for maximum frequency stability) is an air-spaced variable condenser with a screw-driver slot adjustment. The main oscillator tuning condenser can be either dial or screw driver slot controlled. Since this circuit is of the TNT type with a resonant untuned grid coil, it will give maximum results over about only 2MC. The 15 turn coil specified is for use between 58 and 60 MC.

This set is supplied plate voltage by a dynamotor power-generator. A microphone filter must be built into the dynamotor; this consists of a 100 ohm resistor and 20 mfd. 25-volt electrolytic condenser. The latter provides a return path for the voice frequencies, while the 100 ohm resistor acts as an impedance to noise from the common battery supply. The circuit is shown for use in a car with the "plus" terminal grounded to the car frame. If the negative terminal is grounded, the polarity of the 20 mfd. electrolytic condenser must be reversed. Five-meter chokes are also required to be placed at the dynamotor to prevent excessive receiver noise. (The chokes are described in a subsequent paragraph).

A built-in five-inch magnetic loudspeaker is incorporated to eliminate the need of wearing a telephone headset while driving. As can be seen from the pictures, the set is built into a very narrow steel can for the purpose of mounting it above the windshield or on the underside of the car roof. The outside dimensions of this can are 3-in. x 11-in. x 12-in., and the back cover fastens by screws to the cross-ribs in the car roof. The set can also be mounted in any position convenient to the operator. If motor-boating is encountered the detector plate resistor should be center-tapped and this point by-passed to ground with a 1 mfd. condenser.

Two-Tube Transceiver (3): This 5-meter transceiver is small but powerful. Reference to the circuit diagram will show that the 6A6 multi-purpose tube acts as a superregenerative detector and audio-amplifier, transmitting oscillator and speech amplifier; in addition, another tube type-42 functions as a plate modulator when transmitting, and as a second stage of audio amplification for loudspeaker reception. Whatever may be the reproducing medium, loudspeaker or headphones, no DC plate current flows through the windings.

FIG. 25—Showing how the 4PDT switch and small components are mounted under the chassis.

Design Technique: A four-pole double-throw switch with an "off" position in the center of its three positions interconnects the circuits for either transmission or reception. A momentary study of the circuit in either switch position will reveal the electrical function of the various components. In general, better sensitivity can be obtained when the plate RF choke is omitted. The .006 mfd. by-pass condenser must be connected to the approrimate nodal point on the coil which is approximately three turns from the grid end of the 8 turn

FIG. 26—This is the proper way to mount the parts.

coil. This coil is wound with No. 14 wire on a half-inch diameter with less than $\frac{1}{16}$th inch spacing between turns. The coil is soldered directly across the two-plate tun-

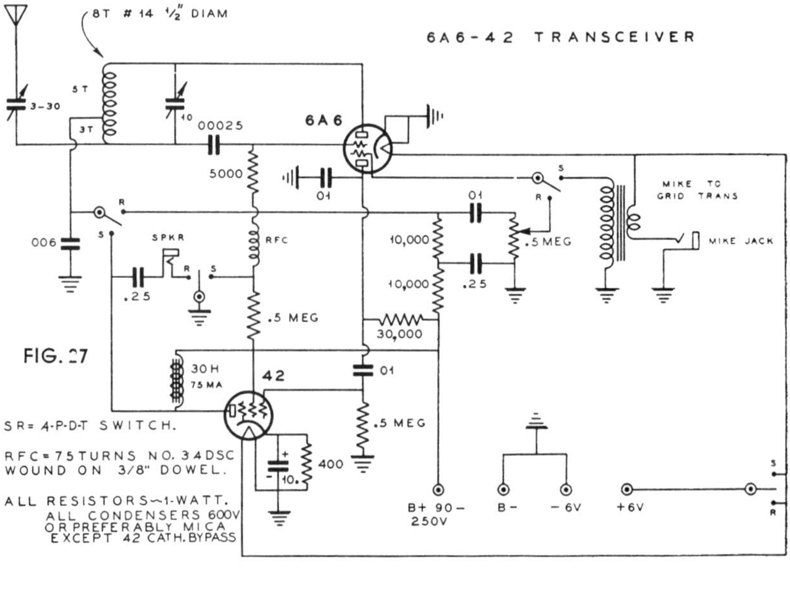

FIG. 27

ing condenser terminals. Probably the same circuit would work on 2½-meters if a small coil was employed since the leads are very short and the 6A6 tube a fine ultra-short wave tube.

The resistance network in the detector plate circuit prevents degenerative amplification (motor-boating). The 10,000-ohm resistor connected from the "plus" B to the 10,000-ohm detector plate resistor can be much higher in value. Better sensitivity and less receiver radiation will take place if the actual plate voltage on the detector is only high enough to allow super-regeneration when the antenna is connected. A 50,000-ohm variable resistor controlled by a knob on the front panel would be a worthwhile addition in place of this 10,000-ohm resistor connecting to the "plus" B supply.

The .01 mfd. condenser connected from the audio plate to chassis-ground functions as both RF and IF by-pass. This hiss level is of such magnitude that it would overload the 42 tube unless a large capacitor, as shown, is connected in the circuit. In addition, the condenser is required as an RF by-pass on account of the close proximity of the triode and RF plates inside the 6A6 envelope.

The plate potential is from 90 to 250 volts; the power output being much higher at higher values of plate voltage. The percentage modulation is quite satisfactory at any plate potential within the limits enunciated, and good speech quality is obtained. Modulation values can be indicated by the usual pilot lamp indicator. The microphone can be of the single-button type.

The antenna coupling condenser is adjusted by a screwdriver to the maximum capacity for a given antenna or feeder system as will just allow the super-regenerative hiss without the appearance of "dead-spots" over the entire tuning range when the control switch is in the "receive" position. The coupling, then, is sufficient for both receiving and transmitting.

The chassis is 6 x 6 x 1½ inches of plate No. 22 gauge steel. The front panel is 7 x 7 inches and is of No. 12 gauge aluminum, large enough so the complete unit can be slipped into a steel cabinet 7-inches high, 7-inches wide, and 7½-inches or 8-inches deep. The bakelite panel is 4-inches by 2½-inches and supports the 6A6 tube as well as the tuning condenser and coil. The tuning condenser has an insulated extension shaft to the dial on the front panel. Horizontal mounting of the 6A6 tube permits very short grid and plate leads in the RF circuit. Connection to the power supplies is by means of a wafer socket, plug and battery cable. The 6-volt switch forms a part of the multiple-leaved send-receive switch since it has an "off" position. The two-plate tuning condenser is made from a three-plate midget condenser.

Portable Transceiver: A mobile or portable transceiver that will stand considerable amount of rough handling and is capable of developing an amazing amount of power for transmitting or receiving can be constructed from the following specifications.

Circuit Details: The oscillator circuit is of the Colpitts type with a split-stator tank condenser, and a tap on the coil for DC feed. The plate power must be fed through an RF choke. The grid blocking condenser is a small 50 mmfd. air-insulated type (no mica condensers are used in the circuit assembly). Leads from the tube to

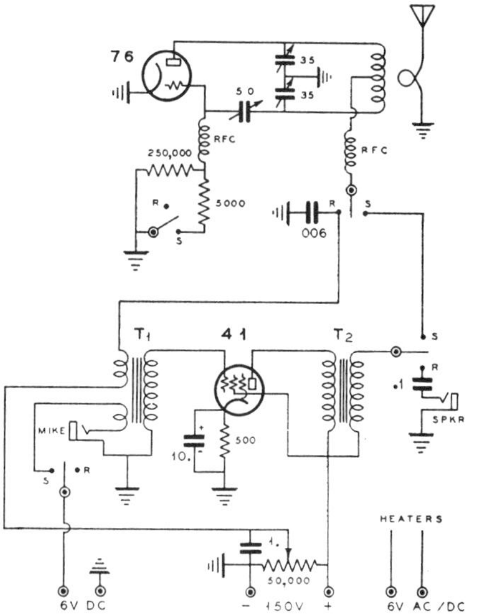

FIG. 28—Switch "S-R" is a 4PDT. RFC—75 turns No. 34 DSC, ⅜" dia. All resistors are 1 watt size. All condensers 600 volt rating, except cathode by-pass which is a 25v., 10 mfd. electrolytic condenser.

FIG. 29—Showing how the antenna loop is coupled around the plate coil.

FIG. 30—"76-41" Transceiver, microphone and portable loud-speaker.

the tank are through ⅜th-inch holes and are not insulated as they are self-supporting. The coil is on G. R. pins, and the jacks are mounted on small slabs of mycalex. The coil center-tap is made by soldering a 3/16th-inch wide strip under the middle of the coil and mounting a small

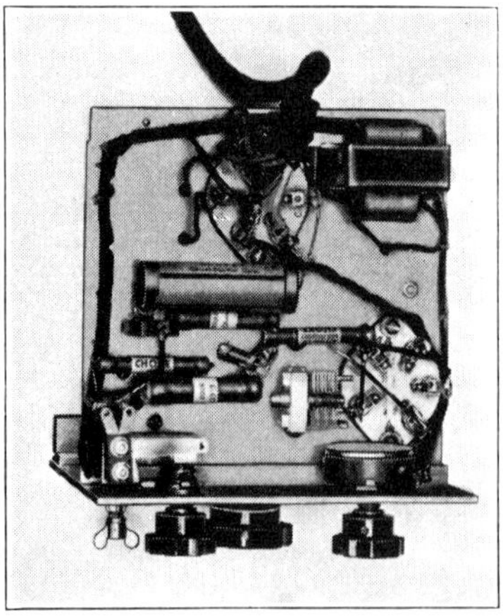

FIG. 31—The components under the chassis are neatly arranged, as shown.

switch-jaw between the pin jacks. The antenna is coupled by means of a one- or two-turn coil wrapped closely around the middle of the tank coil—the exact position and the coil-turns may require some experimenting. If a two-wire feeder is used, it would be well to connect both ends of this coil to insulators, instead of grounding one end. Note: the antenna coil is made of insulated wire to eliminate any danger of shorting the tank coil to ground. The modulator is conventional, differing only from a class A modulator in that it is coupled to its load through a transformer instead of a choke coil.

Front Panel and Case: The accompanying photograph of the transceiver shows the front panel having a machine polish (see "Miscellaneous Notes"); while such a finish enhances the outside appearance it is not necessary. The front panel is six inches square and the base panel is 5¾-inches square. The two pieces are held together by a short length of ⅜th-inch square dural rod, drilled and tapped for 8-32 machine screws having "filister heads."

The case is made of No. 18 gauge black iron, and spot welded. It is recommended that the front panel be taken to a sheet metal worker and have the case fitted around the panel. A handle, called a sash bar lift, may be purchased at any hardware store. After fitting the handle, the case is drilled for the microphone and speaker jacks and for the power cable, then it is painted.

Looking at the center of the front panel, one sees the main tuning dial and the regeneration control and the change-over switch to the left and right, respectively. In the upper right-hand corner is located the antenna lead-through insulator, fitted with a 6-32 wing-nut for convenience. The speaker and microphone jacks are on the left-hand side of the cabinet and the power cable is led through the back and is guarded by a rubber grommet to prevent the cable from chafing.

Note: The transceiver is tuned and adjusted similarly to those previously described.

San Francisco Bay Bridge Equipment: The 5 meter transmitter and receiver here described has been in use by the builders of the great San Francisco-Oakland Bay bridge ever since the first 6A6 tubes were released. The output is between 5 and 10 watts, depending on the plate voltage used. These sets have proved more reliable in service than any of the other units described in previous issues of this book.

The circuit consists of a push-pull 6A6 modulated oscillator using cathode resistor bias. It is loosely coupled to a tuned circuit across a two-wire matched impedance feeder from a half wave antenna. This same antenna tuned circuit is used as a means of tuning the receiver RF stage to the desired band. The resonance curve is sufficiently broad to enable the receiver portion to be tuned over quite a wide band before any great amount of sensitivity is lost.

The receiver is a separate unit, except for a portion of the audio system. A super-regenerative detector is used with two stages of audio amplification, resistance coupled. A type 41 tube is used as a high mu resistance coupled stage working into either the handset receiver through the step-down "mike transformer" or into the

FIGS. 32 and 33—Exterior and Interior Views.

6A6 driver stage for loudspeaker reception. For use in extremely noisy locations, a permanent-magnet dynamic speaker is switched across the output of the Class B audio stage by a slight rearrangement of the 4PDT switch wiring.

A separate microphone battery in the form of an ordinary 4½ volt C battery is used in order that a battery or an AC power pack supply can be plugged into the power socket at the rear of the set.

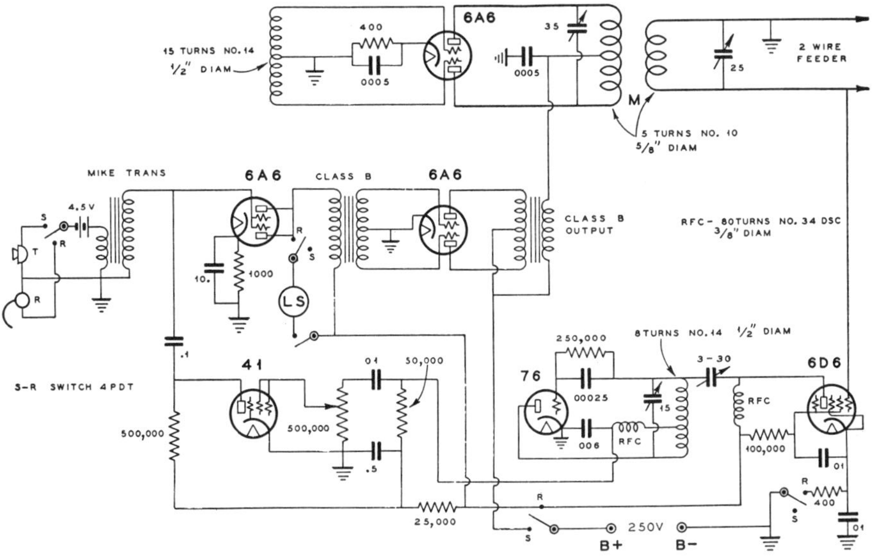

FIG. 34—Complete Circuit Diagram of a San Francisco Bay Bridge 5 Meter Receiver and Transmitter.

Filter Circuits for Automobile Installations: A major problem of 5-meter auto radio confronting users of short-wave equipment is that of securing a suitable and well-filtered plate voltage supply. B-batteries are cumbersome and expensive, if often replaced. A small B-eliminator or dynamotor, operating from the car 6-volt battery, is the solution to this problem. The eliminator or dynamotor occupies little space and the device can be made to supply from 150 to 300 volts.

However, most amateurs who have tried these systems have been troubled by noise in the transmitter or receiver, or both. Additional audio filter in the plus-B leads seem to be of little help. The trouble is caused by RF disturbances which filtrate into the A and B leads to the 5-meter set.

RF disturbances can be confined to the dynamotor or vibrator eliminator itself by means of simple RF chokes. The circuit in Figure A has worked satisfactorily with various dynamotors. The 8 mfd. condenser acts as an audio filter and a low impedance by-pass for the audio or modulator return circuits. The RF choke in the B-plus leads prevents RF from seeking a path up these leads to the set. All RF chokes are mounted as close to the power supply unit as possible.

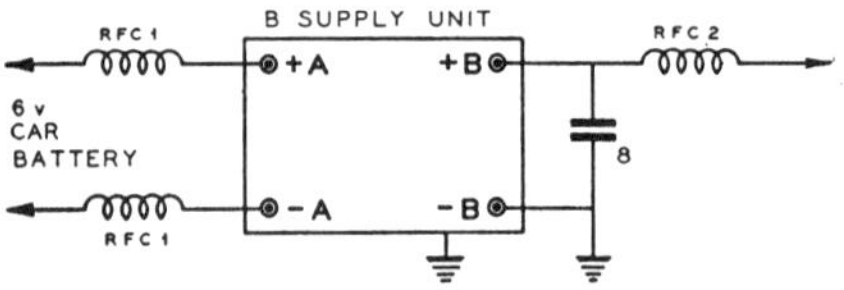

FIG. A

RFC1—25 turns No. 12 wire on 1/2 to 5/8-inch dia. form.

RFC2—75 turns No. 32 to 34 DSC wire on 1/4-inch dia. dowel or bakelite rod.

Either positive "A" or negative "A" battery terminal of car should be grounded.

The RF chokes in the 6-volt leads are wound with a wire size capable of safely carrying the load current, which may be from 2 to 10 amperes, depending upon the rated power input and load to the unit. Usually No. 12 enameled wire, close wound on a 5/8-inch dowel rod for a length of about 2 inches is suitable. The plate RF choke has many more turns than that of the dynamotor power supply, hence, a great number of turns of No. 32 to No. 34 DSC wound on a 3/8-inch diameter rod, for about 1-inch length, will generally suffice.

Occasionally a 1/2 mfd. condenser must be connected from the high-potential side of the battery at the dynamotor terminal to some particular point on the dynamotor frame or housing.

The circuit shown in Figure D will, in many cases remove spurious noises from a 5-meter transmitter when using a dynamotor power supply, or to prevent the clicking noise from a vibrator supply unit. Sometimes these units will be quiet enough for use on a receiver of the super-regenera-

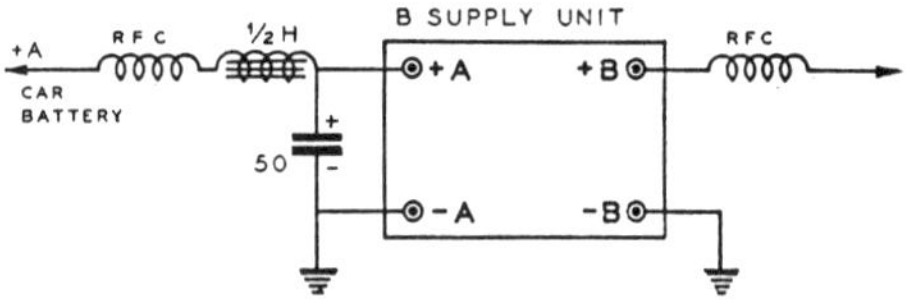

FIG. B

tive type but they will introduce noise in the transmitter due to lack of microphone circuit filtering. A simple filter consists of a 20 to 25 mfd. 25-volt electrolytic condenser to complete the voice frequency circuit, and a 100 to 200 ohm 1-watt resistor in the high-potential side of the 6-volt supply. Care must be taken to correctly polarize the electrolytic condenser with respect to the grounding of the car battery.

The circuits of Figure B and C are practical schemes which can be employed to prevent noises from being injected into either the transmitter or receiver. A low resistance choke of from 0.1 to 1/2 henry having a fractional resistive impedance is inserted in the line. Some small dynamotors are equipped with such a choke but not often with the 50 mfd. condenser or RF chokes. If no audio filter is furnished, at least an 8 mfd. electrolytic condenser must be connected across the plate supply, either in the 5-meter set or at the power supply terminals.

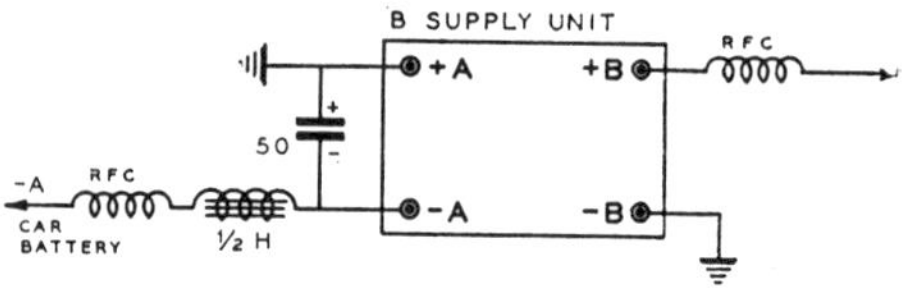

FIG. C

The RF filters must be mounted close to the dynamotor or eliminator to be effective. Ample space can be found inside the dynamotor container for these RF chokes; if not, they must be mounted rigidly in a metal can adjacent to the unit. (Note: The 6-volt supply to the 5-meter set comes from the battery side of the RF filters.)

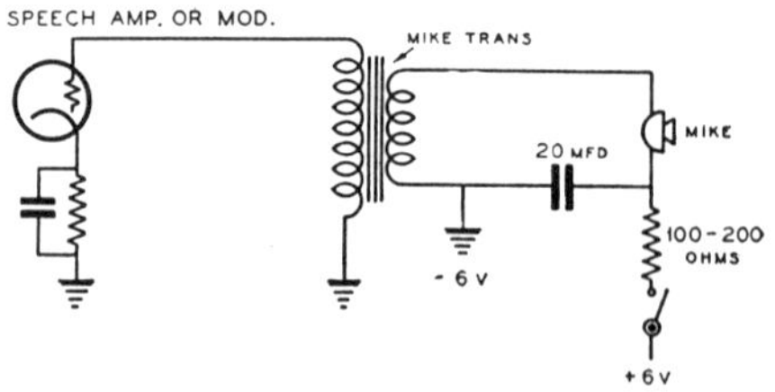

FIG. D

It is good practice to run the power leads directly to the car battery to avoid ignition noises. Frequently, resistor-type spark plug distributor suppressors will eliminate

5-meter interference. The conventional by-pass at the generator is advisable. Fortunately, the car ignition system noise is easily minimized, but the broadcast type of RF choke suppressors will not work on 5-meters. These suppressors are often layer-wound and are not effective at high-frequencies.

Parallel Rod Oscillators: For amateur communication, parallel rod oscillators are extremely desirable for wavelengths between 1 and 10 meters. Parallel rods of the proper diameter and spacing have a high "Q" much greater than can be obtained with tuned coils and condensers at high frequencies; in addition, the efficiency of such a system is nearly twice as high as that of the ordinary oscillator circuits.

A pictorial and schematic diagram of the oscillator circuit is shown in Figures 35 and 36. The circuit consists of two parallel rods of either aluminum or copper pipe, ½ inch in diameter, and separated ½ inch on porcelain stand-off insulators mounted on a long strip of kiln-dry wood; the length of rods are nearly one-half wave long. For five meter operation, two aluminum or copper rods about 7 feet long will suffice, while for work on 2½ meters, the rods will be approximately 30 inches long. One end of the parallel rods is permanently short-circuited and the plate voltage is fed into this point through a 50 turn RF choke. The other end of the rods has a sliding short-circuiting bar for adjustment of the frequency. The load consisting of a resistor and thermocouple, or two-wire matched impedance RF feeders, is bridged across the rods near either end. This distance varies from 3 to 8 inches from the short-circuiting bridge. The antenna feeders may also be coupled inductively by a single "turn" of parallel wires near the rods.

The oscillator acts like a normal "long wave" oscillator in that the plate current is quite low with no load. Under conditions of load, the frequency is mainly determined by the distance between the two short-circuit connections on the rods. Moving the oscillator tube along the rods affects the frequency to some extent, but its main function is to match impedances. Apparently the 45-tube, as shown in the circuit diagram, shortens the half-wave rods from 8 feet to about 6½ to 7 feet, due to the capacity effect of the tube. Based upon this conclusion, the impedance would be maximum at the center where the voltage would be greatest, and diminish at zero at the closed-circuit rod ends. By sliding the 45-tube along the rods, the best impedance match can be obtained for optimum plate load and grid excitation. The latter can also be controlled by means of the external plate-to-filament capacity. Other types of tubes might require this added capacity from grid to filament, instead of plate to filament.

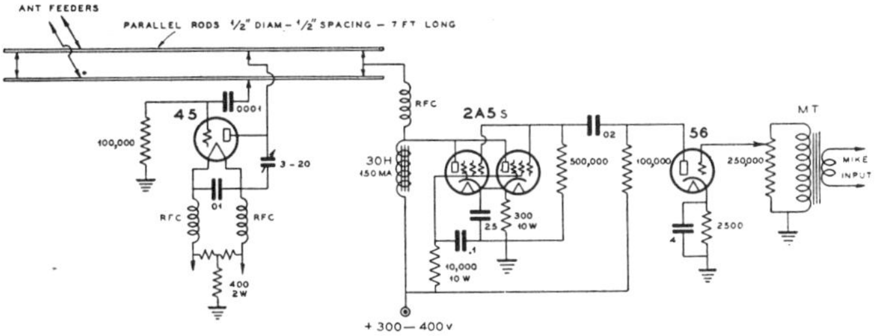

FIG. 35—Parallel-Rod Oscillator and associated speech equipment.

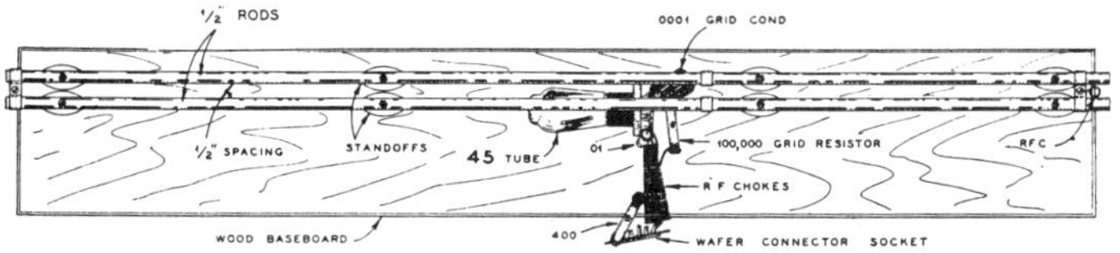

FIG. 36—Bread-board construction of Parallel-Rod Oscillator.

Circuit Details: Filament chokes, placed at right angles to the rods, isolate the circuit. These chokes are wound with 30 turns of No. 14 enameled wire, wound on a ½ inch dowel rod, then slipped off to form a self-supporting coil.

For operation at five meters, a small additional capacity of about 5 to 10 mmfds placed between the plate and cathode or filament is required to maintain oscillation at one frequency of high amplitude. This capacity is not required for 2½ meter operation. The 45-tube is not too effective at 2½ meters; however, the output can be improved by slotting the tube base between the plate and grid prongs, and also by the use of extremely short leads.

Operating the circuit for any great length of time will cause the plate current to drift; this can be eliminated by placing a 400 ohm cathode resistor in the circuit to provide a certain amount of bias to the grid.

The 10,000 ohm 10-watt resistor placed in series with the screen-grids of the pentodes in the modulator drop the applied voltage from 350 to 200 volts; the resistor prevents the temperature of the screen-grids from rising to a very high point.

Notes on Adjustment: The antenna may be properly adjusted by using an ordinary flashlight RF indicator inductively coupled to either end of the closed-circuit rods. The antenna feeders may be moved out, increasing the load, until less upward modulation is obtained when whistling into the microphone. Too much load on any class C stage or oscillator makes it impossible to obtain high precentage amplitude modulation. Too much antenna load will also result in less output and may even stop the circuit from oscillating.

A 50 per cent efficiency can be obtained with closed circuit parallel rod oscillators as compared against 25 to 30 per cent for most other circuits, besides the modulator does not have to deliver nearly as much power output for a given value of carrier output. A carrier from 5 to 10-watts can easily be modulated with one or two pentode tubes without having to resort to a class B system.

Push-Pull Parallel Rod Oscillators: A simple push-pull version of a parallel rod oscillator, similar to the single tube circuit described above, is shown in Figures 37 and 38. In general, the plate rods are the same as the grid assembly, except that the shorting bar is at the far end and the plates are connected to the ends of the rods. The

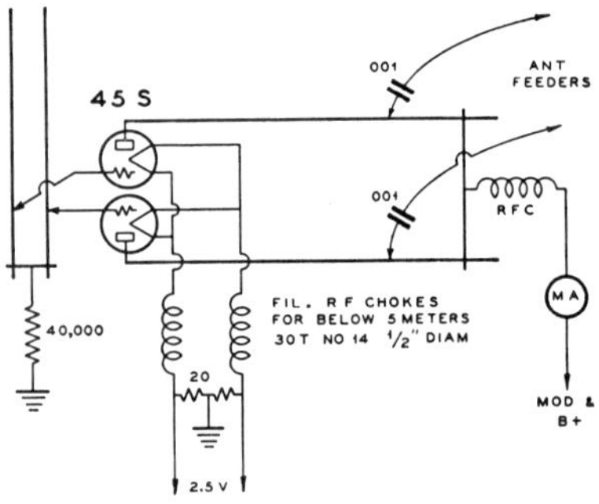

FIG. 37—Parallel-Rod Oscillator Circuit.

plate rods should preferably be extended out at right angles to the grid rods in order to minimize coupling. The grid rods act like a quartz crystal oscillator.

The grids are connected to each rod, ⅕th to ⅓rd the way up from the shorted end. Since the shorted end is a node, the insulation at this point is relatively unimportant, hence, the ends below the shorting-bar can be supported in holes bored into a wooden block. For 5-meter operation, the rods are made 4½ feet long. Air insulation is best for the tops of the grid rods, however, for low or moderate power, glass or isolantite can be used to hold the rods parallel.

A plate milliameter is absolutely necessary if correct tuning is to be accomplished. It will indicate minimum current (with antenna feeders disconnected) when the grid and plate shorting bars are correctly adjusted for oscillation. It will also fluctuate slightly when a screw-driver or pencil is touched to the rods at points other than voltage nodes. With the circuit oscillating properly, the feeders are then connected through a pair of fixed mica condensers

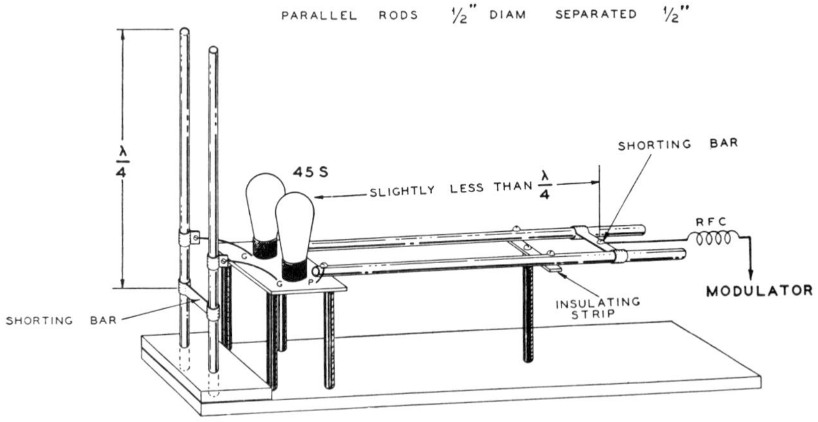

FIG. 38

and gradually slid down the plate bars toward the plate ends of the rods until the correct current is being drawn by the tubes. Some slight readjustment of the grid shorting bar may then be necessary but great care must be exercised or the circuit will stop oscillating. (NOTE: The actual wavelength of oscillation depends on the length of the rods and the length of the tube elements and the connecting wires, plus the capacity between the rods and the internal capacity of the tubes).

Parallel rods can be plate-modulated without excessive frequency modulation. The modulator output is figured as half the DC plate-watts input to the oscillator. The modulation must be kept under 100 per cent. The 45-tubes shown in the diagram can be modulated by a pair of 2A5s in parallel or push-pull.

2½-Meter Transmitter: The circuit shown has both grid leak and cathode bias in order to prevent mishap to the tube in the event of failure to oscillate. The cathode resistor also prevents a tendency for the plate current running away at excessive values of plate voltage and current. It is possible to run the plate voltage up to nearly 500 volts at 100 MA with this system. The type 6A6 tubes are probably more subject to this creeping effect than the RK34s, preliminary tests show, since the former tends to run wild if the cathode current is more than 80 MA.

The 15,000 and 300 ohm resistors should be of the 10 watt size and the grid coil can be made of 5 turns of No. 18 bare wire, about ⅜-in. or $\frac{7}{16}$-in. diameter and one inch long. The coil should be soldered to the socket terminals. The plate circuit consists of a pair of No. 14 bare wires about 12 inches long for 2½ meters with the antenna feeders connected across them about 2 inches from the short-circuited end. These wires are spaced about 1½-in. apart and held rigidly in place by means of small stand-off insulators. The plate RF choke consists of about 30 turns of No. 22 DSC wire, ¼-in. diameter. It is possible to operate the transmitter without any RF chokes because it uses a balanced push-pull circuit, although a simple choke in the plate lead is shown.

A plate voltage of 380 volts at 75 MA gave an input of 28½ watts and an output of 15 watts. The efficiency ran about 53%. Lower plate voltages gave less efficiency, but an average of 50% could be obtained on 2½ meters, which seems to be remarkably high. A 10 or 15 watt carrier on 2½ meters should provide a good signal for most amateur work. Many transceivers put out only a fraction of one watt on 2½ meters.

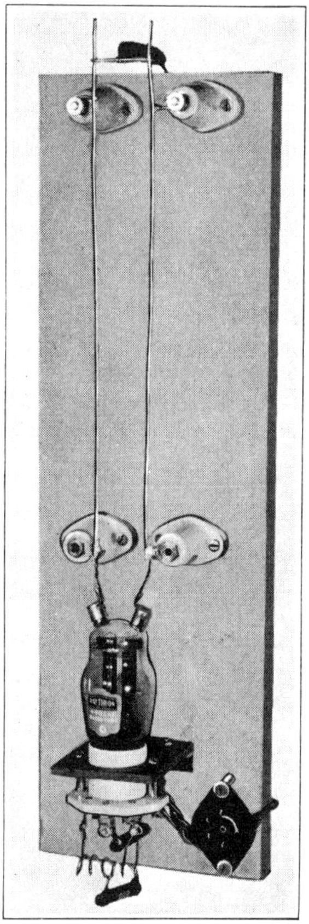

FIG. 39—2½-Meter Transmitter.

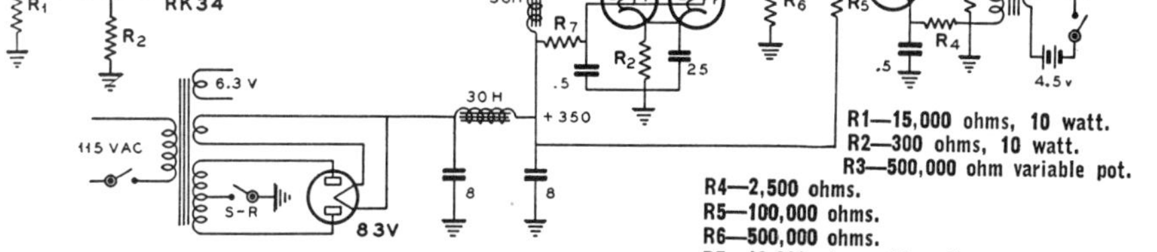

R1—15,000 ohms, 10 watt.
R2—300 ohms, 10 watt.
R3—500,000 ohm variable pot.
R4—2,500 ohms.
R5—100,000 ohms.
R6—500,000 ohms.
R7—10,000 ohms, 10 watt.
L1—5 turns, No. 18, 1" long, $\frac{7}{16}$" dia.
RFC—30 turns, No. 22, ¼" dia., self-supporting.

FIG. 40
Circuit diagram for the complete 2½ meter transmitter.

Stabilized U. H. F. Oscillators: Notes regarding adjustments and the physical makeup of stabilized oscillators for ultra-high frequencies founded upon the circuit of Figure 41 are described herewith.

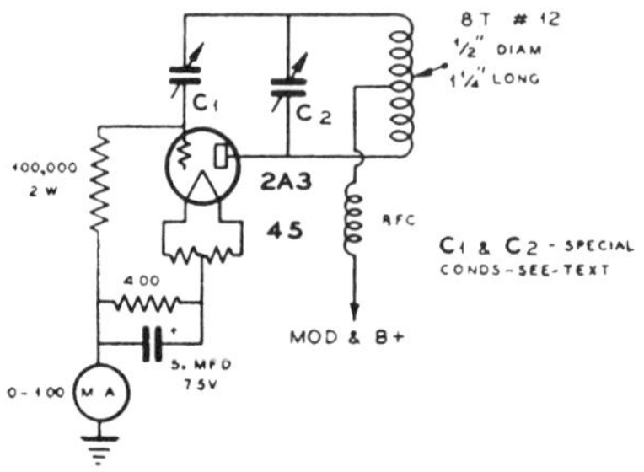

FIG. 41—Circuit for unit shown below.

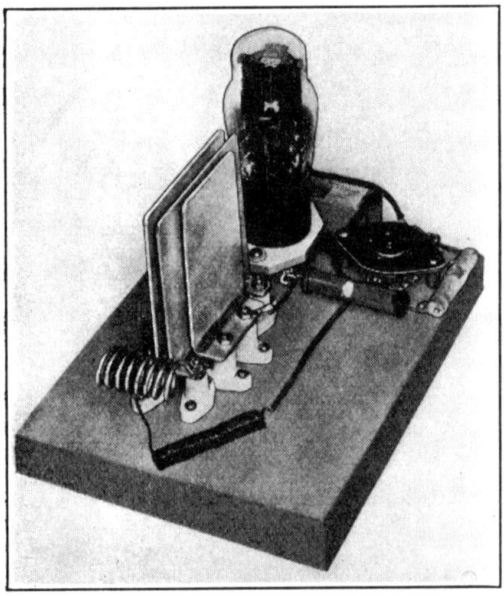

Stable Oscillator shown in FIG. 41.

off insulators. A moderate antenna load, controlled grid excitation, and a combination grid-leak and cathode resistor grid bias make this circuit remarkably free from frequency modulation effects.

A high or low power 2½, 5 or 10-meter oscillator of the stabilized type requires that the grid circuit be tuned for maximum frequency stability and not for maximum antenna output. In construction, the size of the condenser plates and the relative spacings are for tubes having characteristics up to the 801's. With certain tubes the cathode resistor (see Fig. 41) and the audio by-pass condenser can be eliminated. The 400-ohm 10-watt resistor and 5 mfd. condenser are necessary for type 45 tubes to prevent plate current increments from attaining high values when the tube is prevented from oscillating. With this resistor, 50 per cent higher plate voltage can be applied with a margin of safety.

With a carbon-type 100,000-ohm resistor in the grid circuit, no grid RF choke is required. The plate RF choke consists of about 75 turns of small wire wound on a ¼-inch diameter rod, the latter being removed after winding the coil. The oscillator coil should be provided with solder-ing-lug terminals to facilitate band changing.

The condenser C2 is made of two 3-in. x 3½-in. No. 12 gauge aluminum plates, spaced ¼-inch apart. The extra half-inch length is used to provide mounting lugs for fastening to small standoff insulators. The grid condenser C1 consists of a similar plate, 2-in. x 3½-in., spaced about ⅛th-inch from the "floating" plate of C2. Plate spacing is varied by means of oblong slots placed in the mounting lugs, or feet, for rough adjustment. Vernier adjustment is accomplished by bending the upper ends of the plates and by varying the coil spacing between turns.

Excellent frequency stability with a rigid antenna can be obtained if the radiation current is reduced 10 to 20 per cent from the maximum reading. This adjustment is made by varying the link coupling to a tuned circuit connected across the antenna feeders, or by variation of direct electro-magnetic coupling between the oscillator and antenna coils.

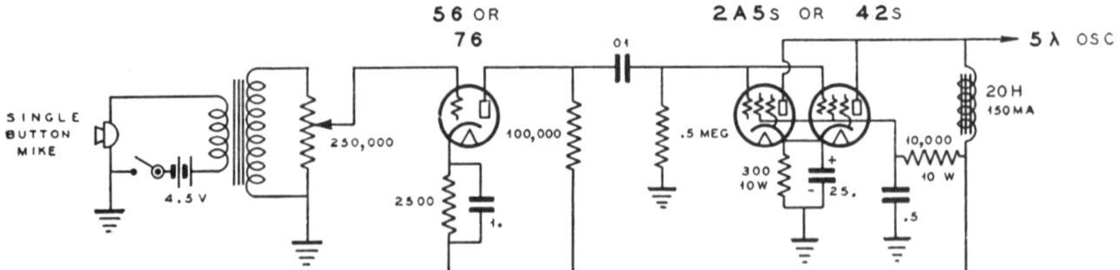

FIG. 42—Modulator for '45 or 2A3 Oscillator.

Referring to Figure 41, the grid RF excitation can be controlled to provide excellent oscillatory stability. The condensers C1 and C2 are formed of three heavy aluminum plates, mounted on small stand-

High Power with 50T Tubes: Outputs of more than a quarter-kilowatt are obtained below 10-meters with 50T tubes in stabilized circuits. Other tubes may not be suitable for this type of service on account of the high interelectrode capacities. A tuned-grid tuned-plate oscillator must necessarily have the proper RF grid excitation if stability and tube life are considered as important factors. The stabilization is accomplished by tapping only across a small portion of the tuned grid circuit, thus allowing the grid tuning circuit to actually become the frequency controlling circuit instead of merely a grid excitation control. Figure 44 shows that the grids are tapped across only enough turns to obtain oscillation, resulting in a relatively sharp tuning of the grid and plate circuits.

FIG. 43—Push-Pull 5-meter transmitter with Eimac 50T tubes.

The principal advantage of frequency control in the grid circuit is that the antenna swinging effect does not directly act upon the frequency control circuit. The power is so low in the grid circuit that double-spaced midget split-stator tuning condensers can be used without danger of flash-over, even at plate voltages as high as 2500 volts. The plate condenser (low-C) is made of a pair of parallel No. 12 gauge aluminum plates, spaced approximately ½-in. to 1-in. apart. The plates are about 3-in. square, mounted at one corner on a standoff insulator. The top end of each plate is connected to the plate cap of the tube closest to it, thus the plate leads are kept very short.

The frequency stability over wide changes in plate voltage is maintained within very close limits when only a small part of the total grid coil is used. Moderate values of antenna load are recommended. The overall stability is sufficient for direct modulation for operation below 10-meters.

The tube shown in the circuit of Figure 41 can be substituted with a 50T, but since higher voltages would be required it will be necessary to redesign the condensers.

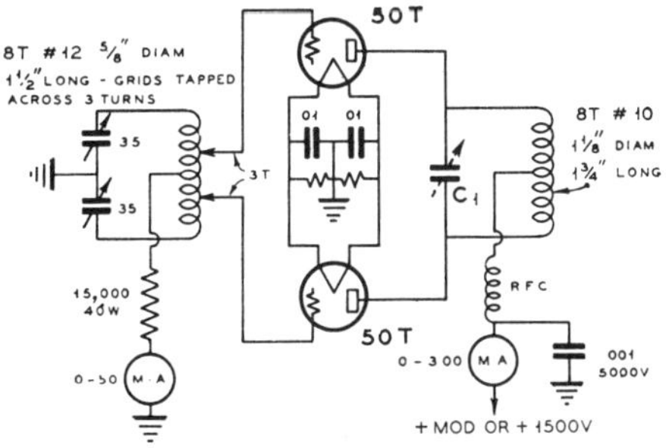

FIG. 44—Grid-controlled, high power stable 5-meter oscillator. The photo above shows the complete unit.

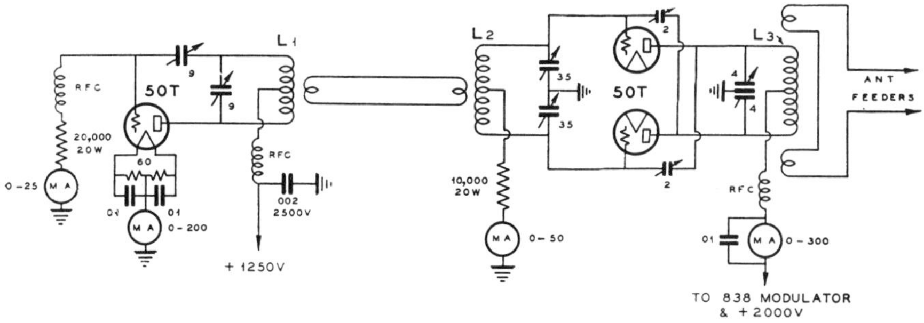

FIG. 45—¼-KW 50 meter M.O.P.A.

RFC—75 turns No. 22 DSC wire, ¼" or ⅜" dia.

L1—7 turns No. 10 wire, 1½" long, ⅞" dia.

L2—8 turns No. 12 wire, 1⅛" long, ¾" dia.

L3—8 turns No. 10 wire, 1¾" long, 1⅛" dia.

The area of the plates will be approximately 4 × 4-inches; the capacity being calculated from C = .225A/d, where A, is the area in square inches and d, the spacing between the plates. About ⅜-in. spacing is required between the two plates of the tuning condenser C1 of Figure 43; in this same Figure, the grid coupling plate is 3-in. × 4-in. in area with nearly one-half inch spacing to the tuning condenser plate.

Figure 45 shows a high-power push-pull neutralized M. O. P. A. The circuit is the same as Figure 2 except for the addition of two neutralizing condensers and the use of a different inductance in the grid circuit for low-C operation. The neutralizing plates are 1-in. x 1½-in. rectangular pieces of No. 12 gauge aluminum, and are separated nearly ½-inch.

This M. O. P. A. will furnish one-quarter KW RF-carrier output and is capable of 90 to 100% modulation without excessive frequency modulation; in addition, the efficiency of this system is nearly 20 per cent higher as compared with push-pull oscillator outputs with the same inputs.

5-Meter Superheterodyne Receiver: This seven tube receiver has a RF stage resonant at the top end of the 5-meter band which actually adds some gain in the form of signal-to-noise ratio. The resistance-coupled IF amplifier allows the reception of 5-meter signals emitted by modulated oscillators. The values of the capacities and grid leaks are such as to prevent amplification at audio-frequencies. The second detector differs from other circuits in that the screen-grid of the 41-tube is connected to the regular input grid thereby acting as a high-mu triode. Actually the detector functions somewhat like a class B tube, in that the grid current starts to flow as soon as the signal is impressed. The rectified grid current is used to obtain semi-AVC action, the voltage of which is fed back to

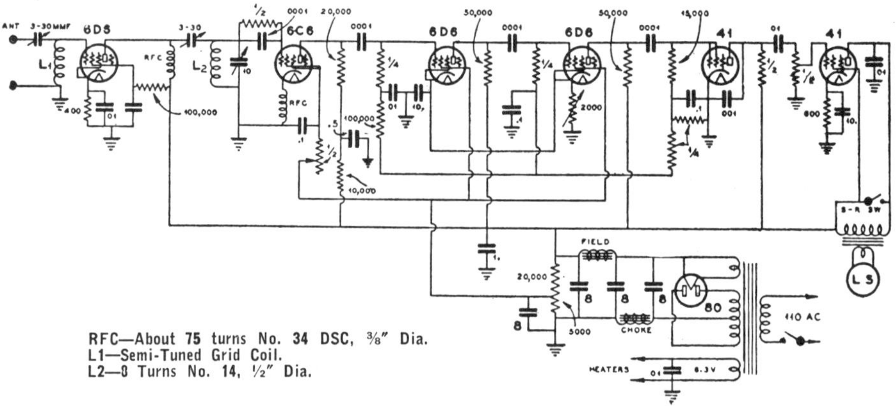

FIG. 46—Frank Jones' 5-meter superheterodyne circuit.

The 5-meter Super should be laid out as shown in the above illustration.

the grid of the two IF amplifiers, through resistive filters, so as to prevent too great a signal being built-up across this detector circuit.

The semi-tuned RF stage greatly simplifies the receiver because no tuning control is necessary for the RF grid circuit, the latter is resonant broadly over the 5-meter amateur band, and since the ratio of L to C is large, the gain of this stage is fairly high. The resonance curve of a non-regenerative tuned circuit at 5-meters is extremely broad, especially if the tuning and stray capacities are small. If this receiver is to be used on any other ultra-short wave bands it will be necessary to change the RF coil as well as the detector coil. This circuit eliminates the need of a two-gang tuning condenser and minimizes common coupling between the RF and the oscillating detector.

Notes: The detector should oscillate weakly for maximum sensitivity and because this degree of oscillation varies slightly over the range of the tuning condenser, a variable screen voltage control is brought out to the front panel. When very strong signals are received, the detector should oscillate a little stronger, in order to prevent an overloading effect which would result in noticeable audio distortion.

The entire set is built on a chasis 8 in. x 17 in. x 1½ in. deep, and fits a 7 in. x 19 in. x ⅛th in. standard relay rack panel.

8-Tube 5-Meter Superheterodyne: Most 5-meter superheterodynes generally produce more noise than signal, but not as much noise as a superregenerative receiver. The high noise-to-signal ratio may be attributed to high IF gain and a lack of RF gain ahead of the first detector. An ordinary resonant or semi-tuned RF amplifier provides a certain degree of gain, but by incorporating regeneration in either the RF stage or detector circuit the amplificatory properties may be increased from five to fifty times. In the ultra-frequency spectrum, regeneration in the detector circuit is preferable because antenna resonance does not effect, or cause "dead spots," in the regeneration control; that is, providing an RF stage precedes the regenerative detector.

In the circuit accompanying this description, image interference is minimized because of regeneration and two pre-selection tuned circuits. The IF frequency being about 2.7 MC, the image is 5.4 MC away from the desired signal. This means that no interference from other amateur signals will be heard in the 5-meter band from 56 to 60 MC.

Delayed AVC is included to prevent overload on strong signals. To provide maximum sensitivity the control voltage is applied to the grid of the two IF stages only.

Technical Details: The 2.7 MC IF transformers are made from the parts of regular IF components. Those used in the design were wound on ⅜-inch diameter tubes. The 450 KC litz coils were removed and two coils each of 120 turns of No. 34 DSC wire put on in "jumble" fashion to cover a winding length of ⅜-inch. Between the adjacent coil edges there is a space of ⅛-inch. These windings, tuned with the mica trimmers of the original transformer, cover from approximately 100 to 120 meters. (If this receiver is to be used close to a 120-meter police radio station, it would be advisable to wind the coils with only 100 turns and tune the transformers to 90 or 100 meters.)

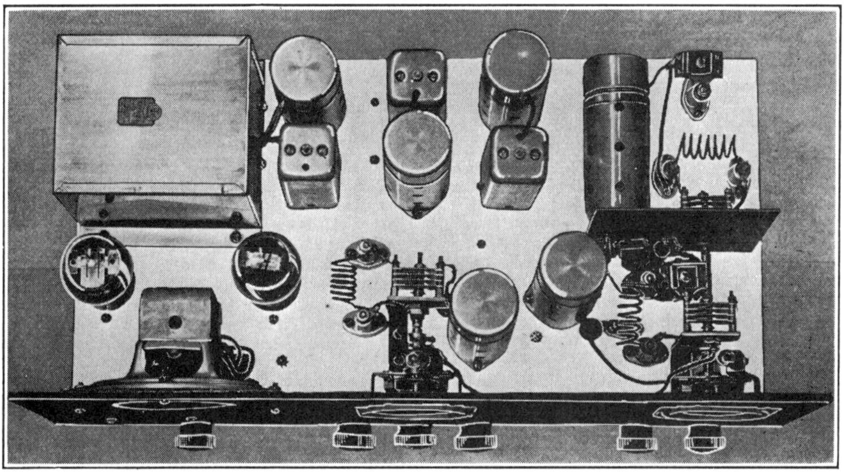

Looking down into the 5-meter Super. The r-f coil is mounted horizontally to permit use of very short leads. An aluminum shield isolates the r-f stage from the detector.

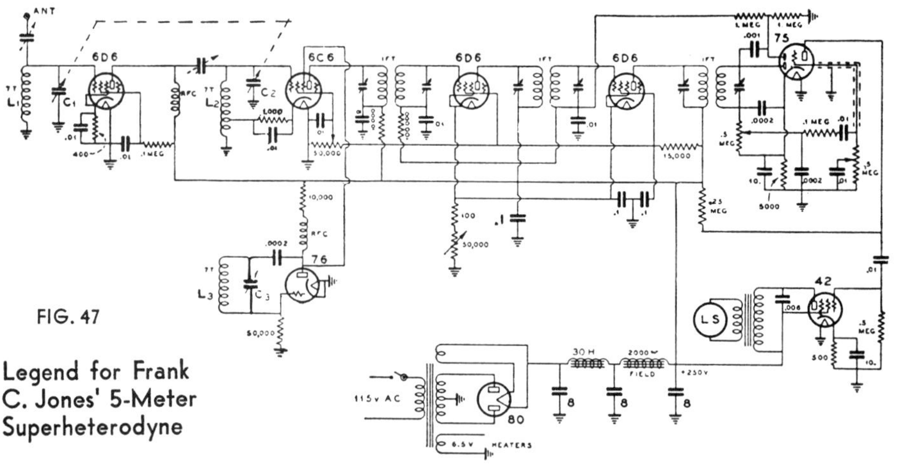

FIG. 47

Legend for Frank C. Jones' 5-Meter Superheterodyne

L1 and L2—Each 1½ in. long, 7 turns, No. 14 enameled wire, ½ in. dia. L3—1 in. long, 7 turns, No. 14 enameled wire, ½-in. dia.

C1, C2, C3—100uufd. double-spaced variable condensers, with only 7 of the original plates remaining. Maximum capacity of these re-built condensers to be about 18uufd.

I.F. Transformers tuned to approximately 2,000 KC.

The two RF chokes are made by winding No. 34 DSC wire on a ⅜-inch diameter bakelite rod, winding to cover about one-inch of the length of the rod. All the other coils are made of No. 14 wire, space wound on a half-inch diameter. These coils are mounted on small stand-off insulators near the tuning condenser terminals.

The IF amplifier is lined-up by means of a modulated oscillator of the all-wave type. Starting from the second detector circuit, each stage is aligned by coupling to the oscillator, then a re-check made of the over-all amplifier by coupling the oscillator into the first detector grid circuit. The latter should connect temporarily through a 1000 ohm resistor to ground instead of to its LC circuit.

To align the detector and RF stages the detector coupling condenser is adjusted until its capacity is low enough to allow the first detector to break into oscillation when the regeneration control is fully advanced at both ends of the tuning range. The RF antenna coupling, or trimmer condenser, is adjusted together with slight coil respacing until the noise level is highest throughout the band. There is usually enough noise from auto ignition to accomplish this, although a harmonic signal from a modulated all-wave oscillator is much superior for this purpose.

Notes: A tone control is provided to reduce automobile ignition interference which is quite serious when using a superheterodyne receiver in most locations.

For receiving, a high half-wave antenna with a transposed two-wire transmission line energizing the receiver through an electro-static screen will minimize practically all man-made noise. If this system is employed, the antenna coupling condenser should then be connected across the tuning condenser and a Faraday electro-static shield placed between the tuned grid and tuned antenna feeder coil.

The receiver is mounted on a 7 x 19 x $\frac{3}{16}$-inch aluminum panel for relay rack mounting. The hole for the loudspeaker opening and the two airplane dials can be cut by means of a flying-bar cutter. The chassis is made of No. 14 guage aluminum, 9 x 17½ x 1¾ inches. The pictures give a good idea of the proper arrangement of the parts.

The Simplest 5-Meter Superheterodyne: A superheterodyne which anyone can easily build in a few hours has recently been developed for use on the short wave bands below ten meters.

The IF amplifier is the really interesting part of this receiver. It gives good amplification over the band of frequencies desired, from 10 KC to a little over 100 KC, and is quite stable. The secret is in using the proper values of resistors and condensers to obtain this resonance characteristic. By using low values of grid resistors, ¼ megohms, and small coupling condensers, .0001 mfd., the response to audio frequencies is practically nil. This value of coupling condenser (.0001 mfd.) and a grid leak of ¼ megohm does not tend to attenuate the higher frequencies such, as for example, 50 KC. This means that the first

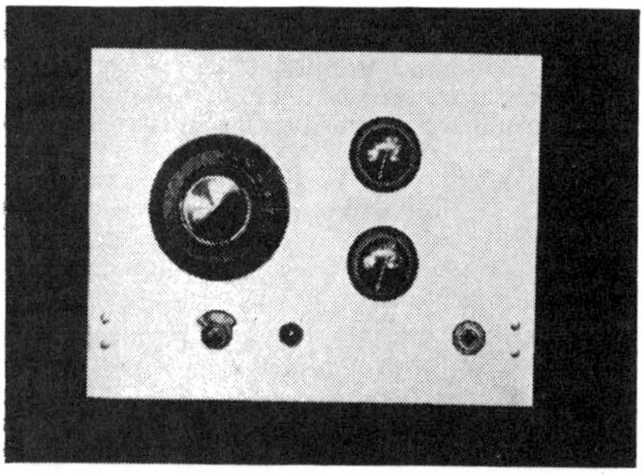

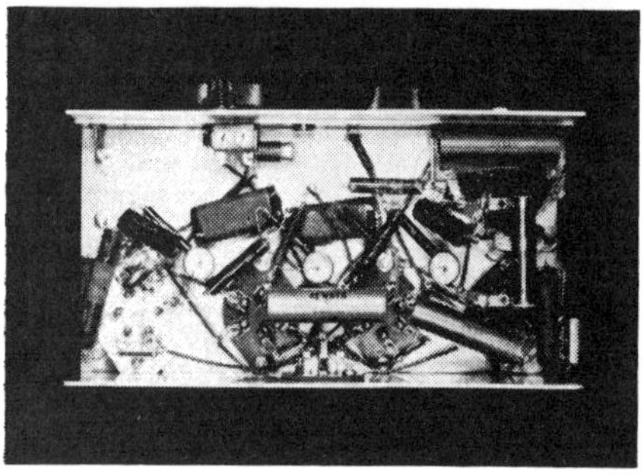

Another constructional design for the simple receiver here described.

detector can be of the autodyne type and act as its own oscillator. By having it oscillate weakly, the tube is in its most sensitive condition, which accounts for the excellent signal to noise ratio obtained in this receiver. It also eliminates the need of tracking two tuned circuits, such as used in most superheterodyne receivers.

The relatively low values of plate resistors of 50,000 ohms tend to even-out the amplification of the IF amplifier for the

Looking into the simple 4-tube Superheterodyne.

Front panel view.

range of 50 to 100 KC in order to be able to receive modulated oscillators. Probably 98% of the phone transmitters on 5 meters use modulated oscillators, so it is necessary

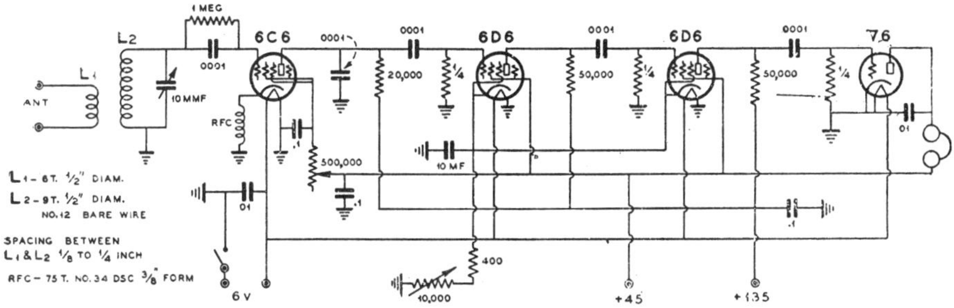

FIG. 48—The extremely simple circuit diagram of the 5-meter Super.

to have a receiver tuned broadly enough to receive these signals. Even if all transmitters were temperature and crystal controlled types, it would still be desirable to have the IF amplifier broadly tuned in order to take care of oscillator drift in the receiver. Two stages of moderate gain per stage give more than enough amplification to bring up the man-made noise level into audibility in the output of the second detector.

When first testing this receiver, the regeneration control should be turned up high enough to insure good oscillation in the first detector. The IF volume control should be turned on full if 180 volt B supply is used, or back just slightly if 135 volts is used. Auto ignition will be heard if an indoor antenna of any convenient length is used—provided cars are passing within a block or so. Below the point of oscillation in the first detector the auto ignition and other noises drop out. It will be found that sensitivity is greatest when the detector is oscillating weakly but on very strong signals, stronger oscillation is desirable to prevent overloading distortion. Coupling between the antenna coil and first detector coil should be adjusted for best weak signal reception, although this is not critical. Too much coupling to a resonant antenna will pull the detector out of oscillation. It will be found that every 5 meter signal will have two points close together where the audio quality is clear, since an autodyne first detector is used.

In building this receiver, good .0001 midget mica condensers and good quality resistors should be used. A noisy resistor or leaky coupling condenser will cause plenty of noise, especially if it is in the first detector or first IF stage. In one receiver built in the laboratory, a noisy plate resistor in the first IF stage caused trouble until it was replaced. The IF units always perform satisfactorily and there is no alignment of tuned circuits to worry about. ± 10% accuracy of values of condensers and resistors is satisfactory.

The screen and suppressor grids of the first detector are connected together. This gives smoother regeneration effects and better conversion gain for 5 meter work. The screen by-pass was made as large as .1 mfd. in order to prevent noise from variation of the regeneration control. Needless to say, this condenser, as well as the heater .01 mfd. by-pass, should be non-inductively wound in order to act as a by-pass for 5 meter purposes.

Ultra-High Frequency Phenomena Below 1 Meter: When an attempt is made to operate a standard triode at increasingly high frequencies it is found that the output and efficiency begin to decrease. The frequency at which this is first observed will depend upon the design of the tube but it will usually be in the 10 to 60 megacycle range. By successive modification of the circuit arrangement and size this decrease in power output and efficiency can be minimized. With optimum circuit arrangements, however, this decrease continues until finally a frequency is reached beyond which oscillations can no longer be produced.

Outside of the various losses tending to decrease the output power and plate efficiency of a vacuum tube with increases in frequency, manufacturers are now constructing miniature tubes of relatively high efficiency. Even with these developments, tubes are still limited to certain frequency limitations, these have to do with the time required for the electrons to travel from the cathode to the anode within the tube structure. This time, the so-called "transit time," is very small in present day commercial types of power tubes, usually much less than one microsecond. Obviously at low frequencies it can be neglected and, in fact, for many tubes it still plays a minor role either in determining the output and efficiency in the high-frequency range or in establishing the limiting frequency for oscillations. When the frequency range of oscillation of a tube is extended by an adequate increase in energy losses and by improvements in

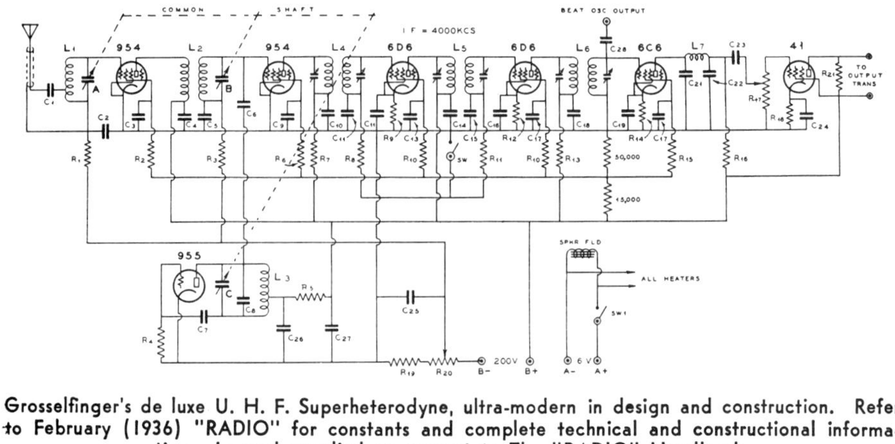

Grosselfinger's de luxe U. H. F. Superheterodyne, ultra-modern in design and construction. Refer to February (1936) "RADIO" for constants and complete technical and constructional information. Legend supplied on request to The "RADIO" Handbook.

electrical design, transit time becomes a dominating factor in the reduction of output power and efficiency in establishing the limiting frequency of oscillation.

This comes about in two ways (as explained by M. Kelly and A. Samuel at the Winter Convention A. I. E. E. Jan. 1935). In the first place, the relative phase of the alternating grid and plate potentials for best operation must be altered to compensate for the time required for the electrons to travel from the region in which the grid has its greatest effect upon their motion to the region in which their motion has the greatest effect upon the plate current. The available control over these phases is usually insufficient to permit a realization of the optimum adjustment. In terms of the measured characteristics of the tube, the transconductance has become complex. But even with the optimum phase adjustment the efficiency is reduced by losses which occur because of the variation in grid and plate potentials during the transit time. Electrons arriving at the plate will in general have velocities greater than the velocity corresponding to the potential of the anode at the instant of their arrival. The excess energy corresponding to the greater velocity is obtained from the oscillating circuit and is dissipated at the plate in the form of heat. Again, in terms of the measured characteristics, the input conductance has been increased above its low frequency value.

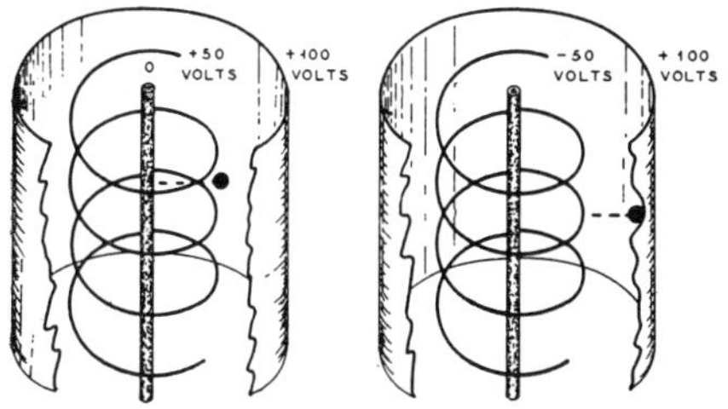

The mechanism which enables electrons to take energy from the oscillatory circuit in their passage across the tube is evident from a consideration of a somewhat simplified case as shown in the Figure above. Assume that the plate is held at a constant potential of 100 volts, and that the grid is held at 50 volts positive just long enough to allow an electron to come from the cathode to the grid plane (very near one of the wires), where its velocity will correspond to a fall of potential of 50 volts. The potential of the grid is then suddenly changed to 50 volts negative. The electron will then fall through an additional potential difference of 150 volts, arriving at the anode with a velocity corresponding to 200 volts, producing just twice as much heat as it would have done had the grid potential not been changed during the transit time. This added energy must come from the source which produced the change in grid potential. In the actual case the change in grid potential is not abrupt but a similar loss occurs. This limits the useful range of a tube to values for which the oscillation period is long compared to the electron transit time.

R.C.A. Acorn Tube

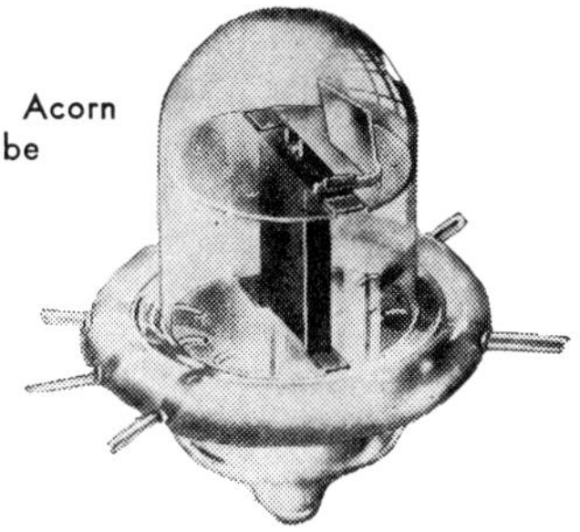

The frequency range in which a given design is near the optimum is limited. Characteristics such as high mutual conductance and a sharp cut-off which make a good tube oscillator at low frequencies, while still of importance at ultra-high frequencies are apt to be secondary to certain special frequency requirements.

The 955 "Acorn" One-Half Meter Tube: In extending the frequency range of vacuum tubes the RCA company has developed a miniature triode capable of operating efficiently on wavelengths lower than one meter. In spite of the small size of the 955-tube it has a low plate resistance, 12,500 ohms, and a high amplification factor of 25 with resulting high mutual conductance of 2000 micro-ohms. Mutual conductance is a fairly-accurate yardstick of tube performance, and regardless of the compromises which have been made to enable this tube to operate as a regenerative oscillator at a wavelength of only 20 inches, the mutual conductance is higher than any of the conventional general purpose triodes, such as the 56 and 76.

In order to keep the capacities low, no base is used on this tube. The leads to the grid, plate and indirectly-heated 6.3-volt cathode come directly out of the glass envelope, and are widely separated. No attempt should be made to solder to these wires because soldering usually results in breaking the glass envelope.

The 954 "Acorn" Pentode: A companion to the 955 tube of the pentode type is now commercially available for amateur experimentation. From the characteristics of this new tube it can be seen that the input and output capacities are only a small fraction of the 6C6 or 57 tube. It therefore becomes very useful as an RF amplifier for wavelengths below 10 meters. In properly designed circuits the tube will give a gain of 3 at one meter, and 10 or

FIG. 49
RCA 955
CIRCUIT DIAGRAM AND
SUGGESTED LAYOUT

L1—8 turns, ½-in. outside diameter, No. 18 wire, spaced ¼-in. between turns.

C1—Tuning condenser; 2 circular brass plates, ¾-in. in diameter; 10/32 thread on adjusting screws.

C2—.00025 mica condenser, postage stamp type.

R1—15,000 ohms, 1 watt carbon resistor.

RFC—¼-in. bakelite rod wound 1½-in. with No. 32 DCC wire.

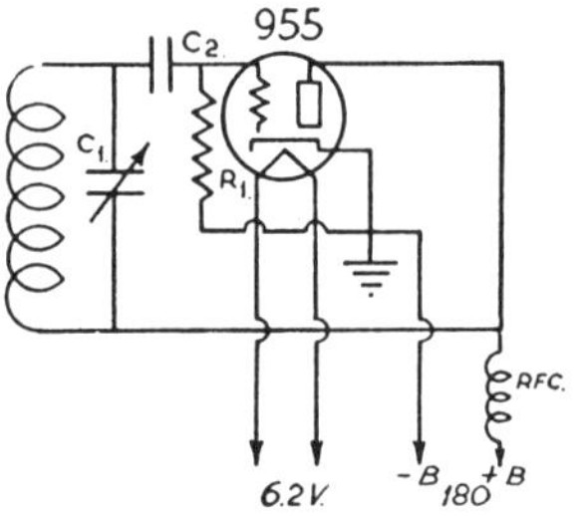

Circuit diagram of transmitter using RCA 955 tube.

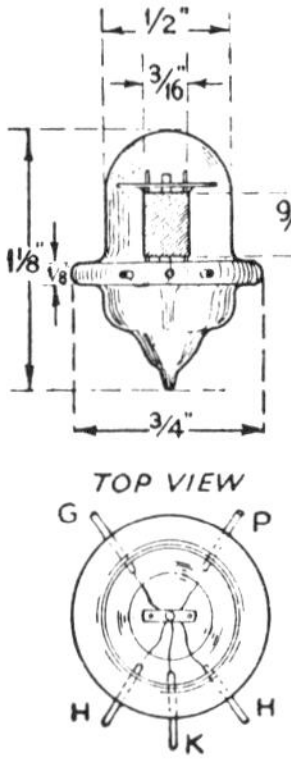

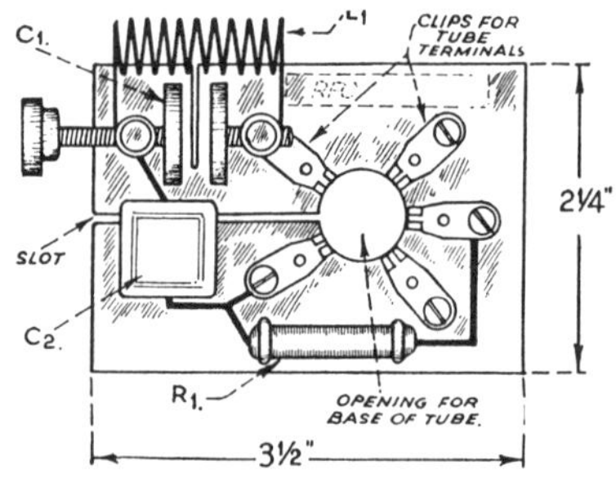

FIG. 50—Plan view of transmitter.

more at 5 meters. Except for its small size it is quite similar to a 57 or 6C6 tube in its other characteristics and may be used as a detector, RF amplifier, AF amplifier, or as a triode tube. The latter use is made in a special vacuum tube voltmeter since the tube is so small and the input capacity so low it can be used for RF measurements. In this case the tube is mounted on flexible leads so that the tube connects directly across the circuit under measurement.

In shielding this tube for RF measurements, the control grid end is inserted through a hole in a metal plate so that the metal edge of the hole is in close proximity to the internal shield. It may be desirable to provide a small metal collar on the baffle hole in order to increase the shielding effect.

RF grounding should be by means of small condensers, right at the tube terminals. For very high frequencies, flat ribbon leads to the clips are preferable to others. The leads must be insulated from the metal shield by means of mica spacers. These connections then act as by-pass condensers. In RF amplifiers, the tuned-grid and plate return by-pass leads should be made to a common point in order to avoid RF interaction.

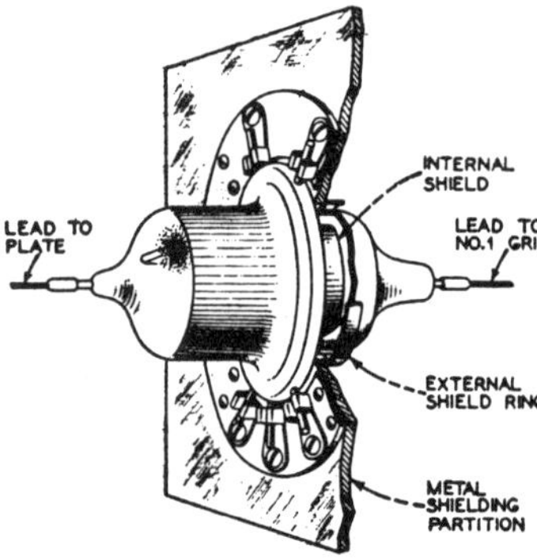

FIG. 51—Terminal mounting template.

As an audio amplifier the plate voltage may be as high as 250 volts; screen voltage 50; control grid 2.1 volts; suppressor grid connected to cathode; plate load resistor 250,000 ohms; and plate current 0.5 ma. With a one megohm grid-leak a voltage amplification of approximately 100 can be obtained.

For detection, the grid bias can be obtained by means of a resistor between 20,000 and 50,000 ohms.

Exerimental ¾-Meter Transceiver: Reference to the circuit diagram shown in Figure 54 will show that the oscillating

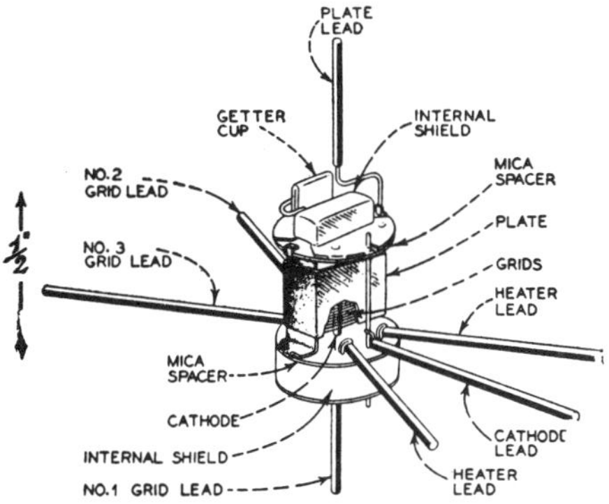

FIG. 52—Internal Structure.

circuit consists of the tube capacities and a parallel-wire LC circuit. At ¾ meters the parallel wire length is slightly over an inch in length and is made by soldering a pair of No. 14 bare copper wires to the tube grid and plate clips. The parallel wire bridge consists of the .0001 grid condenser.

Oscillation is obtained by a cathode RF choke and a 450-ohm cathode resistor by-passed with a .01 mfd. condenser. This condenser by-passes the super-regenerative hiss frequency. The number of turns in the RF chokes is somewhat critical; a variation from 10 to 15 turns will cause the circuit to operate erratically, the optimum winding being about 25 turns of No. 22 DSC wire wound on a ¼-inch diameter form. Oscillation should always be checked by means of a plate milliameter. The plate current should never exceed about 7 milliamperes on the transmit position.

TABLE I				
954				
E_f (a.c. OR d.c.)	6.3 VOLTS			
I_f	0.15 AMP.			
CAPACITY G-P. (WITH SHIELD-BAFFLE)	0.007 MAX. MMFD.			
INPUT	3 MMFD.			
OUTPUT	3 MMFD.			
OVERALL LENGTH	1 1/16" ± 3/16"			
OVERALL DIAMETER	1 3/32" ± 1/16"			
MAXIMUM RATINGS				
E_p (d.c.)	250 V. MAX			
E_{g3} SUPPRESSOR (d.c.)	100 V. MAX			
E_{g2} SCREEN (d.c)	100 V. MAX.			
TYPICAL OPERATION AND CHARACTERISTICS				
	CLASS A AMPLIFIER		BIASED DET.	
E_p	90	250	250 (THROUGH PLATE LOAD)	VOLTS
E_{g2}	90	100	100	VOLTS
E_{g1} (C.G)	-3	-3	-6	VOLTS
SUPPRESSOR (G3) CONNECTED TO CATHODE AT SOCKET				
μ	1100	OVER 2000	—	—
R_p	1	OVER 1.5	—	MEG
G_m	1100	1400	—	μMHOS
I_p	1.2	2.0	.01 MA (WITH NO SIGNAL)	MA.
I_{g2}	.5	.7	—	MA.
PLATE LOAD	—	—	250,000	OHMS

FIG. 53—¾-meter transceiver in metal case, with built-in midget magnetic speaker. The R.C.A. 955 Acorn Tube is plainly visible.

Antenna coupling can be acomplished by connecting the antenna feeder to some point along the parallel wires, or preferably by inductive coupling. For best results some type of beam antenna should be used. An ideal beam antenna system consists of an antenna wire 13¾ inches long and two directors 13 inches long, and one reflector 14¼ inches long. The antenna is spaced a quarter-wave ahead of the reflector wire which will amount to about 7 inches (⅜ of a wavelength spacing between the antenna and director and between the two director wires should be used). This will amount to about 10½ inches spacing. While this system is not very directional, it can be improved by making the antenna arrays more elaborate; however, for experimental purposes the system described will enable the transmitter to be heard for about 2 or 3 miles of air-line distances.

Smaller grid condensers are often desirable in order to obtain higher values of oscillator output.

5-Meter Regenerative R. F. Receiver: On 5-meters R. F. amplifiers are frequently employed to prevent receiver radiation. The sensitivity of such a device would, ordinarily, increase the sensitivity of the receiver, but at 5 meters very little gain is possible. However, the improved circuit shown in the accompanying diagram will give an increase in sensitivity due to the inclusion of controlled regeneration through the medium of a type 954 Acorn Pentode tube.

The circuit diagram shows the simplicity of design; in general, the 954 functions as a regererative screen-grid RF amplifier, the 76 as a super-regenerative detector, the 37 as the interruption-frequency oscillator, and lastly, the 41 as an audio-frequency output tube. Because the 954 tube is capable of giving good amplification on the

FIG. 54

3/4-Meter Circuit

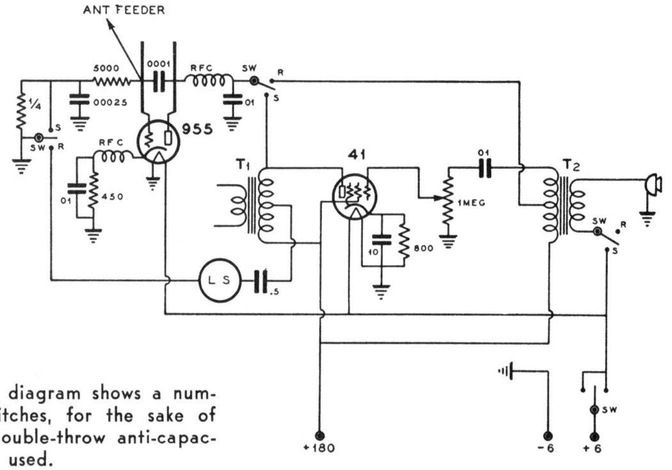

The RF Choke consists of about 25 turns of No. 22 DSC wire, wound on a 1/4-inch diameter form. T1 and T2 are Output transformers. Those used in the Transceiver here shown are of the 2A5 P.P. Output type. Although the circuit diagram shows a number of separate switches, for the sake of simplicity a 4-pole-double-throw anti-capacity switch is actually used.

2½ meter band as well as on 5 meters, it is well to mount this tube very close to the detector to minimize the length of the connecting leads.

General Construction: The receiver is built on a plated steel chassis, 12 in. x 6 in., with a 5 in. x 5 in. shield between the RF stage and the detector. The acorn tube is mounted so that it protrudes through a half-inch hole in the shield, with the plate lead connected to the coupling condenser. This condenser can be set to a value as high as 30 mmfds. for 5-meter operation, and it couples the RF stage to the detector tuned circuit. Tuning is done by two dials because of RF regeneration and the necessity of exactly tuning the RF stage. The detector tuning shaft must be insulated from the chassis and front panel. Plug-in coils cover any bands from 2½ to 10 meters.

The receiver chassis is mounted on a large front panel, 13¾ in. x 18 in., of No. 12 gauge aluminum. Below this deck is

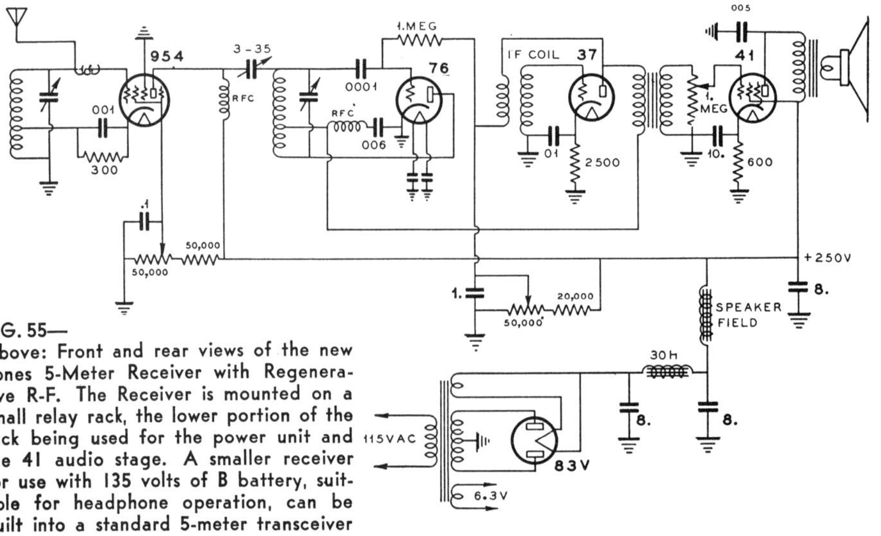

FIG. 55—
Above: Front and rear views of the new Jones 5-Meter Receiver with Regenerative R-F. The Receiver is mounted on a small relay rack, the lower portion of the rack being used for the power unit and the 41 audio stage. A smaller receiver for use with 135 volts of B battery, suitable for headphone operation, can be built into a standard 5-meter transceiver case.

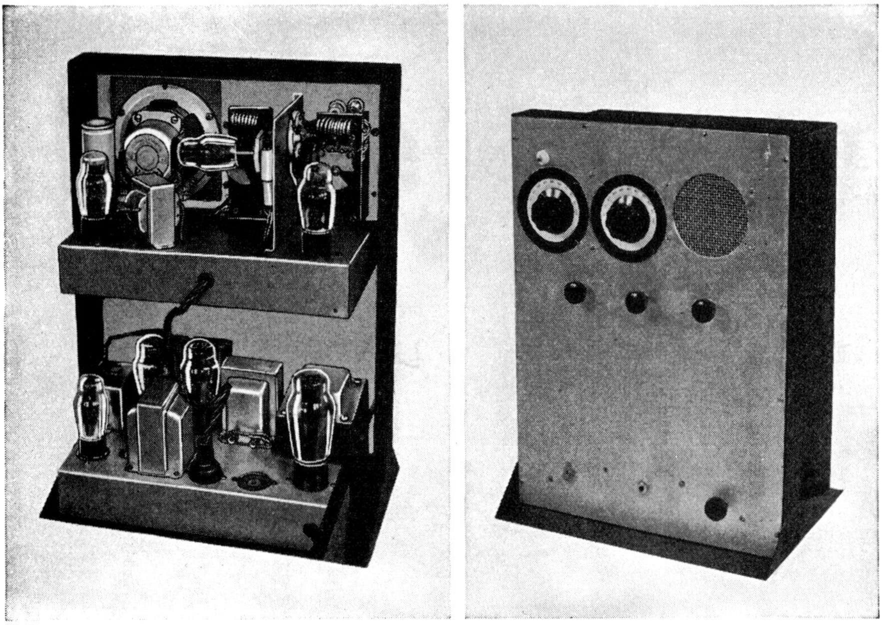

Rear and front views of Jones U.H.F. Receiver with regenerative r-f stage.

mounted another chassis on which are power pack and audio amplifier.

The plug-in coils are wound of No. 14 or No. 12 tinned wire so as to fit into ordinary telephone tip jacks. Three jacks are required for each coil, and are mounted near the tuning condenser terminals to shorten connections. An old hard rubber panel may be used for the 2-in. x 4-in. coil and condenser supports. Hard rubber has less RF loss than most bakelite material and the tip jacks can be mounted on these panels without appreciable signal loss. The RF chokes are made by winding about 75 turns of No. 24 DSC wire on a $\frac{3}{16}$th-in. rod, then slipping the winding off the rod, and mounting between its ends supports.

The detector has a plate voltage control to enable the operator to maintain the lowest degree of hiss without losing super-regeneration or producing a poor tonal quality. The antenna is loosely coupled to the receiver. When the RF stage is tuned to resonance, the noise level will increase appreciably when the RF regeneration is just below the point of RF oscillation. The noise level in this case is caused by auto ignition, neon signs, or other electrical disturbances picked-up by the antenna circuit. Too close antenna coupling will prevent sufficient regeneration. Too little coupling will result in oscillation before the proper amount of screen-grid voltage is applied.

A 3-1 or 4-1 ratio inter-stage audio transformer will provide sufficient drive to excite a loudspeaker. A 175 KC IF transformer, or regular interruption-frequency unit can be substituted for the low-frequency oscillator. When this tube is oscillating, super-regeneration can be obtained at lower plate voltage than with grid-leak super-regeneration. Placing a .005 or .006 mfd. condenser across the output of the pentode amplifier tends to eliminate a certain amount of hiss in the output when no signal is tuned in on the receiver.

Modified 5-Meter Regenerative RF Receiver: This receiver is a modified version of the aforegoing design. In general, the construction is similar, except for the chassis and panel dimensions which are left to the discretion of the builder.

The circuit in Figure 56 shows the 954 tube supplied with cathode regeneration to improve the gain. The detector, a 955 tube, functions as a super-regenerative device with a 6C5 tube acting as a separate interruption frequency oscillator. The 6F6 is an ordinary pentode amplifier, and the remainder of the circuit is conventional practice.

Circuit Notes: The value of the .005 mfd. grid tuning condenser in the 6C5 circuit depends upon the type of interruption-frequency coils. This value, in conjunction with the detector grid-leak, plate voltage and plate by-pass condenser, can be varied to obtain the most satisfactory results. The detector will super-regenerate without the 6C5 oscillator if a .005 or .006 mfd. condenser is connected from ground to the

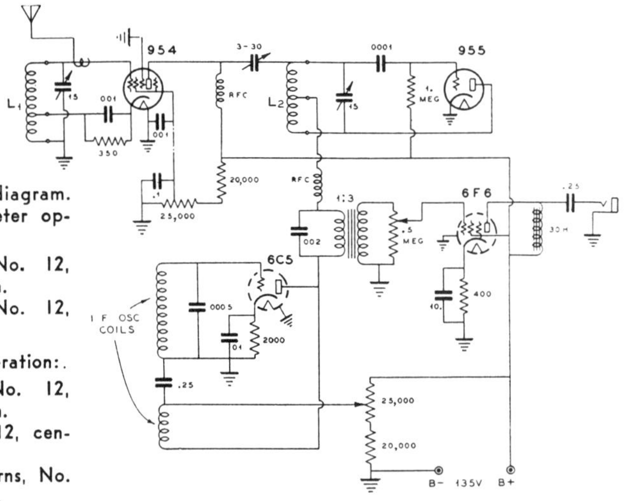

FIG. 56.
Complete circuit diagram. Constants for 5-meter operation:

L1—10 turns, No. 12, tapped at 1 turn.
L2—10 turns, No. 12, center-tapped.

For 2½-meter operation:

L1—4 turns, No. 12, tapped at 1 turn.
L2—4 turns, No. 12, center-tapped.
R-F Chokes: 75 turns, No. 28 DSC, ¼" dia.

top of the audio transformer primary, that is, the plate side; either arrangement is satisfactory.

The detector plate voltage must be adjusted to the optimum value in order to gain full use of the regenerative RF stage. A 25,000-ohm potentiometer provides a constant variable adjustment of this voltage. It should only be set high enough to obtain the super-regenerative hiss when not tuned to a signal. The detector grid leak may be connected as shown, or to ground. The latter may prolong the life of the tube, but the connection to the plus B gives a higher degree of audio fidelity on strong signals. As long as the detector oscillates, the actual grid bias voltage on the detector is about the same in either case.

The RF stage has the cathode connected to the grid coil, one turn above the grounded end on both the 5 and 2½-meter coils. The best position for this tap depends upon the physical arrangement of the apparatus and on the length of the ground leads in this RF circuit.

Front and rear views of U.H.F. Receiver with metal tubes and regenerative r-f stage.

ANTENNAS

An antenna is an electrical conductor supported in free space above or in direct relation with the ground to either radiate or intercept radio-frequency waves. The efficiency of the device depends upon numerous electrical and mechanical factors; those dealing with transmitting antennas are treated in the subsequent paragraphs.

Fundamentals: A wire connected to any source of oscillating electrical energy will radiate electromagnetic waves because of the varying intensity of the electrical field surrounding the wire. The field closest to the wire is called the **induction field** which oscillates to-and-fro; that part of the field which escapes forms the energy in the radiated field which is urged outward and diffused in all directions through space with the speed of light-rays. Any wire supported in space and within range of the radiated field will intercept the energy and will have induced in it a radio-frequency voltage, which is detectable as an incoming signal by receiving apparatus.

An antenna can be compared to any tuned circuit, except that its capacity and inductance are distributed along the wire, instead of being lumped as in a tuned circuit. At resonance, the inductive reactance in an antenna will cancel the capacitive reactance so that only the resistive impedance limits the flow of the oscillating current. By resonating the antenna to the frequency of the induced RF field, the current generated in the wire will be much greater than if it were in a non-resonant condition, the latter offering added inductive or capacitive reactance to the flow of current.

Wavelength, Length, and Frequency: Any tuned resonant antenna circuit must be some multiple of half-wavelengths long in the form of lumped or distributed capacities and inductances. The resonant frequency of an antenna is given by the expression:

$$F = \frac{300,000,000}{\lambda}$$

where F is the frequency in cycles per second; and λ, the wavelength in meters.

By dividing the frequency into the wavelength the actual (exact) length of any half wave antenna can be found in meters (metric measure). To convert the result into (feet) multiply by 1.56 which allows for end effects.

For an antenna consisting of a straight wire electrically one-half wavelength long, the physical length will be approximately 5% shorter than the electrical length; this is due to the fact that it is impossible to secure a wire having a zero diameter supported in space without end insulators.

Impedance and Radiation Resistance: The impedance along a half wave antenna varies from minimum to maximum at the ends. The impedance at the ends can be several thousand ohms, while that at the center would be theoretically about 73 ohms if the antenna was infinitely high above the earth, and was not near any other objects. The actual center resistance varies as shown in Figure 1 for various heights above the ground.

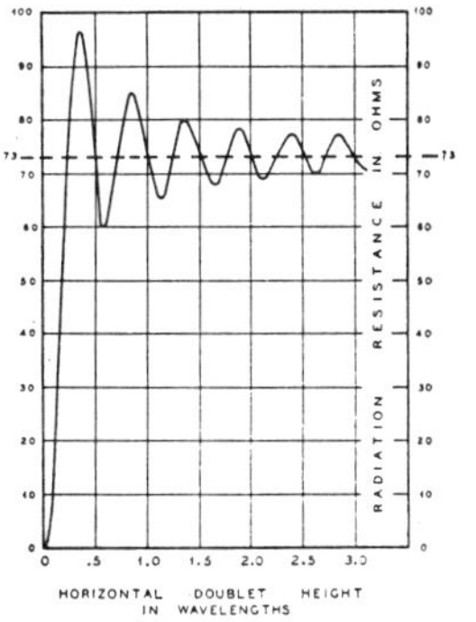

FIG. 1—Radiation resistance of half wave horizontal antenna for various heights above ground.

From this curve it can be seen that the radiation resistance varies with the antenna height above the ground. Radiation resistance is a term which is useful in expressing the power radiated by an antenna. It is that resistance which would consume the same amount of power that is radiated from the antenna. This resistance depends upon the antenna length and construction, and proximity of nearby objects. In addition to the radiated energy, power is also lost as a result of wire and ground resistance, corona and induced power losses in nearby objects. These losses can also be represented as a resistance in series with the antenna consuming an equivalent amount of power. For ultra-high frequency antennas the radiation efficiency can be higher than 90 per cent.

A transmitting antenna usually consists of a wire of definite length which may be grounded, ungrounded or connected to a counterpoise. A ground made by either a direct or capacitive connection acts as a reflector to the aerial wire and so completes the circuit.

With a direct ground connection, the antenna may be either an electrical quarter, or some odd multiple of quarter wavelengths; the ground acts as a subterranean reflector furnishing quarter waves to the antenna to give half waves or multiples of half waves for resonance. A very short wire can be loaded-up to an electrical quarter wave by means of a loading coil; such a device is not as efficient as a higher resonant antenna.

Increasing the antenna length beyond an electrical half length causes a drop in radiation due to non-resonance, normal radiation will be restored, however, when the antenna is tuned to a full wavelength or to some other multiple of half waves.

Directive Properties of Antennas: The directive properties of an antenna are almost entirely dependent upon the length of the radiating portion, the height above the ground and its slope. A short antenna (up to a half wave length long) radiates most of its energy in a circular pattern at right angles to the wire. As the length of wire is lengthened beyond that of a half wave, the radiation pattern changes into cone-shaped loops, one at each end of the antenna. As the length is further increased, four principal radiatory loops appear which radiate from the ends of the antenna. A short antenna, therefore, may be considered as a "broadside radiator," and that of a long antenna an "end-fire radiator." The half wave antenna, such as shown in Figure 1, has a maximum field intensity in the form of a figure "8" at right angles to the wire direction. Adding more and more half wavelengths to the antenna tends to bring the radiatory configurations closer to the end directions of the long wire, and additional field intensity loops of small values are added outwardly at approximately right angles to the wire.

Angle Radiation: Radio waves are reflected or refracted (not ultra-short waves) back to earth from ionized stratospheric layers of atmosphere surrounding the earth. By directing the greater portion of the transmitted wave at certain angles with respect to the horizon, the received signal strength in the direction of the angle of directivity will be augmented in intensity, even when a low-powered transmitter is energizing the antenna. The angle at which the wave should be directed for best results depends upon the distance and direction desired. For extremely long distances a low radiation angle is preferable, or an extremely high angle above the earth's horizon. Intermediate angles will tend to shorten the skip distance.

An antenna placed at certain heights above the ground causes the lower half of the radiated energy to be reflected in or out of phase in an upward direction due to the reflected radiation from the ground. The angle of incidence is a geometrical function between the angles formed by the antenna and ground, and that formed by the antenna and horizon. A half wave horizontal antenna placed between a quarter and a half wave above the ground will have most of its energy reflected upwards at a very high angle of radiation; but, by tilting one end, more energy is sent out in the downward tilt direction.

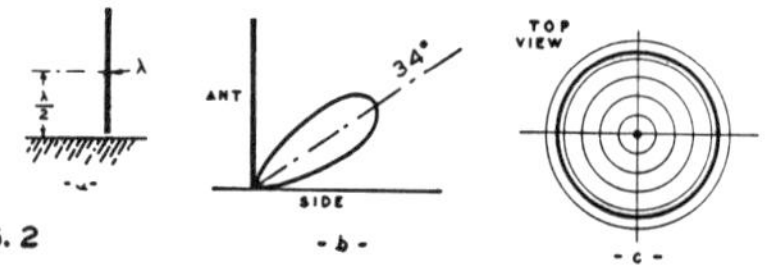

FIG. 2

Full wave antenna gives maximum radiation upward at an angle of 34 degrees from the horizon.

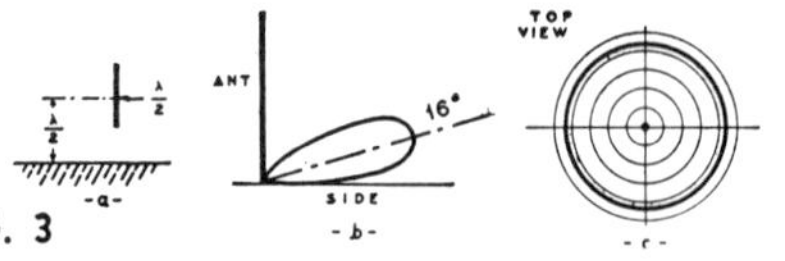

FIG. 3

Half wave antenna, same height as full wave antenna in Fig. 2. Here the radiation is known as 16-degree "low angle" radiation.

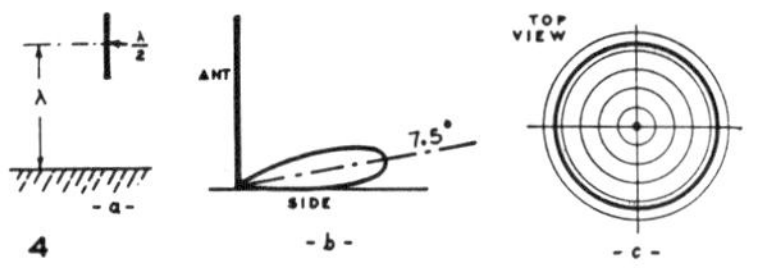

FIG. 4

Increasing height of antenna above ground lowers angle of radiation to 7.5 degrees, as shown.

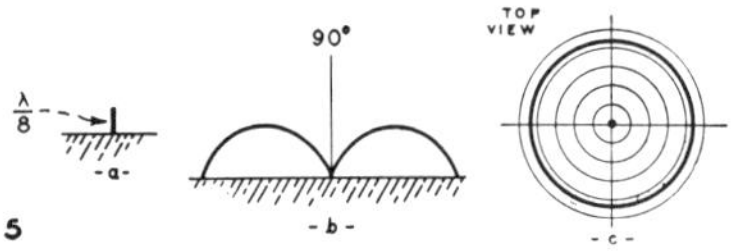

FIG. 5

Short vertical antenna, grounded, gives very low angle radiation.

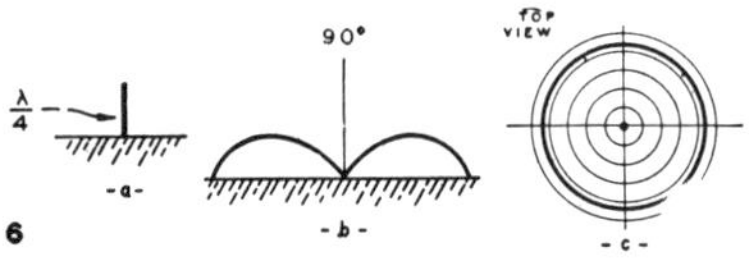

FIG. 6

Quarter wave grounded antenna concentrates more of the radiation at a low angle than the 1/8 wave antenna does.

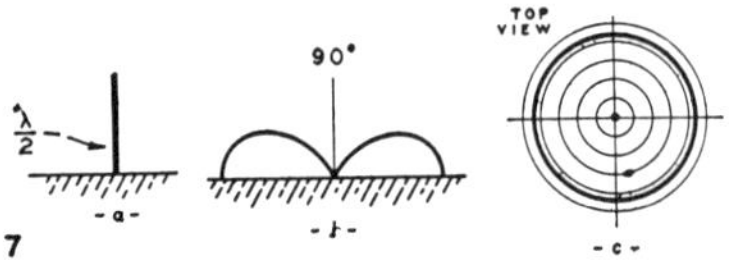

FIG. 7

Grounded half wave antenna.

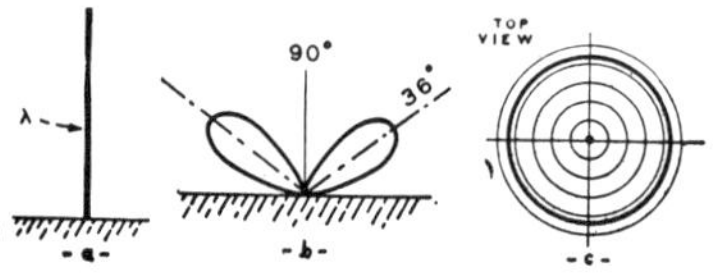

FIG. 8

Full wave grounded antenna.

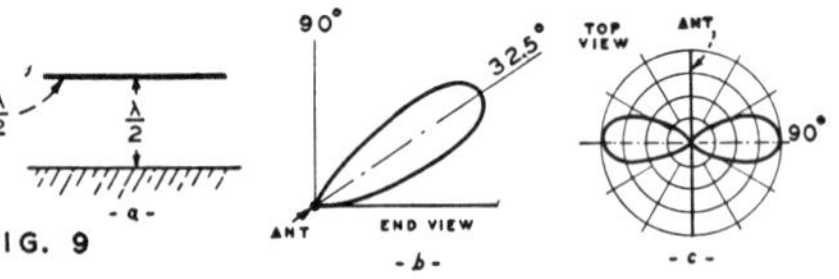

FIG. 9

Horizontal antenna half wave above ground. This is the best height for horizontal antenna for general use.

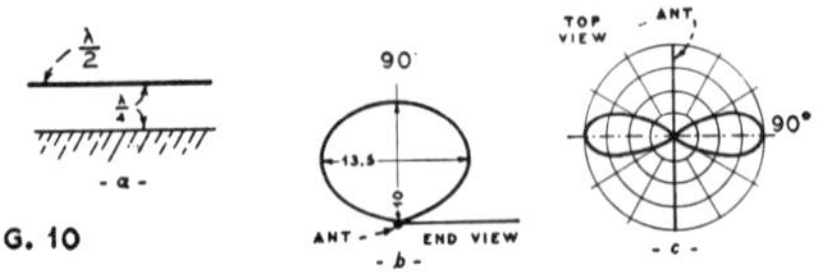

FIG. 10

Horizontal antenna, quarter wave above ground.

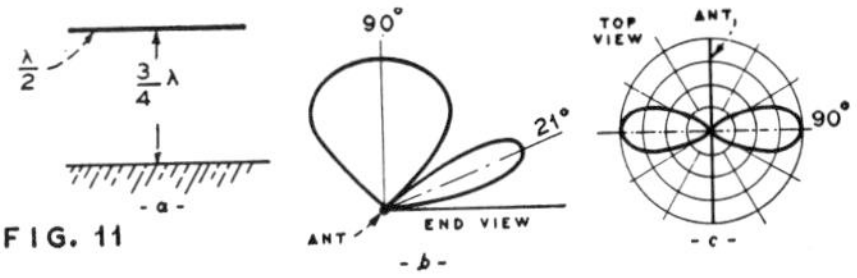

FIG. 11

Horizontal antenna, 3/4 wave high, radiates mainly upward.

Long antennas operated at a harmonic tend to give low angle radiation. An analogy can be made with an ordinary garden hose and nozzle, considering the radiation from one end of an harmonic antenna. As the nozzle is turned from the fine spray position, the cone of water has less angle and is more concentrated. A second or third harmonic antenna is like the fine spray position, while a long one (6 or 8 wavelengths long), projects most of the signal outwards in the form of a very narrow cone, having a radiation much greater at its maximum than a half wave or full wave antenna.

A vertical antenna of quarter or half wave dimensions radiates a very low angle wave which is superb for distance transmission; its maximum radiation is less than a very long horizontal antenna, correctly pointed. Vertical radiators diffuse energy equally well in all directions; however, the energy available in any certain direction is much less than from a concentrated beam, such as given by one of the radiation ellipses from a long horizontal antenna.

Careful antenna design will enable a low-powered transmitter to lay down a powerful signal at certain distant points. Unfortunately, these points change with the time of day, season and other conditions affecting the ionized stratospheric layers.

Hertz and Marconi Antennas: Hertz antennas are those which are characterized by having an even number of quarter waves with voltages present at both end. All doublets are Hertz antennas. The principal advantage of the Hertz antenna is that it has no ground connection and there fore its loss resistance is lower than that of a grounded antenna. It is most widely used for frequencies above 3,000 KC because its physical dimensions become so large as to be awkward at lower frequencies.

Marconi antennas are a quarter wave in electrical length, measured between ground and the far end of the antenna. The electrical length can be adjusted by a tuning condenser, either in shunt or in series with the coupling coil. A shunt condenser increases, and a series condenser decreases, the electrical length of the antenna system. If the Marconi antenna is cut so that its electrical length is exactly one-quarter wave at the transmission frequency, the coupling coil and tuning condensers can be eliminated and some form of a single wire feed line can be used to supply or transfer energy from the antenna to the transmitter. The effectiveness of a Marconi antenna depends, to a great extent, on its height above the ground and upon a very low resistance ground connection; where a sufficiently low resistance ground is not available, a counterpoise is used. Marconi antennas are generally used for frequencies below 3,000 KC. However, of late, the antenna has been employed in mobile and portable transmission apparatus as a quarter wave 5-meter radiator, grounded on the lower end.

Notes: The electrical length of an ungrounded antenna must be approximately equal to one-half the natural wavelength. Series condensers and inductances placed in the system will, respectively, shorten or lengthen the electrical length. Series condensers will only be effective when the wavelength does not exceed that of two-thirds the desired value.

With a grounded antenna, the physical length must be such that the electrical length is equal to one-fourth the natural radiation frequency. The length can be either shortened or extended by applying the same methods described above, providing that the desired length does not exceed one-third the natural radiation frequency.

Directional Antenna Arrays: In some locations directional antennas are used for both transmission and reception. In commercial practice, large "curtains" of wire are used for reception on account of the energy content of the received signal being proportional to the amount of antenna wire exposed to the radio waves. In transmission, directive networks concentrate energy much like reflectors and lenses concentrate light rays. These antenna arrays consist of half wave antenna elements spaced and energized to obtain either a parasitic or reflective characteristic, or combinations of both. Antenna directivity results from phasing the radiation from the adjacent antenna elements to neutralize the radiation in the undesired directions, and to reinforce the radiation in the desired direction. Directivity can be obtained in either horizontal or vertical planes.

Reflector wires are longer than the physical length of the antenna, and when placed behind the antenna, that is, in relation to the desired directivity, are situated one-quarter wave distant; for reflectors placed at the sides, the distance is increased up to one half wavelength. The length of the reflector wires can be obtained by multiplying the wavelength in centimeters by the decimal .485, which equals the physical length in centimeters; the length can also be determined directly in feet by multiplying the wavelength in meters by 1.60.

A reflector wire is required to be longer, that is, to resonate at a lower frequency so that the induced voltage in the reflector will be 180 electrical degrees out of phase with respect to the antenna voltage and also, so that the field surrounding the reflector will be 90 degrees out of phase with respect to that of the antenna. The vectorial addition of the oriented fields will show an increase in the radiated energy in the direction away from the reflector. If the reflector field has too high an inductive reactance, the parasitic field will cancel out the main radiated field in that direction because of one field being 180 electrical degrees out of phase with the other. A reflector is placed one-fourth wave behind the antenna so as to be in phase with the antenna field which is 90 degrees out of phase with the antenna current. Should the reflector be placed very close to the antenna or at a distance so that the inductive reactance of the reflector would vectorially add, the field about the antenna would be materially increased.

Director wires are shorter than those used for reflectors, hence, resonate at a higher frequency so that the capacitive reactance of the wire increases the radiation field in the desired direction. Directive arrays are placed directly in front of the antenna, optimumly spaced at three-eighths wavelength away. The physical length of the director, in centimeters, can be found by multiplying the wavelength in centimeters by the decimal .435. The length can be linearly determined in feet by multiplying the wavelength in meters by 1.425. The sharpness of the directive beam can be increased by adding more director wires in the desired direction.

Some data for ultra-high frequency antenna systems with directive arrays are as follows:

very sharp directivity can be used in the horizontal plane. However, for directivity in the vertical plane, it appears that the best angle above the horizon varies from time to time. For best results the main beam should not be directed at an angle lower than 10 to 12 degrees and not higher than 25 to 30 degrees, and that the vertical directivity should not be too sharp.

The vertical directivity of horizontal antennas depends primarily upon the height of the antenna above the ground rather than upon other characteristics of the antenna. This is because the ground reflects a subterranean image of the antenna and the reflected energy combines with the main energy to reinforce or to cause cancellation, depending upon the vertical angle. The higher the antenna the lower (i. e. the nearer the horizontal) will the reflected energy reinforce the directly radiated energy with the result that the higher the antenna above the ground the closer to the horizontal will be the radiation. This is graphically shown in Figures 9, 10 and 11, from which it is seen that if the height is one wavelength then the greatest portion of the energy will be directed at a vertical angle of approximately 16 degrees, while if the height is one-half wavelength, the angle will be 30 degrees. Horizontal antennas should therefore never be less than one-half wavelength above the ground if they are to be used for long distance communication.

Wave-lengths	Frequency MC	Antenna Length	Reflector Lengths	Director Lengths	Antenna Spacing Directors	Antenna Spacing Parallel Array
5.357	56	8′ 4″	8′ 7″	7′ 7″	4′ 4½″	8′ 9″
5.263	57	8′ 2⅜″	8′ 5½″	7′ 5½″	4′ 3¾″	8′ 7½″
5.172	58	8′ ½″	8′ 3¾″	7′ 4″	4′ 2⅞″	8′ 5¾″
5.085	59	7′ 10¾″	8′ 2″	7′ 2½″	4′ 2″	8′ 4″
5.0	60	7′ 9″	8′ ½″	7′ 1″	4′ 1¼″	8′ 2½″
10.64	28.2	16′ 8″	17′ 1″	15′ 2″	8′ 9″	17′ 8″

Some important factors which enter into directive transmission and reception with antenna arrays are:

(1) The terrain of the intervening country will alter the pattern of directivity, especially if high hills or large structures are in the vicinity of the direction of the transmission.

(2) It is pre-requisite that both the transmitting and receiving systems be in the same geometric plane. At the receiver, the best results will be obtained by rotating the receiving antenna in the signal plane. Rotation is nesessary because of the phase distortion occurring during transmission.

(3) Transmitting antenna, reflectors and directors must not be mounted on metal poles, else the field pattern will be greatly distorted. If metallic masts are used, the ratio between the radiator and the support must **not** approach a half or multiple of half waves.

Vertical and Horizontal Directivity: With either antenna arrays or directional antennas, it is possible to have too much directivity. This is because the waves do not always travel along the same path in reaching the receiver, and that the amount of directivity in the vertical and horizontal planes which can be tolerated is affected by certain factors. In general, it is found that the waves travel very closely along the great circle path to the receiver, and that

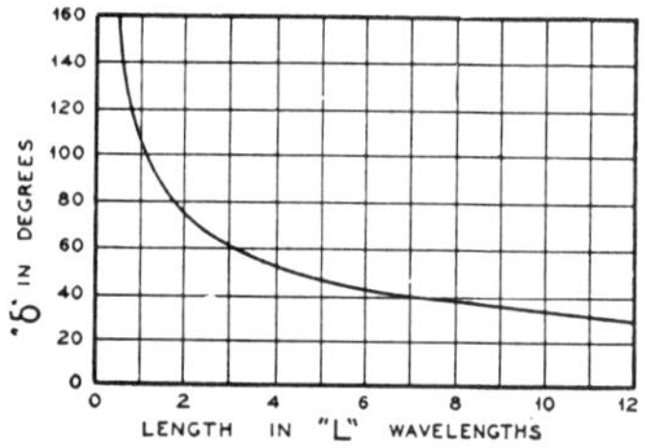

FIG. 12

Terman's Diamond Antenna Charts.

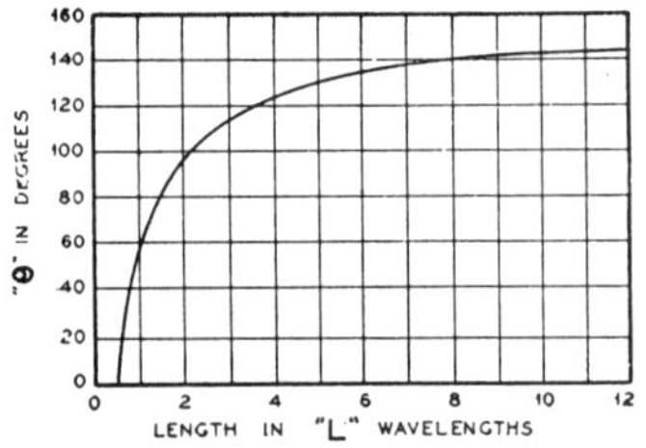

FIG. 13

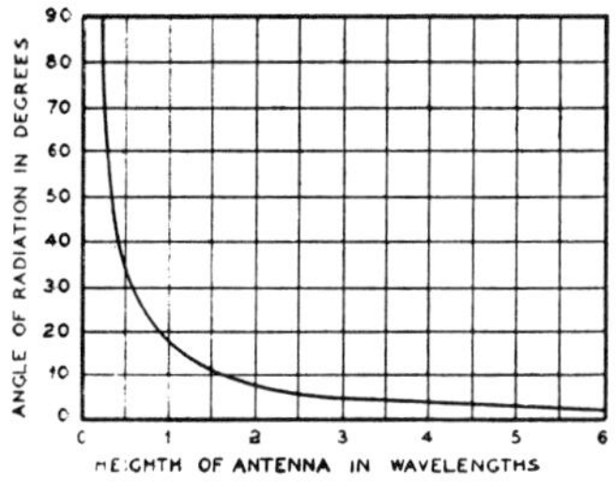

FIG. 14

Horizontal and "V"-type Directional Antenna: The horizontal V and diamond directional antennas, such as shown in Figures 15a and 15b, respectively, are systems of relative simplicity to build and to tune. The principal factor controlling the design of a V antenna is the angle between the wires. This is determined by the length of the wire according to the relation shown in Figure 12; this value is relatively critical. The amount of directivity obtainable is greater the longer the wires, and commercial antennas of this type are commonly made about eight wavelengths long. However, reasonable directivity can be expected of two to four wavelengths.

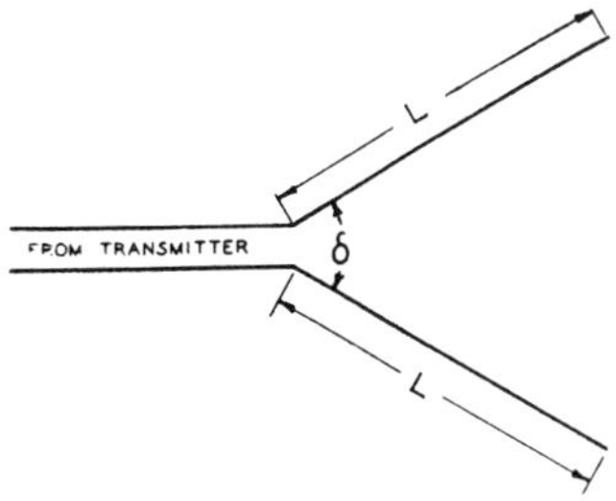

FIG. 15a

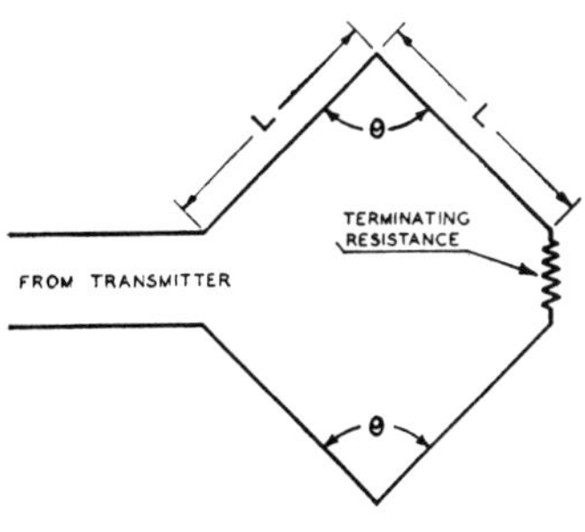

FIG. 15b

A single V antenna is bi-directional. The back-end radiation can be redirected forward by a reflecting antenna (similar to the radiating antenna) located an odd number of quarter-wavelengths behind and faced so that the two antennas are supplied with current 90 electrical degrees out of phase.

The diamond antenna, see Figure 15b, is non-resonant and possesses a current distribution which uniformly dies away from the input corner to the terminating resistance. As a result of this behavior, the diamond antenna is not critical with respect to frequency and can be used without any change or adjustment over a frequency range of at least 2 to 1. The antenna is, furthermore, uni-directional, since the terminating resistance eliminates the radiation which would otherwise take place in the backward direction. These properties make the diamond antenna desirable from many points of view. It can, for example, be used at 20 meters in the daytime and on 40 meters at night without any change. In constructing a diamond antenna it is important to consider the angle θ, which is related to the length of the legs as shown in chart of Figure 13. The terminating resistance has a value of approximate 800 ohms; its use is to eliminate resonances along the line. The antenna also offers a resistance load of about 800 ohms to the transmission line.

Tuned Diamond Antenna: The difference between the tuned diamond and non-resonant diamond antennas is that in the tuned system the resistor at the far end has been eliminated. The tuned diamond is mostly used in the 5 meter band, but is also applicable to 20 and 40 meter operation. The antenna shown in Figure 16 has sides equal to one full wavelength which, for 5 meters, has a physical length of 16 feet 2 inches. The angle in the case shown, that is, for one wavelength a side, is 120 degrees. For sides equal to 2 wavelengths (32 feet 4 inches) the angle must be 87 degrees. The arrow at the top of the diagram shows the direction of wave propagation and also the direction of best reception. The antenna configuration shown in Figure 17 will give a stronger wave in the direction indicated, as compared with the antenna in Figure 16.

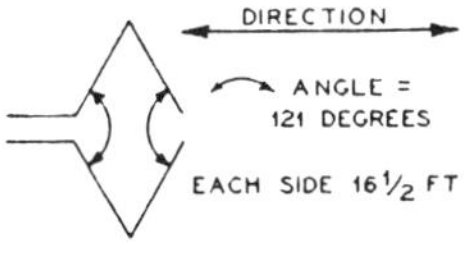

FIG. 16

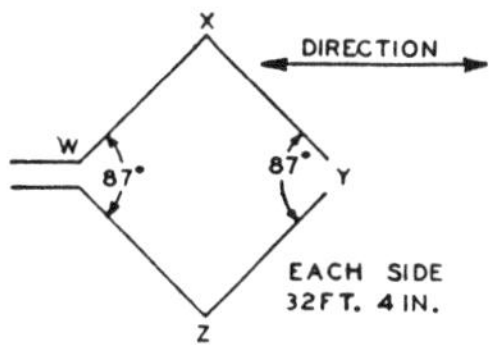

FIG. 17

If radiation in one of the directions is unwanted, the back wave may be eliminated by inserting a 600 to 800 ohm resistor in the open end, as shown in Figure 7; the wattage of the resistor must be such as to safely dissipate one-half of the transmitter output. If the antenna is slanted, so that the "w" end is higher than the "y" end (see Figure 17), the radiation will be stronger towards "y" direction; conversely, if the "y" is made higher than the "w," the propagation will be stronger in the lower pointed direction. For receiving, it will be found the direction in which the transmission is strongest is the direction from which best reception is secured. If the two edges "x" and "z" are not of the same height, the angular direction of transmission with respect to the direction of propagation wil be distorted. For example, if the "x" and "y" are higher than the "w" and "z," the direction of transmission will be as shown in Figure 23. Raising any one of the four corners will cause the radiation to increase away from the side elevated.

Tuned diamond antennas will radiate in an exactly horizontal position provided that the angle of radiation in degrees and the height of the antenna in wavelengths is correctly calculated. These calculations have been simplified, and a curve shown in the chart of Figure 14 will enable a designer to quickly determine the necessary figures; for example, if 36 foot 4½ inch sides are used, which is approximately equal to two wavelengths, 32 feet 4 inches off the ground, the angle of radiation will be 7 degrees, as shown by the chart. Slanting the antenna 6 degrees will cause the energy to be radiated in an exactly horizontal plane.

Note on 5-Meter Antenna Systems: In transmission and reception of 5-meter signals, the direct or ground wave is used. This is because there is little or no reflection from the ionized stratospheric layers of the upper atmosphere. The earth, however, reflects short waves much like a mirror does light rays; for this reason short wave transmitting and receiving apparatus working on the fringes or in wave bands of the quasi-optical frequency spectrum must be in visual range of each other. It is therefore necessary that the antennas at both transmitting and receiving points be placed as high as possible above ground. On account of the direct-wave propagation phenomena, the antennas must have a low angle of radiation. Vertical antennas of the simple half wave type are more effective than horizontal types because of the vertical polarization being greater than that of horizontal polarization; in addition, the greatest radiation pattern is parallel to the earth as well as having a low angle with respect to it; hence, the earth acts as a reflector tending to bend the wave front up and away from the ground. There is less tendency for upward bend with vertical polarization; otherwise a half-wave horizontal antenna would be just as effective.

Grounds: A good ground is essential for satisfactory operation of both transmitter and receiver. Several pipes driven into the earth and spaced a few feet apart will function splendidly. The connections to the

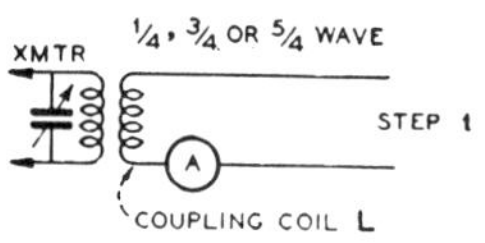

FIG. 18

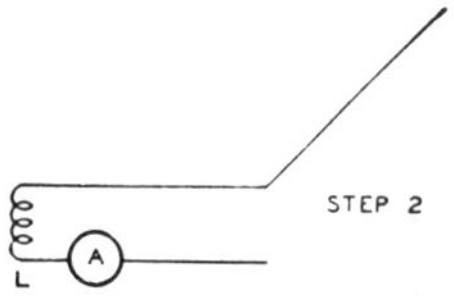

FIG. 19

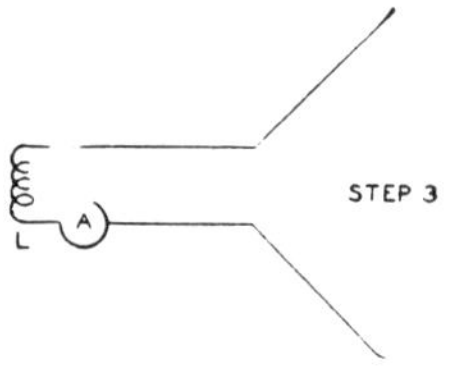

FIG. 20

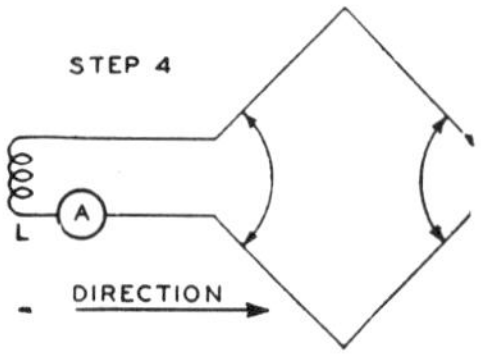

FIG. 21

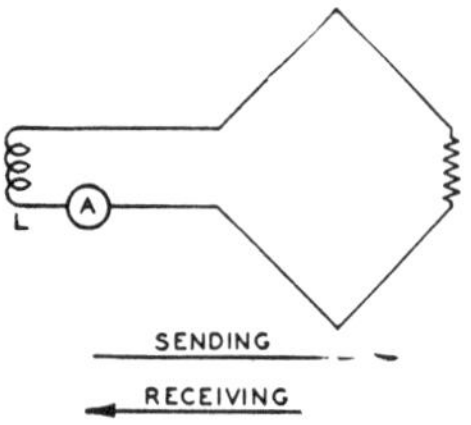

FIG. 22

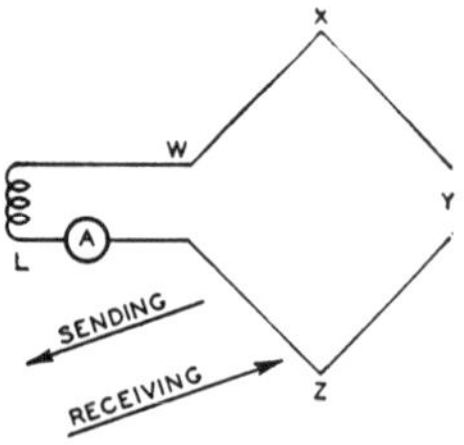

FIG. 23

pipes must be made with well-designed ground-clamps. Ordinary water pipes, such as the mains belonging to the public utility company, do not make good grounds on account of the relatively high resistance of the pipe coupling connections.

When a satisfactory ground cannot be established, a counterpoise must then be used. This device simply consists of one or more wires placed either under, on, or above the ground. For wires above the ground, the relative capacity with reference to the earth is very high. The length of a counterpoise is not critical and is usually made about one-half the length of the flat top portion of the antenna. Counterpoises placed above the ground are insulated and run parallel to the antenna. If the device is not oriented properly, ground currents will flow, causing losses in such places where sand or dry soil is found.

All ground systems perform two primary functions. These are: (1) to serve as an electrical contact with the earth to form one electrode of the radiating system, and (2) a wave-reflecting surface. Power lost in the resistance of the ground system can be reduced by making this resistance as low as possible, or by so adjusting the mode of operation of the antenna so that the antenna resistance is very high with respect to that of ground, and the current entering the ground, for a given power, is relatively small.

Antenna Sites and Locations: The antenna is the most important unit in both transmitting and receiving practice. It must be properly designed, but, above all, it must be given the best possible location if the full efficiency of its radiating properties is to be realized. For long distance communication, the energy must be radiated at the lowest possible angle with respect to the horizon. It is essential that the antenna be as high as local conditions will permit; in addition, be free from nearby objects, such as trees, tin roofs, etc. The comparatively recent development of various kinds of low-loss transmission lines no longer makes it necessary to locate the antenna close to the transmitter. Lines as long as 1,000 feet are being used with very low losses. The antenna **must** be cut to exact length. However, because antenna formulae do not take into consideration the capacity effects, which depend upon the height above ground and the conductivity of nearby objects, the formulas for length will be in slight error; correction must be made for these factors after the antenna is erected.

Dummy Antennas: When a transmitter is being tested, it is often desirable to utilize a dummy antenna instead of the radiating antenna in order to more accurately determine the power outputs, as well as to prevent interference with other stations when the transmitter is being tested.

A dummy antenna consists of a resonant tank circuit whose condenser and coil are selected to resonate at the transmitter frequency. This tank circuit is coupled to the transmitter by either direct, capacitive or link coupling. The latter is most desirable, as it affords a convenient means for varying the degree of coupling between the plate tank of the transmitter and the load tank of the antenna. In order to dissipate power, some resistance must be coupled into the dummy antenna load tank. One method is to connect a series of electric lamp bulbs in a circuit; taps are then taken from the bulbs and are shunted across various turns of the tank coil until the final amplifier draws the proper plate input from its power supply.

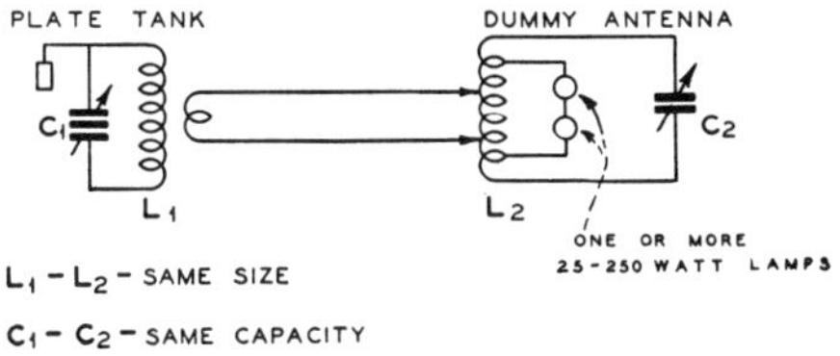

DUMMY ANTENNA

The resistance of electric lamp bulbs varies widely with filament temperature; therefore, it is difficult to accurately determine the power output of the transmitter by the I^2R Law, because R is a variable factor. An approximate estimate of power output can be made by determining the brilliancy of the lamp bulbs when working as a dummy antenna, as compared with the brilliancy of the same bulbs when connected to a source of 60 cycle voltage. The voltage and current for a given degree of brilliancy can be accurately measured at 60 cycles, but not at radio-frequencies. In other words, the power consumed in lighting the lamp to a given degree of brilliancy is the same whether the power is RF energy or that derived from the utility company's power line.

Transmission Lines and Coupling Systems: Transmission lines are devices which electrically convey the power from a transmitter for the energization of a distant antenna; these lines are designed to have minimum radiation so as to prevent the dissipation of RF currents during transit.

The means of coupling the line to the plate tank of the final amplifier in the transmitter, by any one of the known coupling systems, must be carried out to a high degree of accuracy. This is because the characteristic impedance of the line (which depends upon its mechanical dimensions) must be properly matched to the plate circuit for maximum transfer of energy from tube to line. By the same token, the far end of the line must terminate in such a manner that the line impedance is matched to the antenna impedance.

An ideal coupling device is one which must (1) be matched in impedance to the final amplifier at the station end; (2) be matched in impedance to the antenna at the antenna end; and (3) have negligible losses in itself and must not radiate.

Any resonant circuit, of which the plate

tank of a final amplifier is an example, has a rather high impedance (AC resistance) across its ends. Across the two center turns of this same circuit the impedance is low, and the impedance increases as taps are taken out and away from the center point of the tank coil. One of the fundamental electrical laws states that maximum power transfer from one electrical circuit to another occurs when the impedance source of power is exactly equal to the impedance of the received power. Thus, it follows that the coupling device (which may be any one of the various transmission lines or feeders) receives power from the final amplifier most effectively when tapped across just enough turns of the final tank coil so that the impedance across that portion of the tank is equal to the impedance at the station end of the coupling device.

Similarly, an antenna, which is a tuned resonant circuit, varies in impedance along its length. Thus the coupling device must connect to the antenna at a point, or points, where the impedance of the antenna matches the impedance of the coupling device. If the impedances at the station end and at the antenna are not properly matched, power is lost either through radiation from the feeders, which radiation is not effective, or else by loss to ground due to voltages induced in surrounding objects which absorb power, such as house wiring, tin roofs, plumbing and water pipes, etc. Other common losses in coupling devices are due to sharp bends, which have a reflection loss and thus radiate; wire of improper size for the feeder separation; more capacity to ground from one side of the feeders than the other, and high resistance and unsoldered joints.

Zepp Feeders: The portion of the antenna called the Zepp feeder (which is a resonant coupling device and thus forms part of the antenna proper) simply consists of an additional length of antenna which is folded back upon itself in such a way that the standing waves on the two feeders neutralize each other and thus prevent the feeder portion of the antenna from radiating. The first fundamental of Zepp antenna operation is that the flat-top portion which does the actual radiating must be cut to within 10 per cent for the frequency used. No amount of tuning of the folded portion (feeders) of the antenna can properly compensate for a flat-top which is more than approximately 10 per cent too long or too short. The electrical length of a flat-top can be checked by taking down the feeders and disconnecting the antenna from its feeder, and then raising the feeders again. With the feeders raised (with the antenna disconnected) the transmitter is turned on and the feeders are tuned to resonance. Next, the transmitter is turned off; now, without changing the feeder tuning condensers, connect the antenna to its associated feeder in the usual manner. When the transmitter is turned on again, no retuning of the feeder condenser will be necessary if the flat-top portion of the antenna was cut to the proper length. If the feeder tuning condensers must be increased to establish resonance, the flat-top portion of the antenna is too short; if the capacity must be reduced, the flat-top portion is too long.

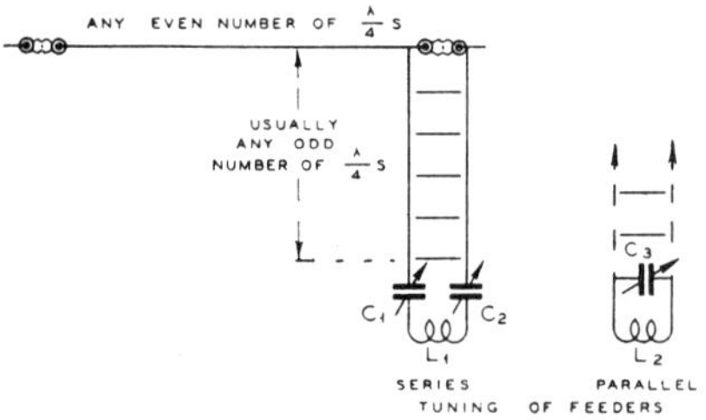

Series tuning of Zepp feeders is used to reduce the electrical length of the resonant system which consists of the two feeders and the pick-up coil. Parallel tuning is used to increase the electrical length of the resonant system, and thus feeders of almost any length can be resonated on practically any frequency by one or the other of these two tuning methods.

The most outstanding feature of a Zepp coupling system is due to its simplicity and ease of adjustment. It is definitely less efficient than the more modern non-resonant transmission lines, such as the Twisted-Pair Feeders, Johnson Q Feeders and the One- or Two-Wire Matched impedance lines, because the Zepp system actually brings a portion of the antenna into the operating room. Theoretically, Zepp feeders do not radiate, but, as a matter of fact, the perfect Zepp feeder only exists on "paper."

Another type of Zepp feeder is one which is attached to the center of the antenna, instead of at one end. This differs from the more common voltage fed type in that it connects to a low impedance point on the antenna instead of at a high impedance point. For this reason it is sometimes known as a "current fed Zepp," or "doublet," whereas the more common type is known as the "voltage fed antenna." If the flat-top portion of the antenna is any odd number of half wavelengths long, it will be found that at the end of each half wave section there is a voltage loop having a very high impedance which a voltage feed may be attached for proper termination.

Zepp Feeder Lengths: All Zepp feeders must be self-resonant to the radiated frequency. The term "Zepp Feeders" includes the coupling coil and coupling condensers, and thus the electrical length of the two feeders and the coupling inductance and capacity must be some multiple of one half wavelength in length. Therefore each feeder, with its associated half of the coupling reactance, must be some odd multiple of one-quarter wave in electrical length. If the mechanical length of the two feeders is either longer or shorter than the length required to produce resonance, the proper amount of electrical length can be either added or subtracted by means of

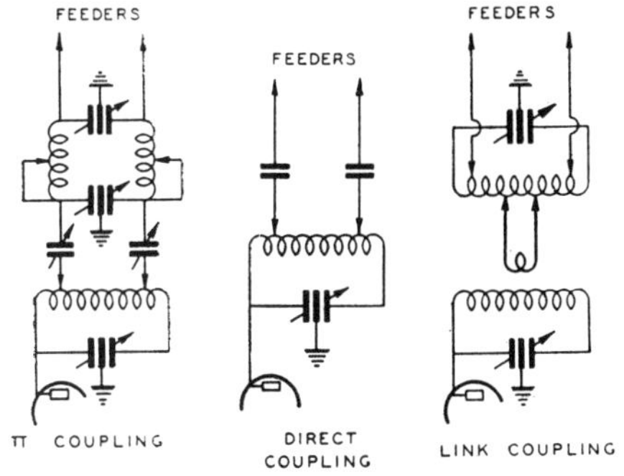

Three methods of coupling the antenna load into the final amplifier. The PI coupling and link coupling circuits can be used with either resonant or non-resonant feeders, whereas the direct coupling circuit can only be used with non-resonant lines.

coils and condensers. Thus when Zepp feeders are tuned, merely their electrical length is varied.

Energy is transferred to the feeders by a small coil usually placed at the station end in inductive relation to the plate tank coil of the final amplifier. The presence of this coil adds electrical length to the feeders. If the electrical length of the feeders, plus the electrical length of this coupling coil, is less than one-quarter wave greater than any multiple of one-half wave, then series condensers must be used to shorten the electrical length of the feeders sufficiently to establish resonance. If, on the other hand, the electrical length of the feeders and the coupling coil is more than one-quarter wave too long, or less than one-quarter wave too short, a condenser shunted across the coupling coil must be used to bring the electrical length back to a multiple of one-half wavelength. Reference to the table shown below will facilitate converting frequency into the mechanical equivalent of full wave, half wave and quarter wave systems. The table will show whether the feeders are longer or shorter than a multiple of one half wave. The length of the two feeders should be added together; for example, the shortest feeders, whose mechanical length would approximate one half wave at 40 meters, would each be 33 feet in length.

For 5 meters one quarter wave is	**4 feet**
For 10 meters one quarter wave is	**8 feet**
For 20 meters one quarter wave is	**16 feet**
For 40 meters one quarter wave is	**33 feet**
For 80 meters one quarter wave is	**66 feet**
For 160 meters one quarter wave is	**132 feet**

Not all Zepp feeders have coupling coils and tuning condensers; instead, a short-circuiting bar is shunted across the two feeders at the lower end to complete the circuit. In this type of construction the feeders are tuned by sliding the bar up and down along the feeders until resonance is established. The final amplifier of the transmitter is then coupled to the feeders by means of a non-resonant, low impedance transmission line which is clipped onto the two feeders slightly above the shorting bar, depending upon the impedance of the transmission line. The impedance across the feeders is lowest at the shorting bar and is highest at the end where the antenna is connected to the feeder. Thus, the impedance of a non-resonant transmission line which is delivering power to this type of Zepp feeder can be matched by merely sliding the clips up and down along the feeders until the standing waves disappear from the non-resonant line. If a neon tube is held against the non-resonant line, its brilliancy should not vary as it is moved along the line. This form of Zepp feeder is quite widely used by the commercial communications companies in coupling a non-resonant line to a directional antenna array. In commercial practice, the losses inherent in all transmission lines are minimized by keeping the lines well up in the air and away from house wiring, tin roofs, etc., and by using high-grade separators and by proper tuning and balancing.

LENGTH OF FEEDERS	Type of Feeder Tuning to Use
Up to One Quarter Wave	Parallel Tuning
Between One and Two Quarter Waves	Series Tuning
Between Two and Three Quarter Waves	Parallel Tuning
Between Three and Four Quarter Waves	Series Tuning
Between Four and Five Quarter Waves	Parallel Tuning
Between Five and Six Quarter Waves	Series Tuning

Zepp Feeder Tuning Data

The principal advantage of the Zepp feeder system is that, no matter how inefficiently it may be built, power will always be drawn from the final amplifier, although the power radiated might be a very small fraction of the energy conveyed in transit. Because a Zepp feeder system draws the greatest amount of power out of the final amplifier and gives the greatest meter indication of RF amperes, is **not** indicative that the system is working efficiently. Other forms of coupling devices usually refuse to draw power from the final amplifier unless the radiating portion of the antenna is actually radiating. Sometimes it is assumed that the non-resonant transmission line is faulty and difficult to adjust because the final amplifier cannot be made to draw enough plate current; however, the fault may be traced to the antenna not having the proper length to draw power from the transmission line. In other words, an effective non-resonant transmission line ordinarily will not draw power from the transmitter unless it can deliver it to the antenna.

Length of Flat-Top Portion for Zepp Feeders: The flat-top portion of a Zepp antenna is not critical as to length. The Zepp feeder tuning system in the radio room will compensate for variations of approximately 10 per cent in the flat-top portion. Thus, the following table of flat-top lengths is suitable for operation on any frequency in the bands listed below:

FOR HALF WAVE ZEPP FLAT-TOPS

Band	Length of Flat-Top
160 meters	250 feet
80 meters	130 feet
40 meters	66 feet
20 meters	33 feet
10 meters	16½ feet
5 meters	8 feet

Spacing Zepp Feeder: If the spacing on Zepp feeders is too great, the standing waves on those feeders do not properly concel out each other. If the spacing is too small, there is excessive heating of the feeder separators and a tendency for them to arc-over. The best compromise for Zepp feeder spacing is about 6 inches. Ceramic feeder separators which resist the absorption of moisture should be used.

Calculating the Length of Any Half Wave Radiator: Antennas which are fed by any type of non-resonant line must be cut to exact length, subject to slight modification due to the presence of nearby objects. For all practical purposes the antenna can be cut to the calculated length and the wire used should be of the kind that will not stretch. Knowing the frequency at which the antenna is to operate, this figure can be converted into wavelengths by dividing it into 300,000,000, thus giving the wavelengths in meters (metric measure); the actual length in feet for a half wave antenna can be obtained by multiplying by 1.56. For example: assume a frequency of 7,200 KC for the transmitter:

$$\frac{300,000}{7,200} = 41.7 \text{ meters}$$

$41.7 \times 1.56 = 65$ feet.

Non-Resonant Transmission Lines: The essential difference between a non-resonant transmission line and the Zepp feeder system is that the non-resonant lines, when properly constructed, have no standing waves on them, and the impedance of the line is the same at both ends, whereas the impedance of Zepp feeders vary uniformly from a low impedance at the transmitter end to a very high impedance at the flat-top end.

The impedance of most non-resonant lines is usually under 800 ohms and is as low as 70 ohms in certain types of twisted-pair or concentric cable lines. In any circuit carrying power, the voltage across the circuit rises as the square root of the impedance. Thus a low impedance line has a very small voltage across it and insulation becomes a minor factor. Because the line is non-resonant there is no circulating current flowing through it, and thus allows the use of smaller wire than that used for a Zepp feeder system. This is not always true of the low impedance lines but holds true for the average lines. Non-resonant lines radiate a negligible amount of power and can therefore be run indoors and close to water pipes and other conductors without losses.

Hertz Single-Wire Matched Impedance Line: This type of non-resonant line is more efficient than those of the resonant class; however, it has some losses. A single-wide fed Hertz antenna consists of a single wire (and phantom return ground) clipped on the antenna 14 to 17 per cent of one-half wave away from any voltage node. In a half-wave antenna the voltage node is at the center. There is some controversy as to the exact point of attachment because it varies slightly with frequency, size of the feeder wire and height of the antenna above ground. Best results can be obtained by moving the point of attachment about four inches at a time around the optimum position until the standing waves disappear, as indicated by a neon bulb held against the feeder and moved along it over a distance of one-quarter wave.

It is highly important that a good ground connection is established on account of the single-wire line being, in reality, a two-wire line, the second wire of which is a phantom return through ground, which is in effect connected to the exact center of the antenna in the case of a one-half wave flat-top. Unfortunately, a perfect ground is never obtained in practice; hence, the single-wire feed line has somewhat higher losses than the better two-wire types of lines. It is essential that the single-wire feeder be placed at right angles to the flat-top radiator which feeds it for a distance of at least one-third of the length of the antenna in order to avoid pick-up from surrounding magnetic fields, which would cause the presence of standing waves to appear on the feeder. One of the characteristics of all non-resonant lines is that they can be extended to practically any length; lines have been successfully built as much as 4,000 feet long.

The End-Fed Hertz Antenna: All types of end-fed antennas do not generally utilize feed lines and can therefore be termed "directly excited antennas." They are useful because they are tunable from the radio room and can be built in less time than practically any other type. Their main disadvantage lies in the fact that the antenna is brought directly into the radio room; therefore a material portion of the radiation is lost by the nearness of immediate metallic surroundings.

An end-fed system is shown in Figures 24 and 25. If some isolation is desired, L1

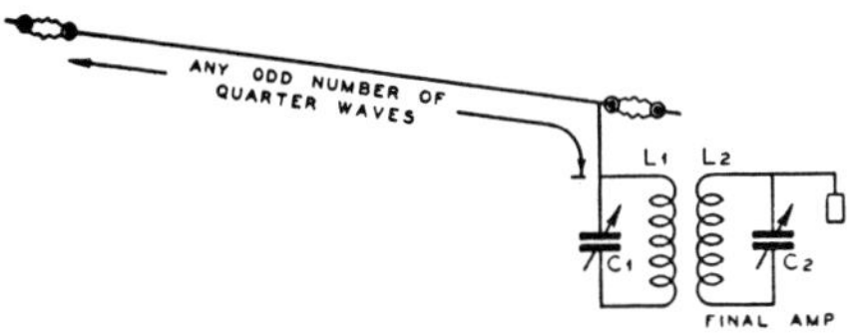

FIG. 24—End fed, current fed Hertz antenna.

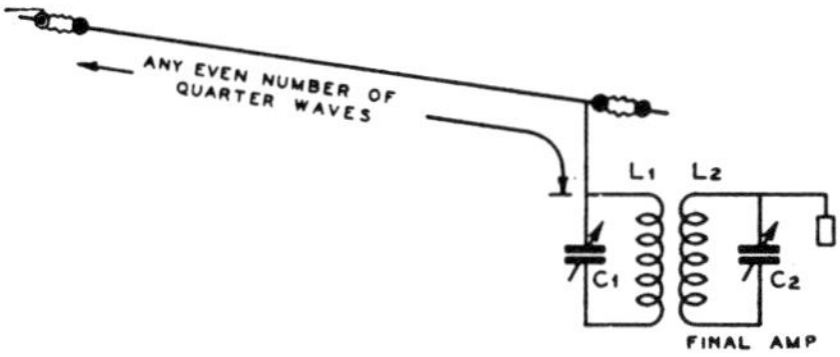

FIG. 25—End fed, voltage fed Hertz antenna.

and C1 may be placed at the top of a pole and a link-coupled feed line coupled between L1 and L2. This antenna is desirable in locations where it is impossible to obtain a good ground connection, a common trouble at higher frequencies when the transmitter is located three or more stories above the ground. No ground should ever be used with an end-fed antenna because the ground tends to make it act as a Marconi antenna, unless care is taken in tuning adjustments.

There are only two types of end-fed antennas: (1) voltage feed, and (2) current feed. Voltage fed systems are those in which the power is fed to the antenna at a point of high impedance, such as the end of a half wave or full wave Hertz. Current fed systems are those in which the power is fed to the antenna at a point of low impedance, such as at the center of a half wave antenna, or near the end of an antenna whose length is any odd number of quarter wavelengths.

Figures 24 and 25 are similar in appearance, yet a difference exists between them. The radiating portion of the antenna in Figure 24 has high voltage present at both ends, and the coil L1 and condenser C1 resonate to the transmitter frequency. In Figure 25 there is a high voltage at the far end of the radiating portion of the antenna, and very little voltage is present where the antenna connects to the top of L1; however, a high current exists at this point, and a high voltage point will be found at the bottom of L1. In Figure 25, L1 has about half as many turns as L1 in Figure 24 for the same transmitter frequency. L1 and C1 in Figure 25 will ordinarily be tuned to approximately twice the transmitter frequency.

Center-Fed Two-Band Vertical Zepp Antenna: Vertical antennas are very useful for long distance transmission in all directions. The antenna shown in Figure 26 is a center-fed half wave Zepp for 40-meter operation and consists of two end-fed half waves excited in phase when operated on 20 meters. This antenna is particularly desirable for 20-meter operation, as it provides low angle radiation not obtainable with an ordinary full wave 20-meter vertical antenna. The latter has two half waves that are out of phase, resulting in high angle radiation.

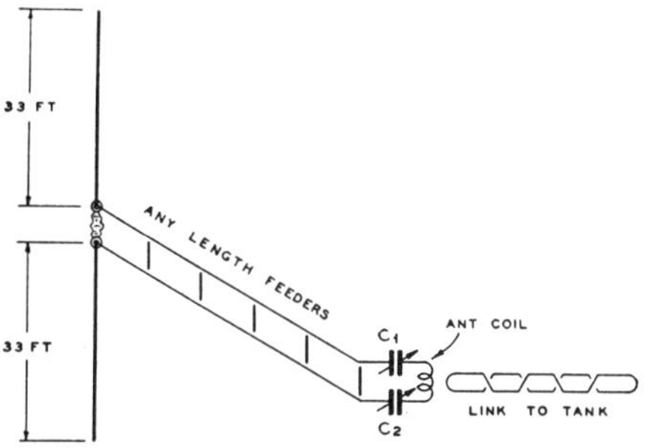

FIG. 26—Two-band vertical antenna.

The feeders may be of any convenient length but isolated from metallic objects on account of the standing waves on the transmission line. The line separators must be able to withstand high voltages, and the circuit must be tuned at the station end to some multiple of half waves. Series tuning is shown, but parallel tuning will suffice if necessary. Series tuning will usually allow more power to be transmitted to the antenna because there is less loss in the large series tuned coil than in a small shunt tuned coil. The antenna ammeter will indicate very low values of current for 20-meter operation, but fairly high values will be indicated on 40 meters.

Two-Wire Non-Resonant Transmission Line: It was previously stated that all non-resonant transmission lines are aperiodic and therefore have no standing waves on them, but have a distributed inductance and capacity which depends on the size of the wires and their spacing. These factors determine the characteristic or surge impedance of the line. This characteristic line impedance is the impedance that must be matched to the antenna at the far end and to the transmitter at the near end, if maximum power transfer is to be realized. It is quite simple to compute the surge impedance of any two-wire line.

$$Z_s = 276 \log \frac{b}{a}$$

where Z_s is the surge impedance; a, the radius of the wire; and b, the distance between the two wires. a and b may be expressed in any units, but the units chosen for the two dimensions must be similar because of the ratio between the two.

Since the impedance of an antenna depends upon the points between which measurement is made and varies from a low value at a voltage node (center of a one-half wave antenna) to a very high value

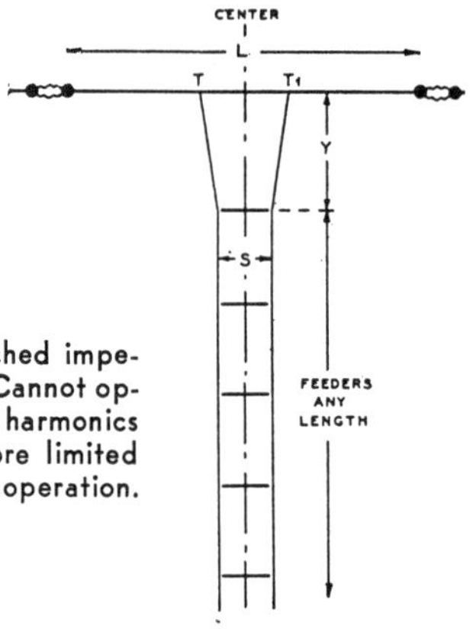

Two-wire matched impedance Hertz. Cannot operate well on harmonics and is therefore limited to one-band operation.

at the voltage loops or ends of the antenna. The line, therefore, must be tapped onto the antenna at points whose impedance is equal to the impedance of the line. With the ordinary type of two-wire matched impedance line it is necessary to fan-out the feeders at the far end to evenly increase or transform the feeder impedance so that it matches the antenna. The details of this matching process between a 600 ohm line and a one-half wave antenna are figured as follows:

$$L \text{ (in feet)} = \frac{492{,}000}{F} \times K$$

or

$$L \text{ (in meters)} = \frac{150{,}000}{F} \times K$$

Where L is the antenna length; F, the frequency in kilocycles; K, .96 for frequencies below 3,000 KC; K, .95 between 3,000 and 28,000 KC; and K, .94 for frequencies above 28,000 KC.

The portion of the antenna between the two taps, T and T1 where the feeders connected is computed as follows:

$$T \text{ and } T1 \text{ (in feet)} = \frac{492{,}000}{F} \times K'$$

or

$$T \text{ and } T1 \text{ in (meters)} = \frac{150{,}000}{F} \times K'$$

Where K' equals the decimal .25 for frequencies below 3,000 KC; K', .24 between 3,000 and 28,000 KC; and K', .23 for frequencies above 28,000 KC.

The fanned Y portion is computed as follows:

$$Y \text{ (in feet)} = \frac{147{,}000}{F}$$

or

$$Y \text{ (in meters)} = \frac{45{,}000}{F}$$

The feeder spacing S for a 600 ohm transmission line is computed approximately as follows:

$$S = 150 \times r$$

Where S is the center-to-center distance between the feeders; and r, the radius of the same wires. These should be expressed in the same units, whether inches or millimeters.

The spacing of the feeders is rather critical and the line must be kept taut. Each side of the line must be of the same length and be symmetrical with respect to ground. The transmission line is connected at right angles to the antenna for a distance at least equal to one-third of the antenna length. Any bends in the feed line should be gradual, as sharp bends cause reflection losses and undesired radiations.

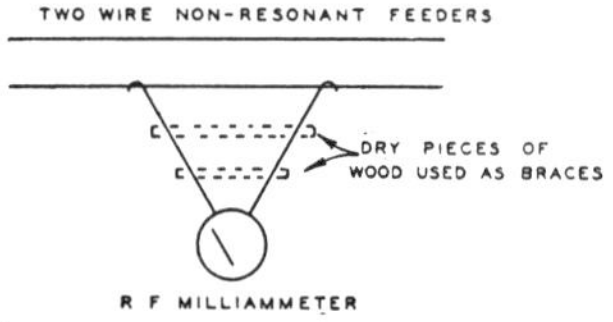

When bare copper wire is used for non-resonant feeders the impedance matching adjustments can be checked by measuring the current along the line, as shown above.

NOTE: If the power output of the final stage is 50 watts or more, and if bare copper wire feeders are used, the RF milliammeter can be inserted as a shunt across approximately 15 inches of one feeder, as shown in the illustration, and the unit moved along one or the other feeder. This method is very useful when adjusting long single or two-wire "non-resonant" feeders. Standing waves with variation in line current is an indication of reflection losses and thus the line current should be made as near constant as possible.

The characteristic impedance of the two-wire line must be matched to the plate circuit of the final amplifier. This can be done by tapping the line on each side of the center of the plate tank coil until the final amplifier operates efficiently at the desired DC input. In most final amplifiers the feeders will be tapped about one-quarter or one-

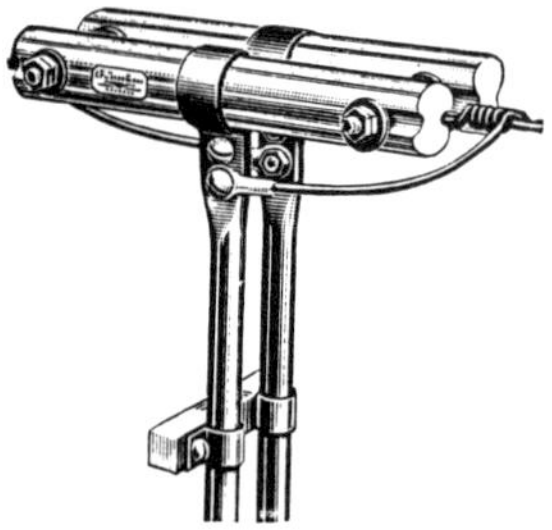

The Johnson "Q"

third of the way out from the center-tap of the coil. Because the tank circuit impedances vary widely, depending on tubes and plate voltages, it is impossible to definitely determine this point in advance. Some form of blocking condensers should be used in the feeders to isolate the DC plate voltage from the antenna and line.

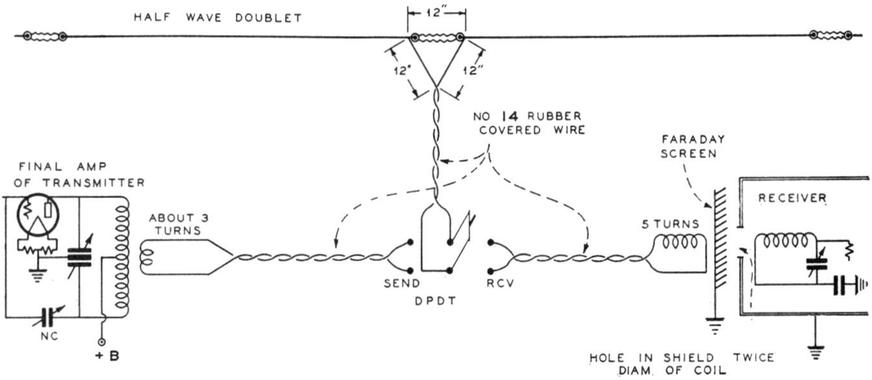

FIG. 27—Practical One-Band Doublet Antenna With Twisted-Pair Feeders.

The familiar Collins Pi Network provides a convenient and exact matching device to most effectively couple a two-wire line to an amplifier.

Transposition blocks are not required if the antenna to which the two-wire line is connected is used for transmission only. Ordinary transposition blocks introduce a certain amount of loss due to the radiation from the sharp bends in the wires. However, certain types of transposition blocks minimize these effects and only cause a minor loss in efficiency.

Twisted-Pair Non-Resonant Line: This type of transmission line as shown in Fig. 27, as well as the Johnson Q Feeders, have low losses. The losses in the twisted-pair feeders are exceptionally low, largely because the small spacing between the wires causes the line to have a very low characteristic impedance. Normally such a transposition line has an impedance in the neighborhood of 175 ohms. This means that for a given amount of power the voltage between the two wires is very low and thus insulation and dielectric losses can be kept to a minimum. Ordinary stranded lamp cord should be avoided because of high losses, but single-conductor No. 12 to No. 18 twisted is satisfactory. (NOTE: The special twisted-pair manufactured by the RCA company and sold in conjunction with the double-doublet all-wave antenna kits, is most excellent for twisted-pair transmission lines; this wire has a special low-loss high-frequency insulation and will transmit power up to as high as 1 KW). The impedance of twisted-pair line is approximately twice the center impedance of a one-half wave antenna and should be fanned out for approximately the last foot of its length where it taps onto the antenna, across the center insulators (see diagram). The taps are about one foot apart, forming an equilateral triangle of approximately one foot on each side out of the two feeders and the center portion of the antenna. Special 80 ohm twisted pair feeders can be connected directly in the center of the antenna, the triangle not being required.

This type of transmission line radiates so little that it can be run around corners, through walls and close to metallic objects with very little loss. Unfortunately, the line will not operate on harmonics.

The only competitor to the twisted-pair non-resonant line is that of the concentric-tube type of two-wire line, which is very expensive to construct. The Johnson Q uses a special quarter-wave matching transformer to couple a more or less conventional 200 to 600 ohm two-wire line to the 72 ohm impedance which exists at the center of a half wave Hertz antenna. This matching transformer consists of two parallel aluminum tubes, each a quarter wave in length. These low-loss resonant Zepp feeders are suspended directly from the center of the antenna. The manufacturers of the Johnson Q System recommend that if their instructions are followed closely, the losses can be held to a very low value. The system is not a Zepp antenna system. The antenna coil should not be tuned because the complete network must be non-resonant for proper operation.

Concentric Transmission Lines: Untuned concentric transmission line (or coaxial cable) is the most satisfactory means for carrying radio-frequency power over any great distance. It has the advantages of low-loss, complete shielding, simple installation without insulation, and easy adjustment. It is weatherproof and may be buried underground or carried up elevator shafts or wire ducts. The small size of the line is so flexible that it may be installed almost as easily as rubber-covered power cable. Objects near the line do not detune it or introduce loss as with an open line. There is no danger from anyone touching the line. No radiation or pick-up can occur which is particularly important in a directional system.

Filling the line with dry nitrogen at slightly greater than atmospheric pressure is a desirable precaution on any outdoor installation, especially when the line is under ground. The dry gas prevents condensation of moisture inside the line.

Coaxial cables can be constructed without the use of nitrogen and function quite satisfactorily. These lines can be designed to match the center-point impedance of a half-wave antenna by considering the following: The surge impedance is given by the expression

$$Z_s = 138 \operatorname{Log}_{10} \frac{D}{d}$$

where D is the inside diameter of the outside tube, and d, the outside diameter of the inside conductor. The outer conductor may be grounded at any point. The inner conductor is insulated from the outside sheath by glass or isolantite beads which are placed at intervals along the line; the beads also furnish the necessary mechanical spacing. See illustration on page 267.

Coaxial feeders are suitable for ultra-short wave automobile or airplane antenna installations besides their application to commercial radio installations. The inner conductor, in the case of a quarter-wave car antenna, connects to the bottom end of the antenna rod while the outer conductor is grounded to the car chassis. The antenna impedance to be matched is approximately 35-ohms in the case cited, and is about 73-ohms for the average half-wave antenna.

Collins Pi Network: This type of network (see Figures 28 to 31) is placed between the final amplifier tank circuit and the load, whether it be single or two-wire, resonant or non-resonant transmission lines, or at the end of an antenna itself. It affords a very flexible means of transforming impedances where non-resonant lines are used and of balancing-out inductive or capacitive reactances when coupling to a resonant feeder system or to an antenna proper. Its principal advantage is that it practically eliminates any impedance mismatch between the transmission line (or end-fed antenna) and the plate tank of the final amplifier. It cannot correct a mismatch between the transmission line and the antenna, and is of no great value where everything is perfectly matched by either inductive or direct coupling. The single-wire type of Collins PI Network is particularly useful when a portable transmitter is used.

The circuit shown in Figure 33 evenly loads each side of the push-pull stage and only causes a very slight capacity unbalance, which is too small to appreciably affect the neutralization. The conventional plate tank condenser C1 of the push-pull stage is shown as a split-stator type, although its use is not essential to the antenna coupling system. L1 is the regular tank coil. L2 has about one-third as many turns as L1 and is exceptionally closely coupled to it. The coil may be wound inside or outside of the plate tank, although the interwound coil shown is to be preferred. It is impossible to obtain close enough coupling by placing L1 and L2 end-to-end, as is done with most Zepp and inductive antenna coupling systems.

Adjusting the PI Network: The PI network acts as a low-pass filter which does not cause an appreciable loss to the fundamental emitted frequency, yet practically eliminates radiation of harmonics. It is illegal to radiate harmonics and for this reason the PI network is of value. For high power operation the spacing of the plates in the tuning condensers should be wide enough to withstand several thousand volts when coupling to certain types of antennas. Normally, the condensers should have sufficient plate spacing to withstand at least 1,000 volts when the network is coupled to transmitters not having more than 100 watts output.

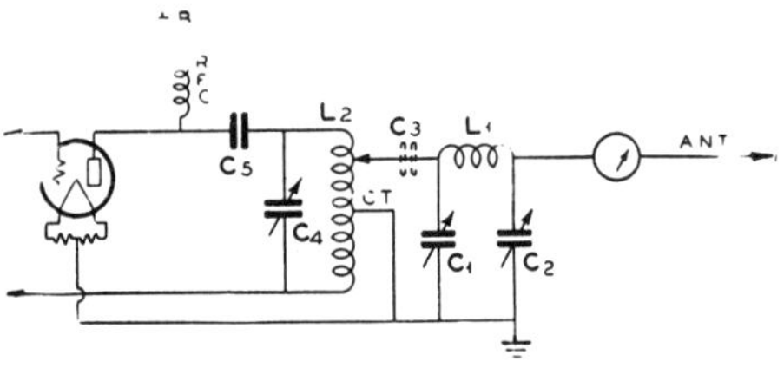

FIG. 28
Single-wire feed line—single section plate tuning condenser—shunt feed. C1 and C2 in all circuits (Fig. 1 to Fig. 3C) are .00035 mfd.

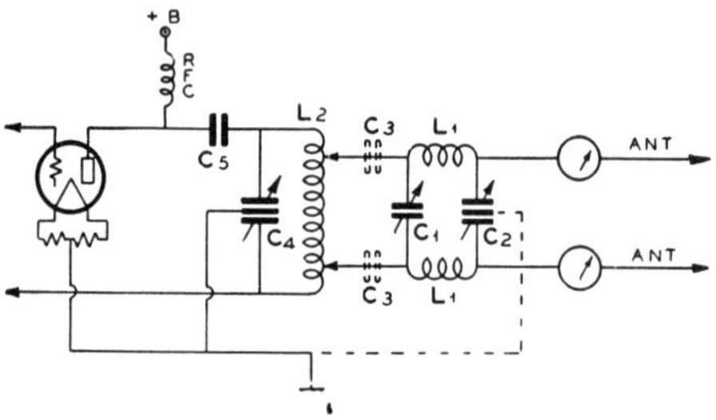

FIG. 29
Two-wire feed line from single-ended amplifier—split-stator plate tuning and optional split-stator used at C2. Shunt feed.

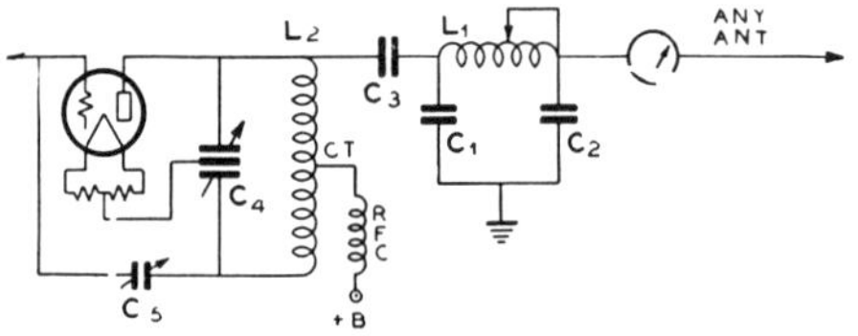

FIG. 30
Single-wire feed from end of low impedance output tube tank. Split-stator tuning and series feed. C1 and C2 should be variable.

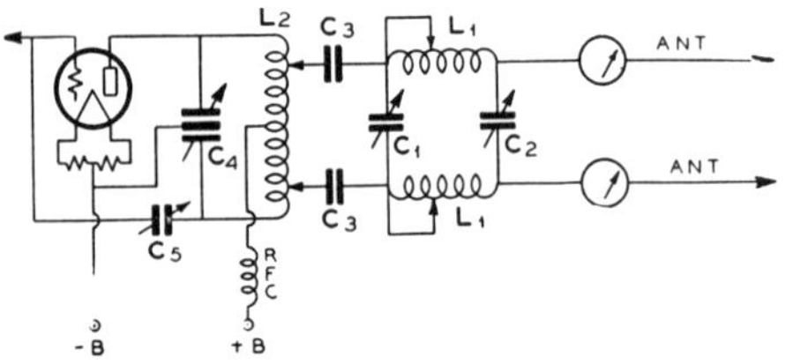

FIG. 31
Two-wire line from single-ended amplifier. Split-stator tuning and series feed.

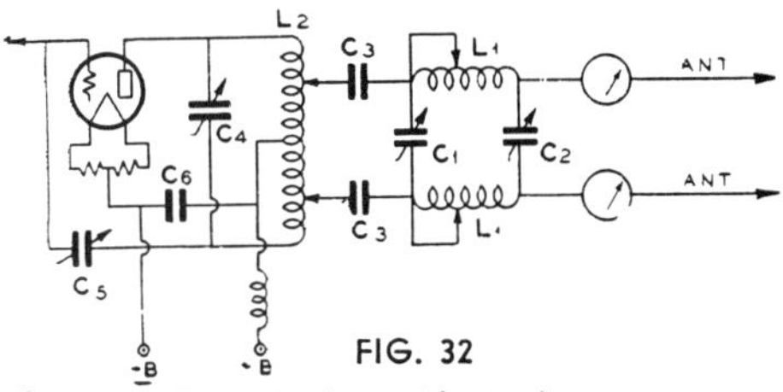

FIG. 32

Same as Fig. 33, but with single-section tuning condenser.

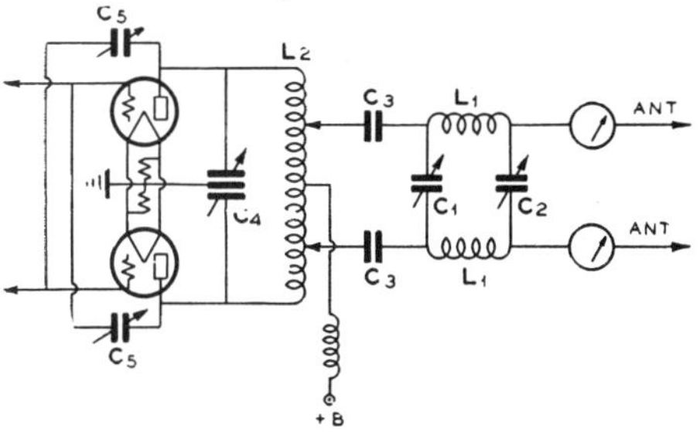

FIG. 33

Coupling a two-wire line to a push-pull final amplifier. Use of single-wire line out of a push-pull final through a PI network is not recommended.

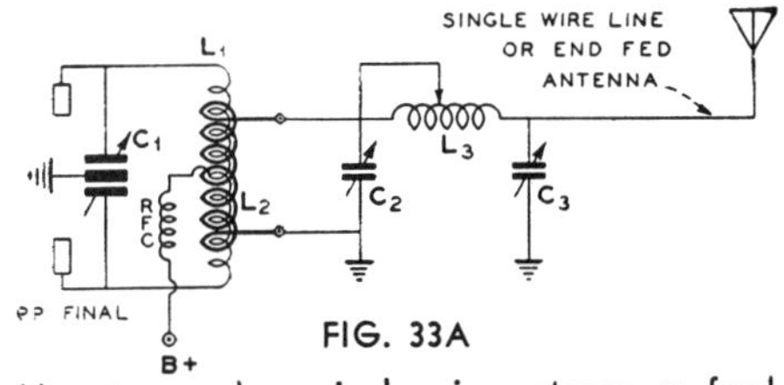

FIG. 33A

How to couple a single wire antenna or feed line to a push-pull final amplifier.

L1 and L2 should be interwound in order to load both tubes equally in a push-pull amplier.
L2—1/3 Tank Turns, interwound or otherwise very closely coupled.
L3—Standard Collins coil.
C2-C3—.00035 mfd. each.

The plate tank of the final amplifier must be tuned to resonance with the PI network disconnected from the final amplifier. The final amplifier must not be retuned thereafter. Then connect the PI network to the final amplifier and to the antenna. Tune the two variable condensers in the PI circuit until maximum antenna current (or feeder current) is obtained at normal values of final amplifier plate current. The PI network condenser which is closest to the final amplifier is used to obtain resonance in the Pi network for any particular setting of the load matching condenser (the nearest one to the antenna). The amount of inductance in the PI network coils must be determined by experiment to obtain best results.

Link Coupling the Final to the Antenna: Link coupling will increase the output in any stage of the transmitter as well as in coupling between the final amplifier tank circuit and the antenna. It is highly desirable to use some form of tuned circuit when coupling the antenna to the final tank circuit. For example, the feeders of a Zepp antenna must be tuned, and link coupling will simplify the mechanical problems involved. Another case is for the single-wire feeder to a Hertz antenna; here the feeder circuit should have an additional tuned circuit in order to minimize harmonic radiation.

Coupling a Zepp antenna ordinarily requires that two coils be placed at either end of the final tank circuit to obtain a balanced condition. This is an awkward method for varying the coupling and it is sometimes difficult to arrive at the proper number of turns in the coupling coils with respect to the shunt or series tuning condensers. However, by link coupling, the antenna tuning parts can be located at some convenient point several feet away from the transmitter. More output can frequently be obtained if impedance matching is correct, this being better accomplished when the loading coils are removed from the plate coil field. Coupling coils are usually wound with heavy copper tubing and thus there is an excess amount of metal with eddy current loss in the field of the plate coil. The use of a two to three-inch diameter plate coil of No. 10 or No. 12 wire, space wound, will in such cases give higher efficiency for low C circuits than heavy copper tubing coils at inputs as high as 1 KW. The Zepp tuned coil should be of copper tubing for high-power operation as the antenna current reaches high values.

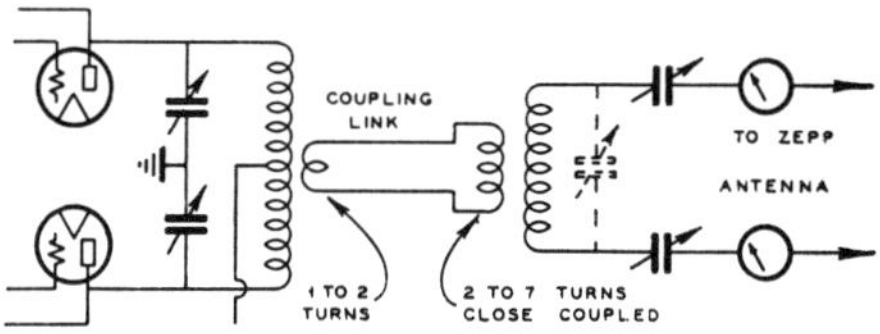

FIG. 34

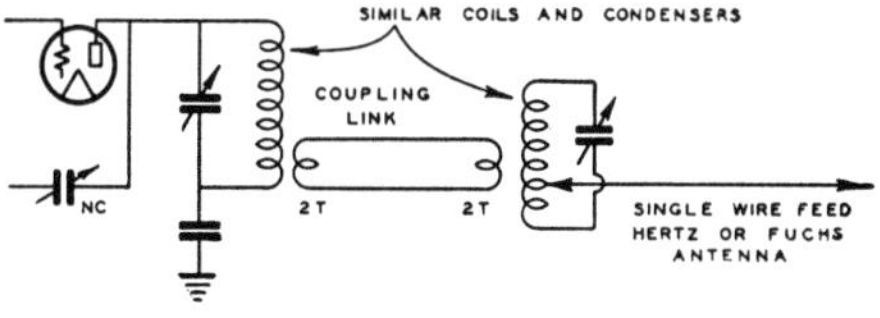

FIG. 35

The link coupling circuit may consist of No. 14 or No. 12 rubber-covered wire, twisted or parallel, with a one turn loop at the plate coil end and two or more turns in the loop at the antenna coil end. Since only inductive coupling is desired, the link coils should be tightly wound over the RF voltage nodal points of the coils, and provided with sufficient insulation to withstand the plate voltage. The number of coupling turns depends upon the closeness of the coupling and the ratio of impedances. The antenna coil is low and the link coil should have more turns, closer coupled, than the plate coil. This system is shown in Figure 34. With a single-ended final amplifier such as shown in Figure 35, the link coil should be at the RF ground potential —sometimes it is necessary to use two turns to secure sufficiently close coupling. It is difficult to obtain proper impedance matching with link coupling unless the link coils are very closely coupled to their tuned coils when the impedances of the tuned circuits are widely different.

Figure 35 shows a method for link coupling to either side of a single-wire fed Hertz antenna or to a Fuchs antenna. The latter is often brought into the station operating room. The system shown permits the antenna coupling circuit to be located near the lead-in point and link-coupled to the transmitter final amplifier. The additional tuned circuit allows impedances to be matched and greatly reduces harmonic radiation. This circuit, when properly adjusted, reflects a pure resistive load on the tube circuit, even for moderate variations of antenna characteristics, such as that due to vibration in the wind. For this reason it should be especially adaptable to self-excited oscillator systems.

The adjustment of the circuit in Figure 35 is simple because the impedances can be made nearly equal. One or two turns in the link coils is sufficient. The latter are located at the RF voltage nodes of the coils. The antenna or antenna feeder should be tapped-up far enough so that normal load is placed on the final amplifier tube when both tuned circuits are exactly in resonance. A simple field strength meter is very useful for indicating when the maximum power is being delivered into the antenna. This maximum power is obtained when the link coupling, antenna tap, and tuned circuit losses are all correct and as low as possible, respectively.

The Collins Multi-Band Antenna: This antenna is suitable for operation on several bands due to the use of a compromise system of r-f feeders. The losses in dry weather are exceptionally low, and even in wet weather this system is comparable to the Zepp. antenna. This new Collins system consists of a half wave antenna at the lowest frequency desired, with parallel copper tubing feeders connected in the center of the antenna. The copper tubing feed line is supported by means of insulating bars. Two copper tubes, each ¼-inch in diameter, are spaced 1½ inches apart and held in position by means of ceramic blocks. These blocks are located on the feed line at intervals of about 20 inches. The characteristic impedance of the feed line is 300-ohms. This is the geometric mean between 75 and 1200-ohms, the center impedance of the antenna when it is used at harmonic frequencies.

By making the feeders a multiple of quarter wavelengths long, the reactance at the station end is negligible and it will provide a resistive impedance of 75 or 1200 ohms. A simple untuned pick-up coil with variable number of turns is suitable for coupling to the transmitter or receiver tuned circuits.

The design formulas are as follows:

$$\text{Antenna length} = L = \frac{(k - .05)\ 492{,}000}{f}$$

$$\text{Feeder length} = l = \frac{234{,}000\ m}{f}$$

L = Feet.
k = No. of half wavelengths.
f = Frequency in kilocycles.
m = Number of quarter wavelengths.
l = feet.

CHART FOR COLLINS MULTI-BAND ANTENNA

Antenna	A	B	C	D	E	F	G
Antenna Length in Feet	136	136	275½	250	67	67	103
Feeder Length in Feet	66	115	99	122	65	98	82½
Frequency Range in Megacycles	3.7—4.0 7.0—7.3 14.0—14.4	3.7—4.0 14.0—14.4	1.7—2.0 3.7—4.0 7.0—7.3 14.0—14.4	1.7—2.0 3.7—4.0	7.0—7.3 14.0—14.4 28.0—29.0	7.0—7.3 14.0—14.4 28.0—29.0	3.7—4.0 7.0—7.3 14.0—14.4
Nominal Input Impedance in Ohms	1200 all bands	75 all bands	1200 160—80—20M., and 75 on 40M.	1200 all bands	75 on 40M. 1200 on 20 and 10M.	1200 all bands	1200 all bands

The impedance mismatch at the antenna is not very serious, being 4-to-1, but by the use of resonant lengths of feeders the actual efficiency of the feeders runs above 97% for moderate lengths. The feeders weigh about 10 pounds (an average for the vertical portion), and they hang from the center of the antenna. The antenna wire should therefore be of hard-drawn copper or steel core wire under tension, in order to prevent undue antenna sag.

160-Meter Coupling Systems: A simplified PI coupling system is shown in the Figure below. The 150 mmfd. and the 500 mmfd. variable condensers are effectively in series, through the common chassis ground connection. The advantages of this arrangement are: (1) there is no DC on the tuning condensers and the condensers will not flash-over on modulation peaks; (2), there is freedom from filter and rectifier trouble; (3) closer spaced tuning condensers can be used; (4) ample leeway for the tuning circuit because large variable condensers are used. The plate coil L1 consists of 60 turns of No. 20DCC wire, close wound. on a 2-inch diameter form, tapped at the 40th, 50th and 60th turn.

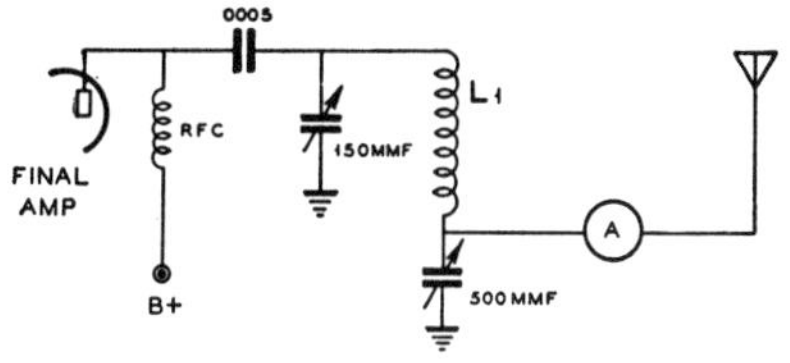

FIG. 36—A simplified antenna coupling system. The 500 mmf. condenser is an ordinary receiving type variable condenser; the 150 mmf. condenser is of the high-voltage type.

Figure 38 shows the common inverted-L Marconi antenna using parallel tuning of the pick-up coil. Figure 39 shows the same antenna in a T-form, instead of an inverted-L. Practically all 160-meter antennas are of the quarter wave type and are similar to those used in the broadcast band for either transmission or reception.

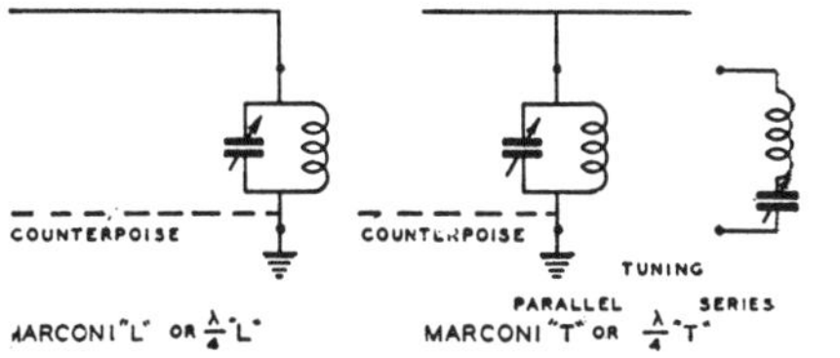

FIG. 38

FIG. 39

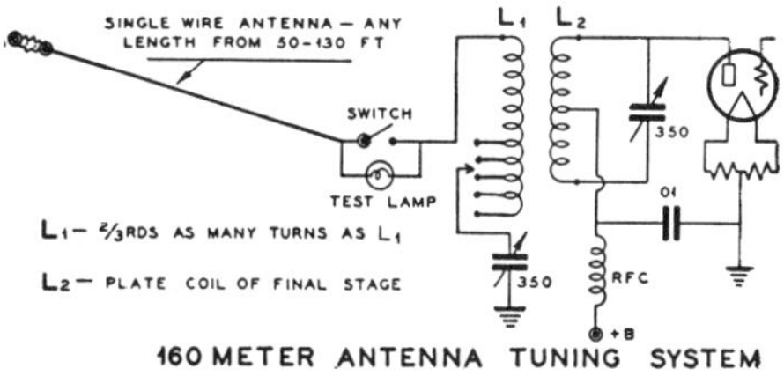

FIG. 40—The circuit shows the use of a tuning lamp in series with the antenna and a shorting switch for bridging the lamp after the antenna is tuned. A better method is to merely wrap a turn or two of wire around the lead-in wire and connect the ends of the loop to the lamp. The lamp can then be left permanently in the circuit.

FIG. 37—How to Use a 40 Meter Zepp Fed Hertz on 160 Meters

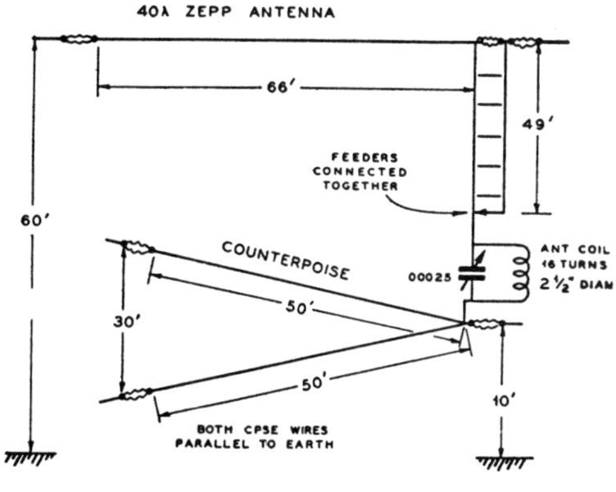

The illustration shows a 40 meter Zepp fed Hertz antenna for operation in the 160 meter band. A counterpoise, about 10 feet above earth, completes the circuit to ground and makes a Marconi, or quarter wave grounded antenna out of the combination. The Zepp feeders are connected together and attached to the tuning condenser and to one end of the antenna coupling coil, as shown. The other end of the coupling coil connects to the counterpoise. If the feeders are not of the same length as those shown in the diagram, the number of turns on the coupling coil must be changed in order to establish resonance. The coupling coil should be loosely coupled to the tank circuit of the transmitter. For 40 meter operation, the Zepp feeders are adjusted in the usual manner with the coil and condensers, and the counterpoise is not used.

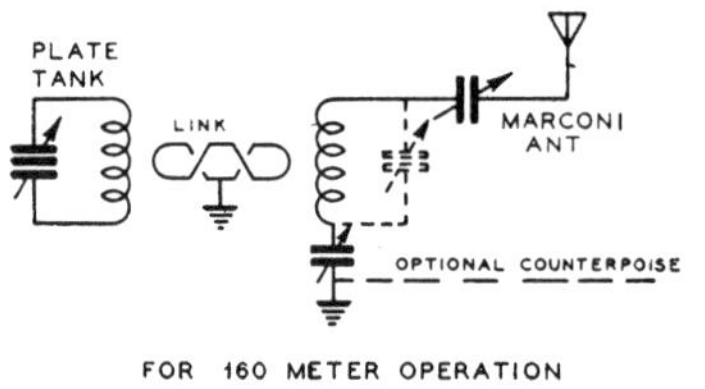

FIG. 41—Link-coupling the plate tank to a 160-meter Marconi antenna.

Figures 40 and 41 show other methods of adjusting and coupling the quarter wave, or Marconi antennas. The choice depends largely upon the individual location. It is always desirable to keep the lead-in and the coupling coil remote from all house wiring and metal objects to minimize losses. Figure 41 shows a feeder system which can be used to isolate the lead-in and coupling coil from the transmitter and metallic objects. The system shown in Figure 41 has probably the lowest losses. Any grounds can be replaced by a counterpoise.

Notes: A Marconi antenna for 160-meters can be adjusted by using series tuning to ground or counterpoise. This requires a tapped antenna loading coil, and a series condenser of from .00025 to .0005 ufd. maximum. Resonance is obtained by switching taps and varying the condenser until the antenna loads the final stage plate current to its normal values. If this value is more or less than the rating of the tube, the coupling between the loading coil and the final tank coil should be increased or decreased.

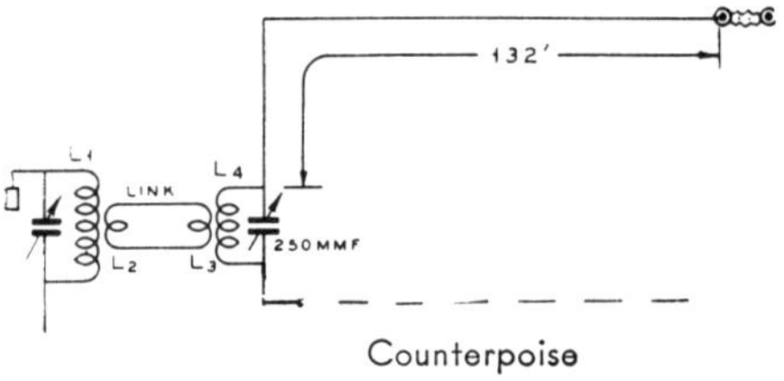

FIG. 42—160-meter antenna system with short counterpoise. L4 should have fewer turns than L1. L2 and L3 each have 2 to 4 turns.

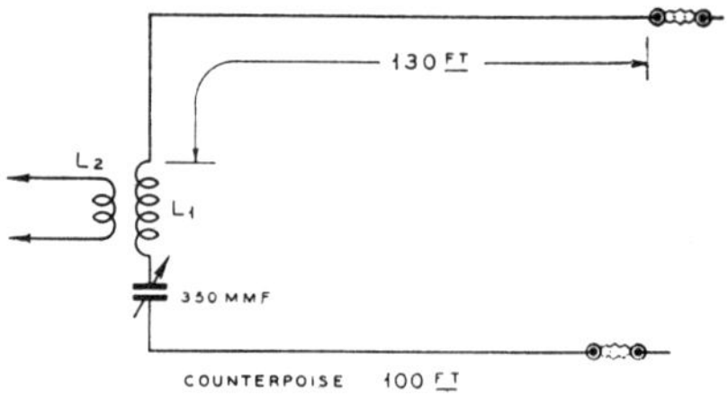

FIG. 43—160-meter series tuned system. L1 should have fewer turns than the plate coil L2, to which L1 is loosely coupled.

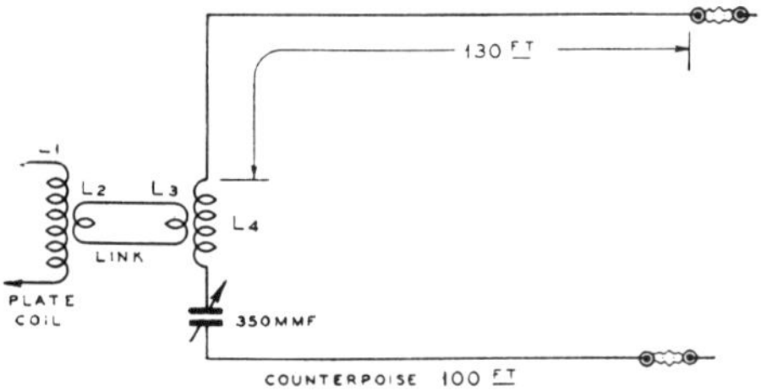

FIG. 44—160-meter series tuned system with link coupling of 2 to 4 turns between L1 and L4.

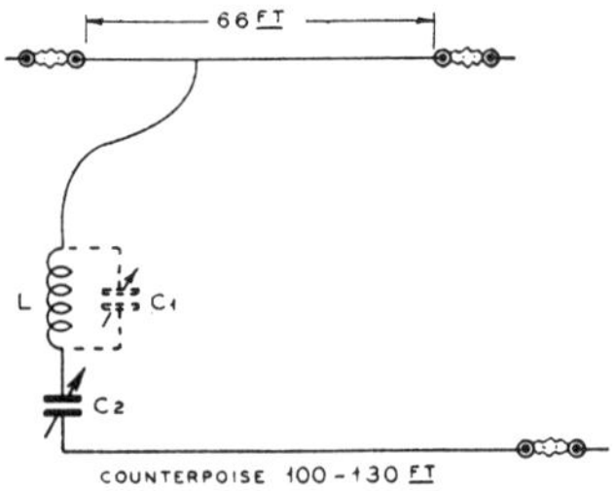

FIG. 45—40-meter single wire fed Hertz for 160 meter operation. If "L" has sufficient turns, C1 is not required. "L" is coupled to the plate tank circuit.

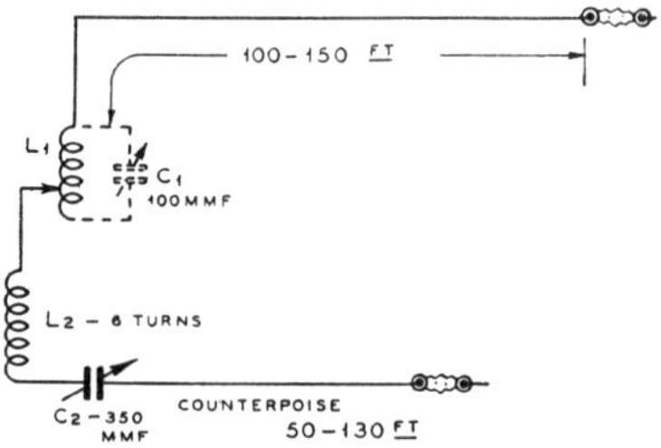

FIG. 46—160-meter system with separate tuned circuit L1 and C1. L2 is loosely coupled to the plate tank circuit.

Single Wire Feeder Antenna for All-Wave Operation:

The time-honored value of 18¾ feet off center for attaching the feeder to a single wire antenna for 80 meter operation, and the value of 9 feet 4 inches off center for a 40-meter antenna are suitable for operation on one frequency only. (See "Hertz Single Wire Matched Impedance Line" previously described). These values, on the other hand, are very poor if the antenna is to be used on harmonics. The figures are based on the approximate formula that the single wire feeder will have no standing waves on it if it is tapped to the half wave flat top at a point which is 1/7th of the antenna length, from center. The figure of 18¾ feet from center for a 132-foot antenna places the feeder tap 15¼ feet from the voltage node for 40 meter operation, and 1¾ feet for 20 meter operation.

SINGLE WIRE FED ANTENNA FOR ALL BAND OPERATION

ANTENNA FOR 80, 40, 20 & 10 METERS IS 134 FEET LONG, WITH SINGLE WIRE FEEDER TAPPED 22 FT. OFF CENTER. THE FEEDER CAN BE EITHER 66 OR 132 FEET LONG.

ANTENNA FOR 40, 20 & 10 METERS IS 67 FT. LONG, WITH FEEDER TAPPED 11 FT. OFF CENTER. THE FEEDER CAN BE 33 66 OR 99 FEET LONG.

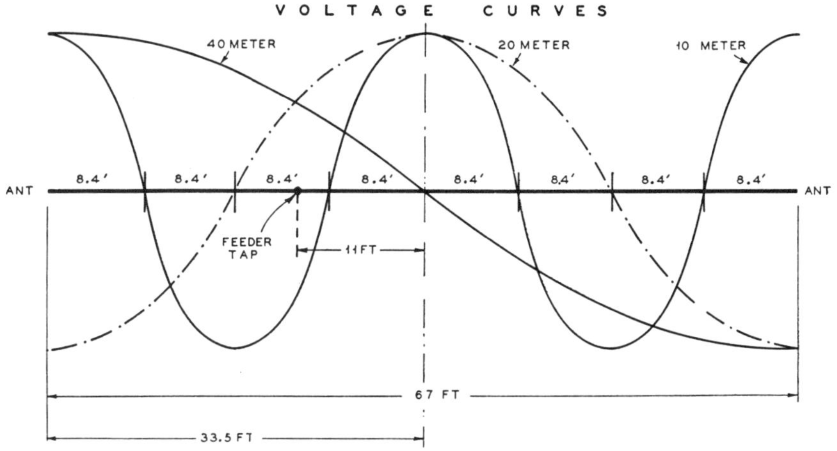

These values are nearly double, and less than half the correct values respectively, consequently the standing waves would be very bad, poor impedance matching and considerable loss of power will result.

A compromise of approximately ⅛th, instead of 1/7th, will give 22 feet on 80 meters for a 135-foot antenna, 12 feet on 40 meters, and 5 feet on 20 meters. These values are such that good impedance match will result, and satisfactory all-band operation will be obtained. For a 67 foot 40 meter antenna, the feeder should be tapped on the flat top at a point about 11 feet off center, which gives 6 feet from the node on 20 meters and 2½ feet from the node on 10 meters. The former value of 9 feet 4 inches, now proved erroneous, gives a distance of only 0.8 feet from a node for 10 meter operation, which explains the generally poor results obtained when such 40 meter antennas are used for 10 meter operation.

The reactive impedance effect of the feeder, being slightly off its point of correct impedance match, can be practically eliminated by making the feeder some multiple of quarter wavelengths long. At the station end, the impedance would then be purely resistive. The formula for calculating the feeder length in feet would be:

$$\frac{234{,}000}{f_1} = l,$$

where f_1 = lowest frequency used in kilocycles. The antenna length should be cut so that it will resonate at the middle frequency band desired, from the formula:

$$\frac{(k - .05)\,492{,}000}{f_2} = L \text{ in feet}$$

k is the number of half wavelengths and f_2 is the frequency in kilocycles. The slight error in length for the lower and higher frequencies must be tolerated because the actual length is a compromise.

A typical antenna for 80, 40, 20 and 10 meters would be 134 feet long, with the feeder tapped to the antenna at a point 22 feet off center, the feeder being either 66 or 132 feet long. Similarly, a 40, 20 and 10 meter antenna would be 67 feet long, the feeder tapped to the antenna at a point 11 feet off center, the feeder being either 33, 66 or 99 feet long. Two similar antennas, placed at right angles to one another, will provide reception or transmission in nearly all directions.

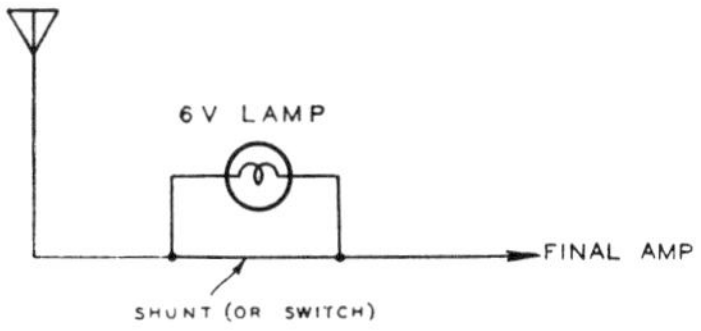

Inexpensive antenna current indicator for single wire fed Hertz antenna.

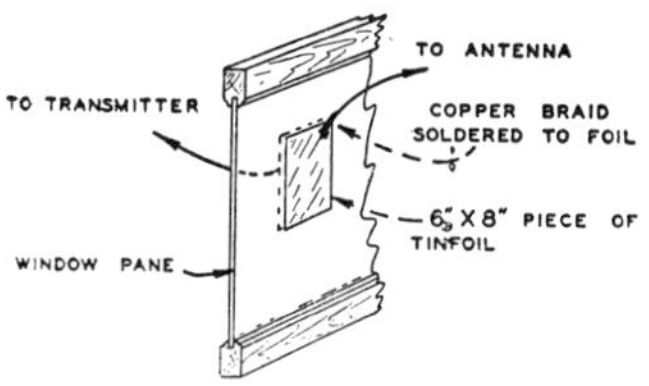

Method for use with receiving antenna or low-power transmitter lead-in. A sheet of tin foil on each side of the window pane acts as a condenser of low impedance.

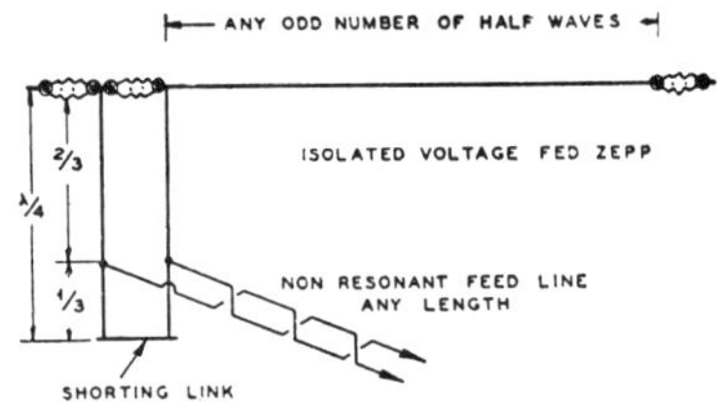

System for matching a non-resonant long RF feeder to the antenna by means of a quarter wave resonant stub line. The location of the non-resonant feed line taps on the quarter wave stub should be adjusted for equal currents each side of the connection to the stub. For transmitting, the non-resonant line should not be transposed. It should have no standing waves; i.e., unequal values of current along the line. The quarter wave stub should have the shorting link adjusted for exact resonance at the desired transmitting frequency.

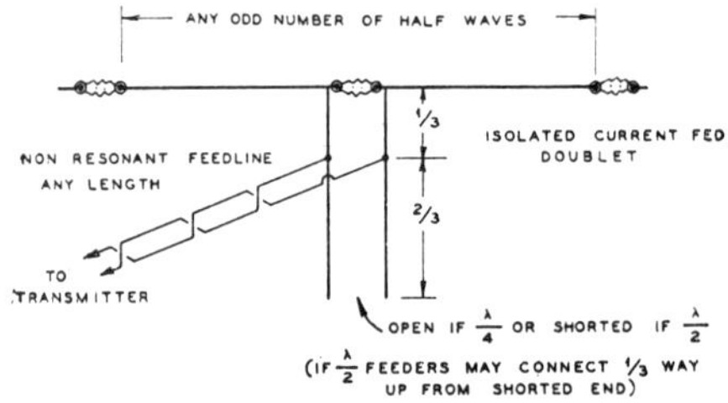

This system is similar to the one shown above, and the adjustments are made in like manner. The length of the stub (or building-out portion) is critical. The flat-top, plus the building-out section, must be in exact resonance with the transmitting frequency. The transposed non-resonant line reduces noise pick-up if the antenna is used for receiving, but it introduces irregularities when the system is used for transmitting.

10-Meter Vertical Antenna with ¼-Wave Matching Transformer: The present active solar cycle has brought the 10-meter band to the fore. Many experimenters prefer to use specially designed antennas for 10-meter operation. The antenna shown in the diagram is particularly adaptable for this type of work. It consists of a 25-foot pole, supported on the roof or to one side of a building or other structure, with a 16.5-foot vertical antenna wire run up along side the pole and insulated from it with a number of small insulating strips or rods. At the bottom of the 16.5-foot section, is another section of two wires, called "the matching transformer." These wires are 8-feet long, one of them being a portion of the antenna proper. A shorting

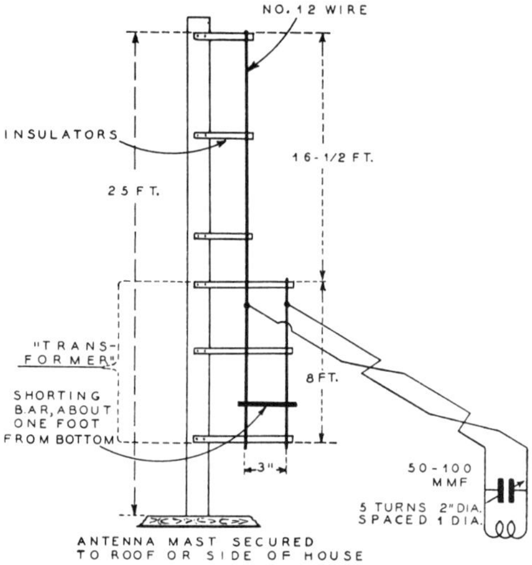

10-METER VERTICAL ANTENNA WITH QUARTER WAVE MATCHING TRANS.

bar, connected across the bottom of the two wires, see diagram, is moved upward or downward for antenna tuning. Likewise, the feed line, tapped on the two wires at a point about one-third the way down from where the two-wire portion begins, is also later adjusted and readjusted to tune the system.

Tuning Procedure:

(1) Place transmission line ⅓rd the way down from the point where the two wires begin, that is, ⅓rd the way down from the top of the "matching transformer".

(2) Adjust the shorting bar by placing it approximately 1 foot or 18 inches from the bottom of the "matching transformer".

(3) Turn "on" transmitter, and loosely couple the antenna coil to the final amplifier plate coil.

(4) Place a "field strength meter" (described in this section) somewhere where it can be seen from the roof, or let someone else watch the reading of the meter.

(5) Never re-adjust the field strength meter once it is set, while the antenna is being tuned.

(6) Take readings on the field strength meter and adjust the antenna coupling to the instrument so that half scale readings are obtained.

(7) Return to the roof, put on a pair of gloves, and adjust the shorting-bar until the field strength meter denotes maximum reading.

(8) Next, adjust the position of the feed-line to a point where maximum indication is again had on the field strength meter.

(9) Lastly, re-adjust the shorting-bar so that a more accurate position can be found, as again denoted by still greater reading of the field strength milliammeter.

A Field Strength Measuring Set: A field strength meter is a very useful instrument for determining maximum efficiency from a transmitting set and antenna. It actually gives an indication of the power in the antenna and thus it is more reliable than reading values of RF current in the antenna or feeders. A method for using the field strength meter is to connect a short antenna wire to the coil, L1, as shown in the circuit diagram shown here.

Coil L1 is a standard plug-in coil, wound like an ordinary short-wave receiver coil, but with a tap for the antenna taken at a point one-third the way up from the bottom end of the winding. This coil is tuned with a midget variable condenser of about 100 mmfd. The tube used in this field strength meter is a type 30 (2 volt filament) with the plate and the grid of the tube tied together. The MA meter is a 0-1 MA DC milliammeter.

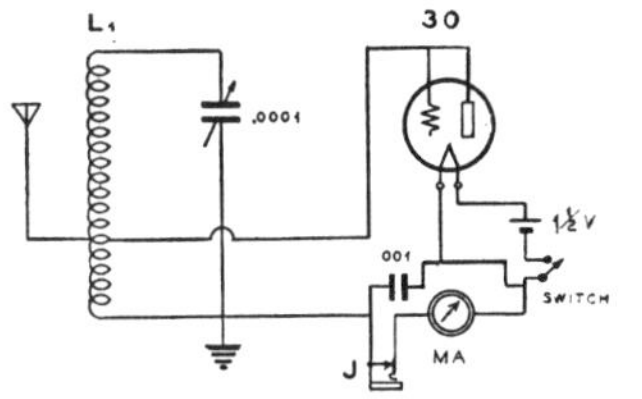

The extremely simple circuit of the Field Strength Measuring set.

Circuit diagram of Field Strength Test Set, a most useful and practical piece of equipment. It is far better to rely on this test set for measuring RF current in the antenna than to resort to the use of the conventional RF thermo-ammeter in the antenna circuit. L1 is an ordinary receiving type plug-in coil, 1½-in. dia. Here is the coil-winding data: For 80 and 160 meters—Wind 63 turns of No. 22 DCC or DSC, close wound; tap at 20 turns from the bottom end of the winding. For 40 and 20 meters—Wind 12 turns, space wound to occupy a winding space of ¾-in.; tap at 4th turn from bottom. For 5 and 10 meters—Wind 2 turns, spaced ½-in. apart, and tap at the exact center. The test set and battery must be encased in a shield can.

To make field strength measurements, the instrument is placed either in the radio room or somewhere in the immediate vicinity. An antenna a few feet above the ground is entirely suitable for use with this instrument. The transmitting system is tuned for maximum reading on the milliammeter. The field strength meter should be tuned to resonance with the transmitter frequency. This meter is also useful as a phone monitor, or as a check for key clicks. For these purposes an ordinary headset is plugged into the telephone jack J. The type 30 tube acts as a diode and it provides sufficient emission from a single 1½-volt dry cell to actuate the meter, or a headset at low volume. The diode is only connected across part of the tuned circuit and thus the actual selectivity is considerably higher than the usual connection where the diode is connected across the entire tuned circuit.

Field Strength Measuring Set, front and interior views.

Field-Strength Meter for 10-Meter Adjustments: The simple field-strength meter shown in the circuit diagram (Page 264) is equipped with coils for 10-meter operation. To adapt the instrument for other wavelengths necessitates a change in coils. In addition to its regular use, the meter will function as an neutralizing indicator.

A 0-1 MA DC milliammeter is connected in series with coil L1 and a carborundum detector. The maximum deflection of the meter denotes maximum antenna radiation.

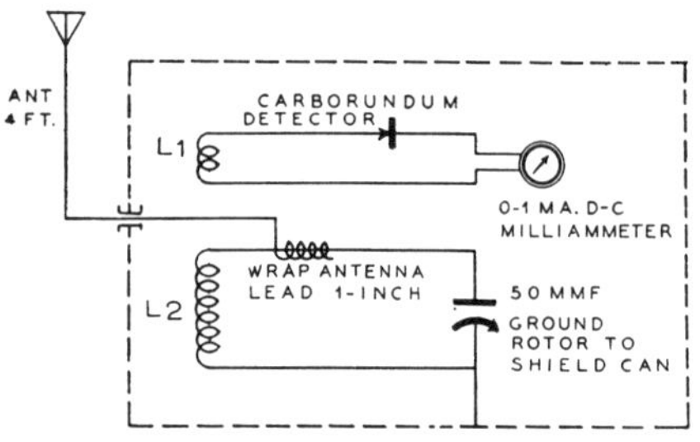

FOR 10-METER OPERATION, L1 HAS 2 TURNS, 1-1/2" DIA. L2 HAS 5 TURNS, 3/4" DIA. L1 IS SLIPPED OVER L2 OR LOOSELY COUPLED.

10-METER FIELD STRENGTH MEASURING SET

The length of the antenna wire, usually about 4 feet, and the proximity of the instrument to the antenna determines the amount of meter deflection. The antenna can be shortened or lengthened, as required, so that the indicator will be about half scale position when preliminary readings are taken.

The instrument should be encased in a metal can for best results. The antenna wire is twisted around the top lead of coil L2 for a distance of about 1 inch. The range of the meter deflection is varied by increasing or decreasing the number of turns.

Choosing Antennas for Various Locations: Several types of antennas will give satisfactory results in a given location. In brief, the suggestions offered below will be of help to the experimenter in choosing a good antenna for his particular location.

For those who reside in densely populated areas, such as apartment house districts, hotels, and so forth, a satisfactory antenna for operation from 80-meters down, would be a half wave antenna with either single wire or Zepp feeders. If the transmitter is located on the top floor it is often possible to use an end-fed half, or full wave antenna. For 160-meter phone operation a quarter wave (from 75 to 132 feet) antenna and a one or two wire counterpoise is desirable. If the radio room is on the lower floor of an apartment house, hotel, etc., and if one band operation only is desired, the problem merely resolves itself into the use of a twisted-pair RF feeder to a half wave doublet antenna. This twisted pair reduces losses in transmission and the feeders can be run close to walls, down light-wells, around corners, etc., without appreciable effect on performance. If more than one band operation is desired, two antennas with twisted-pair feeders should be used, one for each band, because a half wave doublet antenna with twisted-pair feeders operates successfully only on one band.

For the operator who wishes to work on both the 75 and 20 meter phone bands from a single antenna, a satisfactory compromise is to use a very long antenna operated on one of its harmonics, end or single wire feed at the transmitter. Some multiple of half wavelengths is the proper antenna length to use.

For the operator who is to operate only in the 20-meter band, either a vertical half wave antenna or two separate horizontal antennas, placed at right angles to each other, with some form of two-wire feeder, will be satisfactory.

Antennas For Receiving

Noise Reduction Systems: The only principle which has been successfully employed for the reduction of man-made static is to locate the antenna in a comparatively noise free area and to employ a lead-in of such a type that pick-up on the lead-in is eliminated. To place the antenna in a noise free location is a unique problem for each installation. However, the type of lead-in is an important design problem. There are two general types—the shielded lead-in and the balanced transposed line. The shielded line is unsuitable for high frequencies because to be effective, the shielding must be grounded every few feet with short ground wires. This is obviously impossible in most installations.

The balanced line, however, is eminently suitable for many reasons. When used in conjunction with a well-designed transformer at the set, pick-up on the line is almost completely eliminated. No grounding is necessary. Losses are lower than in a shielded line.

In designing the line the space between the wires and the size of the wires is important. The farther apart they are, and the smaller they are, the higher is the characteristic impedance of the line. If a line is terminated at each end with its characteristic impedance, its transmission is nearly constant at all frequencies. However, when the terminating impedances are widely different from the proper value, the transmission varies greatly with frequency, the curve passing through a series of peaks and valleys corresponding to resonance points in the line.

For the RCA World-Wide Antenna System a line having 180 ohms impedance was chosen because this value is about the average input impedance of most short-wave receivers and because it is about the average impedance of the double doublet antenna over the short-wave frequency spectrum.

Because the antenna does not represent an impedance exactly equal to the line impedance at all frequencies, the transmission curve does have a series of minor peaks and valleys, varying in efficiency two or three-to-one. The line length was adjusted experimentally by throwing lengths in and out of the circuit, until a length was found such that a transmission peak occurred at each one of the important short-wave broadcasting bands.

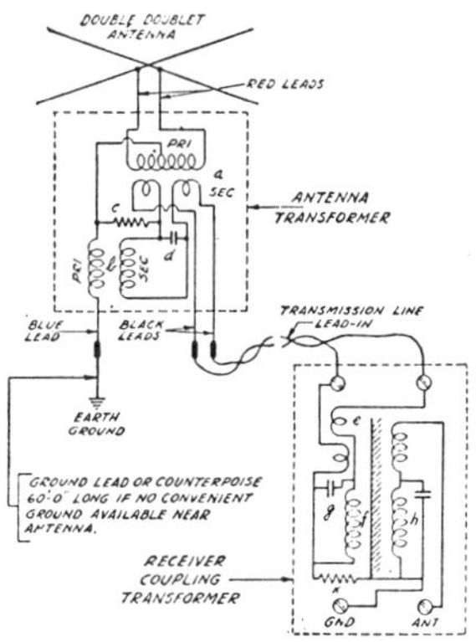

FIG. 46—Improved R.C.A. World-Wide antenna system.

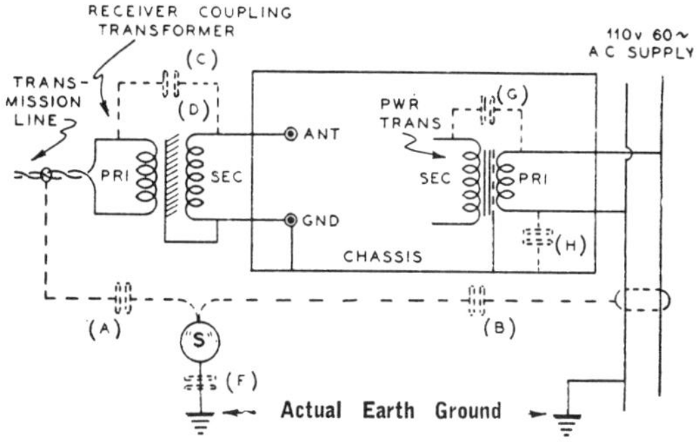

FIG. 47

Mechanically, the line consists of a rubber-covered twisted-pair with stranded tinned copper wire for each conductor. After exhaustive tests submarine cable rubber was specified for insulation of the transmission line due to its low losses and high natural rubber content. The life of this transmission line is materially increased by the use of this high quality rubber insulation. While twisted-pair was indicated to produce a line of the proper impedance, it is also important that the wires be close together with frequent transpositions to avoid picking up out of phase signals.

In order to keep the losses low when the line is wet, it is important that no cotton be used as insulation. Even when a cotton wrap is well impregnated, the impregnating material soon evaporates away and moisture then gets in, increasing the line losses.

It is very important to note that the noise eliminating feature of the system depends entirely on the design of the transformer which couples the line to the set.

The purpose of this transformer is to eliminate in-phase signals while transmitting out-of-phase signals. The expression "in phase" means that the voltages of the two sides of the line go positive together and then go negative together. Obviously, this type of signal will produce no current in the primary of the transformer, it simply changes its potential. "Out-of-phase" signals are those which cause one side of the line to go negative when the other goes positive and then the reverse. This type of signal does not produce primary current. The mere presence of a transformer does not eliminate the in-phase signals (or noise), because if there is capacity coupling, the noise will be transmitted to the set through that capacity.

In the transformer under discussion a special and highly efficient static shield is used, completely eliminating capacity coupling. As a result, the in-phase signals and noise picked up by the line are eliminated while the out-of-phase signals picked up by the antenna are transmitted to the receiver.

The circuit diagram of the complete antenna system is shown in Figure 46.

When choosing a noise-free area to locate the "double doublet" antenna, it is well to consider the accepted theory that the strength of noise interference varies inversely as the square of the distance from the source of noise.

The receiver coupling transformer of the system completely eliminates automobile ignition noise. This is explained by referring to Figure 47.

"S" represents a signal generator such as a source of auto ignition noise; (a) the capacity coupling from "S" to the transmission line; (b) the capacity coupling from "S" to the power supply line; (h) the capacity coupling from one side of the power supply line to the metal chassis; (f) the capacity coupling from "S" to actual earth ground.

(A) The noise voltage that would be induced by capacity coupling (a) into the transmission line would correspond to an "in-phase" signal and therefore would be coupled or fed through to the secondary of the receiver coupling transformer by the capacity (c) if this capacity (c) were not eliminated by the special and highly efficient electrostatic shield (d). If it were not for shield (d) a noise voltage would be developed across "ANT" and "GND" of the receiver due to a completed circuit from "GND" to chassis frame through "h" to the power supply line which is usually grounded on one side and thence back to "S" through (f).

(B) The noise voltage that would be induced by capacity coupling (b) causes current to flow through the power transformer and develop a noise from ground to the chassis through capacity (h). If no receiver coupling transformer was used this voltage would occur across ("ANT" and GND") the input terminals of the receiver and hence cause noise. When the RCA antenna system is used including the receiver coupling transformer this voltage occurs between the primary and the electrostatic shield, since capacity (c) has been eliminated.

However, this does not produce primary current. Therefore this noise voltage does not induce a voltage in the transformer secondary.

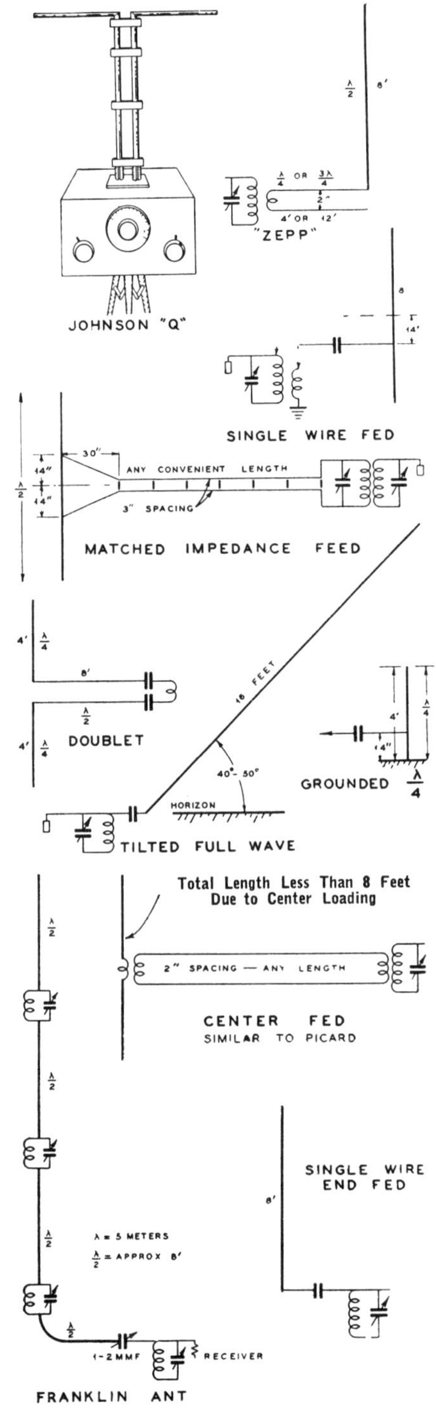

5-Meter Antennas.

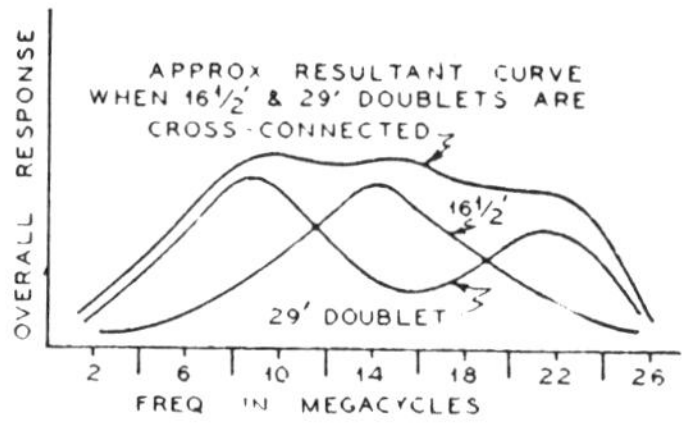

(C) The electrostatic shield provided with most power transformers serves to offset the capacity coupling (g) and thus prevents the introduction of RF noise voltages into the voltage supply of the receiver directly.

No doubt the above reasons (A) and (B) contribute a very real improvement in signal-to-noise ratio to be had with this system on auto ignition interference.

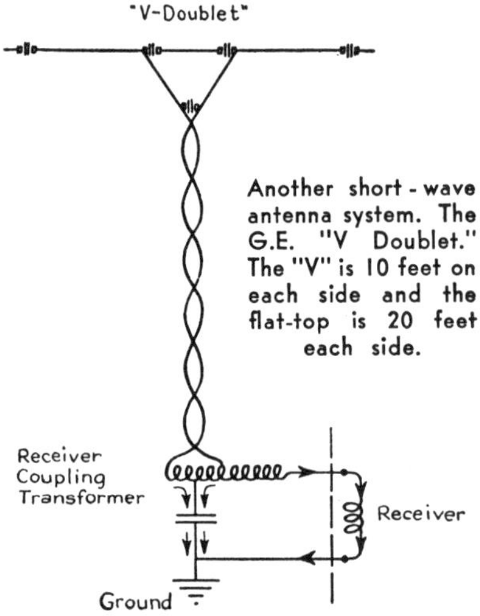

Another short-wave antenna system. The G.E. "V Doublet." The "V" is 10 feet on each side and the flat-top is 20 feet each side.

The Franklin Antenna System: A very effective ultra-short wave antenna is the Franklin 5-meter type which consists of a number of half wave sections with a resonant circuit between each section.

The Franklin antenna is very interesting in that the received signal can be doubled or more with a three section wire 24 feet long with two tuned circuit cut in at 8 foot intervals. These tuned circuits can be 6 turns of No. 10 or 12 wire, wound on a five-eighths inch form with little spacing between turns and tuned by a three or four plate midget tuning condenser. These coils are soldered directly across the condenser terminals and the 8 foot antenna sections terminate on these connections. These circuits are resonanted to five meters by previous adjustment when the coil and condenser are coupled like a wavemeter to a transmitter or receiver circuit. With an outside antenna, these tuned trap circuits must be protected against moisture. Quarter wave matching stubs or sections about four feet long can be used instead of small tuned circuits.

The purpose of these tuned circuits are to prevent phase reversal of standing waves of voltage and current in an antenna of several half wave lengths. The "phasing coils" reverse the phase without themselves radiating to any extent; the desired effect of a number of antennas all radiating in phase is obtained.

A full wave antenna, 16 feet long, without a phasing coil and condenser trap circuit has a radiation pattern like a shamrock or four leaf clover, without much energy being projected at right angles to the antenna. This radiation pattern should have a maximum in a direction parallel to the earth for 5-meter transmission or reception, so a 16 foot antenna can be used if it is tilted at an angle of 40 or 50 degrees toward or away from the desired directions. It will be more effective if tilted towards the desired direction since its upper "loop" would be parallel to the earth, and the effective height would be greater.

Any form of antenna can be used for five-meter work, even a wire several hundred feet long, but best results are obtained if the antenna is designed for five-meter use. The vertical halfwave antennas mounted on roof tops with two-wire matched RF feeders, or the simple Franklin antennas are by far the best for non-directional transmission and reception.

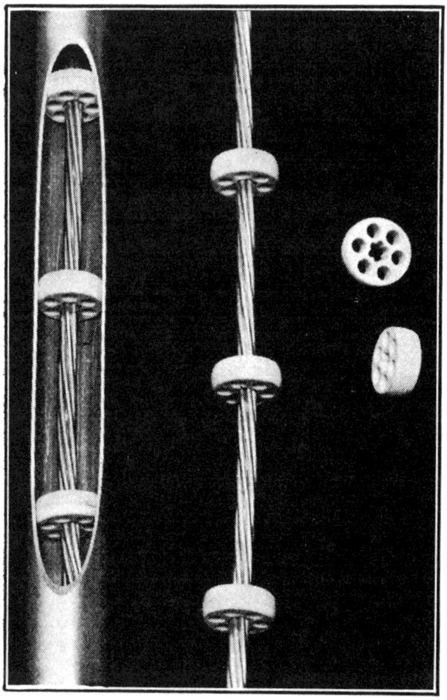

Concentric transmission line components, Heintz & Kaufman type.

5 Meter Receiving Antenna: A surprisingly good receiving antenna consists of an eight foot wire with its lower end coupled through a very small capacity to the grid circuit of the receiver. This type works well in any type of building that has not been constructed with too much steel or reinforcement wire, such as used in stucco coated exteriors. Moving the antenna a few feet in a room will often increase the signal several fold due to reflective or directive effects of nearby objects, such as house wiring. If the antenna can be raised in a vertical position, or nearly so, excellent results will be obtained.

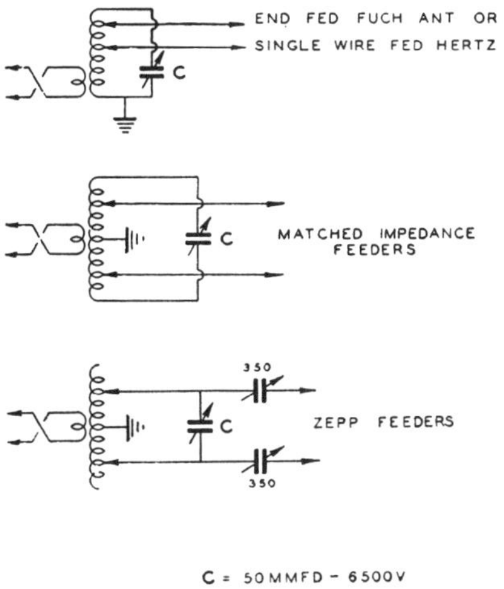

Three methods of link coupling the final tank coil of a transmitter to antenna tuning or impedance matching circuits.

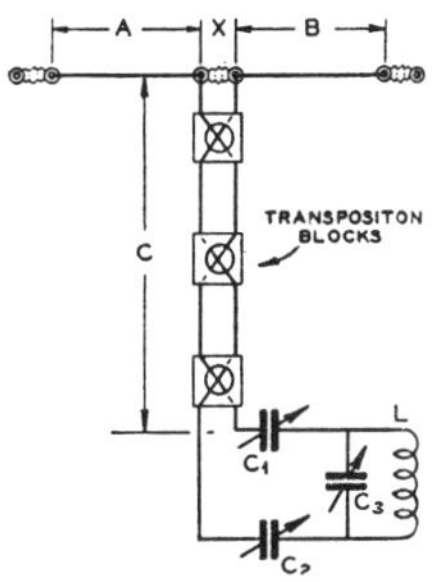

Doublet receiving antenna with transposed feeders for noise reduction. Average lengths for "A" and "B" are 33 feet each. "C" can be any length over 33 ft. "L" is loosely coupled to the receiver and the system is tuned to resonance with condensers C1, C2 and C3. The capacity of C1 and C2 is 350 mmf. each. C3 is 100 mmf. Coil "L" has 15 turns of No. 22 DSC wire, on a 1½-in. dia. coil form.

Guage No. B&S	Cross Sectional Area: Dia. in Mils	Cross Sectional Area: Cir. Mils	Cross Sectional Area: Sq. Inches	Turns per Linear Inch: DCC.	Turns per Linear Inch: SCC.	Turns per Linear Inch: DSC.	Turns per Linear Inch: SSC.	Turns per Linear Inch: Enam.	Turns per Linear Inch: Enam. and SCC.	Turns per Linear Inch: Enam. and SSC.	Turns per Square Inch: DCC.	Turns per Square Inch: SCC.	Turns per Square Inch: DSC.	Turns per Square Inch: SSC.	Turns per Square Inch: Enam.	Turns per Square Inch: Enam. and SCC.	Turns per Square Inch: Enam. and SSC.	Ft. per Pound: DCC	Ft. per Pound: SCC.	Ft. per Pound: Bare	Res per 1000 Ft.: Copper	Carrying Capacity, Copper: 1000 CM Per Amp.	Carrying Capacity, Copper: 1500 CM Per Amp.
0000	460.0	211600	.1662																	1.561	.0499	211.6	140.7
000	409.6	167800	.1318																	1.968	.0629	167.8	111.3
00	364.8	133100	.1045																	2.482	.0793	133.1	88.9
0	324.9	105500	.08289																	3.130	.1000	105.5	70.3
1	289.3	83690	.06573																	3.947	.1260	83.7	55.7
2	257.6	66370	.05213																	4.977	.1592	66.4	44.1
3	229.4	52640	.04134																	6.276	.2004	52.6	35.0
4	204.3	51740	.03278																	7.914	.2336	41.7	27.7
5	181.9	33100	.02600																	9.980	.3192	33.1	22.0
6	162.0	26250	.02062	5.44	5.60															12.58	.4028	26.3	17.5
7	144.3	20820	.01635	6.08	6.23															15.87	.5080	20.8	13.8
8	128.5	16510	.01297	6.80	6.94													19.6	19.9	20.01	.6405	16.5	11.0
9	114.4	13090	.01028	7.64	7.68													24.6	25.1	25.23	.8077	13.1	8.7
10	101.9	10380	.008155	8.51	8.55													30.9	31.6	31.82	1.018	10.4	6.9
11	90.74	8234	.006467	9.58	9.60													38.9	39.8	40.12	1.284	8.2	5.5
12	80.81	6530	.005129	10.62	10.80	11.8	12.1	12.1	11.4	11.8	121	136	139	146	146	130	139	48.9	50.2	50.59	1.619	6.5	4.4
13	71.96	5178	.004067	11.88	12.06	13.2	13.5	13.5	12.8	13.2	153	171	173	183	183	162	173	61.5	63.2	63.80	2.042	5.2	3.5
14	64.09	4107	.003225	13.10	13.45	14.7	15.1	15.2	14.2	14.7	187	213	216	229	239	201	216	77.3	79.6	80.44	2.575	4.1	2.7
15	57.07	3257	.002558	14.68	14.90	16.4	16.9	17.0	15.8	16.5	229	264	268	287	290	250	271	97.3	100	101.4	3.247	3.3	2.2
16	50.82	2583	.002028	16.40	17.20	18.2	18.9	18.7	17.6	18.4	280	327	333	358	350	309	338	119	126	127.9	4.094	2.6	1.7
17	45.26	2048	.001609	18.10	18.80	20.3	21.2	21.4	19.5	20.5	340	404	412	448	458	381	421	150	155	161.3	5.163	2.0	1.3
18	40.30	1624	.001276	20.00	21.00	22.6	23.6	24.0	21.7	22.9	412	498	510	559	575	469	524	188	196	203.4	6.510	1.6	1.1
19	35.89	1288	.001012	21.83	23.60	25.4	26.8	27.2	24.2	25.8	508	629	644	715	739	587	665	237	247	256.5	8.210	1.3	.86
20	31.96	1022	.0008023	23.91	26.40	27.8	29.5	30.1	26.5	28.4	596	752	773	867	904	701	805	298	311	323.4	10.35	1.0	.68
21	28.46	810.1	.0006363	26.20	29.70	30.8	32.8	33.6	29.6	31.5	752	949	949	1078	1129	878	991	370	389	407.8	13.05	.81	.54
22	25.35	642.4	.0005046	28.58	32.00	34.1	36.6	37.7	32.7	35.0	899	1161	1161	1337	1419	1071	1227	461	491	514.8	16.46	.64	.43
23	22.57	509.5	.0004002	31.12	34.30	37.6	40.7	42.3	36.1	39.0	1070	1416	1416	1656	1785	1306	1518	584	624	648.4	20.76	.51	.34
24	20.10	404.0	.0003173	33.60	37.70	41.5	45.3	47.2	39.7	43.1	1266	1722	1722	2048	2225	1575	1858	745	778	817.7	26.17	.41	.27
25	17.90	320.4	.0002517	36.20	41.50	45.7	50.3	52.9	43.7	47.9	1491	2085	2085	2525	2800	1907	2289	903	958	1031	33.00	.32	.21
26	15.94	254.1	.0001996	39.90	45.30	50.2	55.7	59.0	47.8	52.8	1745	2515	2515	3108	3484	2281	2788	1118	1188	1300	41.62	.25	.17
27	14.20	201.5	.0001583	42.60	49.40	55.0	61.7	65.8	52.1	58.1	2029	3019	3019	3811	4328	2713	3381	1422	1533	1639	52.48	.20	.13
28	12.64	159.8	.0001255	45.50	54.60	60.1	68.3	73.9	57.0	64.4	2347	3611	3611	4666	5456	3250	4141	1759	1903	2067	66.17	.16	.11
29	11.26	126.7	.00009953	48.00	58.80	65.5	75.4	82.2	61.9	70.6	2696	4294	4294	5688	6761	3830	4988	2207	2461	2607	83.44	.13	.084
30	10.03	100.5	.00007894	51.10	64.40	71.3	83.1	92.3	67.4	77.9	3076	5081	5081	6911	8527	4547	6075	2534	2893	3287	105.20	.10	.067
31	8.928	79.70	.00006260	56.80	69.00	77.3	91.6	103.0	72.8	85.3	3489	5981	5981	8389	10568	5305	7267	2768	3483	4145	132.70	.079	.053
32	7.950	63.21	.00004964	60.20	75.00	83.7	101.0	116.0	79.1	93.9	3931	7003	7003	10101	13363	6250	8818	3137	4414	5227	167.30	.063	.042
33	7.080	50.13	.00003937	64.30	81.00	90.3	110.0	130.0	85.6	103.0	4398	8143	8143	12130	16952	7326	10672	4697	5688	6591	211.00	.050	.033
34	6.305	39.75	.00003122	68.60	87.60	97.0	120.0	145.0	91.7	112.0	4883	9407	9407	14499	20967	8403	12610	6168	6400	8310	266.00	.039	.026
35	5.615	31.52	.00002476	73.00	94.20	104.0	131.0	164.0	98.8	123.0	5391	10817	10817	17247	26745	9766	15186	6737	8393	10480	335.00	.032	.021
36	5.000	25.00	.00001964	78.50	101.00	111.0	143.0	182.0	105.0	133.0	5917	12346	12346	20408	33051	11080	17777	7877	9846	13210	423.00	.025	.017
37	4.453	19.83	.00001557	84.00	108.00	118.0	155.0	206.0	113.0	146.0	6452	13996	13996	24015	40766	12758	21295	9309	11636	16660	533.40	.020	.013
38	3.965	15.72	.00001235	89.10	115.00	126.0	168.0	235.0	120.0	157.0	6978	15763	15763	28106	54990	14291	24685	10666	13848	21010	672.60	.016	.010
39	3.331	12.47	.000009793	95.00	122.50	133.0	181.0	261.0	128.0	172.0	7524	17630	17630	32690	68120	16308	29412	11907	18286	26500	848.10	.012	.008
40	3.134	9.888	.000007766	102.50	130.00	140.0	194.0	290.0	134.0	184.0	8045	19589	19589	37779	84246	18041	33727	14222	24381	33410	1069.00	.009	.006
41	2.800	7.841	.000006160	112.00	153.00													17920	30610	42130	1323.00	.008	.005
42	2.494	6.220	.000004885	124.00	168.00													22600	38700	53100	1667.00	.006	.004
43	2.221	4.933	.000003873	140.00	192.00													28410	48600	66970	2105.00	.005	.003
44	1.978	3.910	.000003073	153.00	210.00													35950	61400	84460	2655.00	.004	.0025

Copper Wire Table

Power Supplies

A source of high voltage direct current must be applied to the plate electrode of a vacuum tube in order that the tube may properly function. The power is usually obtained from the AC mains. It is then transformed, rectified and filtered to produce a uni-directional current.

All vacuum tubes are, in principle, rectifiers and can be used for converting alternating currents into direct currents. In general, there are two types of rectifiers known as the half-wave and full-wave types. In a half-wave rectifier, the tube passes one-half of the wave of each alternation and blocks the other half; thus current flows for half the time and drops to zero the other half of the time. This causes a very uneven voltage output, because it varies from zero to maximum 60 times per second. Half-wave rectifiers produce a pulsating uni-directional current having a large undulatory DC characteristic.

To minimize the pulsations, and to make the current flow in a more continuous manner, a full-wave rectifier is used. This type of rectifier consists of two tubes, each connected to one end of the secondary of a transformer with a grounded center-tap connection. When one end of the transformer winding is going through the most positive part of the cycle, with respect to the center-tap, the other end of the transformer is most negative. Therefore, when one tube is conducting, the other is inoperative (plate negative with respect to the cathode); one-half cycle later, the other tube conducts and the first is non-conducting, such is the process of full-wave rectification. The output voltages from the tubes are connected together through a common rectifier tube filament circuit, and thus the tubes alternately supply current to the output circuit which is connected between the filament winding of the tubes, and the center-tap of the high voltage transformer. The rectifier filaments are always positive in polarity.

Full-wave rectifiers deliver a direct current pulsation 120 times per second instead of one pulsation for each cycle, therefore, the variation in output voltage is less. However, the pure direct current required for the amplifier tube is not yet available, so that some form of a low-pass filter must be placed between the rectifier and load.

A low-pass filter is a device which selects and passes certain types of electric currents and rejects and by-passes other unwanted types. It should be remembered that a pulsating DC voltage might be considered as that of a pure DC voltage which has a somewhat smaller alternating voltage superimpressed on it. In other words, the combination of the two voltages is, in every respect, exactly similar to the pulsating DC voltage output of the rectifier.

All filters and filter operations are designed to select and reject alternating currents. The characteristic of the alternating current which enables the filter to function in the aforesaid manner is its frequency. A low-pass filter offers little impedance to the passage of alternating currents of low frequency, but materially impedes the flow of such currents of high frequency. DC can also be considered as AC of zero frequency, thus passes straight on through a low-pass filter to the load with little or no impedance. On the other hand, the pulsations, or ripple consists of an AC current having a frequency of 120 cycles per second (twice the mains frequency) which the low-pass filter prevents from reaching the load where it would make its presence known as hum. A low-pass filter generally consists of two elements: an inductance, or choke coil, placed in series with the load, and one or more capacities (filter condensers) shunted across the load. An inductance or choke coil is a device which resists any change in that current that flows through it, and it offers a relatively high resistance to the flow of a varying current. The more variations there are per second, the more resistance it offers to the flow. Because it is in series with the load, the AC component (or ripple) passes with only the greatest of difficulty.

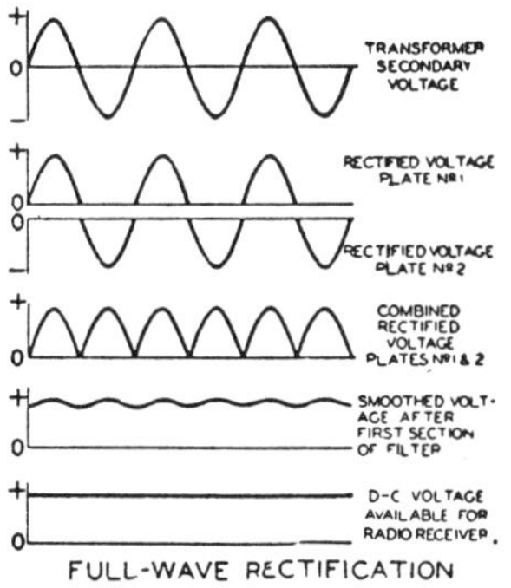

FIG. 1—Showing effects of rectification and filtering of an a-c current.

On the other hand, a capacity (or condenser) is a device which has exactly the opposite action to that of a choke. It offers a relatively low impedance path to the flow of alternating or pulsating current and yet represents a very high or often an infinite resistance to the flow of direct current.

Electricity always follows the path of least resistance. Thus the DC will choose to travel through the choke and back to ground through the load serving the useful function of attracting electrons from the filament over to the plate of the amplifier tube on its way. The AC component, or ripple, on the other hand, faces a high resistance in the choke, but a very easy path back to ground, where it seeks to go, through the condenser. It also chooses the path of lowest resistance, and consequently is by-passed directly to ground. The choke prevents the AC ripple from reaching the plate circuit of the amplifier tube, where it would cause undesirable hum.

The first three circuits, a, b, and c, indicate parallel types of filters, and d, indicates the resonant filter connected across the supply line direct from the rectifier. On account of the effectiveness of the filter it is rather bulky and requires large values of inductance and capacity for successful operation. The resonant-type choke requires much less equipment, yet more accurate adjustment. Its application is generally limited to high power equipment, or to installations where proper equipment is available to determine effective filtering efficiency. For simple amateur installations, the "brute force" filter in some of its many forms will prove to be highly practicable.

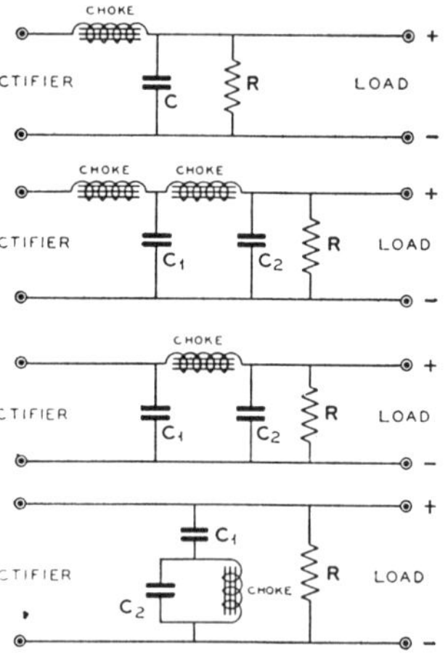

FIG. 2—Common types of filter circuits.

For amateur applications, the ripple voltage must not exceed about 1 per cent for radio telephone service, in the audio and preliminary amplifier stages. With crystal-control the power supply to the final stage (often isolated) should have no more than 5 per cent ripple, and preferably much less. The simplest way to determine ripple voltage is to measure it with a voltmeter, and from reliable reports from other stations, as to whether or not the "carrier" is pure DC.

The diagram, Figure 3, shows a simple scheme for measuring the ripple voltage with the aid of a fixed condenser (about ¼ to 1 ufd.) and a high-resistance copper-oxide alternating current voltmeter. To make the measurement, connect the apparatus as shown. AFTER THE TRANSMITTER IS IN OPERATION, insert the voltmeter plug in the jack and read the voltage. This will be the approximate RMS

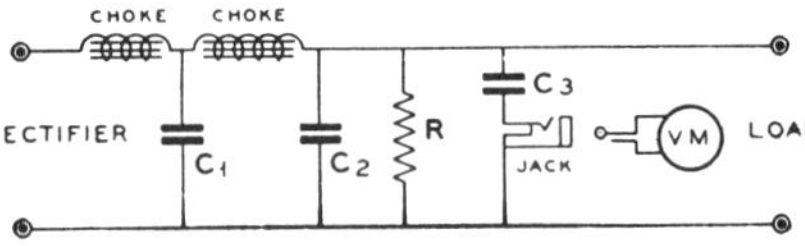

FIG. 3—Measuring a-c ripple.

ripple voltage (approximate because it may be altered by the wave shape, and condenser capacity to a certain extent). Before turning off the plate current **remove the voltmeter** from the circuit. If this precaution is not observed when the transmitter is started or stopped, the rush of current caused by the condenser charging or discharging may burn-out or damage the meter. The meter must be connected in the circuit in such a way that it is on the low potential or negative side of the condenser. **Contact with the high voltage can cause a fatal shock!**

Generally, an input filter system, such as shown in Figure 2b, is advisable, especially with mercury-vapor tubes. The choke absorbs energy, similar to a fly-wheel. Due to the high inductance it resists any change in the current flowing through it. The choke coil selected should be large enough to carry the maximum direct current load without heating, and without losing too much inductance. Any coil wound on an iron or steel core has a certain amount of inductance, determined by the size of the core and the size and number of turns of wire wound on the core. The direct current which is impressed on it magnetizes the core, and this reduces the value of inductance. It is quite possible to raise the direct current up to magnetic saturation, in which case the core will cease to exist magnetically, and the effect will be that of a pure resistance. Closed magnetic circuits of steel or iron will saturate quite easily. All cores on properly designed chokes are fitted with a break in the magnetic circuit. Usually this is in the form of a piece of fibre, bakelite, or other material which is inserted between the ends of the laminations; any method of breaking the magnetic continuity will suffice. This gap is commonly called an "air-gap," but for mechanical reasons it is better to use a non-magnetic substance, instead of air as the spacing. Magnetic saturation can be avoided in chokes by liberal design, and by the use of plenty of copper and iron. Iron core material is often cheaper and easier to obtain than the employment of a large number of turns of wire; therefore, chokes should be designed with large cores, the dimensions of which should be kept within certain limits for practicability.

Choke coils are easily built as they contain only a single winding. Care, however, must be exercised in insulating the windings from the core as the winding must often stand the full plate voltage, plus the "peak," or 1.41 times the output voltage which is delivered from the rectifier system.

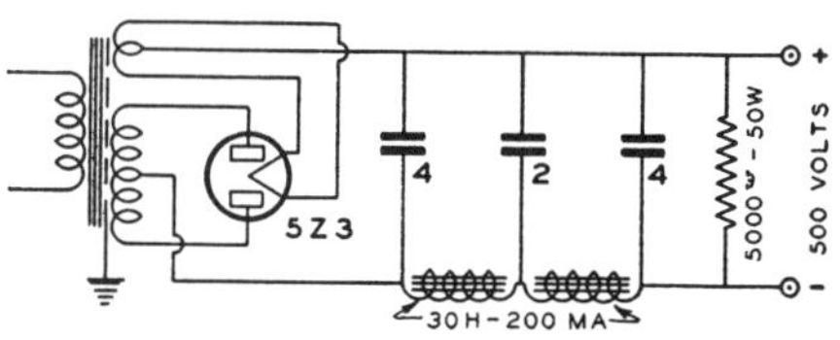

Standard full-wave rectifier using 5Z3 tube. This tube handles more current than the 80.

The other portion of the "filter" is the capacity, or condenser. The latter consists of two types, paper and electrolytic. Paper condensers consist of two strips of tinfoil, separated by high-voltage waxed-paper, and are available in capacities up to about 2 mfd. for voltages up to nearly 5000 volts. The electrolytic types are available in several voltage ranges of about 450 to 600 volts, maximum, per section, which is usually about 8 mfd. capacity. The action of an electrolytic condenser is dependent upon the fact when pure aluminum is immersed in a solution of sodium borate (other solutions are also used) a very thin film of oxide is formed on the surface of the metal. This film, which is apparently of molecular dimensions, forms the dielectric of the condenser. Because the capacity of any condenser is inversely proportional to the thickness of the dielectric, and directly proportional to the dielectric constant, it will be seen that the very thin film of dielectric will give remarkably high capacities for extremely small areas of aluminum.

Electrolytic condensers have the following disadvantages: single units cannot be made for operation at more than about 450 volts; they draw an appreciable current (a few milliamperes), and, after a few years of service often break down. On voltages higher than 350 to 450 volts, it is necessary to connect several condensers in series, or in series-parallel to obtain increased capacity. Under these conditions it is sometimes necessary to connect an equalizing resistor across each condenser as shown in Figure 4.

Some types of paper condensers are impregnated with wax, and some with oil, especially the higher voltage types. The oil type is most desirous, although more expensive than the ordinary wax impregnated types. Paper condensers are rated for "flash" and "normal operating voltage" test ratings; the first refers to a test, usually about twice or three times the normal operating voltage of the condenser, and is only a manufacturers' test rating. The normal operating voltage, or working voltage, is the maximum voltage the condensers will be required to stand in service; this value is often the square root of 2, or 1.41 times the direct current voltage. For reasons of safety, it is good practice to use condensers of at least 1.5 times the normal working or operating direct current voltage as read on the output voltmeter across the filter terminals.

All mercury-vapor tubes are rated by their "maximum peak" current and "maximum inverse peak voltage." The "maximum peak" current rating is a measure of the ability of a tube to stand extremely high transient currents. This rating is intended to form a basis for set design in limiting the abnormal currents that occur during short circuit conditions or with certain types of filters. In addition to this

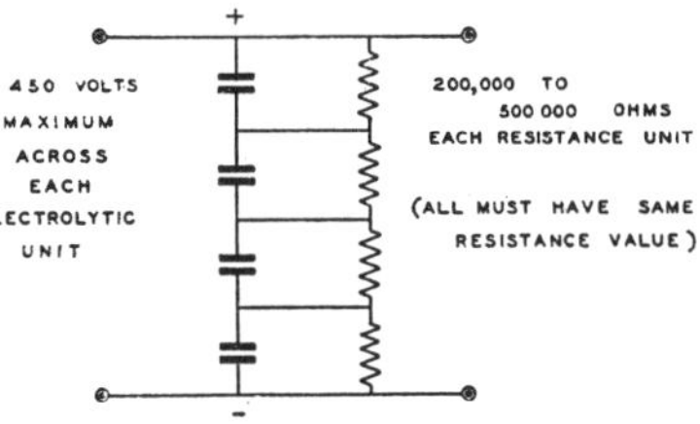

FIG. 4—Electrolytic condensers connected in series plus to minus, with 500,000 ohm, 1 watt, resistor connected across each condenser in order to equalize the leakage and to prevent excessive strain on any one section of condenser

Typical low voltage power supply.

High voltage full-wave power supply using 866 or 866A rectifier tubes.

rating can be included the "maximum average plate current," this is based upon tube heating. It is the plate current as measured on a DC meter and represents the highest average current which can be continuously carried through the tube.

The "maximum peak inverse voltage" is the highest instantaneous voltage that a tube will safely stand in the direction opposite to that which it is designed to pass current. In other words, it is the safe arc-back limit with the tube operating within the specified temperature range. The relations between peak inverse voltage, the direct voltage, and the r.m.s. value of alternating voltage depend largely upon the individual characteristics of the rectifier circuit and the power supply. The presence of line surges, keying surges, or any other transient wave form distortion may raise the actual peak voltage to a value which is higher than that calculated from the sine wave voltages in the transformer. It should, therefore, be emphasized that the maximum rating of the tube refers to the actual inverse voltage and not to the calculated values. A cathode-ray oscillograph or spark gap connected across a tube is useful in determining the actual inverse peak voltage. In single-phase circuits, the peak inverse voltage on a rectifier tube is approximately 1.4 times the r.m.s. value of the plate voltage applied to the tube. In poly-phase circuits the peak inverse voltage must be determined vectorially.

Clarifying some points in the above paragraph it can be said that the maximum inverse peak voltage depends for its value on the peculiar qualities of alternating currents, where the usual type of thermocouple, dynamometer, or similar common types of meters actually give the "square root or mean square," or r.m.s. value of the current, or voltage in a circuit. This means that ordinary meters read the effective current or voltage, or that which would cause the same heating effect by an equivalent direct current. In an AC circuit the maximum peak voltage or current is the square root of 2, or 1.4 times that indicated by the meter reading. In other words, in an AC circuit, with say 100 volts indicated, the actual peak voltage is 141 volts. In a simple half-wave rectifier system, therefore, with 1000 r.m.s. volts across the transformer secondary, there will be 1,410 volts peak voltage, and a single half-wave rectifier tube would have this voltage impressed on it, either positively when the current flows, or "inverse" when the current does not flow. The inverse peak may be twice this value if condenser input filter is used in a half-wave rectifier. With a full-wave system with a center-tap transformer, the voltage across the entire secondary will be twice that of a similar half-wave system, or 2,000 volts, applying the above example. The maximum peak inverse voltage across each tube when not conducting (negative half of the cycle) will be 2,000 x 1.41, or 2,820 volts. Obviously, care must be taken in the choice of rectifier tubes and associated equipment, because it is the peak voltage which breaks down the insulation and causes failure.

In the rectifier output circuit the two half-waves combine to form pulsating direct current, and the peaks of this current are also 1.41 times the indicated, or average value. This means that all units, such as condensers, etc., must be arranged to withstand this voltage.

Power supply for c-w operation. Low capacity filter condensers are used.

The voltage regulation of a rectifier and filter system must be given careful attention in the design of a power supply. It depends on the selection of a power transformer of substantial size and a reasonable overload capacity; the secondary should be of low resistance, and the transformer so designed that the voltage, will not drop appreciably when the secondary load is increased. The selection of the proper chokes. of low resistance and high inductance and of low saturation, all contribute to the maintenance of good regulation in the power supply unit. The use of the so-called "swinging choke" which changes its inductance with variation in load, is also a help in this direction.

A heavy-duty resistor should be connected across the filter output so as to draw an appreciable load. This "bleeder" resistor should normally draw about 10 per cent of the full-load current. The resistor places a load on the system so that a chance open-circuit will not allow the condensers to build up to the full 1.4 times the normal voltage which, obviously, would place a strain on the entire system. A bleeder resistor must be wire-wound, preferably of the 50 or higher-watt dissipating variety. This resistor also helps keep the voltage constant, and to prevent "chirpy" signals when keying a CW or telegraph transmitter.

Technical Note: Bleeder resistors must be mounted so as to allow free circulation of air, and placed as far away as possible from transformers, condensers, and other equipment, or the latter may be damaged from the heat radiated. Ohmic values are given in the table below for bleeder resistors satisfactory for most any amateur transmitter power supply.

Output Voltage	No. of Units in Series	Resistance Ohms	Current Ma
500	1	25,000	20
1000	2	50,000	20
1500	3	75,000	20
2000	4	100,000	20
2500	5	125,000	20
3000	6	150,000	20

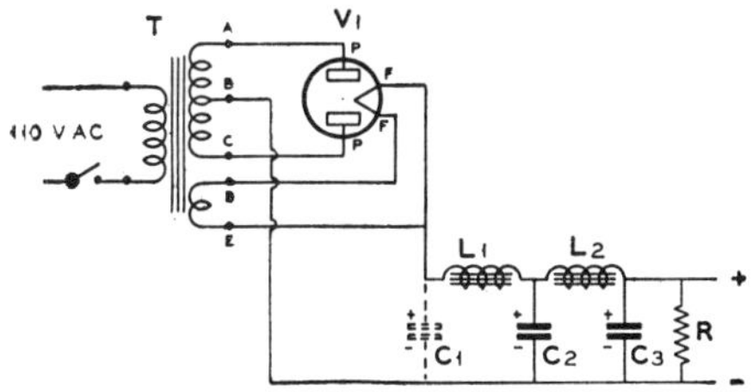

When "C1" is connected in the circuit, the filter is termed "Condenser Input." If "C1" is omitted, the filter is called "Choke Input."

A further method of obtaining good voltage regulation is to use choke input to the filter. This is essential for all types of circuits having mercury-vapor tubes. If a condenser is connected directly across the output of a mercury-vapor rectifier system (except in some form of voltage doublers where a condenser is necessary) the condenser will draw nearly the peak 1.4 times the normal current from the rectifier at all times, and will also change the output voltage considerably; thus regulation will be poorer. Except in small units, vacuum type rectifers with choke input to filters is strongly recommended, both for increased tube and condenser life, and for better regulation. A fuse placed in the power supply system may save a tube, condenser, or other piece of equipment, which costs many times the value of the fuse. It is desirable to mount the chokes in a position of minimum inductive field of the power transformer.

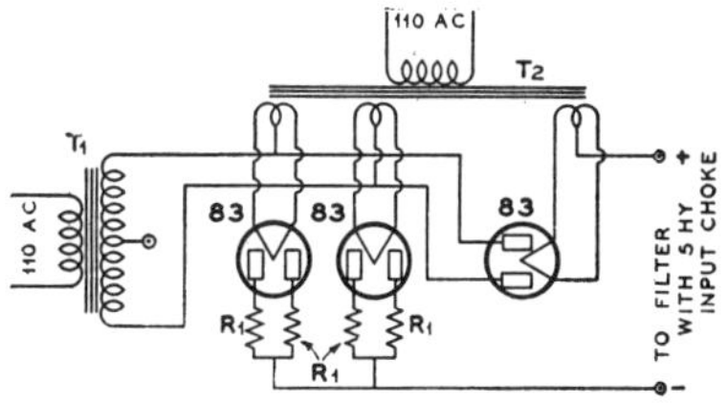

Bridge rectifier suitable for 1000 volt supply. R1 are equalizing resistors, 100 ohms each, 10w.

Plate Supply Circuits and Ratings: Inasmuch as practically all amateur transmitter plate supplies use mercury-vapor rectifier tubes, the data compiled herein concern this type of tube only. Tubes of this type are rated on the basis of peak inverse voltage, and peak plate current.

Condenser Input: Where a filter circuit is used having condenser input, the peak plate current per tube in a full-wave circuit may rise to values as great as four times the DC load current depending on the value of input capacitance. This naturally results in poor tube economy. In the case of 866 tubes, for example, the peak plate current of .6 ampere might be reached when the DC output obtained is only .15 ampere. A second factor which limits the application of condenser input filters for amateur work is the poor regulation obtained in such circuits.

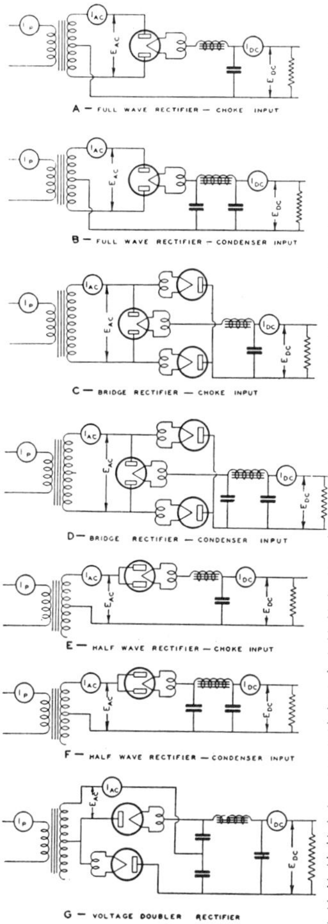

A: E_{DC}-435 V, I_{DC}-100 MA, E_{AC}-1100 V, I_{AC}-71 MA, I_{PRI}-.6 MA

B: E_{DC}-675 V, I_{DC}-100 MA, E_{AC}-1100 V, I_{AC}-103 MA, I_{PRI}-.9 MA

C: E_{DC}-860 V, I_{DC}-100 MA, E_{AC}-1100 V, I_{AC}-96 MA, I_{PRI}-1.1 A

D: E_{DC}-1200 V, I_{DC}-100 MA, E_{AC}-1100 V, I_{AC}-148 MA, I_{PRI}-1.65 A

E: E_{DC}-480 V, I_{DC}-100 MA, E_{AC}-550, I_{AC}-120 MA, I_{PRI}-.9 A

F: E_{DC}-580 V, I_{DC}-100 MA, E_{AC}-550 V, I_{AC}-190 MA, I_{PRI}-1.45 A

G: E_{DC}-1080 V, I_{DC}-100 MA, E_{AC}-550 V, I_{AC}-310 MA, I_{PRI}-1.55 A

Typical voltage and current readings in various types of power supplies.

Choke Input: With a filter circuit having a choke input, the peak plate current per tube in a full-wave circuit will generally be about 50 per cent greater than the DC. With a saturated reactor this peak current will be increased as the load current is increased to as high as 2½ times the DC.

With the knowledge of the peak inverse voltage, and the peak plate current of the rectifier tubes, it is apparent that the proper tube or tubes and associated components can be readily determined for any plate supply output. These values for given tubes are enumerated below:

Tube Type	Peak Inv. Volts	Peak Plate Current (amp.)
82	1,400	.40
83	1,400	.80
66	7,500	.6
66A	10,000	.6
72	7,500	2.5
72A	10,000	2.5
869	20,000	5.0

Standard Rectifier Circuits: Figures 1 to 6 on the facing page illustrate typical rectifier circuits applicable to amateur use. The single-phase half-wave circuit of Figure 1 is not very popular due to the fact that the ripple is of greater magnitude and being of lower frequency than other systems is more difficult to filter. With choke input, the DC voltage will be approximately .45 that of the r.m.s. voltage E. Figure 2 illustrates the full-wave single-phase circuit which every amateur is familiar with. Figure 3 is identical in nature with Figure 2, except that four tubes (more if desired) are used to obtain higher current output. The resistors shown in the plate circuits of these tubes are very essential, otherwise one tube will generally take most of the load with the natural result that the tube life is greatly decreased; a drop of about six volts across these resistors will insure stability. Figure 4 shows a bridge circuit with four tubes, its advantage is that high DC voltages can be secured without expensive (high peak inverse voltage) tubes and with low voltage transformers. For full-wave rectification the DC voltage can be increased by using the entire secondary output of the plate transformer, in fact, the voltage will be exactly doubled; of course, this halves the current output due to the transformer current carrying limitations. Figures 5 and 6 are similar to that of Figure 2, except that they apply to three-phase circuits. In the circuit of Figure 5, each tube carries current for one-third cycle. The circuit of Figure 6 is very commonly employed in high power transmitters where three-phase power is available due to the high DC output voltage attained. This circuit has the added advantage that the ripple frequency is high, being six times the supply frequency, allowing simple filtering.

Analyzing these rectifier circuits have given the values indicated above as the maximum operating and output values for any of the tubes described.

As an example in applying these figures to the 866 tube, it is found that in a full-wave circuit (Figure 2), the maximum transformer voltage E, each side of the center-tap is .35 x 7500, or 2650 volts. This gives a DC voltage at the input to the filter of 2650 × .9, or 2400 volts. The maximum DC output is .66 times the peak plate current of .6 ampere, or 400 MA. Hence, voltages and currents lower than these values can be used. With a saturated input reactor, the allowable DC is reduced. However, as these saturated reactors are normally used in conjunction with a class B amplifier load, the high DC and peak plate currents are normally of short duration, reducing the tube life by an amount which is not excessive.

Predetermining DC Voltage: An examination of Figure 7 will show that it can be reduced to the more simpler form of Figure 8. Here, the ratio of transformation is such that E volts are induced in the transformer secondary. From the theoretical DC output which is .9 × E, and subtracting all the voltage drops, which include the drop across R_t (the transformer resistance), across L_t (the transformer leakage inductance, across V (the tube drop), across RL1 and RL2 (the choke resistances) it will be found that the output voltage can be accurately estimated. If the transformer regulation is known, a value of E can be obtained which already incorporates the transformer losses. The DC output is then (.9 × E) minus 15 (the normal voltage drop across a mercury-vapor tube, minus I_{dc} x (RL1 plus RL2). This gives a definite means of predetermining the DC output voltage from a rectifier using a choke input filter.

Notes on Operating Mercury-Vapor Rectifier Tubes: In respect to the operation of the tube it is considered good practice to allow the filament of the tube to come up to full temperature before the plate voltage is applied. Otherwise, the active material may be knocked off the filament. The precaution given is merely this: that the space charges around the cathode reduce the fall of potential at the surface of the cathode so that the positive ions which are drawn to the hot cathode or filament do not have energy enough when they strike the filament to knock off the active material. It is the positive ions of mercury accelerated in the high field around the filament when there is a large space charge that is attracted to the negative filament and injure its surface.

Every precaution should be taken to keep the mercury tube out of intense radio-frequency fields, because radio-frequency oscillations introduce potentials into the gaps

Figure No.	Transformer Volts "E"	DC Output Volts at Input to Filter	DC Output Current in Amperes
1	.7 x Inv. Pk. Vtg.	.45 x E	1.33 x Pk. Plate
2	.35 x Inv. Pk. Vtg.	.9 x E	.66 x Pk. Plate
3	.35 x Inv. Pk. Vtg.	.9 x E	1.32 x Pk. Plate
4	.7 x Inv. Pk. Vtg.	.9 x E	.66 x Pk. Plate
5	.43 x Inv. Pk. Vtg.	1.12 x E	.83 x Pk. Plate
6	.43 x Inv. Pk. Vtg.	2.25 x E	1.0 x Pk. Plate

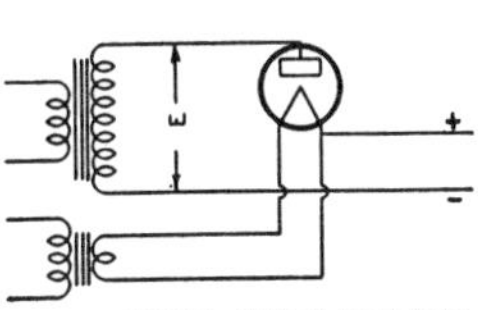

SINGLE PHASE HALF WAVE

FIG. 1

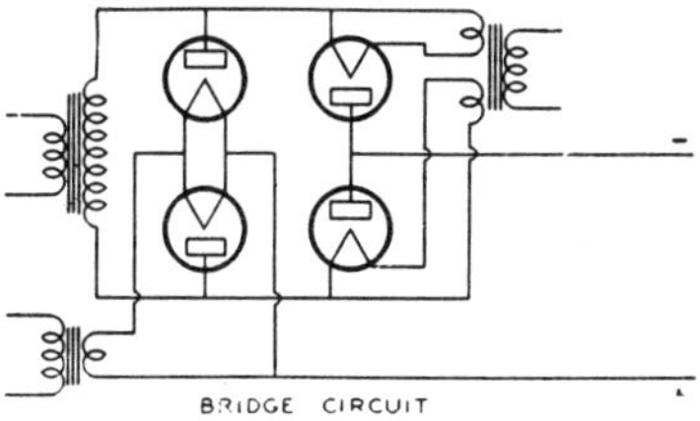

BRIDGE CIRCUIT

FIG. 4

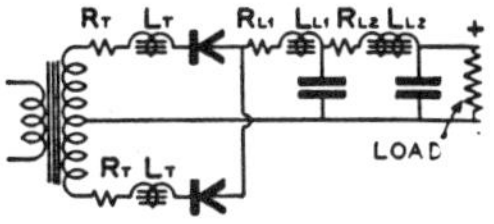

SIMPLIFIED FORM FIG. 7

FIG. 8

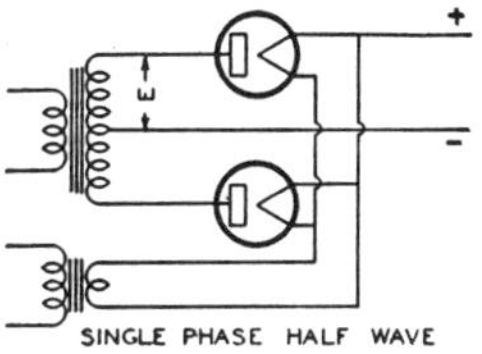

SINGLE PHASE HALF WAVE

FIG. 2

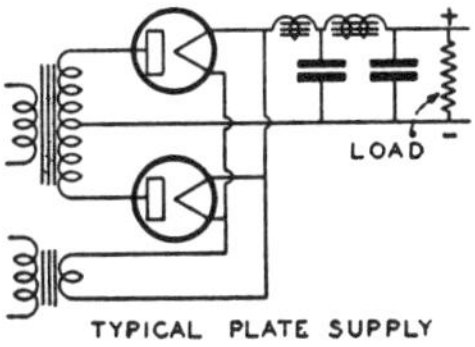

TYPICAL PLATE SUPPLY

FIG. 7

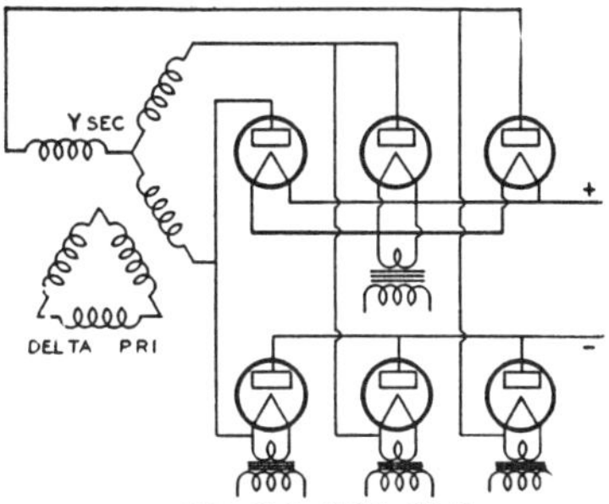

FULL WAVE THREE PHASE

FIG 6

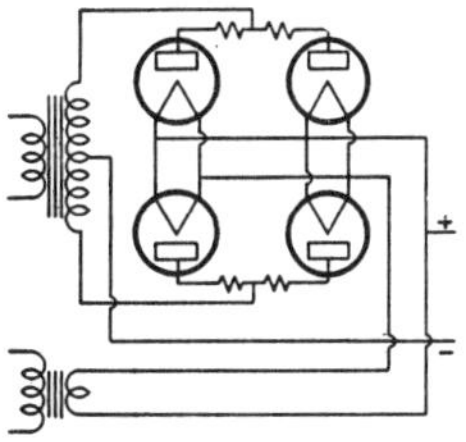

FULL WAVE PARALLEL TUBES

FIG. 3

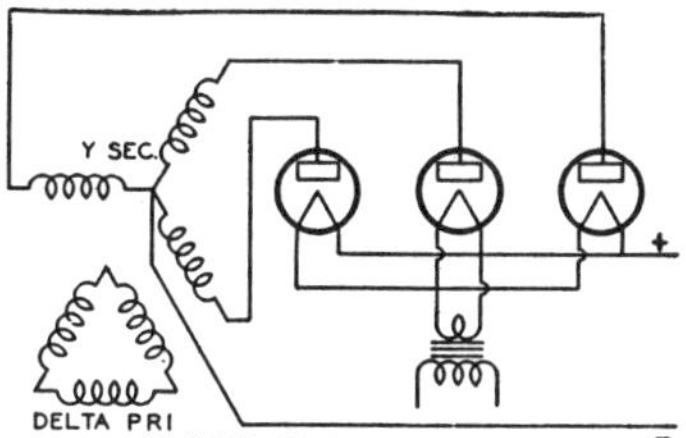

HALF WAVE THREE PHASE

FIG. 5

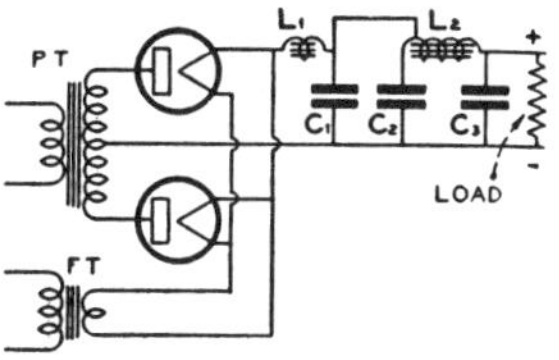

FIG. 9

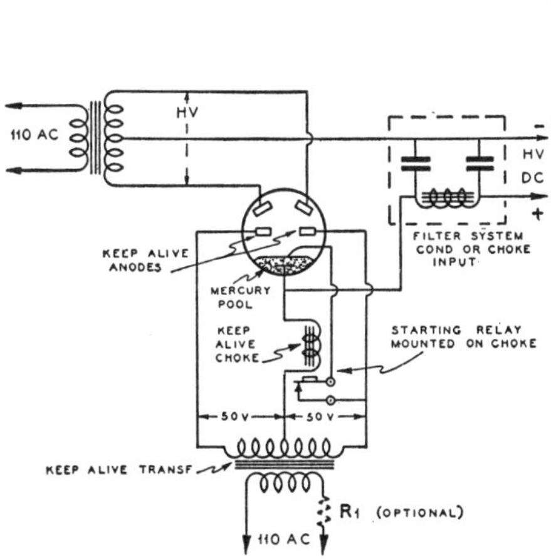

2C MULTI-ARC

Circuit diagram showing Multi-Arc starting, keep-alive circuit and filter supply. The Mercury Arc is capable of rectifying extremely high voltages and currents. Its normal operating life is long; its cost is not excessive.

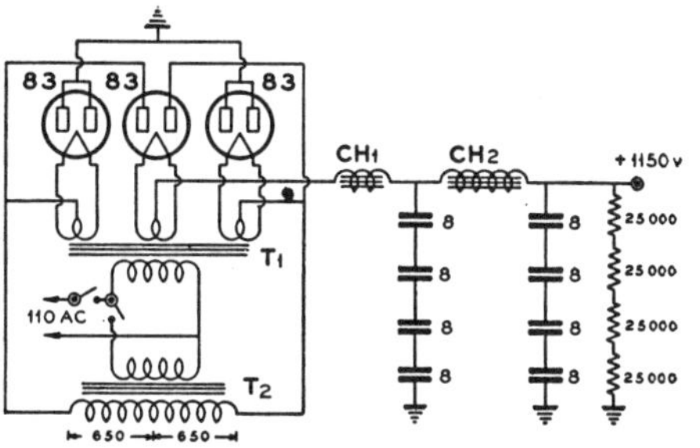

Bridge rectifier using inexpensive parts. Output 1150 volts at 250 MA. Ordinary 8 mfd. 450 working-volt electrolytic condensers are connected in series, as shown.

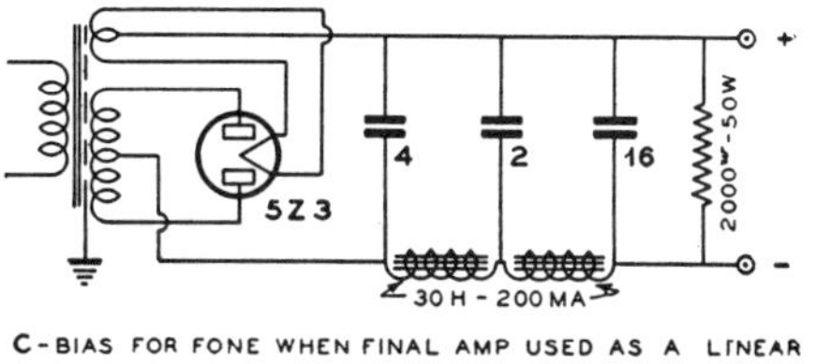

C-BIAS FOR FONE WHEN FINAL AMP USED AS A LINEAR

between the cathode and the anode of the tube which, superimposed upon conditions already existing, leads to ionization and changes of current when they are not wanted. If the mercury tube is operating critically, it takes little to produce the necessary ionization which is needed for its operation. This added potential introduced by means of RF currents playing between the electrodes is sufficient to start ionization that is not desired. The tube should likewise be kept out of magnetic fields as such fields have the effect of changing the energies of the electrons in the atoms of mercury-vapor by distorting their hypothetical orbits and making certain direction of motion easier than others. Much difficulty in filtering the output of the rectifier may be eliminated by isolating the power supply from the transmitter and shunting the power supply with a fairly high mica condenser preceded by an adequate RF choke. A condenser of .002 ufd. may be used advantageously for 40 meter operation.

Power Supply Components

The design of a good power supply is as equally important as the design of the transmitter or receiver; too much emphasis cannot be placed upon the proper selection of the various electrical components going into the power supply assembly. The design specifications and other complementary features are outlined in the subsequent paragraphs.

Transformer Design: A common problem in radio and allied work is to determine how a transformer can be built to supply certain power requirements for a particular application, or how to calculate the windings needed to fit a certain transformer core which is already on hand. These problems can be solved by a small amount of calculation.

The most important factor in determining the size of any transformer is the amount of core material available. The electrical rating, as well as the physical size, is determined almost entirely by the size of the core. The core material is also important, but the present practice is to use high-grade silicon-steel sheet. It will be assumed that this type of material is to be employed in all construction herein described. Soft sheet-iron, or stovepipe iron is sometimes substituted, but transformers made from such materials will have about 50 to 60 per cent of the power rating, pound for pound of core, as those made from silicon-steel. The core size determines the performance of a transformer because the entire energy circulating in the transformer (except small amounts of energy dissipated in resistance losses in the primary) must be transformed from electrical energy in the primary winding to magnetic energy in the core, and reconverted into electrical energy in the secondary. The amount of core material determines quite definitely the power that any transformer will handle.

Transformer cores are often designed so that if the losses per cubic inch of core material are determined, these losses can be used as a basis for calculating the rating of the transformer. These losses exist in watts, and are divided between the eddy current loss and the hysteresis loss. The eddy current loss is the loss due to the lines of force moving across the core, just as if it were a conductor, and setting-up currents in it. Induced currents of this type are very undesirable and they are merely wasted in heating the core, which then tends to heat the windings, increase the resistance of the coils and reduce the overall power handling ability of the transformer. To reduce such losses, transformer cores are made of thin sheets, usually about No. 29 gauge. These sheets are insulated from each other by a coat of thin varnish, shellac, or japan, or by the iron-oxide scale which forms on the sheets during the manufacturing process, and which forms a good insulator between sheets.

"Hysteresis" means "to lag," and hysteresis in an iron-core means that the magnetic flux in the core lags behind the magnetizing force that produces it, which is, of course, the primary supply. Because all transformers operate on alternating current, the core is subjected to continuous magnetizing and demagnetizing force, due to the alternating effect of the AC field. This force heats the iron, due to molecular friction caused by the iron molecules reorienting themselves as the direction of the magnetizing flux changes. The higher the field strength, the greater the heat produced. A condition can be reached where a further increase in magnetizing flux does not produce a corresponding increase in the flux density. This is called "saturation" and is a condition which would cause considerable heat in a core. In practice, it has been found that all core material must be operated with the magnetic flux well below the limit of saturation.

Core losses manifest themselves as heat and these losses are the determining factor in transformer rating. They are spoken of as "total core loss," generally used as a single figure, and for common use a core loss of from .75-watts to 2.5-watts per pound of core material can be assumed for 60 cycles. The lower figure is for the better grades of thin sheet, while the higher loss is for heavier grades. About 1-watt per pound is a very satisfactory rating for common grades of material. This rating is also dependent on the manner in which the transformer is built and mounted, and in the ease with which the heat is radiated from the core. Transformers with higher losses may be used for intermittent service.

The transformer core loss can be assumed to be from 5 to 10 percent of the total rating for small transformers. Thus, if the core loss is known, the rating of the transformer can be easily determined. If the figure of 1-watt per pound is assumed, the problem is further simplified. To determine the rating of the transformer, weigh the core. If, for example, the core weighs 10 pounds, the transformer will handle from 100 to 200-watts. Such a transformer core can be assumed to have about 150-watts nominal rating. If the weighing of the core is inconvenient, the weight can be calculated from the cubic contents, or volume. Sheet-steel core laminae weigh approximately one-fourth pound per cubic inch.

Transformer cores are generally made of two types, shell and core. The shell-type has a center leg which accommodates the windings, and this is twice the cross-sectional areas of the side legs. The core-type is made of strips built-up into a hollow box-like affair of uniform cross section. For the shell-type core, the area is taken as the square section of the center leg, in this case 2¼ in. x 4½ in. and in the core-type, this area is taken as the section of 1 leg, and is also 2¼ in. x 4½ in., or an actual core area in both cases of 10.1 square inches, which is large enough for a comparatively large transformer.

To determine the number of turns for a given voltage, apply the following formula:

$$E = \frac{4.44\,N\,B\,A\,T}{10^8}$$

Where E equals the volts of the circuit; N, the cycles of the circuit; B, the number of magnetic lines per square inch of the magnetic circuit; A, the number of square inches of the magnetic circuit; and T, the number of turns.

The proper value for B, for small transformers, and for ordinary grades of sheet-iron, such as are now being considered, is 75,000 for 25 cycles and 50,000 for 50 or 60 cycles.

Rewriting the above formula:

$$T = \frac{E \times 10^8}{4.44\,N\,B\,A\,T}$$

and since N and B are known

$$T = \frac{10^8}{4.44 \times 60 \times 50{,}000} \times \frac{E}{A}$$

from which

$$T = 7.5 \times \frac{E}{A}$$

That is, for a transformer to be used on a 60 cycle circuit, the proper number of turns for the primary coil is obtained by multiplying the line-voltage by 7.5 and dividing this product by the number of square inches cross-section of the magnetic circuit.

On a 25 cycle circuit, the 7.5 becomes 12, and on 50 cycles it becomes 9.

Tentative Design: Assume a transformer core that is to be used on a 115 volt, 60 cycle circuit for supplying power to two rectifier tubes, each of which takes 1,000 volts on the plate. The rectifier is of the full-wave type. The core measures 2½ inches x 4½ inches; hence,

$$T = \frac{7.5 \times 115}{2.25 \times 4.5} = 85 \text{ (to the nearest turn),}$$

and the volts per turn equals

$$\frac{115}{85} = 1.353 \text{ which is the same for all coils.}$$

Now, the secondary coil must have two windings in series, each to give 1,000 volts, and with a middle-tap. The secondary turns will be $\frac{2000}{1.353} = 1478$ with a tap taken out at the 739th turn.

Allowing 1,500 cm per ampere, the primary wire should be No. 12. The size of the wire on the plate coils may be No. 22 or 24 for a 400 to 300 ma. rating.

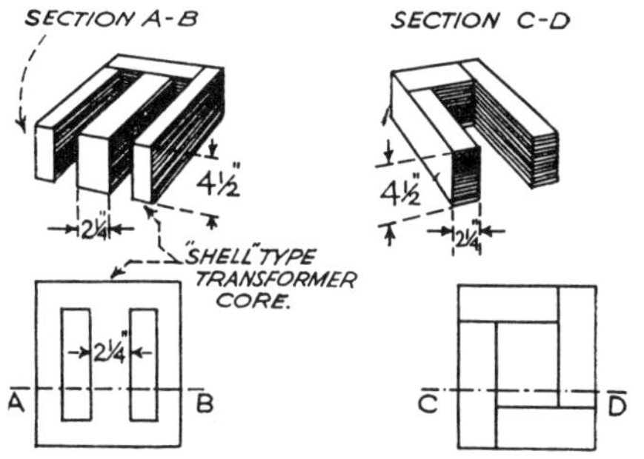

Types of transformer cores.

To determine the quantity of iron to pile up for a core, it is well to consider 1 to 1.5 volts per turn as a conservative range. For trial assume 1.25 volts. Then by transforming the first equation:

$A = 7.5 \times \frac{E}{T}$ or, the area required is 7.5 times the volts per turn; in this case, 7.5 × 1.25 = 9.38 sq. in.

The magnetic cross section must be measured at right angles to the laminations that are enclosed by the coil; the center leg when the core is built up around the coil; and either leg where the core is built up inside the coil, that is, between the arrows in the sketches shown above.

It should be kept in mind that there is a copper, or resistance loss in all transformers. This is caused by the passage of the current through the windings, and is commonly spoken of as the "IR" loss. It manifests itself directly as heat, and varies as the load is varied; the heavier the load, the more heat is developed. This heat, as well as other heat losses, must be removed, or the transformer will burn up. Most transformers are so arranged that both the core and windings can radiate heat into the surrounding air, and thus cool themselves. Large transformers are mounted in oil, for cooling, and also for the purpose of increasing the insulation factors.

In any transformer, the voltage ratio is directly proportional to the turns ratio. This means that if the transformer is to have 110 volts input, and 250 turns for the primary, and if the output is to be 1,100 volts, 2,500 turns will be needed. This may be expressed as:

$$\frac{E_p}{E_s} = \frac{T_p}{T_s}$$

It is often more convenient to take the figure obtained for the primary winding, and by dividing by the supply voltage the number of turns per volt is calculated. This accomplished, the number of turns for any given voltage can be calculated by simple multiplication.

Radio transformers are generally of small size. The matter of power factor can therefore be disregarded, more especially because they work into an almost-purely resistive load. In the design of radio transformers, the power factor can be safely assumed as unity, in which case the apparent watts and the actual watts are the same. Admittedly this is not always a correct assumption, but it will suffice for common applications.

The size of the wire to be used in any transformer depends upon the amperage to be carried. For a current of 1-ampere as a continuous load, at least 1,000 circular mils per ampere must be allowed. For transformers which have poor ventilation, or intermittent heavy load service, or where price is not the first consideration, 1,500 circular mils per ampere will suffice. If, for example, a transformer is rated at 100-watts primary load on 110-volts, the current will be

$$I = \frac{W}{V} = \frac{100}{110} = 0.90 \text{ amperes,}$$

and if the assumption is 1,000 circular mils per ampere, it will be found that this will require 1,000 × .90, or 900 circular mils. The wire table shows that No. 20 wire for 1,200 mils, is entirely satisfactory. If it is desired to use 1,500 circular mils, instead of 1,000, this will require 1,500 × .90 of 1,350 mils, which corresponds to approximately No. 19 wire. The difference seems to be small, yet it is large enough to reduce heating and to improve overall performance. Assume, for tentative design, a 600-volt, 100MA high-voltage secondary; a 3-ampere 5-volt secondary; and 2.5-volt 7.5-ampere secondary. Simple calculation will show a 60-watt load on the high-voltage secondary; 15-watts on the 5-volt winding; and 16-watts on the 2.5-volt winding, a total of 91 watts. The core and copper loss is 10-watts. The wire sizes for the secondaries will be for 100 mils current, No. 30 wire, 3-amperes at 5-volts, No. 15 wire; No. 11 wire for the 7.5-ampere secondary.

For high voltage secondary windings, a small percentage should be allowed to overcome the resistance of the small wire used, so that the output voltage will be as high as anticipated. The figures given in the table include this percentage which is added to the theoretical ratio of turns, and consequently the number of turns shown in the table can be accepted as the actual number of turns to be wound on the core of any given transformer.

Allowance should always be made for the insulation and size of the windings. Good insulation should be provided between the core and the windings, and also between each winding and between turns. Numerous materials are satisfactory for this purpose; varnished paper or cloth, called "empire," or paper is very satisfactory, although costly. Good bond paper will serve well as an insulating medium for small transformer windings. Insulation between primary and secondary and to the core must be exceptionally good, as well as the insulation between windings. Thin mica, or "micanite" sheet is very good. Thin fibre, commonly called "fish paper" is also a good insulator, Bristol Board, or strong, thin cardboard may also be used. In all cases, the completed coil should be impregnated with insulating varnish, and either dried in air or baked in an oven. Common varnishes or shellac are unsatisfactory on account of the moisture content of these materials. Air-drying insulating varnish is practical for all-around purposes; baking varnish may be substituted, but the fumes given off are inflammable and often explosive. Care must be exercised in the handling of this type of material. Collodion and banana oil lacquer is positively dangerous, and in the event of a short-circuit of transformer burn-out, a serious fire may result.

If it is desired to wind a transformer on a given core, it is much better to calculate the actual space required for the windings, then determine whether there is enough available space on the core. If this precaution is not observed, the designer may find that only about half the turns can actually be wound on the core, when the work is about three-fourths finished. From 15 to 40 per cent more space than is actually required must be allowed. The winding of transformers by hand is a space consuming process. Unless the builder is an experienced coil-winder, there is every chance that a sizable portion of the space will be used-up by insulation, etc., not sufficient space remaining for the winding. Calculate the cubical space needed for the total number of turns, and allow from 15 to 40 per cent additional space in the core "window." Thereby much time and labor will be saved.

Filter Chokes: A choke is a coil of high inductance. It offers an extremely high impedance to alternating current, or to current which is substantially alternating, such as pulsating DC delivered at the output of a rectifier.

Choke coils are used in power supplies as part of the complete filter system in order to produce an effectively-pure direct current from the pulsating current source, that is, from the rectifier. The size of the choke must be such that the current flowing through it does not cause an appreciable voltage drop due to the ohmic resistance of the choke; at the same time sufficient inductance must be maintained to provide ample smoothing of the rectified current.

Smoothing Chokes: The function of a smoothing choke is to discriminate as much as possible between the AC ripple which is present and the desired DC that is to be delivered to the output. Its air-gap should be large enough so that the inductance of the choke does not vary materially over

Choke Table for Transmitter Power Supply Units

Current M.A.	Wire Size	No. Turns	Lbs. Wire	Approx. Core (Area)	Air Gap	Wt. Core
200	No. 27	2000	1.5	1½"x1½"	3/32"	4 lbs.
250	No. 26	2000	1.75	1½"x2"	3/32"	5 lbs.
300	No. 25	2250	2	2"x2"	⅛"	6 lbs.
400	No. 24	2250	3	2"x2½"	⅛"	7 lbs.
500	No. 23	2500	4	2½"x2½"	⅛"	10 lbs.
750	No. 21	3000	6	2½"x3"	⅛"	14 lbs.
1000	No. 20	3000	7.5	3"x3"	⅛"	18 lbs.

NOTES: These are approximately based on high-grade silicon steel cores, with total airgaps as given. Airgaps indicated are total of all gaps.

The use of standard "E" and "I" laminations is recommended. If strips are used, and if an ordinary square core is used, the number of turns should be increased about 25%. Choke coils built as per the above table will have an approximate inductance of 10 to 15 henrys. Because considerable differences occur due to winding variations, allowable flux densities of cores, etc., the exact inductance cannot be stated; these chokes will, however, give satisfactory service in radio transmitter power supply systems.

The wire used is based on 1000 circular mils per ampere; this will cause some heating on long runs, and if the chokes are to be used continuously, as in a radio telephone station in continuous service, it is good practice to use the next size larger choke shown for such loads.

the normal range of load current drawn from the power supply.

Swinging Chokes: In certain radio circuits the power drawn by a vacuum tube amplifier can vary widely. Class B audio amplifiers are good examples of this type of amplifier. The plate current drawn by a class B audio amplifier can vary a thousand per cent, or more. It is desirable to keep the DC output voltage applied to the plate of the amplifier as constant as possible, and the voltage should be independent of the current drawn from the power supply. The output voltage from a given power supply is always higher with a condenser input filter than with a choke-type input filter. When the input choke is of the **swinging** variety, it means that the inductance of the choke varies widely with the load current drawn from the power supply. Thus, at low load currents the inductance of the swinging choke is high and the filter acts as a choke input filter, with a relatively low output voltage. When the load current increases, the inductance of the swinging choke decreases and the filter circuit begins to act more and more like a condenser input filter. This causes the output voltage to rise somewhat, although the rise is usually adjusted so that it just offsets the voltage drop caused by the transformer and choke resistance, plus the drop across the rectified tubes. A swinging choke does not have much smoothing effect, but it is valuable in improving the voltage regulation of the power supply. The use of a swinging choke is desirable in a CW transmitter to reduce keying thumps which occur when a condenser input is used.

Design and Construction of Chokes: A choke is made up from a silcon-steel core which consists of a number of thin sheets of steel, similar to a transformer core, but wound with only a single winding. The size of the core and the number of turns of wire, together with the air-gap which must be provided to prevent the core from saturating, are factors which determine the inductance of a choke. The relative sizes of the core and coil determine the amount of DC which can flow through the choke without reducing the inductance to an undesirable low value due to magnetization.

The same core material which is used in ordinary radio power transformers, or from those which are burned-out, is satisfactory for all general purposes.

In construction, the choke winding must be insulated from the core with a sufficient quantity of insulating material so that the highest peak voltages which are to be experienced in service will not rupture the insulation. It is good practice to

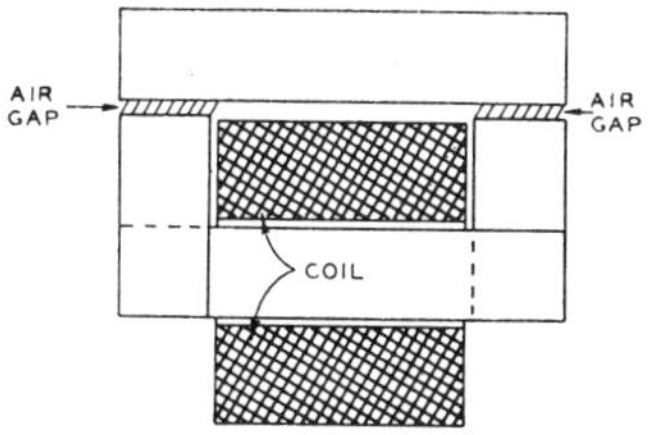

Two types of choke coil construction. The air-gap is approximately 1/32 inch.

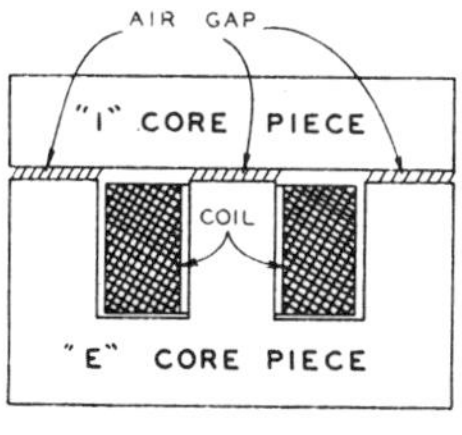

The air-gap can be filled with non-magnetic material, such as brass, bakelite, etc.

operate chokes with the cores grounded; in addition, the choke may be placed in the negative high voltage lead, in order to minimize breakdown and to keep the filtering properties at high efficiency. If the choke is mounted on a breadboard, the core need not be grounded. In some cases where extremely objectionable hum is introduced it may be necessary to completely shield and ground the entire choke assembly.

Design of Voltage Dividers: The calculation of the correct resistance values and the power ratings of voltage dividers can be determined by the following procedure:

Determine the voltage to be required at each tap and the current to be drawn from it. Vacuum tube manuals can be referred to for tube data.

Determine the bleeder current desired. This value will depend upon the total current drawn by all the tubes plus what the power supply can deliver without over-heating.

Determine the current flowing in each section of the divider.

Calculate the resistance of each section by Ohm's law.

Determine the power rating from the equation:

$$\text{Watts} = \frac{I^2R}{1,000,000}$$

(I = Milliamperes)

The value I represents the highest current any section is required to carry. If the divider is to consist of several resistors the wattage of each section should be calculated separately and the actual current in that section used for the calculation.

Voltage dividers offer a common impedance to several circuits and so may give rise to regeneration and degeneration. These effects may be eliminated by employing by-pass condensers and extra filters in individual supply leads. (Hint: to constitute an efficient by-pass, the condenser reactance must be considerably lower than the circuit resistance, generally .1 the resistance being by-passed will suffice.)

Resistor-Capacity Filters: Those excellent resistor-capacity filters that are featured in custom-built and very expensive equipment can be designed from very simple formula. In general, these filters function to stabilize, reduce hum and to prevent common coupling between stages. The effectiveness of the properties of an RC network is proportional to the ratio of the resistance to the reactance of the circuit; a resistive ratio of 50 to a capacitive ratio of 1 is satisfactory for all general purposes. In other words, the resistance in the filter must have a value about 50 times greater than the reactance of the condenser at the lowest frequency to be filtered.

Figures 10, 11 and 12 show how to connect RC filters. By-pass condensers such as those placed across C-bias resistors should have such a value that the impedance of the C-bias circuit is small in comparison to the ohmage of the resistor.

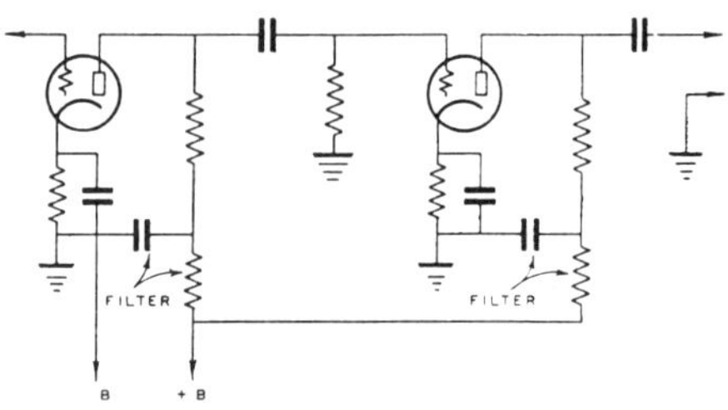

FIG. 10

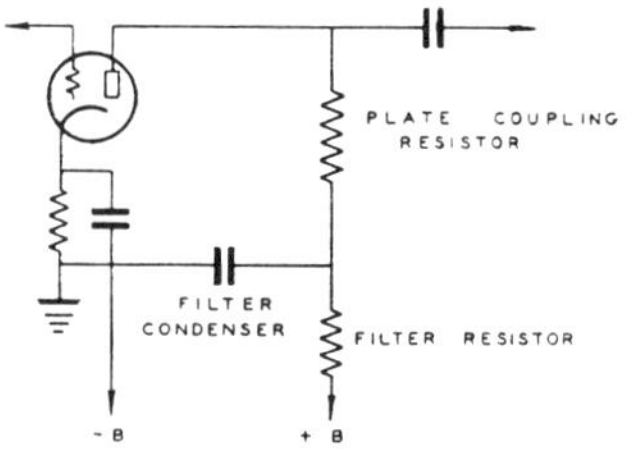

FIG. 11

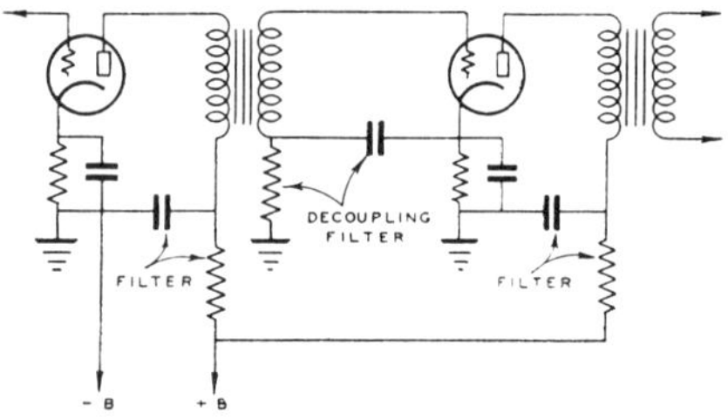

FIG. 12

Winding Turns Per Inch

B & S Gauge No.	D.C.C.	S.C.C.	Enamel	B & S Gauge No.	D.C.C.	S.C.C.	Enamel
6	5.44	5.60		26	39.90	45.30	57.00
7	6.08	6.23		27	42.60	49.40	64.00
8	6.80	6.94		28	45.50	54.00	71.00
9	7.64	7.68		29	48.00	58.80	81.00
10	8.51	8.55		30	51.10	64.40	88.00
11	9.58	9.60		31	56.80	69.00	104.00
12	10.62	10.80		32	60.20	75.00	120.00
13	11.88	12.06		33	64.30	81.00	130.00
14	13.10	13.45	14.00	34	68.60	87.60	140.00
15	14.68	14.90	16.00	35	73.00	94.20	160.00
16	16.40	17.20	18.00	36	78.50	101.00	190.00
17	18.10	18.80	21.00	37	84.00	108.00	195.00
18	20.00	21.00	23.00	38	89.10	115.00	205.00
19	21.83	23.60	27.00	39	95.00	122.50	215.00
20	23.91	26.40	29.00	40	102.50	130.00	230.00
21	26.20	29.70	32.00	41	112.00	153.00	240.00
22	28.58	32.00	36.00	42	124.00	168.00	253.00
23	31.12	34.30	40.00	43	140.00	192.00	265.00
24	33.60	37.70	45.00	44	153.00	210.00	275.00
25	36.20	41.50	50.00				

						SECONDARY WINDINGS (Turns for Voltages Given)																	
											HIGH-VOLTAGE WINDING												
WATTS	Section of Core (inches)	Area of Core (Square inches)	Primary Turns	Primary Wire Size	Turns per Volt	2.5 volts	5.0 volts	6.3 volts	7.5 volts	10 volts	250 volts	300 volts	350 volts	400 volts	450 volts	500 volts	600 volts	700 volts	800 volts	900 volts	1000 volts	1250 volts	1500 volts
10	½ x ½	.25	3500	31	32	80	160	205	240	320													
10	½ x ⅝	.31	2800	31	24.2	61	122	147	182	242													
12	½ x ¾	.37	2300	30	20.0	50	100	126	150	200													
12	⅝ x ⅝	.38	2280	30	19.6	48	96	124	147	196													
15	⅝ x ¾	.46	1875	29	16.1	42	84	105	124	161													
22	⅝ x 1	.62	1400	28	12.2	31	61	77	92	122													
20	¾ x ¾	.55	1570	28	13.6	34	68	86	102	136													
25	¾ x 1	.75	1150	27	10.0	25	50	63	75	100	2620	3150	3700	4200	4750	5250							
30	¾ x 1¼	.93	930	26	8.1	21	42	52	62	81	2100	1500	3140	3400	3800	4200							
50	¾ x 1½	1.12	770	24	6.7	17	34	43	50	67	1860	2100	2500	2840	3150	3500	4200	5000					
50	1 x 1	1.0	860	24	7.5	19	38	48	57	75	1950	2400	2700	3150	3600	3900	4700	5500					
60	1 x 1¼	1.25	690	23	6.0	15	30	38	45	60	1600	1900	2200	2500	2800	3150	3800	4400					
65	1 x 1½	1.50	575	23	5.0	13	25	32	38	50	1300	1575	1850	2100	2400	2650	3150	3700					
75	1 x 1¾	1.75	490	22	4.2	11	21	27	31	42	1100	1320	1550	1750	2000	2200	2650	3150	3800	4000	4400		
110	1 x 2	2.0	430	21	3.7	9	18	23	28	37	980	1170	1370	1550	1750	1960	2300	2750	3100	3500	3900		
105	1¼ x 1¼	1.56	550	21	4.8	12	24	31	36	48	1260	1510	1770	2050	2240	2510	3050	3500	4100	4500	5020		
100	1¼ x 1½	1.87	460	21	3.8	9	19	25	29	38	1000	1200	1400	1600	1800	2000	2400	2720	3200	3560	4000		
120	1¼ x 1¾	2.18	400	20	3.5	9	18	21	26	35	920	1100	1315	1470	1650	1840	2200	2560	2940	3300	3700	4620	5500
140	1¼ x 2	2.5	350	19	3.2	8	16	20	24	32	840	1020	1180	1340	1510	1680	2050	2350	2680	3000	3380	4200	5050
125	1½ x 1½	2.25	380	20	3.3	8	16	21	25	33	870	1040	1210	1400	1560	1730	2100	2420	2800	3120	3500	4400	5250
150	1½ x 1¾	2.64	330	18	2.9	7	14	19	22	29	760	910	1130	1220	1360	1530	1840	2100	2450	2750	3050	3800	4650
200	1½ x 2	3.0	290	17	2.42	6	12	15	18	24	630	765	890	1020	1150	1265	1522	1780	2050	2380	2350	3200	3840
300	2 x 2	4.0	215	15	1.87	5	9	12	14	19	490	590	690	780	880	980	1180	1360	1570	1760	1950	2350	2940
400	2 x 2½	5.0	175	14	1.52	4	8	10	12	15	395	470	550	640	710	790	950	1110	1265	1420	1590	1980	2400
500	2 x 3	6.0	145	13	1.26	3	6	8	9	12	330	395	455	530	595	660	790	920	1060	1200	1330	1650	2000

Rules of the Board of Underwriters

Receiving Stations: Owners of insured residences and buildings are compelled to comply to the following Underwriter's rules:

a—Outdoor antenna and counterpoise conductor sizes shall not be less than No. 14 if copper or No. 17 if of bronze or copper-clad steel. Antenna and counterpoise conductors outside of buildings shall be kept well away from all electric light and power wires or any circuit of more than 600 volts, and from railway, trolley or feeder wires, so as to avoid the possibility of contact between the antenna or counterpoise and such wires under accidental conditions.

b—Antenna and counterpoise where placed in proximity to electric light or power wires of less than 600 volts, or signal wires, shall be constructed and installed in a strong and durable manner, and shall be so located and provided with suitable clearances as to prevent accidental contact with such wires by sagging or swinging.

c—Splices and joints in the antenna span shall be soldered unless made with approved splicing devices.

d—The preceding paragraphs a, b, and c, shall not apply to power circuits used as receiving antenna, but the devices used to connect the light and power wires to radio receiving sets shall be of approved type.

e—Lead-in conductors, that is, conductors from outdoor antennas to protective devices, shall be of copper, approved copper-clad steel or other metal which will not corrode excessively and in no case shall they be smaller than No. 14, except that bronze or copper-clad steel not less than No. 17 may be used.

f—Lead-in conductors from the antenna to the first building attachment shall conform to the requirements for antennas similarly located. Lead-in conductors from the first building attachment to the building entrance shall, except as specified in the following paragraph, be installed and maintained so that they cannot swing closer to open supply conductors than the following distances:

Supply wires 0 to 600 volts....... 2 feet
Supply wires exceeding 600 volts.10 feet

Where all conductors involved are supported so as to secure a permanent separation and the supply wires do not exceed 150 volts to ground, the clearance may be reduced to not less than 4 inches. Lead-in conductors on the outside of buildings shall not come nearer than the clearances specified above to electric light and power wires unless separated therefrom by a continuous and firmly fixed non-conductor which will maintain permanent separation. The non-conductor shall be in addition to any insulating covering on the wire.

g—Each lead-in conductor from an outdoor antenna shall be provided with an approved protective device (lightning arrester) which will operate at a voltage of 500 volts or less, properly constructed and located either inside the building at some point between the entrance and the set which is convenient to a ground, or outside the building as near as practicable to the point of entrance. The protector shall not be placed in the immediate vicinity of easily ignitable material, or where exposed to inflammable gases or dust or flyings of combustible materials.

m—The grounding conductor from the protective device may be bare and shall be of copper, bronze or approved copper-clad steel, and if entirely outdoors shall not be smaller than No. 14 if of copper nor smaller than No. 17 if of bronze or copper-clad steel. If wholly indoors or with not more than ten feet outdoors it need not be larger than No. 18. The protective grounding conductor shall be run in as straight a line as possible from the protective device to a good permanent ground. The ground connections shall be made to a cold-water pipe where such pipe is available and is in service connected to the street mains. An outlet pipe from a water tank fed from a street main or a well may be used, providing such outlet pipe is adequately bonded to the inlet pipe connected to the street water main or well. If water pipes are not available, ground connections may be made to a grounded steel frame of a building or to a grounding electrode, such as a galvanized pipe or rod driven into permanently damp earth or to a metal plate or other body of metal buried similarly. Gas piping shall not be used for the ground.

i—The protective grounding conductor shall be guarded where exposed to mechanical injury.

An approved ground clamp shall be used where the protective grounding conductor is connected to pipes or piping.

j—The protective grounding conductor may be run either inside or outside the building. The protective grounding conductor and ground, installed as prescribed in the preceding paragraphs h and i may be used as the operating ground.

It is recommended that in this case the operating grounding conductor may be connected to the ground terminal of the protective device.

If desired, a separate operating grounding connection and ground may be used, this operating grounding conductor may be either bare or provided with an insulated covering.

k—Wires inside buildings shall be securely fastened in a workman-like manner and except as provided in paragraph m of this section shall not come nearer than two inches to any electric light or power wire not in conduit unless separated therefrom by some continuous and firmly fixed non-conductor, such as porcelain tubes or approved flexible tubing, making a permanent separation. This non-conductor shall be in addition to any regular insulating covering on the wire.

1—Storage battery leads shall consist of conductors having approved rubber insulation. The circuit from a filament "A," storage battery of more than 20 ampere-hours capacity, NEMA rating, shall be properly protected by a fuse or circuit-breaker rated at not more than 5 amperes. The circuit from a plate, "B," storage battery or power supply shall be properly protected by a fuse.

Transmitting Stations: The following paragraphs apply to amateur stations only:

a—Antenna and counterpoise conductors outside buildings shall be kept well away from all electric light or power wires or any circuit of more than 600 volts, and from railway, trolley or feeder wires, so as to avoid the possibility of contact between the antenna or counterpoise and such wires under accidental conditions. Antenna and counterpoise conductors when placed in proximity to electric light or power wires of less than 600 volts, or signal wires, shall be constructed and installed in a strong and durable manner, and shall be so located and provided with suitable clearances as to prevent accidental contact with such wires by sagging or swinging.

b—Antenna conductor sizes shall not be less than given in the following table:

Material	Stations to which power supplied is less than 100 watts and where voltage of power is less than 400 volts	Stations to which power supplied is more than 100 watts or voltage of power is more than 400 volts
Soft copper	No. 14	No. 7
Medium drawn copper	No. 14	No. 8
Hard drawn copper	No. 14	No. 10
Bronze or copper-clad steel	No. 14	No. 12

c—Splices and joints in the antenna and counterpoise span shall be soldered joints unless made with approved splicing devices.

d—Lead-in conductors shall be of copper, bronze, approved copper-clad steel or other metal which will not corrode excessively and in no case shall be smaller than No. 14.

e—Antenna and counterpoise conductors and wires leading therefrom to ground switch, where attached to buildings, shall be firmly mounted five inches clear of the surface of the building, on a non-absorptive insulating support, such as treated pins or brackets, equipped with insulators having not less than five inches creepage and air-gap distance to inflammable or conducting material, except that the creepage and air-gap for continuous wave sets of 1,000 watts and less input to the transmitter shall not be less than 3 inches.

f—In passing the antenna or counterpoise lead-in into the building, a tube slanting upward toward the inside or a bushing of non-absorptive insulating material shall be used, and shall be so insulated as to have a creepage and air-gap distance in the case of continuous wave sets of 1,000 watts and less input to the transmitter, not less than three inches, and in all other cases not less than five inches. Fragile insulators shall be protected where exposed to mechanical injury. A drilled window pane may be used in place of a bushing, provided the creepage and air-gap distance, as specified above, are maintained.

g—Adequate lightning protection either in the form of a grounding switch or suitable lightning arrester shall be provided. The grounding conductor for such protection shall be at least as large as the lead-in and in no case smaller than No. 14 copper, bronze, or approved copper-clad steel. The protective grounding conductor need not have an insulating covering or be mounted on insulating supports. The protective grounding conductor shall be run in as straight a line as possible to a good, permanent ground suitable for the purpose. The protective grounding conductor shall be protected where exposed to mechanical injury.

h—The operating grounding conductor where used shall be of copper strip not less than ⅜ inch wide by $\frac{1}{32}$ inch thick, or of copper, bronze or approved copper-clad steel having a periphery, or girth, of at least ¾ inch, such as No. 2 wire, and shall be firmly secured in place throughout its length.

i—The operating grounding conductor shall be bonded to a good, permanent ground. Preference shall be given to water piping. Other permissible grounds are grounded steel frames of buildings or other grounded metal work in the building, and artificial grounding devices such as driven pipes, rods, plates, cones, etc. Gas piping shall not be used for the ground.

j—The transmitter shall be enclosed in a metal frame, or grill, or separated from operating space by a barrier or other equivalent means, all metallic parts of which are effectually connected to ground.

k—All external metallic handles and controls accessible to the operating personnel shall be effectually grounded.

No circuit in excess of 150 volts should have any parts exposed to direct contact. A complete dead-front type of switchboard is preferred.

l—All access doors shall be provided with interlocks which will disconnect all voltages in excess of 750 volts when any access door is opened.

m—Under the conditions noted in paragraphs 1 and 2, below, wiring may be grouped in the same conduit armored cable, electrical metallic tubing, metal raceway, pull-box, junction box or cabinet.

1. Power-supply wires are introduced solely for supplying power to the equipment to which the other wires are connected.

2. Wires other than power-supply wires run in conduit, armored cable, electrical metallic tubing, metal raceways, pull-box, junction box or cabinet with power supply wires are insulated individually or collectively in groups by insulation at least equivalent to that on the power-supply wires or the power and other wires are separated by a lead sheath or other continuous metallic sheathing.

Common Word Abbreviations

ABT	about	FB	fine business	MI	my	TKS	thanks
AHD	ahead	FM	from	MG	motor-generator	TNX	thanks
AHR	another	FR	for	MK	make	TK	take
AMT	amount	FRK	frequency	MO	more	TMW	tomorrow
ANI	any	GA	go ahead	ND	nothing doing	TNK	think
AY	any	GM	good morning	NG	no good	TR	there
BD	bad	GN	good night	NL	night letter	TS	this
BK	break	GT	get, got	NM	no more	U	you
BUG	speed key	GG	going	NR	number	UR	yours
BT	but	GND	ground	NW	now	UD	it would
BN	been	HA	laughter	OB	old boy	UL	you'll
B4	before	HI	laughter-high	OL	old lady	TT	that
BI	by	HR	hear—here	OM	old man	V	from
BCUZ	because	HW	how	OP	operator	VB	very bad
BTWN	between	HV	have	OT	ought	VT	vacuum tube
BTR	better	HF	high frequency	OW	old woman	VY	very
BRT	brought	HLO	Hello	PLS	please	WA	word after
BIZ	business	HM	him	PSE	please	WB	word before
CK	check	HP	high power	PT	put	WD	would
CN	can	HQ	headquarters	PX	press	WF	word following
CUD	could	I	"ok"	R	ok	WK	work
CUL	see u later	IC	I see	RCD	received	WL	will—would
CUM	come	K	go ahead	RT	right	WN	when
CNT	can't	LID	poor operator	RI	radio inspector	WL	wavelength
C	see	LFT	left	SA	say	WT	what
CW	continuous wave	LIL	little	SM	some	WX	weather
DE	from	LST	last	SS	single signal		
DLD	delivered	LTR	letter	SIG	signal	X	interference
DX	distance	MI	my	SINE	sign	XMTR	transmitter
DNT	don't	MA	milliamp.	SUM	some	YF	wife
DA	day	MSG	message	SEZ	says	YL	young lady
DH	dead-head	MILL	typewriter	SINE	op's initia	YR	your
DC	direct current	MST	must	STICK	pencil	30	finish
ES	and	MNI	many	SKED	schedule	73	regards
EI	that is			TFC	traffic	88	love and kisse

Conversion Table

Factors for conversion, alphabetically arranged

MULTIPLY	BY	TO GET
Amperes	×1,000,000,000,000	micromicroamperes
Amperes	×1,000,000	microamperes
Amperes	×1,000	milliamperes
Cycles	×1.000,000	megacycles
Cycles	×.001	kilocycles
Farads	×1,000,000,000,000	micromicrofarads
Farads	×1,000,000	microfarads
Henrys	×1,000,000	microhenrys
Henrys	×1,000	millihenrys
Kilocycles	×1,000	cycles
Kilovolts	×1,000	volts
Kilowatts	×1,000	watts
Megacycles	×1,000,000	cycles
Mhos	×1,000,000	micromhos
Microamperes	×.000,001	amperes
Microfarads	×.000,001	farads
Microhenrys	×.000,001	henrys
Micromhos	×.000,001	mhos
Micro-ohms	×.000,001	ohms
Microvolts	×.000,001	volts
Microwatts	×.000,001	watts
Micromicrofarads	×.000,000,000,001	farads
Milliamperes	×.001	amperes
Millihenrys	×.001	henrys
Millimhos	×.001	mhos
Milliohms	×.001	ohms
Millivolts	×.001	volts
Milliwatts	×.001	watts
Ohms	×1,000,000,000,000	micromicro-ohms
Ohms	×1,000,000,000	micro-ohms
Volts	×1,000,000	microvolts
Volts	×1,000	millivolts
Watts	×1,000,000	microwatts
Watts	×1,000	milliwatts
Watts	×.001	kilowatts

Radio Symbols

The following symbols are commonly used in radio work and many of these symbols are used in the pages of this book:

E_F.............Filament (or heater) terminal voltage
E_B.............Average plate voltage (DC)
I_B.............Average plate current (DC)
E_P.............AC component of plate voltage (effective value)
I_P.............AC component of plate current (effective value)
E_C.............Average grid voltage (DC)
I_C.............Average grid current (DC)
E_G.............AC component of grid voltage (effective value)
I_G.............AC component of grid current (effective value)
E_{FF}...........Filament (or heater) supply voltage
E_{BB}...........Plate supply voltage (DC)
E_{CC}...........Grid supply voltage (DC)
U...............Amplification factor
r_P.............Plate resistance
s_M.............Grid plate transconductance (also mutual conductance, gm)
R_P.............Plate load resistance
Z_P.............Plate load impedance
DC.............Direct Current (as adjective)
AC.............Alternating Current (as adjective)
RMS...........Root Mean Square
U.P.O..........Undistorted power output
C_{GK}...........Grid-cathode (or filament) capacitance
C_{PK}...........Plate-cathode (or filament) capacitance
C_{G1P}..........Effective grid-plate capacitance in a tetrode (cathode [or filament] and screen grounded)
$C_{G1}(k+g_2)$......Direct interelectrode capacitance of grid to cathode (or filament) and screen
$C_P(k+g_2)$......Direct interelectrode capacitance of plate to cathode (or filament) and screen

Test Instruments and Calculations

The technique of making electrical measurements and the use of measuring equipment encountered in the problems of amateur radio practice are outlined in this section.

Voltage Multipliers: In practically all radio measurement work a 0-1.5 DC milliammeter has been found to be sufficiently sensitive for average amateur service. To use this instrument for the measurement of voltage requires that a resistor be placed in series with the meter, and the value of which depends upon the highest voltage to be measured and equals the range of the meter in milliamperes times the series resistance: expressed,

$$R = E_{max} / I_{max} \text{ and } E = IR$$

Current Shunts: To increase the range of the above instrument up to say, 10 amperes, requires the use of a current shunt, whose function is to carry part of the total current thereby lowering the flow of current through the meter. For any current reading the value of the shunt resistance is found by dividing the resistance of the meter by the maximum range of the meter, minus 1.

Resistance Measurements: Resistances can be measured with a precision comparable to that of the meter accuracy with the aid of the following resistance formula:

$$R_x = \frac{R_m (R_o + R_m)}{\dfrac{E}{I_m} - R_o}$$

where, in the diagram, R_x is the unknown resistance; R_m, the internal resistance of the meter; R_o, the limiting resistor; E, the battery voltage; and I_m, the current through the meter.

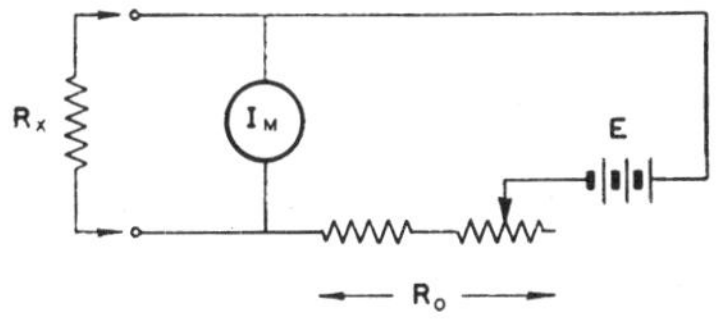

FIG. 1—Ohmmeter circuit.

In setting up the circuit, full scale deflection on the meter is first obtained. The values E and R_o are considered to be standard and do not affect the meter reading by any appreciable amount.

Vacuum-Tube Voltmeters: This vacuum-tube voltmeter will measure AC voltages from 60 cycles up to any frequency. The instrument will determine the audio response of an amplifier, a radio receiver, or the gain of an audio or radio-frequency amplifier; in addition, it can be made to measure the percentage modulation of a phone transmitter or indicate frequency characteristics. In conjunction with an auxiliary full-wave linear rectifier, the peak voltmeter herein described will measure both positive and negative modulation loops of a modulated carrier signal.

FIG. 2—Compact vacuum tube voltmeter.

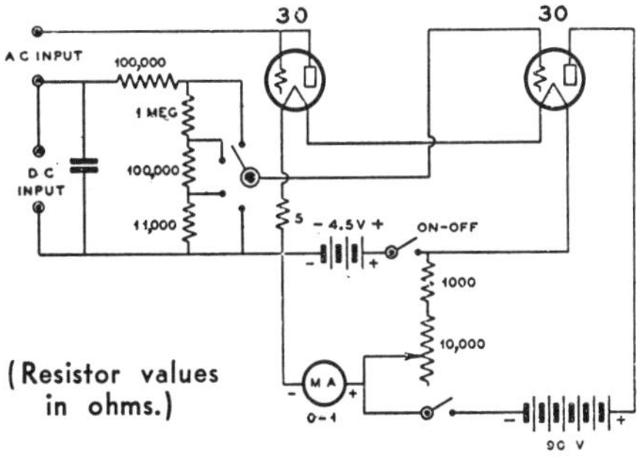

Vacuum tube voltmeter circuit.

In the circuit diagram it will be seen that the VT voltmeter consists of a peak voltage rectifier of the diode type and a DC amplifier incorporating a 0-1 milliammeter for reading low voltages. It is interesting to note that the rectifier is of the linear type, and not square law indicating, hence, the meter scale is direct reading. The diode measures peak voltage if the DC load is at least one megohm. This load can be divided by a series of resistance units for multi-range purposes. That is, the voltmeter can be made to cover from a small fraction of a volt to over 100 volts by means of a multiplier switch with-

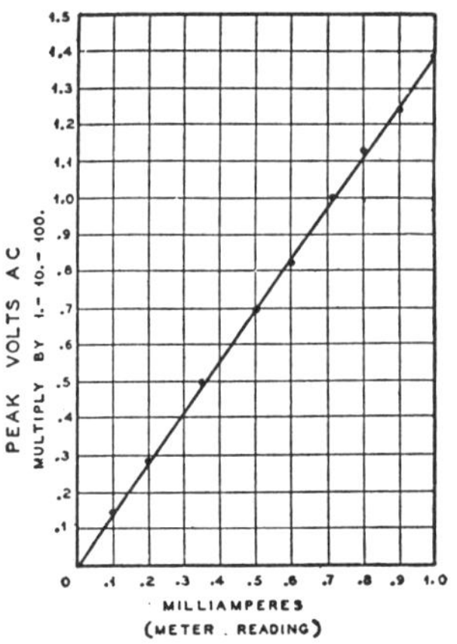

FIG. 3—Vacuum tube voltmeter chart.

out affecting the input circuit. In this circuit, an ordinary milliammeter has been substituted for the more expensive microammeter.

The connection for measuring DC voltages draws less than 0.1 ma. for voltages up to 100, and in the 1 volt range the load is less than 1 microampere. This feature makes the instrument ideally suitable for measuring DC voltages where no appreciable load is permissible, such as the voltage developed by AVC in a radio receiver.

The resistances, tubes and battery voltages are of such values that for all practical purposes the meter scale is direct reading. The multiplier may be marked 1, 10, and 100 and the scale deflection multiplied by these values for either DC or RMS voltages. The error is not over 6 per cent. For actual peak values of AC voltage, the scale reading must be multiplied by 1.4. Over a period of time this form of voltmeter should not be relied upon for better than plus-or-minus 10% accuracy because of tube and battery voltage. For any individual set of measurements a calibration can be made, although generally the relative ratio of voltages is of more importance. The voltmeter is accurate for this purpose.

The 10,000 ohm resistor is adjusted so that the meter needle is brought to zero when no AC voltage is impressed. The 5 ohm filament resistor provides about ¼-volt negative bias on the diode to prevent current from flowing in the multiplier resistors when no AC voltage is present in the circuit. This makes the 10,000 ohm resistor control setting practically constant for any step on the multiplier. (Hint: If the meter cannot be set to exact zero a vernier adjustment may be added by placing a 200 ohm variable resistor in series with the regular control resistor.) The battery switch should preferably be a DPST type in order to open-up the plate circuit at the same time the filament and balancing circuit is opened.

In designing the meter it will be found that a bakelite front panel will simplify all insulation problems. It measures 5 x 8 x $\frac{3}{16}$-inches, and is mounted at right angles to a wooden baseboard. The latter hold the tubes and batteries, as shown. The aluminum cover is 5 x 9 inches and serves to protect the instrument from dust and breakage.

In using this VT voltmeter, it is necessary to have a DC path through the circuit under measurement. Most circuit measurements provide such a path, although occasionally it must be provided by means of the secondary of an audio transformer for AF measurements, or an RF choke for RF measurements. In such cases the choke should shunt the AC input binding posts. The DC resistance path of the circuit under measurement can be as high as 50,000 ohms without affecting the calibration of the VT voltmeter. Since the input is a diode this V.T.V. will have a loading effect on the circuit under test.

High Impedance V.T. Voltmeter: This V.T. voltmeter is suitable for making measurements which require an instrument having a very high input impedance. This instrument can be placed across grid circuits, or even across RF coils in some cases. Its range is from a very small fraction of a volt to slightly over one volt, r.m.s. value, or approximately 1.5-volts peak value. It is a peak type voltmeter.

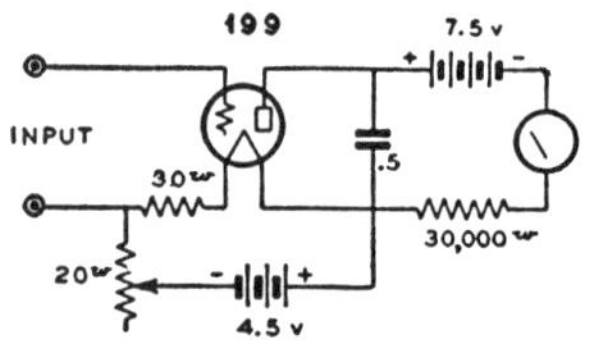

FIG. 4—Circuit diagram of portable VT voltmeter.

The circuit is of simple design, employing C batteries for the power supply and a 0-50 micro-ammeter. The entire voltmeter is mounted in a card filing case about 6-in. x 8-in. x 5-in. A small bakelite panel supports the meter, rheostat and input binding posts. The UX199 tube is mounted on brackets under the panel. The two batteries lie flat in the bottom of the case; the 4½-volt unit supplies the C-bias and the filament current, while the 7½-volt unit acts as a B battery.

The micro-ammeter requires no opposing voltage networks as the B battery voltage is low and a series protective resistor of 30,000 ohms is in the circuit. This resistor has little effect at low readings, but tends to drop the plate voltage at higher values of current, resulting in a somewhat linear calibration curve of input voltage versus meter deflection. A .5-mfd. low-leakage condenser by-passes all AC current from plate to filament, improving the detection characteristic.

A 30-ohm resistor provides C-bias due to filament current voltage drop which amounts to about 1.5 volts. The 20-ohm filament rheostat allows an initial reading adjustment and also provides an off-switch for the filament circuit. The rheostat is adjusted to give an initial deflection of 5 micro-amperes, one main scale division on the meter. The input leads must be short-circuited at all times when the instrument is not in use.

The unit may be calibrated by using a potentiometer and a low-reading AC voltmeter across a filament winding of a transformer. The AC voltmeter reads r.m.s. values; consequently 2.5 volts r.m.s. equals 3.53 volts peak. The potentiometer can be employed to impress a known ratio of this voltage across the V.T. voltmeter. The actual impressed peak voltage should never exceed the bias voltage of about 1.5 volts.

This same method can be utilized with higher C-bias and plate voltage to read greater values of RF or AF voltage. In such cases, the micro-ammeter should have an opposing current network with a variable resistor arranged to allow a zero meter reading at the desired initial plate current.

C.W. Frequency-Meter Monitor: This CW monitor and frequency-meter can be used to calibrate a receiver or to check the frequency of a transmitter as well as monitor the tone and keying of a CW transmitter.

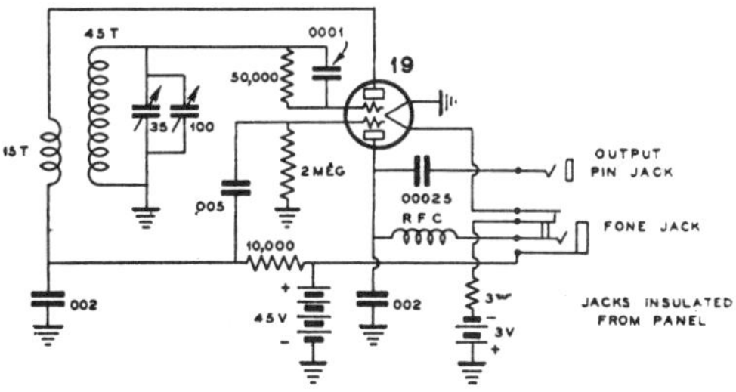

FIG. 5—Circuit diagram of frequency meter-monitor.

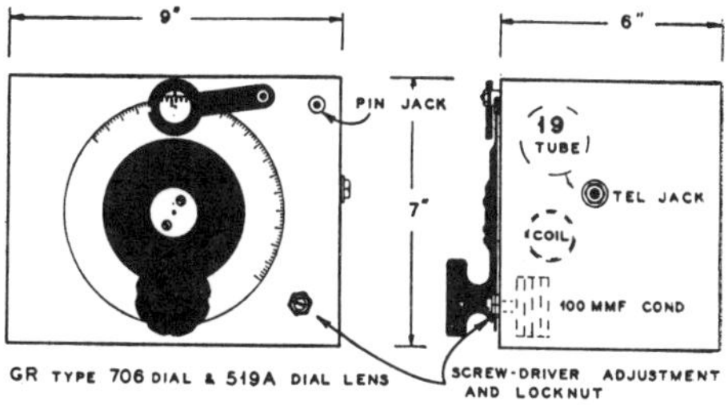

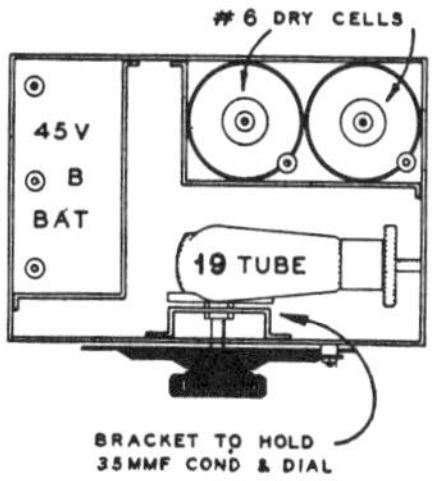

FIG. 6
CONSTRUCTIONAL DETAILS

The filament and B batteries are inside the case. The proper location for the coil is shown in the upper illustration.

The instrument in its air-tight aluminum case.

The oscillator circuit is of the simple regenerative type. Band-spread is obtained by two condensers, a 100-mmfd. variable which has a screw-driver slot adjustment and locking nut, and 35-mmfd. variable driven by a large General Radio Co. vernier dial; this dial can be read accurately to within one part in 1500. It is equipped with a magnifying glass enabling the operator to read scale divisions to a very small fraction.

The instrument is built into an aluminum box about 7 inches high, 9 inches long and 6 inches deep; the thickness of the metal being approximately ⅛th inch. The corner posts are made of ¼-inch square rod. Each aluminum sheet is fastened to the square rod along the edge with three of four machine screws. Aluminum strips of No. 12 gauge, one inch wide are used to form brackets to hold the two No. 6 dry cells and a portable type 45-volt B battery in place. This same strip also provides a bracket for rigidly mounting the 35 mmfd. tuning condenser to the front panel. Since the dial is supported by this condenser rotor, the latter should be of sturdy construction and the bracket should permit at least three points of suspension mounting to the front panel. The coil, jack, and tube socket are mounted horizontally on the end of the panel and the resistors and fixed condensers are supported by direct connections on the wiring. The jack is a type which has an extra set of contacts, closing the filament circuit when the telephone plug is inserted into the jack. This eliminates the possibility of forgetting to turn off the filament switch when the unit is not in use.

The second triode unit of the Type 19 functions as an audio amplifier in order to augment the strength of the beat note signal when the meter is monitoring a CW transmitter. Because the two plates are capacitively coupled inside of the tube, it is necessary to provide an RF by-pass of .001 or .002 mfd. from plate to filament, and a well-designed 2½-millihenry RF choke in series with the telephone jack. The grid resistor of 2 megohms is so high in value that a .005 mfd. mica condenser is sufficient for the audio coupling capacitor.

A tip-jack, insulated from the front of the panel allows pick-up of an external signal. A wire from a few inches to a few feet long can then be used as an "antenna." This wire must be coupled very loosely into the oscillator circuit so that variations in the external length will not produce any deleterious effects on the frequency calibration; variations in length, however, change the intensity of the beat note signals, but without affecting the stability of the instrument.

Technical Notes: If the instrument is to oscillate over the 80-meter band, and the harmonics used on 10, 20, and 40-meters, the coil must have a secondary inductance of 20 microhenrys. A winding one-inch long of No. 22 DSC wire on a one-inch diameter will give this value; or No. 22 Enameled wire wound on a ⅞-inch diameter, one-inch long. The tickler is wound over the ground end of the main coil with a layer of tape or paper between, approximately 12 or 15 turns of No. 30 DSC wire will produce the desired result.

If the meter is to be used in the 160-meter band, the coil is wound with 72 turns of No. 26 Enameled or No. 28 DSC wire on a one-inch diameter bakelite tube. This coil will have about 80 microhenrys inductance and nearly complete bandspread tuning. In both cases the band-setting condenser is placed at abount two-thirds capacity, and once the band is located over the dial range, the condenser should be set with a locknut.

This instrument may be calibrated from standard frequency transmissions (see next sub-topic) which can be picked-up by means of any good short wave receiver. These transmissions are given periodically and are a convenient method of calibration, unless one has another oscillator and wishes to calibrate from harmonic frequencies of local broadcast stations. Then the extra oscillator should be set to give zero-beat (point of no signal between two beat notes) with the carriers of broadcast stations as heard in a BCL receiver, and the harmonics hetrodyned to zero-beat with the frequency-meter (see "Making Zero-beat Adjustments" in this section). Broadcast stations whose frequencies lie between 1000 and 850KC are convenient for this purpose, especially those stations which maintain carriers within 10 or 15 cycles of their assigned frequencies.

An AC Frequency Meter-Monitor: The device here shown uses a type 24 tube as an electron-coupled oscillator, tuned over the 160 meter band. Harmonics of the 160 meter range are used in the 80, 40 and 20 meter bands, and the actual calibration curve shown in Fig. 2 was plotted for the 80 meter band. For 160 meter use, the curve readings would be halved, because frequency in kilocycles is used instead of meters of wavelength.

The oscillator circuit uses a small tuning condenser shunted by a large band-setting

Front view of frequency meter-monitor.

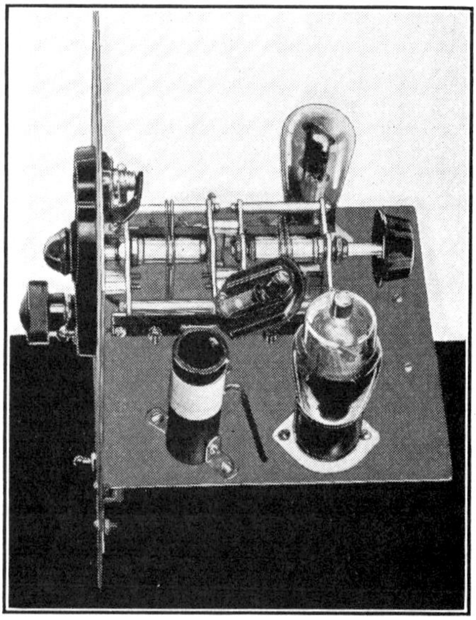

Interior view of frequency meter-monitor. Note placement of grid condenser and wide spacing of coil from metal chassis.

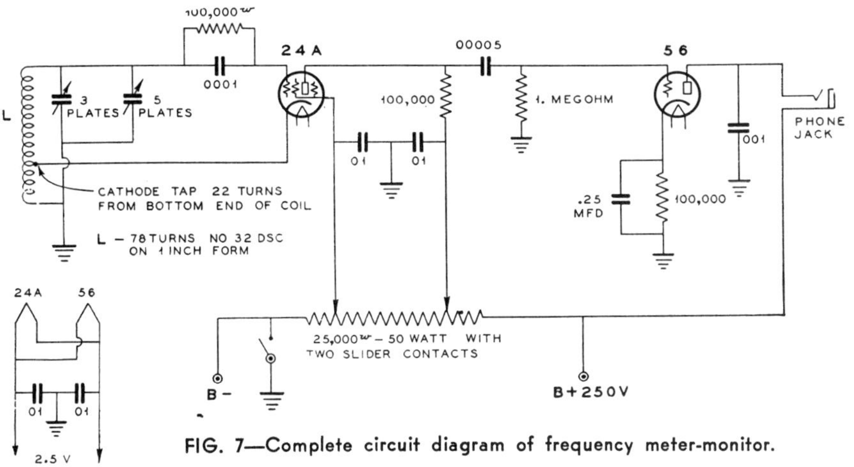

FIG. 7—Complete circuit diagram of frequency meter-monitor.

condenser. The latter is adjusted only when calibrating the frequency meter. The smaller condenser has a vernier dial which can be read accurately for frequency determination.

The 56 tube is used as a detector, beating the external signal against the 24 oscillator fundamental or one of its harmonics. A slight external coupling through a pin jack to the grid of the 56 can be used to pick-up low power transmitter signals. The same coupling can be used to provide a beat note into a receiver for checking the frequency of received signals.

The pictures show the constructional details. Of paramount importance is good rigidity and well-soldered connections. The coil is made of 78 turns of No. 32 DSC wire, wound to cover about one-inch of space on a one-inch diameter bakelite tube. The cathode tap is made at a point 22 turns up from the ground end of the coil winding.

The filament and B supply can be taken from the receiver power supply. A switch is provided which opens the negative B lead to the frequency meter ground, making the instrument inoperative, when desired. The filaments or heaters should be turned on during all of the time the receiver or transmitter is to be operated. About a half-hour warm-up period should be allowed before calibration is made, in order to minimize frequency creepage.

This instrument can be calibrated either from standard frequency transmissions in the amateur bands, or by means of broadcast station transmissions. The latter are required by law to operate within 50 cycles of their assigned frequency, and most of the higher power or better stations operate within 10 to 15 cycles of their exact assignment. A broadcast receiver can be used to pick up these stations and a small oscillator using a B battery, a couple of dry cells and a type 30 tube can be used to "zero beat" any particular received broadcast station. An electron-coupled oscillator capable of tuning across the broadcast band will give stronger harmonics than a type 30 tube oscillator in this range, therefore a serviceman's test oscillator should be used if possible. The local oscillator with zero beat to a station on 880 KC, for example, will have a second harmonic on 1760 KC in the 160 meter band. The fourth harmonic would be 3520 KC in the 80 meter band, the 8th harmonic on 7040 KC in the 40 meter band, and the 16th harmonic, if audible, on 14,080 KC in the 20 meter band.

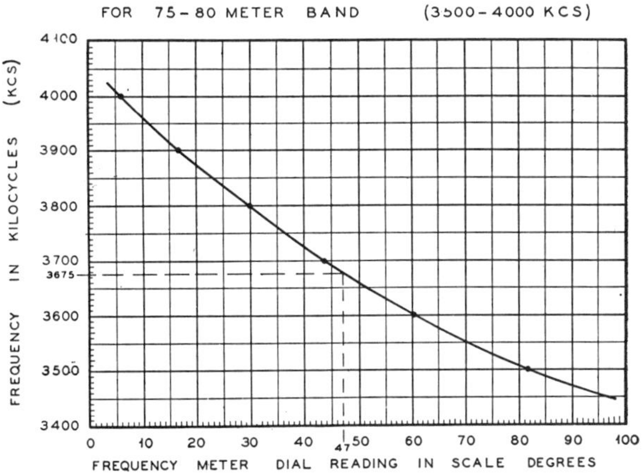

FOR EXAMPLE — SAY FREQ METER DIAL READING OF SIGNAL WAS 47; PROJECT THE VERTICAL LINE "47" TO CALIBRATION CURVE, THEN PROJECT THIS POINT HORIZONTALLY TO THE KC READING WHICH, IN THIS CASE, IS 3675 KCS.

FIG. 8—FREQUENCY METER CURVE.

The frequency meter is used to beat-note against this frequency and at zero beat, as heard in the broadcast band receiver and also in the frequency meter. An exact calibration point can be obtained. Several broadcast stations can be used to provide "harmonics" in this manner for several calibration points of the frequency meter.

Graph paper with 10 divisions to the inch, both horizontally and vertically, (100 squares to the square inch) is used for drawing the curve. Fig. 8 shows a typical 75-80 meter (3500 KC to 4000 KC) curve as plotted from the frequency meter here described. The horizontal line at the bottom of the graph denotes the tuning dial scale divisions from 1 to 100, but marked only in units of 10 on the graph paper. The vertical portion of the graph is used to denote frequency in kilocycles, beginning with 3400 KC and ending with 4000 KC. Similar charts can be plotted for the other amateur bands.

The first requirement is to find a certain frequency, such as 3500 KC, so that the curve-plotting process can begin. Standard frequency transmissions are sent on the amateur bands at regular intervals.

When the standard frequency station announces (in telegraphic code) that it will transmit on 3500 KC, the receiver in the amateur station is then tuned to 3500 KC in such a manner that 3500 KC falls at the extreme end of the tuning dial . . . at the 100 degree scale indication on the dial. Then, with the chart at hand, and the standard frequency of 3500 KC known, the amateur receiver is tuned to zero-beat. This found, the frequency meter dial is rotated until a point is found where the frequency meter also zero-beats with the receiver. The next step is to observe the setting of the frequency meter dial, also the setting of the amateur receiver dial. If the receiver dial is at 100° for 3500 KC, and if the frequency meter dial is at 80°, a dot is placed on the graph paper at a point where the vertical line which corresponds with No. 80 on the horizontal line crosses this line, as shown in Fig. 8. This point of intersection will be the 3500 KC point on the graph curve.

If a station asks you to check its frequency, you first zero-beat the signal on your receiver, then you zero-beat the frequency meter against the receiver. You next observe the setting of the frequency meter tuning dial and by means of the graph you can quickly find the frequency of the station which asks you for a frequency check. Suppose this signal is found at 600 on the frequency meter tuning scale; running your finger UP on the chart, you find that 60 on the horizontal line intersects with 3600 on the vertical line. Thus the frequency of the station you are checking is 3600 KC.

To check the frequency of your own transmitter, zero-beat the frequency meter against your transmitted signal and find the frequency from the curve.

Because there are 10 dividing lines for each unit of the graph, the frequency can be quite accurately shown, perhaps within 2 or 3 KC of the exact frequency. On the other hand, there are 100 points on the horizontal line, so that the entire dial sweep from 0° to 100° can be easily followed.

Schedule of Radio Emissions of Standard Frequency: The National Bureau of Standards has a regular schedule of standard frequency emissions from its station WWV, Beltsville, Md., near Washington, D. C. These broadcasts are available to transmitting stations for adjusting their transmitters to exact frequency, and to the public for calibrating frequency standards and transmitting and receiving apparatus.

These broadcasts are given on two days a week on three single frequencies 5000, 10,000 and 15,000 KC. Those transmissions on 5000 KC are particularly useful at distances within a few hundred miles from Washington, those on 10,000 KC are useful for the rest of the United States, and those on 15,000 KC are useful in the United States and other parts of the world as well. Each Tuesday and Friday (except legal holidays) three frequencies are transmitted as follows: noon to 1 p. m., Eastern Standard Time, 15,000 KC; 1:15 to 2:25 p. m., 10,000 KC; 2:30 to 3:30 p. m. 5000 KC.

The transmissions consist mainly of continuous, unkeyed carrier frequency, giving a continuous whistle in the phones when received with an oscillating receiver. For the first five minutes the general call (CQ de WWV) and the announcement of the frequency are transmitted. The frequency and the call letters of the stations (WWV) are given every 10 minutes thereafter.

The accuracy of the frequencies transmitted is at all times better than a part in five million. From any of them, using the method of harmonics, any frequency may be checked.

Making Zero-Beat Adjustments: Methods for making accurate zero-beat adjustments between two oscillators or carrier frequencies are given herewith:

To make a zero-beat adjustment between a transmitter and an oscillating receiver the first requirement is that a continuous wave (CW) signal be received, properly identified, then the receiver set to zero-beat. Second, the regeneration is reduced in the receiver until oscillation nearly stops. The receiver is next tuned slightly and the beat note heard (the strength of the note will be somewhat weaker) should be carefully reduced to zero frequency. When the condition is recognized, the receiver will be in EXACT zero-beat with the received signal. Unless a beat-frequency indicator is used, a telephone headset will match the two frequencies to within one cycle. Precaution must be exercised during an audible adjustment due to the fact that the zero state may be more than one cycle per second.

To zero-beat two carrier frequencies such as a local oscillator and transmitter signal, all that is required is the use of an oscillating receiver. First, the transmitter signal is picked-up and the oscillating receiver made to zero-beat with it. Next, the beat oscillator is turned on and made to zero-beat with the receiver. The frequency of the receiver is NOT varied during this procedure. Now, says J. K. Clapp in an issue of the I. R. E. Proceedings, "If the

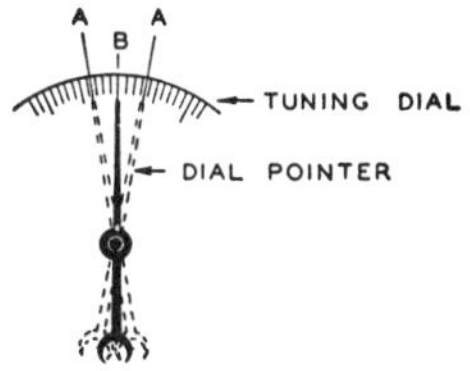

FIG. 9

"A-A" are the two sides of the signal having the same "beat note" (tone).
"B" is the "zero beat" point (no tone).

frequency of the oscillating receiver is moved away from the zero audible beat setting an audible beat tone of, say 1000 cycles is heard, the difference frequency between the signal and the local oscillator will be heard in the form of a waxing and waning audio-frequency tone. If the frequency of the receiver is varied slightly, thereby changing the audio-frequency, no change in the rate of waxing and waning occurs, showing that the beat is between the signal and the local oscillator. If the waxing and waning does **change** when the receiver frequency is varied, the beat note is between the wrong pair of oscillators, and the adjustment should be made again with more care—after the waxing and waning beat is heard, the oscillator may be re-adjusted to bring the rate of waxing and waning to one, or less cycles after which the two frequencies will be matched to within one cycle."

Modulated Oscillators: There are numerous types of modulator-oscillator circuits, all of which have certain advantages and disadvantages but, for ordinary service around an amateur's station or laboratory, one of the more simple types of modulator-oscillator is satisfactory.

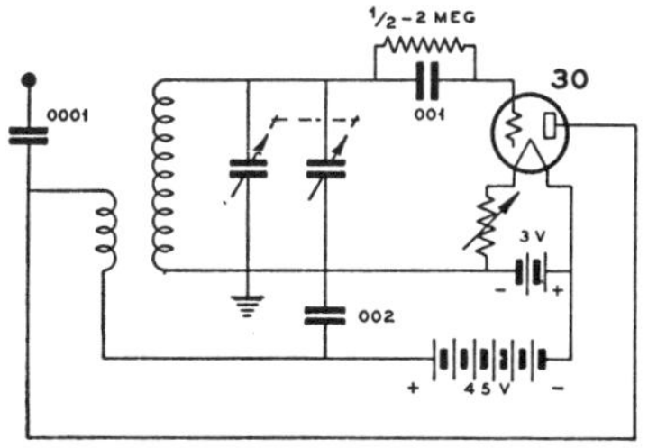

FIG. 10—Circuit diagram of Modulated Oscillator with type 30 tube for battery operation.

In Figure 10 is shown a schematic diagram of an instrument employing either a 199- or type 30-tube arranged to perform the dual function of oscillator and modulator. Modulation is obtained at an audio frequency of 500 or 1000 cycles by a grid blocking action. The tone may be adjusted by changing the grid-leak or grid condenser, and also by rheostatic filament

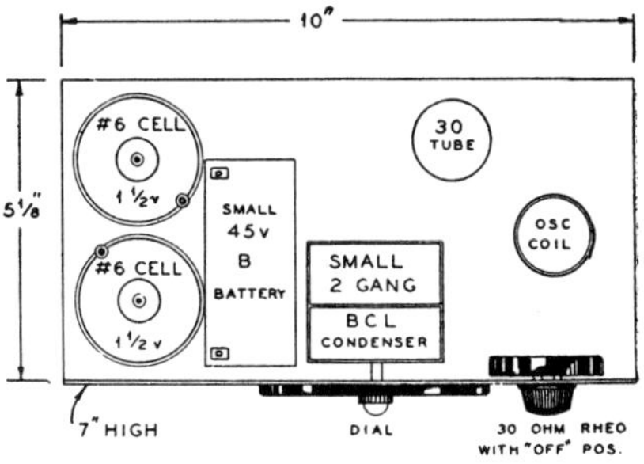

Parts layout for battery model.

control. A two-gang broadcast receiver type variable condenser is used as the main tuning element by connecting both sections in parallel. The coils may be taken from an old BCL receiver where the primary (tickler in this case) has between 20 and 30 turns. The oscillator may be built into a metal box with sufficient space allowed to include a portable 45-volt B battery and two No. 6 dry cells.

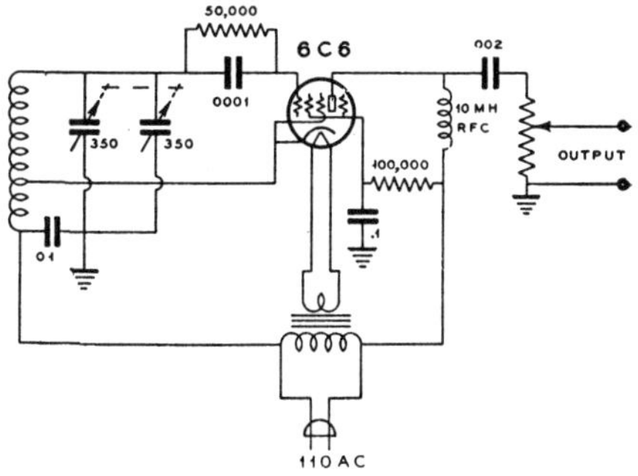

FIG. 11—Circuit diagram for A-C Oscillator.

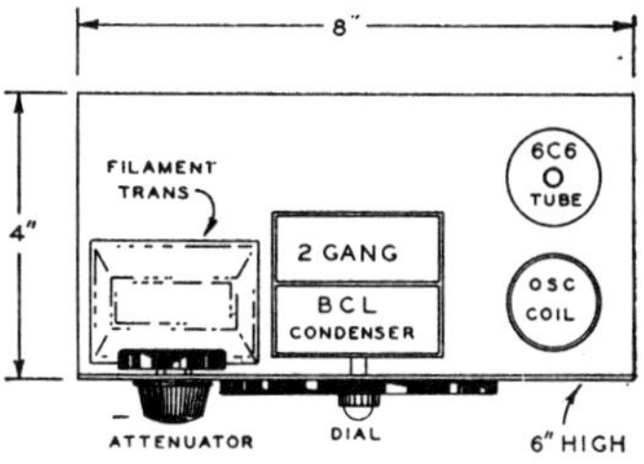

Parts layout for A-C Oscillator.

The oscillator circuit shown in Figure 11 is of the electron-coupled type modulated by the 60 cycle AC plate voltage. In this circuit, as well as in the one previously described, the same type tuning condensers are used. The tuning coil has about 120 turns of No. 30 DSC wire wound on a 1¼-in. diameter tubing. The cathode winding must have an optimum value of turns to obtain the correct amount of modulation from the AC plate supply. It would be more desirable to wind a special coil with taps every five turns from the lower end in order to find the most suitable connection. The filament supply to the circuit is furnished by an old bell-ringing transformer. Small mica condensers isolate the instrument housing and output circuit from the 110-volt AC mains. The disadvantage of this circuit is in the very low-frequency modulation.

Calibration of either of these oscillators can be accomplished by means of a broadcast receiver. The high frequency range of the oscillator down to 550 KC can be calibrated by means of direct or beat-note reception of known frequency broadcast stations and the oscillator signal. The upper range may be roughly calibrated by extending the curve, or more accurately, by employing the second harmonic which will be audible in the broadcast range in the receiver. Dividing this reading in each case will give the fundamental frequency of the oscillator. The latter should tune to about 350 KC, which makes it useful to line up 450 KC superheterodyne receivers.

If a careful calibration of the fundamental frequency is made on the oscillator, the harmonics may be used to locate short wave stations, either amateur or broadcast.

The frequency of a quartz-crystal (a component in a single signal receiver) can be determined very closely by setting the quartz plate on, or leaning it against the grid of the oscillator. At resonance the oscillator will suddenly change, as listened to in a broadcast receiver tuned to the oscillator second harmonic. This test requires the manipulation of both the oscillator and the BCL receiver, but once the crystal frequency is found, the IF amplifier in the single signal receiver can be lined up to that frequency by means of the oscillator.

Note: Lining up an IF amplifier should always begin at the grid of the tube preceding the last stage of IF transformer. After that transformer is aligned (by ear or output meter), the next preceding stage may be lined up using less coupling to the grid of the next preceding stage. It is emphasized that one must always work with a fairly weak signal, because many sets have AVC which would introduce errors with a strong signal peaking, unless a meter is used.

The Harmonic Oscillator: A simple multivibrator with an auxiliary control oscillator capable of supplying a series of calibration points for the calibration or check-

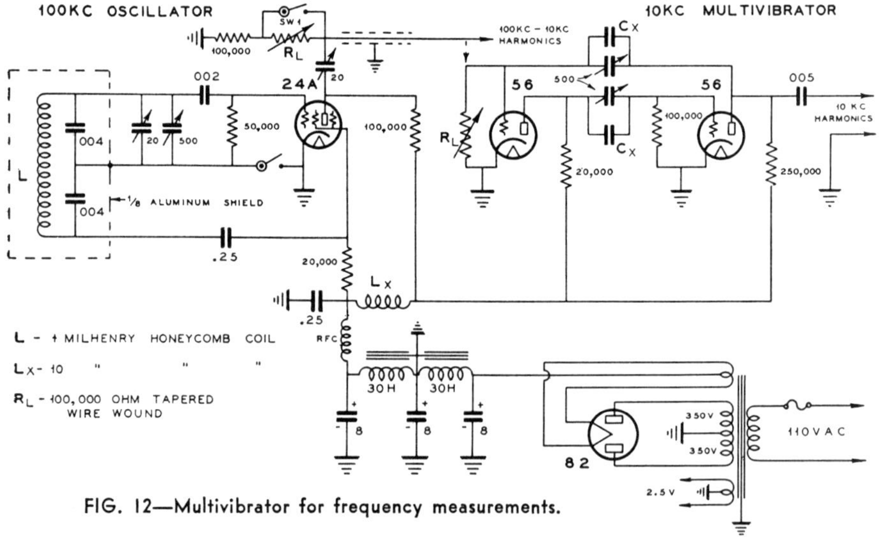

FIG. 12—Multivibrator for frequency measurements.

For example, if the short-wave station is listed at 6.01 megacycles, the oscillator can be set at 1502.5, 1202, 1001.66, 858.6 KC etc., which will all give harmonics on 6.01 MC. By checking at least two fundamental points, it is possible to ascertain which harmonic is heard in the short wave receiver. The fundamental of 1502.5 has an harmonic on 4.5 MC which may cause an error, but by swinging the oscillator setting over to 1202, the next fundamental having a 6.01 MC harmonic would give no harmonic at 4.5 MC.

ing of transmitters, oscillators and radio receivers at fixed intervals in frequency can be constructed by simply following the wiring diagram appearing in Figure 12. This schematic is practically self-explanatory, the parts complement together with the electrical specifications are listed.

The oscillator circuit is a modified Colpitts to which electron-coupling has been added to insure high-stability. To facilitate identifying harmonics of either the multivibrator or control oscillator, provision has been made by the use of switch sw_1 to short out a portion of the grid-leak R_L. When the switch is in the "open" position the harmonic of both systems will have equal strength. When the switch is "closed" the relative strength of the multivibrator harmonics will be weaker.

One of the chief uses of the harmonic oscillator is in the measuring of frequency of any broadcast station; the technique follows:

By utilizing the 5000 kilocycle standard frequency transmissions of WWV, broadcast and amateur stations can accurately check their frequencies and standardize calibrated apparatus. The following steps are required in making the measurements:

(1) Tune the control oscillator to zero-beat with a standard signal or one of known accuracy. (See next sub-topic for making zero-beat adjustments.)

(2) Turn "on" the multivibrator and adjust the relaxation frequency to 10 KC. The device now supplies a series of 10 KC intervals which correspond to the broadcast channels.

(3) Tune the receiver used in (1) to the station frequency to be checked. A beat-note will now be heard between the oscillator harmonic and station frequency. If no beat is detected, the station frequency is as accurate as that of the standard frequency. Usually this will not be the case; hence, the beat frequency heard will be indicative of the number of cycles the station is off from standard. If the beat is too fast to count, it is only necessary to change the control oscillator and read the difference frequency directly from the dial which, of course, has been pre-calibrated for this purpose. The dial must be returned to its original setting after each frequency reading else the instrument will be in error the next time it is used.

To calibrate a receiver or oscillator at 10 KC intervals, it is only necessary to zero-beat the frequency going under calibration with the output of the harmonic oscillator. It is to be realized that the instrument has been previously set to zero-beat with the standard signal.

Beat-Frequency Audio Oscillator: A simple variable audio oscillator can be made by using a 2A7 or 6A7 tube in the circuit shown. The oscillator is remarkably stable, practically free from frequency drift, and is capable of holding its calibration over long periods of time.

The circuit consists of two RF oscillators electronically coupled to each other and to the output. The oscillatory circuits 0-1 and 0-2 oscillate at frequencies near 500 KC. The generated frequency is appreciably stable, and any changes in electrode voltages similarly affect adjacent electrodes in the tube envelope in a like manner. To reduce the frequency drift to minimum due to ambient room temperatures, it is best to place the oscillator coils and condensers in a small balsa-wood box. The use of large capacity and small coils in the oscillatory circuits minimizes radio frequency resistance, reduces the amplitude of harmonics and assures good wave form. The resistances R_1 in series with the oscil-

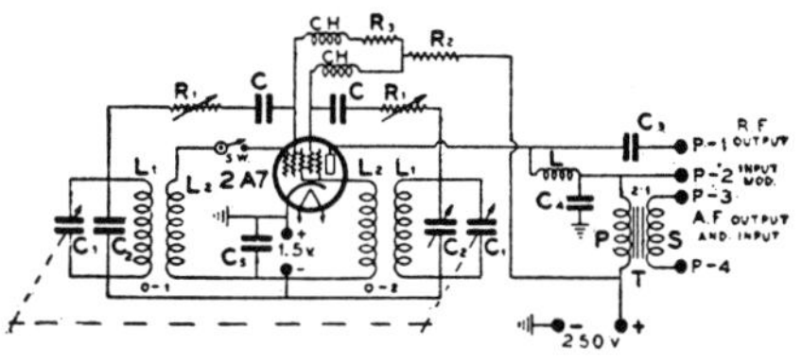

FIG. 13

Complement of parts:

C—.002 mfd. C1—50 mmf. C2—.0001 mfd.
C3—.0005 mfd. C4—.001 mfd. C5—.01 mfd.
R1—0-100,000 ohms. R2—40,000 ohms.
R3—10,000 ohms. Transformer ratio is 2:1.
L—L1—L2—100 mh. CH—20 mh.

lator "plate" leads provide resistance stabilization and limit the oscillation to the straight line part of the tube characteristic.

If switch "sw" is used to open the grid circuit of one of the oscillators, the other can be used as an RF test oscillator, since a .001 mfd. variable condenser provides a wide frequency coverage. The two 50 mmfd. condensers C_1 should be ganged together so that as the capacity of one condenser is increasing the other is decreasing; this causes a variation of frequency both up and down from approximately 500 KC to produce a beat-note of from a few cycles to as high as 10,000 cycles per second.

The triode grid and plate circuit are connected similar to a superheterodyne HF oscillator, with a one-to-one ratio coil such as would be employed to tune a normal receiver over the 100 to 200 meter band. The other circuit incorporates the screen-grid as the oscillator grid and the control-grid as the oscillator "plate" element. The regular plate functions as a demodulator-amplifier circuit with an audio transformer for coupling to an external circuit; the output can also be electron-coupled through capacitor C_3.

A few uses for a beat-frequency oscillator of this type are:

(1) Bridge measurements of all kinds.

(2) Measurements in electrical communication apparatus.

(3) Studies of response curves of loudspeakers and transformers.

(4) Characteristic analysis of filters, cables and dielectrics.

(5) RF signal generator for testing and lining-up receivers.

(6) Audio frequency measurements on speech amplifiers and modulators of a phone transmitter.

Over-Modulation Indicator: One of the Federal Communication Commission requirements are that every phone station is required to have a means for determining the limit of modulation. Thus, an over-modulation indicator must be in constant service when the phone transmitter is in operation.

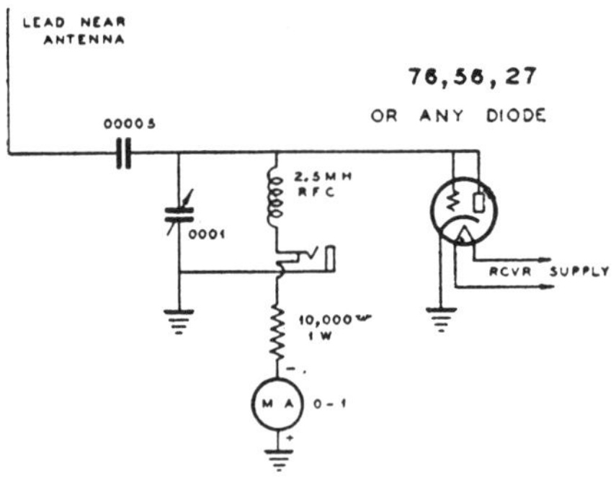

FIG. 14—Circuit diagram of Overmodulation Indicator.

A piece of No. 12 gauge aluminum is bent in one piece to provide a mounting stand for the Overmodulation Indicator.

The Linear rectifier shown in Figure 14 will indicate the slightest amount of carrier shift, which means over-modulation. The instrument is simply adjusted by varying the capacity of a small condenser; thus, the optimum meter deflection is obtained without resorting to coil-coupling schemes. The 50 mmfd. fixed condenser and 100 mmfd. variable condenser form a variable attenuator for RF voltage. The instrument is capacitively coupled to the antenna by running a line from the unit close to the antenna lead-in or feeder.

The indicator is built into a 4-in. x 12-in. x 14-guage aluminum strip, bent as shown in the photograph. The wiring details appear in the schematic. The RF choke is of the pie-wound type and is connected in the circuit so as to allow monitoring the modulated signal; monitoring CW signals will show up key-clicks in the headphones, and the meter can be used to show relative radiation.

The needle on the meter should remain stationary at some fixed reading on the scale, such as half or two-thirds maximum deflection.

This form of over-modulation indicator cannot be used with controlled carrier modulated transmitters. For such transmitters, a 45 or 80-tube, or 879 should be connected in such a manner as to indicate negative peaks, and the transmitter monitored by a selective superheterodyne receiver with crystal filter, in order to test for voice transitions outside the channel in use. Such an indicator acts as a half-wave rectifier with its plate connected to the filament center-tap of the modulated class C stage and its filament connected through a 0-5 or 0-10 Ma DC meter and a 10,000 ohm resistor to the plate RF return circuit of the class C stage before it reaches the plate modulator.

A cathode-ray oscillograph is the best indicator for over-modulation. The trapezoidal figures readily show distortion and modulation capability even more clearly than the sine wave figures. The trapezoidal figure requires only a simple form of oscillograph.

Cathode-Ray Oscilloscope

A sketch of a modern cathode-ray tube is shown in Figure 15. The device functions as follows: A filament heats a tube called a cathode, and negatively charged particles of electricity are emitted in all directions. These electrons are attracted by a positively charged plate called an anode. This anode has a perforation in its center and a stream of electrons shoots out through this opening and impinge upon a chemically treated surface (willemite or calcium-tungstate) called a screen, which is at the top end of the tube. Electrons striking this surface produce a glow or fluorescence which in turn varies with the intensity of the element controlling the flow of electrons. A negatively charged cylinder concentrates the electrons emitted by the filament so that practically all pass through the small hole in the anode. The percentage of electrons passing through the hole in the anode depends upon the size of the hole, the field between the cathode and the anode, the shape of the surrounding electrodes, electrical conditions about the anode, and other factors dependent upon the temperature of the anode which are too numerous to mention. Possibly the fraction passing through the hole can vary from one-ten millionth or less to possibly one per cent of the electrons. The energy given to the electron to pass to the chemically treated screen can only be approximated by direct experiment; however, the electrons gain their energy from the electrical accelerating field.

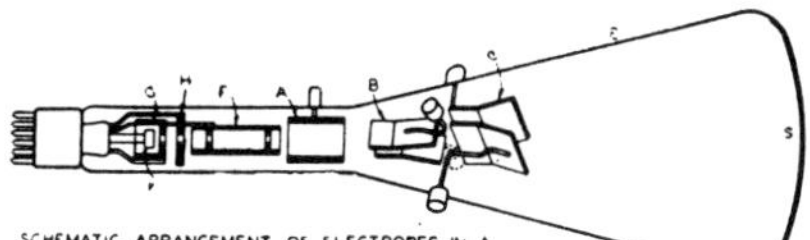

SCHEMATIC ARRANGEMENT OF ELECTRODES IN A CATHODE-RAY TUBE OF THE ELECTROSTATIC-DEFLECTION TYPE

FIG. 15.

Phenomenon of Flouorescence: The phenonenon of fluorescence may best be explained as follows; electrons striking the fluorescent material at the screen-end of the tube expel electrons from that material. In the course of return of these electrons to their parent atoms, light is emitted. The cathode-ray in striking the end of the tube gradually spends its energy ionizing atoms. They ultimately end with zero velocity and are picked up by some positive ion or else slide along the glass walls of the tube envelope to one of the electrodes and escape. Most of the cathode rays penetrate the screen and are ultimately absorbed. A few are reflected with loss of energy. The rays that were not absorbed or reflected are scattered obliquely and go forth with lesser energy. The energy that must be expended to cause fluorescence is at present unknown. Fluorescence is caused by an electron with as little energy as possible, perhaps one or two volts. However, **it is invisible.** To **see** fluorescence, enough energy must be liberated per unit volume so that the volume of density of light emitted causes a visual effect on the eye of the observer. Taking into consideration the not altogether improbable assumption, many investigators are inclined to believe that when an observer's eye is dark adapted he could just see the density of ionization in a gas in which there are one thousand million ions per cubic centimeter. In other words, the energy of the electrons in the region in which the cathode ray strikes, must be sufficient to give a thousand million ions per cubic centimeter or more. The fluorescent material gets rid of the energy in the form of light.

Practical Aspects of the Tube: The high vacuum cathode-ray tube is in many respects like a high-vacuum amplifier tube. No special technique is required in installing and operating the tube other than that normally employed in the handling of high vacuum tubes. The cathode is indirectly heated from the winding of a transformer and is operated at rated voltage without any adjustments. The currents to the electrodes of the "electron gun" usually total 0.1 or 0.2 milliampere or less than the current in the voltage divider which (see the accompanying diagram, Fig. 16) may be as low as one milliampere. Under these conditions, the ripple in the rectified voltage is small and a condenser of 0.5 to 2.0 mfds. supplies adequate filtering. The DC power required is low so that a small transformer with proper insulation for the high voltage will suffice. A few one-watt carbon resistors and potentiometers serve as the voltage divider. A half-wave rectifier or a voltage

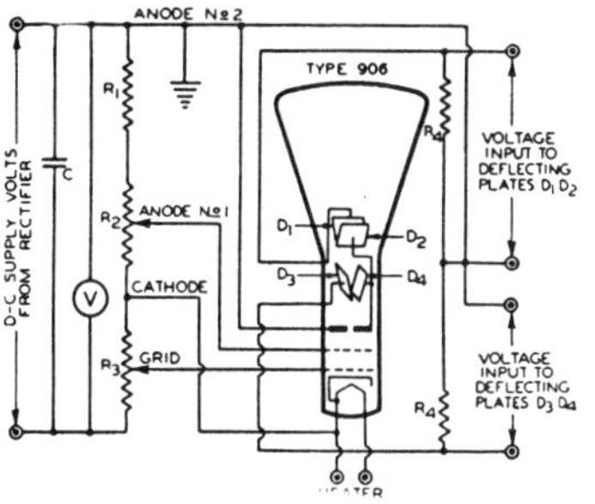

FIG. 16—Cathode-ray tube circuit. (R.C.A.)

doubler circuit is suitable. The rectified DC voltage is approximately equal to the peak of the AC voltage for the half-wave circuit or twice this value for the voltage doubler circuit is suitable. The rectified DC voltage is approximately equal to the peak of the AC voltage for the half-wave circuit or twice this value for the voltage doubler circuit. The rectifier tube carries only a small current but must withstand a peak inverse voltage of twice the peak AC voltage of the transformer.

The control electrode is normally operated with a bias voltage negative with respect to the cathode. It is used for controlling the beam current and hence the brilliancy of the fluorescent pattern. The control electrode is frequently made adjustable by means of a potentiometer tap in the voltage divider in order to permit a range of voltage from zero to a voltage sufficiently negative to completely cut off the beam current.

When this tube has no accelerating electrode, it is normally connected to a fixed tap on the voltage divider corresponding to a rated voltage.

The focusing electrode (1) is used to focus the beam current to a sharply defined spot on the screen. The voltage of this electrode is equal to approximately ⅓rd of the voltage on anode 2. A range of adjustment upward to about ⅓rd of the voltage on anode 2 is allowed.

When the brilliancy of the fluorescent pattern is adjusted, considerable de-focusing sometimes results due to the regulation of the high-resistance voltage divider. This is readily corrected by adjusting the focusing voltage, or it may be eliminated, if desired, by increasing the current in the voltage divider.

If a range of anode voltages is required, a potentiometer on the line voltage side of the high-voltage transformer will change the voltage on all electrodes simultaneously. This method will keep the pattern approximately in focus. Proper location or shielding of the tube is advisable in order to avoid the influence of stray electric and magnetic fields.

Scanning or Deflecting the Cathode-Ray Beam: The scanning of cathode rays upon the fluorescent screen is done by deflecting the beam of electrons vertically and horizontally, that is, in a zig-zag fashion until the whole surface of the screen has been

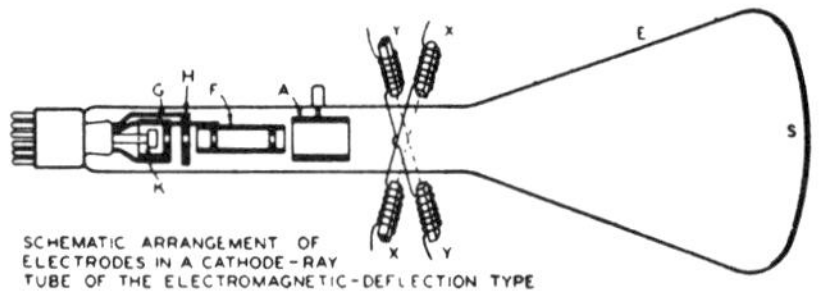

FIG. 17

irradiated. The deflecting action may be accomplished by either electro-static or electro-magnetic fields. The fields must be at right angles to each other and must intersect at the tube axis. In practice, one field is controlled by the current or voltage under observation; the other is controlled by an alternating voltage to give a desired timing control. The field serves to spread the rays or tracing over the fluorescent screen. Whatever method is employed for deflecting the electronic beam, it need only be remembered that the rays are attracted or repelled by the charges in the electric field.

How Cathode-Ray Wave Patterns Are Developed: The electron stream in passing between the lower deflecting plates (see Figure 15) is deflected toward the positive plate to an amount depending upon the momentary electric field set up by the potential difference between the plates. A second deflection at right angles to the first will take place when the second pair of plates is reached if a difference of potential exists between them. The result is that at any instant the recording point forming the end of the electron beam occupies a position on the viewing screen, which, both in direction and distance from its normal position is the resultant of the deflecting forces due to the differences of potential acting at that instant. In linear measurement, this distance amounts to about one millimeter for each volt of the resultant potential difference, according to the Western Electric Company bulletins. If the variations of the two intensities are cyclic and the frequencies of the cycles bear some simple integral ratio to each other, the two components will be the same each time for any point in the cycle of the lower frequency and, therefore, the spot will travel over the same path repeatedly and produce a stationary pattern. The pattern may be considered as a plot in rectangular coordinates of the relation between the two intensities or, if one of the fields is made to vary with time in some known manner, the variations of the other field with respect to time may be studied.

Sweep Circuits and Auxiliary Apparatus: In order that the wave form of phenomena producing a vertical deflection may be viewed, a horizontal deflection is applied to sweep across the screen at a uniform rate. The linear time sweep may traverse the screen only once when observing a non-recurrent wave form, or it may be arranged to be returned rapidly to its starting position and to repeat the linear time sweep.

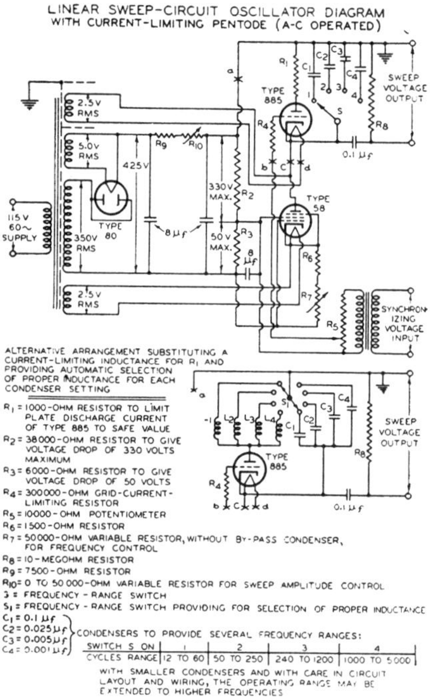

FIG. 18

When the frequency of the wave form to be observed is a multiple of the repetition frequency of the linear time sweep, the wave form remains stationary on the screen. The number of cycles of the wave appearing throughout the sweep on the screen shows the ratio of the frequency of the wave form to the frequency of the linear sweep voltage. Thus, a sweep frequency of 3,000 cycles per second shows four cycles of a 12,000 cycle per second wave-form.

A linear time sweep generator with good linearity, short-return sweep time, excellent frequency stability, and adjustable in frequency over the complete audio-frequency spectrum, is available with the present tubes. The type RCA-885 gas-filled triode tube was especially designed for this service. With suitable circuits, a sweep of 200 volts amplitude (or by special arrangements, 400 volts amplitude) can be obtained. The linear time sweep is generated by charging a condenser at a constant current rate. The constant current characteristic of the plate circuit of a pentode amplifier tube is used preferably as the constant-current controlling device. A diode operated at low filament voltage to produce saturation may also be substituted for this purpose. Another means consists in using only a small portion of the initial charging curve of the condenser. Since the exponential charging characteristic of a condenser is initially linear, the placement of a high resistance in series with the condenser and a high charging voltage will permit an appreciable amount of voltage to develop across the condenser before it departs appreciably from linearity. Either the charging or the discharging voltage of the condenser may be arranged to produce the linear time sweep. The return sweep occurs on discharge or charge according to the circuit arrangement. The time constant of the circuit causing the return sweep should be low with respect to the time sweep. Generally, the arrangement is used in which the condenser charging produces the linear time sweep and the discharge the return sweep. Any one of these methods properly employed is capable of a high degree of linearity.

The return sweep on the discharge of the condenser may be accomplished electrically by a gas-filled triode, or in certain applications it may be done mechanically by a rotating contact, a tuning fork, or other means. The gas-filled triode permits a large frequency range and locked synchronization with the wave-form being observed. Synchronization is locked by means of. a small amount of wave-form voltage coupled to the grid circuit of the gas-filled triode. The mechanical method either controls the phenomena being viewed and is, therefore, self-synchronized, or it provides a standard with which the frequency of the wave-form can be compared.

Other times bases are used for various applications. A 60 cycle per second wave of approximately sine-shape voltage from the power line is often useful as a sweep voltage. When the amplitude is made large enough to cause the end portions to sweep beyond the limits of the screen, the central portion is nearly linear. If the frequency of the wave-form being viewed is a multiple of 60 CPS, the wave-form will appear stationary on the screen and will have an approximately linear time distribution. Since the sweep and the return sweep of the 60 CPS voltage are the same, the wave-form is spread twice across the screen. One method for preventing confusion of the wave-form pattern consists in applying some of the 60 CPS voltage with a 90-degree phase shift to the deflection in the vertical direction. The result is that the sweep and the return sweep appear as 2 separate lines on the screen since the 60 CPS voltages sweep out elongated ellipses instead of a line. Another method consists in making the return sweep invisible by applying some of the 60 CPS voltage with a 90-degree phase shift to the brilliancy control electrode. The beam current is cut off during the return sweep by the negative half of the 60 CPS voltage. Where exact linearity is not necessary, this 60-C. sweep method is convenient. Since there are a large number of 60 CPS intervals over the audio-frequency range, a stationary wave-form is readily obtained. The wave-forms of different frequencies can be spread to convenient proportions by increasing the 60 CPS sweep voltage within the limits permissible in the deflecting circuits of the cathode-ray tube.

TYPICAL CATHODE-RAY OSCILLOGRAPH CIRCUIT
USING EITHER RCA-905 OR RCA-906

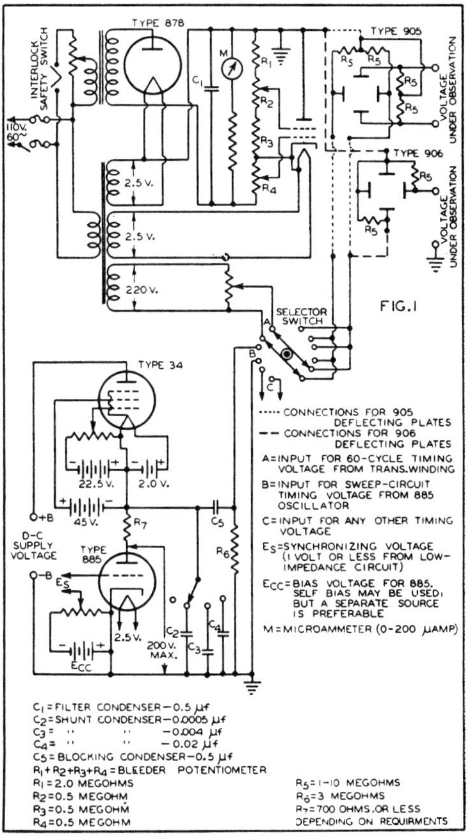

FIG. 19

A circular time base is often useful for frequency comparisons. In general, an ellipse results with axes parallel to the deflecting directions when voltage of the same frequency but with 90-degree phase relation are applied to the deflecting plates. When the deflection amplitudes in the 2 directions are equal the ellipse becomes a circle. The voltage with a 90-degree phase relation is readily obtained by means of a condenser and resistance. If the voltage source supplying the circular time base is changed rapidly from zero to maximum, the successive circles produced by each cycle are swept into a spiral.

The spiral affords a convenient time sweep of known variable velocity and of considerable length on the screen.

Other auxiliary apparatus frequently brought into play are amplifiers and current transformers. Resistance coupled amplifiers are always adaptable when a uniform response over a wide range is required. For extremely low frequencies and DC voltages, directly coupled amplifiers are used. For high frequencies where a wide range is needed, resistance coupled amplifiers having screen-grid or pentode tubes and low plate-load resistances are employed. The resistance coupled amplifier increases voltages to the proper level for applying them to electro-static deflecting plates or to electro-magnetic deflecting coils. If a low resistance is introduced into a circuit where current is to be observed, a small voltage drop in the resistance is produced which may be amplified to a level suitable for deflecting the cathode-ray beam.

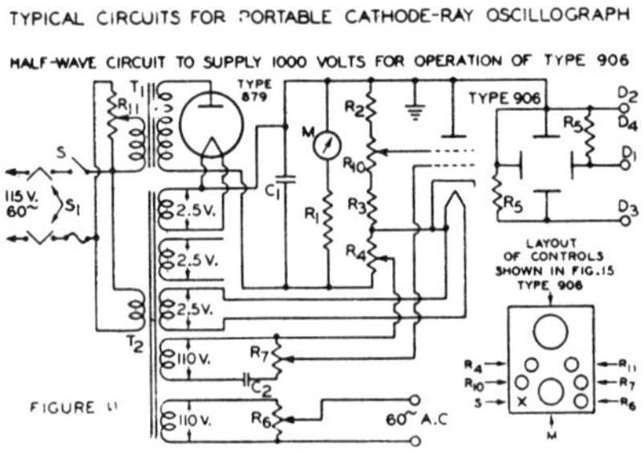

FIG. 20

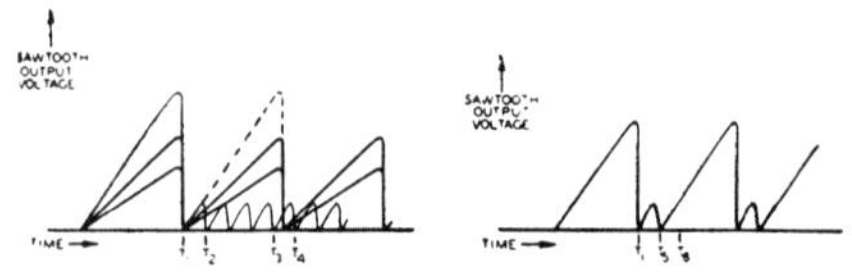

FIG. 21—Wave-form of a saw-tooth oscillator.

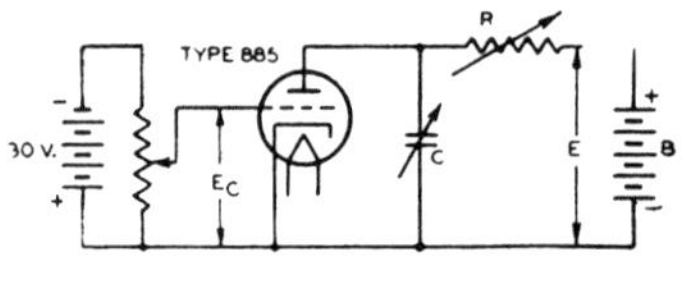

FIG. 22—Simple sweep-circuit oscillator with current-limiting resistor.

Current transformers provide a means for converting current into voltage suitable for direct connection to the electro-static deflection plates. When a uniform response over a frequency range is desired, the primary of the current transformer is connected across a low resistance in the circuit in which the current to be observed flows. The inductance in the frequency range used will be large with respect to the resistance. The voltage developed across the primary is stepped-up by the turn ratio in the transformer, the secondary of which is connected to the deflection plates. For example, with a turn ratio of 100:1 and the primary shunted with 10 ohms, a current of 10 milliamperes will produce 10 volts on the secondary or a peak-to-peak amplitude of about 0.5-inch on the screen of a cathode-ray tube operated at 1000 volts.

Applications of Cathode-Ray Tubes: Fundamentally, the cathode-ray tube may be regarded as an electron pointer or the movement of an uncalibrated electrical indicating device. The calibrating scales and circuit are provided by the user to suit the conditions under which measurements or indications are being made. Only a few practical applications are given here, others of more wider scope are found in treatises solely devoted to this phase of the subject.

One of the simplest uses of the cathode-ray tube is in the measurement of voltage. This is most conveniently done with the electro-static deflection type of tube. The displacement of the spot on the screen is directly proportional to the applied voltage. When a DC voltage is applied, the polarity as well as the magnitude is indicated by the displacement of the spot. When an alternating voltage is applied, the spot sweeps back and forth with an amplitude proportional to the peak-to-peak value of the applied voltage. For example, a 10-volt r.m.s. sine wave produces a sweep with an amplitude equal to 28-volts. At frequencies above 8 cycles per second, the sweep of the spot appears as a line, due to the persistence of vision. There is no error due to frequency until extremely high-frequencies are reached. Overvoltage on the deflecting plates, which is not excessive, merely sweeps the spot off the screen. Thus, the cathode-ray tube, being rugged, having a high-impedance, and being independent of frequency, is useful as a peak voltmeter.

Since the electron beam can respond to several deflecting fields simultaneously or in rapid succession, the cathode-ray tube can be admirably applied to time or frequency comparisons. The feature of being able to combine in the cathode-ray tube the effects of two or even more factors makes it possible to obtain graphical results directly on the screen.

Application Notes: The high voltage anode of the cathode-ray tube requires 1000 volts DC for proper operation. Also DC voltages are required for the amplifier. The RCA-879 tube (rectifier) is used in a half-wave rectifying circuit for providing the necessary anode voltages for the RCA-906. The 80-type tube, connected in a full-wave rectifying circuit is used for both rectifiers, individual filter circuits are provided. The transformer is over-size to prevent stray magnetic leakage that would otherwise affect the operation of the cathode-ray tube.

In applications involving extremely accurate measurements, the current to anode (2) should be reduced to the minimum value consistent with the desired brilliance of pattern. Where brightness is an important consideration, it may be obtained by increasing the voltage applied to anode (2) up to the maximum value. This practice, however, is not always desirable since the greater acceleration of the electrons in the beam causes reduced, deflection sensitivity.

It should be noted that the beam producing a spot of high intensity will burn the fluorescent screen if it is allowed to remain stationary even for a short interval. Such operation may cause excessive heating of the glass with resultant puncture. To prevent this possibility, it is recommended that the beam be kept in motion. It is well to apply controlling voltage to the deflecting system before permitting the electron stream to flow. Stopping of the electron beam may be accomplished by removing the voltage on anode (2) or by increasing the bias on the control electrode to cut-off.

Users of cathode-ray tubes are cautioned to strictly observe the technical information printed on bulletins packed in the tube cartons. Emphasis must also be stressed on the fact that extremely high voltages are applied to the tube and every precaution must be observed to keep from coming in contact with these potentials—fatalities can occur.

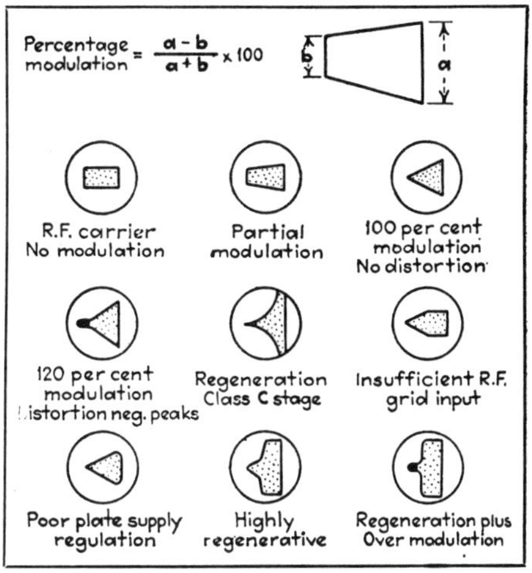

FIG. 2—Oscilloscope chart, showing typical trapezoidal readings of a phone transmitter when sweep circuit is not used.

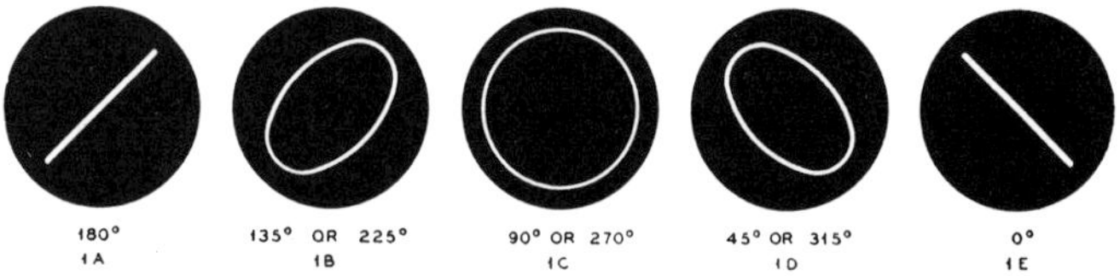

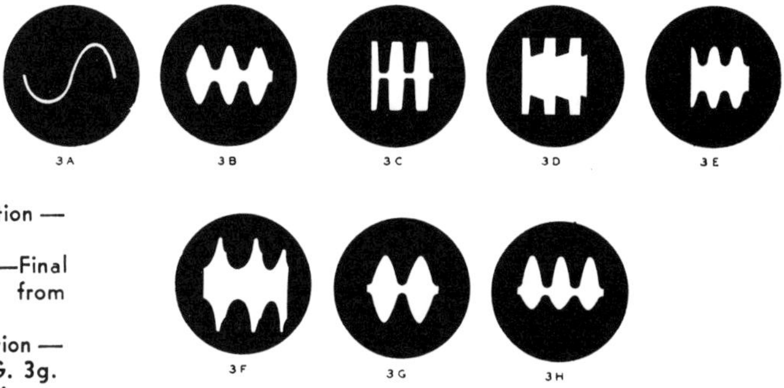

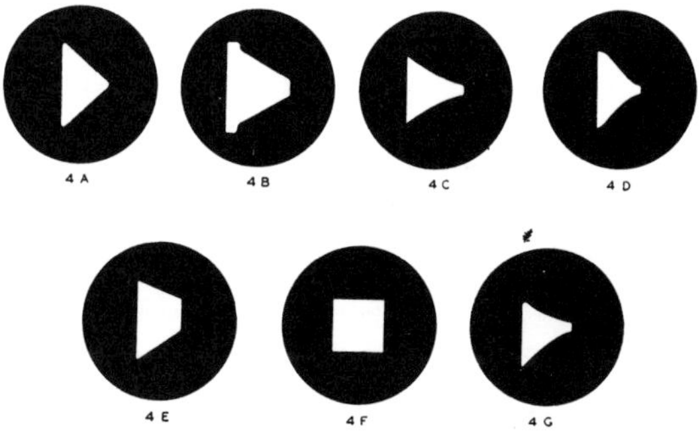

FIG. 1 — Phase relation between two AC voltages in an audio amplifier system.

FIG. 3a—400 cycle audio waveform.

FIG. 3b—Grid Modulation—ideal condition.

FIG. 3c—Grid Modulation—over modulated.

FIG. 3d—Grid Modulation — with too much r.f. grid excitation.

FIG. 3e — Grid Modulation—partially modulated.

FIG. 3f—Grid Modulation — Too much positive grid current.

FIG. 3g — Grid Modulation — 100% modulation.

Fig. 3h—Grid Moduation—Final tank circuit detuned from resonance.

FIG. 4a — Grid Modulation — Same adjustment as FIG. 3g.

FIG. 4b — Grid Modulation — Typical adjustment, heavily modulated with excessive r.f. excitation and output.

FIG. 4c — Grid Modulation — Same as 4b with no grid current when unmodulated.

FIG. 4d—Grid Modulation—R.F. excitation very low and overmodulated.

FIG. 4e — Grid Modulation — about 50% modulated.

FIG. 4f — Grid Modulation — Zero per cent modulation.

FIG. 4g — Grid Modulation — Same as FIG. 3h plate tank detuned.

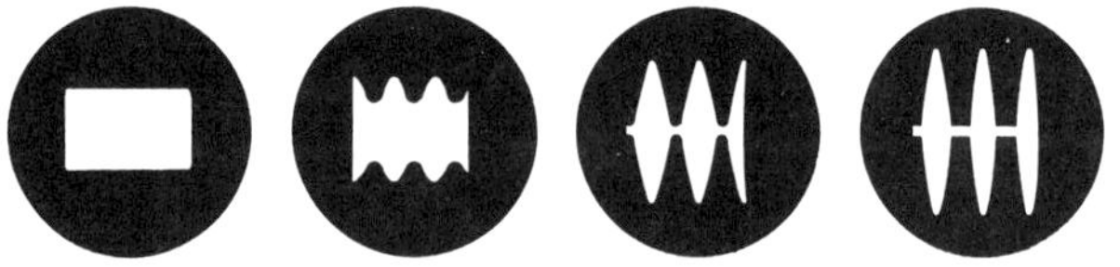

5 A 5 B 5 C 5 D

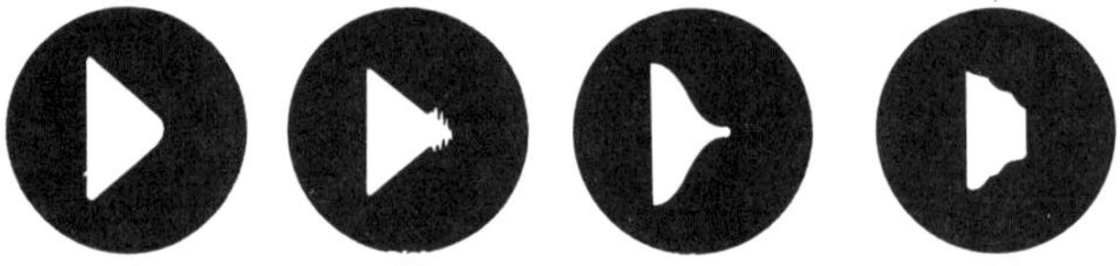

6 A 6 B 6 C 6 D

6 E 6 F 6 G

7 A 7 B 7 C 7 D

FIG. 5a — Plate Modulation—Carrier only.
FIG. 5b — Plate Modulation—Partially modulated.
FIG. 5c — Plate Modulation—100%.
FIG. 5d — Plate Modulation—Over modulated.
FIG. 6a — Plate Modulation—100%.
FIG. 6b — Plate Modulation — Best adjustment obtainable with insufficient Class B audio by-pass condenser (only 2 mfd).
FIG. 6c — Plate Modulation—Over modulated before reaching normal 100% condition.
FIG. 6d — Plate Modulation—Same as 6c at point of best modulation obtainable. Regeneration, lack of r.f. shielding, Class B trouble and too heavy antenna loading.
FIG. 6e — Plate Modulation—Maximum modulation with Class B greatly mismatched to Class C load.
FIG. 6f—Plate Modulation—Maximum modulation with too heavy antenna load.
FIG. 6g—Plate Modulation — With insufficient grid r.f. excitation.
FIG. 7a—CW carrier with one section filter on final stage and saturated choke.
FIG. 7b—CW carrier with one section using larger choke.
FIG. 7c—CW carrier with two section filter, both filter chokes too small.
FIG. 7d—CW carrier 2 section filter with large chokes.

LISSAJOU'S FIGURES

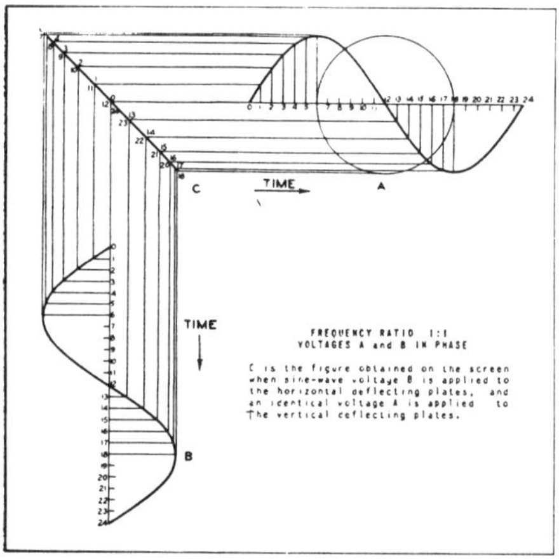

FIG. 23

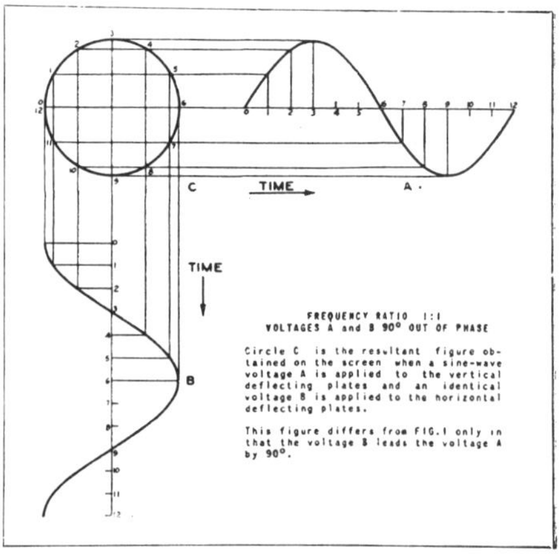

FIG. 24

Oscilloscopic Studies of Grid and Plate Modulation

Interesting comparisons between grid and plate modulation can be revealed through the medium of the cathode-ray oscilloscope. The tracings accompanying these paragraphs portray the conditions under which the average phone transmitter operates.

In general, both grid and plate modulation systems can be made to produce nearly 100 per cent distortionless modulated output. The difficulties encountered are about equal; for the average amateur station, grid modulation is more easily adjusted for 20-meter operation, and plate modulation is easier adjusted on 80 and 160 meters, due to excitation problems. A correctly adjusted plate modulated phone presents a number of problems, the magnitude of which can be verified by listening-in to several radio-telephonic transmissions.

Oscilloscopic Patterns: When two AC voltages of the same magnitude are impressed on the vertical and horizontal plates of a cathode-ray tube, the result will be as shown in one of the patterns of Figure 1 (Page 299) these determine the degree of phase shift in an audio amplifying system.

For checking audio amplifiers, the output of a sine wave audio oscillator is impressed on the vertical plates, using an interruption (saw-tooth) oscillator to produce a sine wave picture on the fluorescent screen at the end of the tube. By connecting the vertical plates across the output of the amplifier through a suitable attenuator, it is possible to compare the wave-form for distortion. Class B amplifiers will sometimes show a high-frequency oscillation superimposed on the main sine wave which tends to increase the breadth of certain portions of the tracing. Circuit changes can be made while studying the figures and trouble-shooting is thereby simplified.

There are two methods for studying the modulation of a phone transmitter, these are: (1), by a sweep circuit oscillator giving solid sine wave envelopes and (2), by a solid trapezoidal figure. Each of these have certain advantages, the latter probably being best for voice tests. The trapezoidal method requires very little equipment, only a 60 cycle tone need be impressed on the horizontal plates as well as for the tone into the audio system of modulator. The vertical plates are connected to a RF pick-up coil coupled to the antenna or final tank circuit. With no modulation, the figure is rectangular, at 100 per cent modulation the figure will be triangular, provided the transmitter is correctly designed and adjusted. The percentage can be calculated by means of the formula in Figure 2 (Page 299). The various patterns portrayed will serve as a guide in analyzing transmitter troubles.

To obtain quantitative data for this discussion, two transmitters were tested, one a grid-modulated set having a combination of cathode-bias and fixed cut-off grid bias with four type 150T tubes in the final stage; the other was a plate-modulated set with two 211 tubes modulated by a pair of 838 tubes in class B.

Figure 3 (Page 299) shows some of the figures outlined for different adjustments of the grid-modulated set, these were taken from an RCA oscilloscope. The audio input was practically a pure sine wave of about 400 cycles per second. The figures appear to have good form except when too much RF grid excitation is applied. Detuning the final tank circuit shortens the modulated envelope on one side or the other.

Figure 5 (Page 300) depicts the same set studied by a trapezoidal configuration. Here, undesirable modulation characteristics are more clearly shown than in the sine wave figures. Inability to modulate 100 per cent, linearly, is readily indicated when grid current flows. For definitive patterns, it is necessary to have a heavy antenna load, zero grid current and steady plate current, otherwise grid-modulated phone tracings will appear like the figure noted as "typical heavy grid-modulated phone." The grid current in the latter case was a few milliamperes and the plate current about 330 ma. Linear modulation was not obtained with more than 275 ma. To properly vary the current the grid excitation should be reduced, rather than changing the antenna load. Too low a value of C in the final tank circuit will also cause non-linearity, and values as high (or higher) than needed for plate modulation are necessary.

The plate modulated phone required a rather light antenna load to secure 100 per cent modulation in a linear form. Figure 5 (Page 300) shows various degrees of modulation when using a sweep circuit oscillator. Figure 6 (Page 300) portrays the trapezoidal figures arrived at under various conditions. A common plate voltage supply as employed for the modulator and class C stage and at least 6 ufd. by-pass condenser was needed from the Class B output center-tap connection to ground in order to allow over 70 per cent modulation without excessive distortion. Two ufd. was insufficient, even though the class C stage was isolated by an additional section of filter. The oscilloscope was connected so as to include the characteristic of the speech amplifier, modulator and class C stage. The antenna output was about one-half of what it would have been for CW; this was because the value of C in the class C stage was not as high as it should have been. If the tuning capacity was increased by a factor of 2, the output would have improved, even though the circuit was no longer low-C. Many amateur phones overmodulate, a condition attributable to low-C in the final stage which precludes the attainment of 100 per cent modulation at high efficiency.

Overmodulation can occur long before 100 per cent modulation is reached, unless careful design and adjustment is made. A constant monitoring device for indicating overmodulation and carrier shift is necessary in every phone station; one of excellent design is described elsewhere in this text.

The oscilloscope is an invaluable device for studying the output wave forms from the speech amplifier to the class B stage, as a whole or any part. Insufficient by-pass across the class B plate supply immediately becomes apparent, as well as amplitude dis-

R. C. A. Type TMV-122B Cathode-Ray Oscilloscope.

tortion in any stage. A frequency check can be made if a beat-frequency or variable-frequency audio oscillator is available, although in most transmitters the amplitude distortion is far worse than frequency distortion. The exploring or pick-up coil connected to the vertical plates will show how much modulation is "leaking back" into the buffer and crystal oscillator stages. Lack of sufficient neutralization will be indicated as a steady carrier in the output. Under such conditions, 100 per cent modulation is obviously impossible as indicated by a point on the triangle. A check should therefore be made to determine that this condition does not exist when the plate voltage is removed from the modulated stage.

Keying surges in CW transmitters or ripple on the carrier, due to insufficient power supply filter, can be studied on the oscilloscope. Figure 7 (Page 300) shows some carrier oscillograms obtained with various filter combinations on the final stage. Keying surges are difficult to picture without a rapid camera shutter and fast lens, this is because the clicks are in the form of transients which die away rapidly. As viewed on the screen, violent peaks and irregularities occur in the otherwise-smooth carrier band when keying.

Notes: An oscilloscope is easily affected by stray AC fields due to power transformers or filter chokes not being electro-magnetically shielded. This condition shows up as a curvature or ellipse of the straight line on the screen when an AC voltage is impressed on one set of the plates. Stray RF fields from a relatively high-powered buffer stage will sometime prevent a thin line from appearing on the screen when preparations are being made to test a final amplifier stage. The best method is to amply shield the transmitter, although in some cases shielded leads and a separate ground connection to the oscilloscope will allow tests to be conducted.

Radiotelephone Transmitter Circuit Connection to Cathode-Ray Oscilloscope: The diagrams show how to measure the modulated r-f wave of a radiotelephone transmitter with the aid of an Oscilloscope. This connection does not measure the audio-frequency characteristic of the speech amplifier and modulator.

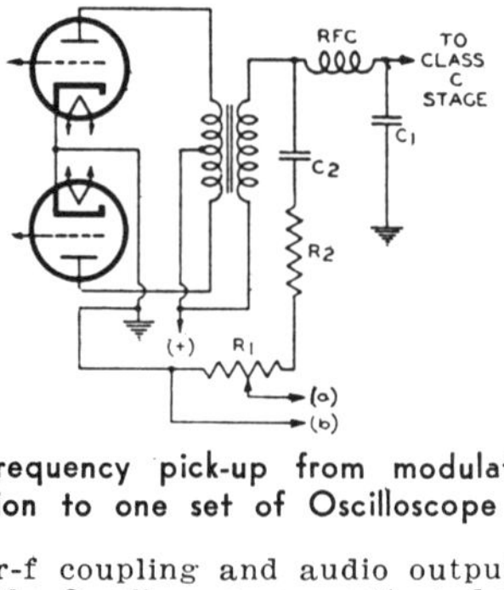

Audio-Frequency pick-up from modulator for connection to one set of Oscilloscope plates.

The r-f coupling and audio output voltage to the Oscilloscope are adjusted to give the desired size of figure on the cathode-ray tube. The circuit connection here shown will provide trapezoidal figures for testing or monitoring the percentage of modulation.

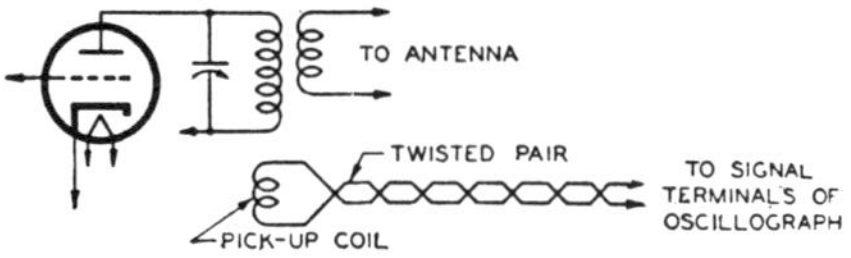

R-F pick-up for connection to plates of Cathode-Ray Oscilloscope.

Absorption Type Wavemeter: Known as "the old stand-by," this simple wavemeter is useful in every amateur station. It consists of a tank circuit which, when tuned to resonance, gives a visual indication of resonance by means of a glow from a flashlight globe or neon lamp. Peak resonance is indicated by peak brilliance of the lamp. This wavemeter will not operate on harmonics or beats. The principle of operation is simple. Several methods of indicating devices and a suggested design for the wavemeter are seen in the diagram, Fig. 25.

A shows the tuned tank system with an aperiodic circuit (looped lamp). The lamp is lighted by means of r-f induced in the tank circuit.

In **B** resonance is indicated by a flashlight globe shunted across approximately one inch of the lead. This is one of the best methods to use because it tunes very sharply and the indication of resonance is sharper on the condenser tuning scale than the system shown in **A.**

C is a variation of **B**, except that the coil is tapped instead of being of the plug-in type.

D uses a neon lamp as the medium of indication.

A r-f meter can be substituted for the lamp in any of the systems shown.

The wavemeter can be calibrated by checking it against a frequency meter. Or the "click" method can be used. Couple the wavemeter to the detector of a regenerative circuit, or to the oscillator of a superheterodyne and by tuning the wavemeter to resonance a distinct "click" is heard in the headphone connected to the receiver. There will be an error in the calibration due to the value of frequency of the i-f amplifier used in the receiver. This value can be rechecked and a curve can be plotted, so that ultimately the wavemeter can provide calibration within 10 KC, depending upon the type of circuit used.

A high C tank is best for sharp tuning. With a .00035 mfd. condenser, the following coils will cover the amateur bands with fair accuracy:

160 Meters: 36 turns, No. 18, close wound on a 4-inch diam. form.

80 Meters: 18 turns, No. 18, spaced one diameter on a 4-inch diam. form.

40 Meters: 8 turns, No. 18, spaced ¼-inch between turns on a 4-inch diam. form.

20 Meters: 4 turns, No. 18, spaced ½-inch between turns on a 4-inch diam. form.

10 Meters: 8 turns, No. 14, spaced one diameter on a one-inch diam. form.

5 Meters: 4 turns, No. 14, spaced one diameter on a one-inch form and tuned with a 3-30 mmf. condenser.

The coil turns should be secured by applying small drops of household cement or "coil dope" to the turns. Clear lacquer is also suitable for this purpose. The wavemeter should be housed in a metal can and the variable condenser rotor should be grounded to the can. A wavemeter of this type for 5 and 10 meter operation can be made by winding a small coil of bare wire, "air-wound" and supported directly to a midget variable condenser. This unit is then mounted on a support strip and the variable condenser is equipped with a bakelite rod, about 1 foot long, with the control knob at the end farthest from the tuning unit. Another rod is attached to the support which holds the condenser and coil, so that the wavemeter can be held in one hand by means of this rod, the other hand being used to turn the long bakelite rod which varies the capacity of the tuning condenser. A 5 and 10 meter wavemeter of this type is **not** housed in a shield can.

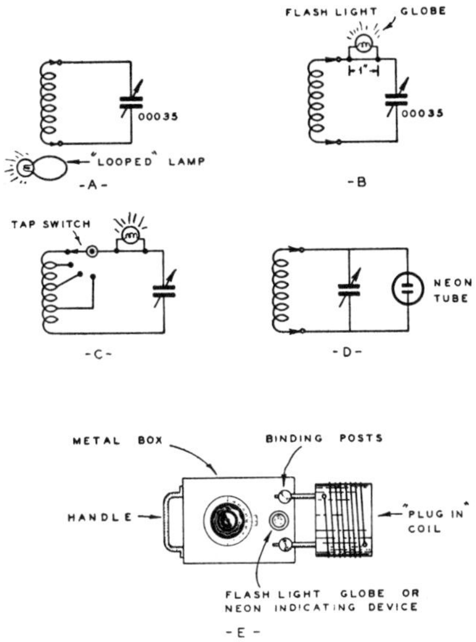

FIG. 25

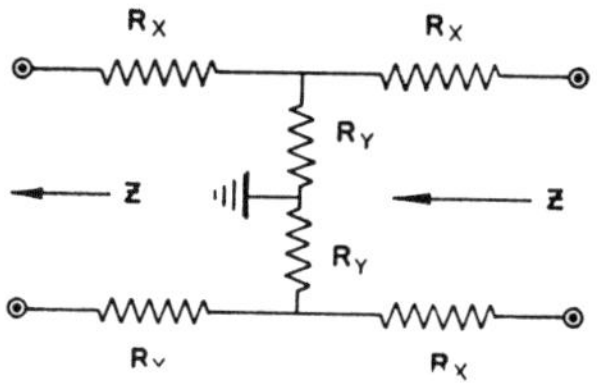

FIG. 26—"H" Pad

Design of Fixed Networks (T&H Pads): The "T" section, which is an unsymmetrical network (unbalanced) is most frequently used where small unbalances in the line or to ground are of little importance. Fixed type networks are chiefly employed in circuits where it is desired to limit the

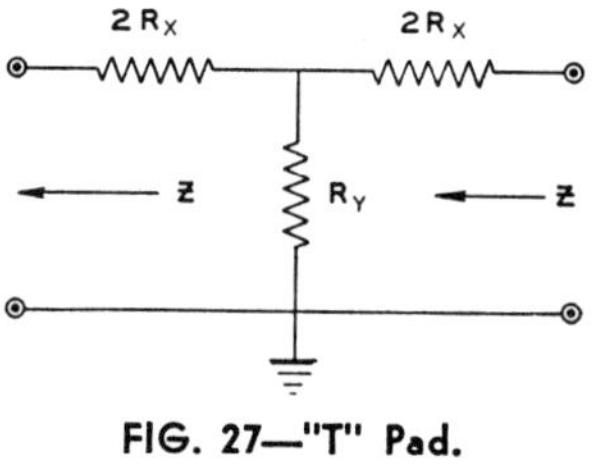

FIG. 27—"T" Pad.

amount of input voltage available to excite amplifiers, thus precluding the possibilities of overloading certain components in the amplifying system. A resistive network which functions as an absorption device loses its identity as a "pad" and is most often referred to as an "insertion loss," because the section has been inserted to attenuate a known and definite quantity.

To design a fixed "T" or "H" type section for some predetermined loss in DB, the following equations are given. These equations only hold good when the line impedances terminating each end of the network are equal; therefore from Figure 26,

$$R_x = \frac{Z}{2}\,\frac{(K-1)}{(K+1)}$$

$$R_y = 2Z\,\frac{K}{(K^2-1)}$$

$$K = \text{antilog}\,\frac{N_{db}}{20}$$

Where R_x equals the series resistor (this value must be multiplied by 2 for "T" sections); R_y, the shunt resistor; Z, the line impedance; and K, a constant derived by taking the inserted attenuation in DB and dividing by 20, then extracting the antilog.

Impedance Matching Networks: At audio frequencies an impedance matching network comprised of resistive impedances can be substituted for an impedance matching transformer or like device. Unfortunately, this type of network introduces a small loss; however, this loss is of little consequence because it can be counteracted by simply working the input or output circuits at a higher level.

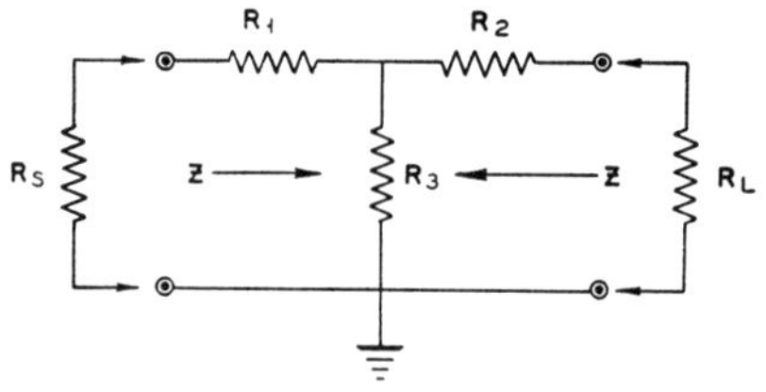

FIG. 27—Impedance Matching Network.

In Figure 27 it is very important that the resistors R-1 and R-2 be placed correctly in the configuration otherwise impedances will be mismatched and reflections will occur in the system. In Figure 28, resistor R1 must face the highest terminal impedance.

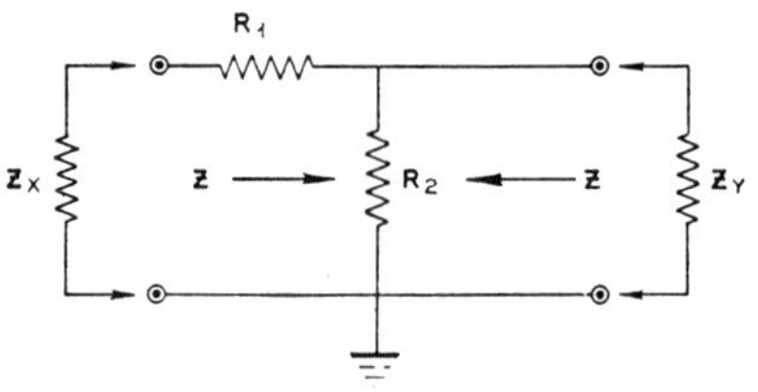

FIG. 28—"L"-Type Network.

To design an impedance matching network of the "T" type requires the use of the following equations:

$$R\text{-}1 = \frac{(R_S + R_L)\,K_1 + (R_S - R_L)}{2}$$

$$R\text{-}2 = \frac{(R_S + R_L)\,K_1 - (R_S - R_L)}{2}$$

$$R\text{-}3 = \frac{(R_S + R_L)}{2K_2}$$

Where R_S is the input impedance; R_L, the output impedance; and K_1 and K_2 are constants taken from the table shown below. These constants appear directly opposite the amount of attenuation in the N_{db} column

N_{db}	K_1	K_2
1	.057	0.115
2	.114	0.232
3	.171	0.352
4	.226	0.477
5	.280	0.609
6	.331	0.747
7	.382	0.897
8	.430	1.055
9	.476	1.233
10	.519	1.422
11	.560	1.634
12	.598	1.863
13	.634	2.122
14	.667	2.404
15	.697	2.720
16	.726	3.075
17	.752	3.468
18	.776	3.907
19	.798	4.398
20	.818	4.952
21	.835	5.555
22	.852	6.262
23	.867	7.013
24	.880	7.868
25	.893	8.870
26	.904	9.977
27	.914	11.188
28	.923	12.484
29	.931	14.091
30	.938	15.734
31	.945	17.744
32	.950	19.810
33	.956	22.339
34	.960	24.939
35	.965	27.121
36	.968	31.393
37	.972	35.397
38	.975	39.515
39	.978	44.555
40	.980	50.237
41	.982	56.079
42	.984	63.230
43	.985	70.583
44	.987	78.792
45	.988	88.836
46	.990	100.165
47	.991	111.813
48	.992	126.070
49	.993	140.729
50	.994	158.672

To design an impedance matching network of the "L" type requires this set of equations:

$$R_1 = Z_x(Z_x - Z_y)$$

$$R_2 = \frac{Z_x Z_y}{\sqrt{Z_x(Z_x - Z_y)}}$$

Since the insertion loss is a function of the impedances terminating the network it can be calculated as follows:

$$K = \sqrt{\frac{Z_x}{Z_y}} + \sqrt{\frac{Z_x}{Z_y} - 1}$$

Where loss in DB = 20 $\text{Log}_{10}K$.

Decibels—Technique and Practical Application

The decibel unit used in radio engineering and virtually universal in all power and energy measurements is actually a unit of amplification expressed as a common logarithm of a power or energy ratio. One decibel is 1/10th of a bel. One bel or 10 decibels indicates an amplification by 10, the common logarithm of 10 being 1. Similarly, 2 bels or 20DB means amplification by 100; 30DB means amplification by 100 and so on. The power ratio for one decibel is expressed as

$$\frac{P_1}{P_2} = 10^{.1} \text{ .... (1)}$$

where P_1 is the power input; P_2, the power output. The number of decibels represents a power gain or loss depending upon whether the relation P_1/P_2 is greater or less than 1.

Expressions for various power ratios are now commonly employed in communication engineering at audio and at radio frequencies. To express a ratio between any two amounts of power, it is convenient to use a logarithmic scale. A table of logarithms facilitates making conversions in positive or negative directions between the number of decibels and the corresponding power, voltage and current ratios.

The Logarithmic Table: A table of logarithms is presented in Figure 29. This table does not differ essentially from any other similar table except that here no proportional parts are given and the figures are stated to only three decimal places; this arrangement has been found to be satisfactory for all practical purposes. A complete exposition on logarithms is without the scope of this HANDBOOK, however, the very essentials together with the practical use of the tables and their application to decibels is given herewith. Thus, a person need not be concerned with the study of logarithms other than their direct employment to decibels.

The logarithm of a number usually consists of two parts; a whole number called the characteristic, and a decimal called the mantissa. The characteristic is the integral portion to the **left** of the decimal point (see examples below), and the mantissa is the value placed to the right. The mantissa is all that appears in any table of logarithms. In the logarithm the mantissa is independent of the position of the decimal point, while on the contrary, the characteristic is dependent only on the position of the number with the relation to the decimal point. Thus in the following examples:

	NUMBER		LOGARITHM
(a)	4021.	=	3.604
(b)	402.1	=	2.604
(c)	40.21	=	1.604
(d)	4.021	=	0.604
(e)	.4021	=	—1.604
(f)	.04021	=	—2.604

It will be seen that the characteristic is equal, algebraically, to the number of places minus one, which the first significant figure occupied to the left of the decimal point. In (a) the characteristic is 3; in (b) 2; in (d) 0; in (e) —1; and in (f) —2. The following should be remembered: (1) that for a number greater than 1, the characteristic is **one less** than the number of significant figures in the number; and (2), that a number wholly a decimal, the characertistic is negative and is numerically **one greater** than the number of ciphers immediately following the decimal point. Notice (e) and (f) in the above examples.

Finding a Logarithm: To find a common logarithm of any number simply proceed as directed herewith: Suppose the number to be 5576. First, determine the characteristic. An inspection will show that this number will be three. This figure is placed to the **left** of a decimal point. The mantissa is now found by referring to the logarithm table in Figure 29. Proceed by selecting the first two numbers which are 55, then glance down the N column until coming to these figures, advance to the right until coming in line with the column headed 7, the number will be 745. (Note that the column headed 7 corresponds to the **third** figure in the number 5576). Place the mantissa 745 to the **right** of the decimal point making the number now read 3.745. This is the logarithm **of 5576. Important:** do not consider the last figure, 6, in the number 5576 when looking for the mantissa; in fact, disregard all figures beyond the first three when determining the mantissa, however, be doubly sure to include **all** figures when ascertaining the magnitude of the characteristic.

Practical applications applying the logarithm to decibels will follow. Other methods using the logarithm will be discussed as the subject develops. See page 306 for Figure 29, Logarithm Table.

Power Levels: In the design of radio devices and amplifying equipment the power level is taken at six milliwatts (.006w). This corresponds to the arbitrary reference level of zero decibels. All power levels above the reference level are designated as "plus" quantities, and below as "minus." The figure always being prefixed by a plus

(+) or minus (—) sign commanding the direction in which the quantity is to be read.

Power to Decibels: The power output (watts) of any amplifier may be easily converted into decibels by the following formula, assuming that the input and output impedances are equal:

$$N_{db} = 10 \text{ Log}_{10} \frac{P_1}{P_2} \qquad (2)$$

where N_{db} is the desired power level in decibels; P_1, the output of the amplifier; and P_2, the reference level of 6 milliwatts. The subnumeral, 10, affixed to the logarithm indicates that the Log is to be extracted from a table to which 10 must be raised in order to produce a number.

By substituting values for the letters shown in the above formula, take the following illustration:

An amplifier using a 2A5 tube is said to deliver an undistorted output of three watts. How much is this in decibels?

Solution by formula (2):

$$\frac{P_1}{P_2} = \frac{3}{.006} = 500$$

and Log 500 = 2.69

therefore 10 × 2.69 = **26.9 DECIBELS**

By placing other values for those shown in the solution any output power may be converted into decibels **provided** that the decibel equivalent is **above** the zero reference level or the power is **not less** than 6 milliwatts.

To solve most all problems to which the solution will be given in minus DBs, a simple understanding of algebraic adding is required. To add algebraically, it is necessary to observe the plus and minus signs of expressions. (Do not confuse these signs with decibels.) In the succeeding illustrations notice that the result was caused sometimes by addition and at other times by subtraction.

(a)	(b)	(c)	(d)
+2	—4	—4	+4
—4	—2	+2	+2
—2	—6	—2	+6

The terms used in (c) are those that apply to decibel calculations.

When a solution to a problem involving logarithms will be in minus DBs, note particularly that the characteristic of the logarithm will be prefixed by a minus sign (—). This sign only effects the characteristic while mantissa remains positive. The mantissa always remains thus, no matter the direction the solution brings the decibel. A prefix —1 to a logarithm means that the first figure of the number will be the **first place** to the **right** of the decimal; —2, will occupy the second place to the right, while a cipher fills the first place; —3, the third place with two ciphers filling the first and second places, and so on.

To multiply a **minus** characteristic and a **positive** mantissa by 10, each part must be considered separately, multiplied by 10, and then the products added algebraically; thus, in the following illustration:

An amplifier using a 199 tube has an output of 5 milliwatts. How much is this in decibels?

Solution by formula (2):

$$\frac{P_1}{P_2} = \frac{.005}{.006} = .83$$

Log .83 = —1.9 (actually —1.920)

Therefore 10 × —1.9 = **—1 DECIBEL** (10 × —1 = —10; and 10 × .9 = +9, hence, adding the products algebraically = —1).

By substituting other values for those in the above solution, any output power

FIG. 29

Three Place Logarithms

N	0	1	2	3	4	5	6	7	8	9
00	000	000	000	000	000	000	000	000	000	000
10	000	004	008	012	017	021	025	029	033	037
11	041	045	049	053	056	060	064	068	071	075
12	079	082	086	089	093	096	100	103	107	110
13	113	117	120	123	127	130	133	136	139	143
14	146	149	152	155	158	161	164	167	170	173
15	176	179	181	184	187	190	193	195	198	201
16	204	206	209	212	214	217	220	222	225	227
17	230	233	235	238	240	243	245	248	250	252
18	255	257	260	262	264	267	269	271	274	276
19	278	281	283	285	287	290	292	294	296	298
20	301	303	305	307	309	311	313	316	318	320
21	322	324	326	328	330	332	334	336	338	340
22	342	344	346	348	350	352	354	356	358	359
23	361	363	365	367	368	371	372	374	376	378
24	380	382	383	385	387	389	390	392	394	396
25	397	399	401	403	404	406	408	409	411	413
26	415	416	418	420	421	423	424	426	428	429
27	431	433	434	436	437	439	440	442	444	445
28	447	448	450	451	453	454	456	457	459	460
29	462	463	465	466	468	469	471	472	474	475
30	477	478	480	481	482	484	485	487	488	490
31	491	492	494	495	496	498	499	501	502	503
32	505	506	507	509	510	511	513	514	515	517
33	518	519	521	522	523	525	526	527	528	530
34	531	532	534	535	536	537	539	540	541	542
35	544	545	546	547	549	550	551	552	553	555
36	556	557	558	559	561	562	563	564	565	567
37	568	569	570	571	572	574	575	576	577	578
38	579	580	582	583	584	585	586	587	588	599
39	591	592	593	594	595	596	597	598	599	601
40	602	603	604	605	606	607	608	609	610	611
41	612	613	614	616	617	618	619	620	621	622
42	623	624	625	626	627	628	629	630	631	632
43	633	634	635	636	637	638	639	640	641	642
44	643	644	645	646	647	648	649	650	651	652
45	653	654	655	656	657	658	659	659	660	661
46	662	663	664	665	666	667	668	669	670	671
47	672	673	673	674	675	676	677	678	679	680
48	681	682	683	683	684	685	686	687	688	689
49	690	691	692	692	693	694	695	696	697	698
50	699	699	700	701	702	703	704	705	705	706
51	707	708	709	710	711	712	713	713	715	715
52	716	716	717	718	719	720	721	722	722	723
53	724	725	725	726	727	728	729	730	730	731
54	732	733	734	734	735	736	737	738	738	739
N	0	1	2	3	4	5	6	7	8	9

FIG. 29

Three Place Logarithms

N	0	1	2	3	4	5	6	7	8	9
55	740	741	741	742	743	744	745	746	747	747
56	748	749	749	750	751	752	752	753	754	755
57	755	756	757	758	758	759	760	761	761	762
58	763	764	764	765	766	767	767	768	769	770
59	770	771	772	773	773	774	775	776	776	777
60	778	778	779	780	781	781	782	783	783	784
61	785	786	786	787	788	788	789	790	791	791
62	792	793	793	794	795	795	796	797	798	798
63	799	800	800	801	802	802	803	804	804	805
64	806	806	807	808	809	810	810	811	811	812
65	813	813	814	814	815	816	816	817	818	818
66	819	820	820	821	822	822	823	824	824	825
67	826	825	827	828	828	829	829	830	831	831
68	832	833	833	834	835	835	836	837	837	838
69	838	839	840	840	841	842	842	843	843	844
70	845	845	846	847	848	848	849	849	850	850
71	851	851	852	853	853	854	854	855	856	856
72	857	857	858	859	859	860	860	861	861	862
73	863	863	864	865	865	866	866	867	868	868
74	869	869	870	871	871	872	872	873	873	874
75	875	875	876	876	877	877	878	879	879	880
76	880	881	882	882	883	883	884	884	885	885
77	886	887	887	888	888	889	889	890	891	891
78	892	892	893	893	894	894	895	896	896	897
79	897	898	898	899	899	900	900	901	902	902
80	903	903	904	904	905	905	906	906	907	907
81	908	909	909	910	910	911	911	912	912	913
82	913	914	914	915	915	916	917	917	918	918
83	919	919	920	920	921	921	922	922	923	923
84	924	924	925	925	926	926	927	927	928	928
85	929	929	930	930	931	932	932	933	933	934
86	934	935	935	936	936	937	937	938	938	939
87	939	940	940	941	941	942	942	943	943	944
88	944	945	945	946	946	946	947	947	948	948
89	949	949	950	950	951	951	952	952	953	953
90	954	954	955	955	956	956	957	957	958	958
91	959	959	960	960	960	961	961	962	962	963
92	963	964	964	965	965	966	966	967	967	968
93	968	968	969	969	970	970	971	971	972	972
94	973	973	974	974	975	975	975	976	976	977
95	977	978	978	979	979	980	980	980	981	981
96	982	982	983	983	984	984	985	985	985	986
97	986	978	987	988	988	989	989	989	990	990
98	991	991	992	992	993	993	993	994	994	995
99	995	996	996	996	997	997	998	998	999	999
00	000	004	008	012	017	021	025	029	033	037
N	0	1	2	3	4	5	6	7	8	9

below 6 milliwatts or the zero reference level may be converted into decibels.

Determining DB Gain or Loss: In using amplifiers it is a prime requisite to know the decibel gain or loss when the input and output powers are known. To determine the gain or loss in DB employ the following formula:

$$\text{(gain) } N_{db} = 10 \text{ Log} \frac{P_o}{P_i} \quad (3)$$

$$\text{(loss) } N_{db} = 10 \text{ Log} \frac{P_i}{P_o} \quad (4)$$

where N_{db} is the number of DB gained or lost; P_i, the input power; and P_o, the output power.

Applying, for example, formula (3): Suppose that an intermediate amplifier is being driven by an input power of .2 watts, and after amplification, the output is found to be 6 watts.

$$\frac{P_o}{P_i} = \frac{12}{.2} = 60$$

$$\text{Log } 60 = 1.77$$

Therefore 10 × 1.77 = **17.7 DB POWER GAIN.**

Amplifier Ratings: The technical specifications or rating on power amplifiers must contain the following information: the overall gain in decibels; the power output in watts; the value of the input and output impedances; the input signal level in DB; the input signal voltage; and the power output level in decibels.

If the specifications on any one particular amplifier had included only the input and output signal levels in DB, it then would be necessary to know how much these values represented in power. The methods employed to determine power levels are not similar to those used in previous calculations. Caution should therefore be taken in reading the following explanations with particular care and attention being paid to the minor arithmetical operations.

The Anti-logarithm: To determine a power level from some given decibel value, it is necessary to invest the logarithmic process formerly employed in converting power to decibels. Here, instead of looking for the log of a number it is now necessary to find the anti-logarithm or number corresponding to a given logarithm.

In deriving a number corresponding to a logarithm it is important that these simple rules be committed to memory: (1) that the figures that form the original number from a corresponding logarithm depend entirely upon the mantissa or decimal part of the log; (2), that the characteristic serves only to indicate where to place the decimal point of the original number; and (3), that if the original number was a whole number the decimal point would be placed to the extreme right.

The procedure of finding the number corresponding to a logarithm is explained as follows: Suppose the logarithm to be 3.574. First, search in the table under any column from 0 to 9 for the numbers of the mantissa 574. If the exact number cannot be found, look for the next **lowest** figure, which is nearest to, but less than, the given mantissa. After the mantissa has been located simply glance immediately to the left to the N column and there will be read the number, 37. This number comprises the first two figures of the number corresponding to the antilog. The third figure of the number will appear at the head of the column in which the mantissa was found. In this instance the number heading the column will be 5. If the figures have been arranged as they have been found, the number will now be 375. Now since the characteristic is 3, there must be four figures to the **left** of the decimal point; therefore, by annexing a cipher the number becomes 3750; this is the number that corresponds to the logarithm 3.574. If the characteristic was 2 instead of 3, the number would be 375. If the logarithm was —3.574 or —1.274 the antilogs or corresponding numbers would be .00375 and .375 respectively. After a little experience a person can obtain the number corresponding to a logarithm in a very few seconds.

Converting Decibels to Power: It is always convenient to be able to convert a decibel value to a power equivalent in order to determine the ratio difference. The formula used for converting decibels into watts is similar in many respects to equation (2), the only difference being that the factor P_1 corresponding to the power level is not known. Usually the formula for converting decibels into power is written as

$$N_{db} = 10 \text{ Log} \frac{P_1}{.006} \quad (5)$$

In practice it has been found that it is too difficult to explain the solution to the above equation on account of the expression being written in the reverse. However, by re-arranging the various factors, the expression can be simplified to permit easy visualization, thus

$$P = .006 \times \text{antilog} \frac{N_{db}}{10} \quad (6)$$

where P is the desired power level; .006, the reference level in milliwatts; N_{db}, the decibels to be converted; and 10, the divisor.

To determine the power level, P, from a decibel equivalent simply divide the decibel value by 10, then take the number comprising the antilog and multiply it by .006, the product gives the power level of the decibel value.

Note: In all problems dealing with the conversion of **minus** decibels to power it often happens that the decibel value $—N_{db}$, is not always equally divisible by 10. When this is the case, the numerator in the factor $—N_{db}/10$ must be made evenly divisible by the denominator in order to derive the proper power ratio. Note that the value $—N_{db}$ is negative, hence, when dividing by 10, the negative signs must be observed and the quotient labeled accordingly.

To make the numerator in the value $—N_{db}$ equally divisible by 10, proceed as follows: Assume $—N_{db}$ to be the logarithm —38 with a zero mantissa, hence, in order to make —38 divisible by 10 simply annex as many units as is necessary from the zero mantissa and add them to the —38 until the figure can be equally divided. An examination will show it was only necessary to add two units to bring —38 up to —40. CAREFULLY NOTE that every unit borrowed from the zero mantissa must be returned to it as a positive quantity multiplied by 10. Thus, the two units borrowed to bring —38 up to —40 is returned as 20, making what was a zero mantissa now have a value of 20. The numerator $—N_{db}$, now becomes —40.20; this figure can now be equally divided by 10.

While the above discussion applied strictly to negative values the following examples will clearly show the technique to be followed for most all practical problems.

(a) The output level of a popular velocity ribbon microphone is rated at —74DB. What is this equivalent in milliwatts?

Solution by equation (6)

$$\frac{—N_{db}}{10} = \frac{—74}{10} \quad \text{(not equally divisible by 10)}$$

Routine:

```
   —74    mantissa
   +6 60  (6×10)
  ------
   —80 60
```

$$\frac{—N_{db}}{10} = \frac{—80.60}{10} = —8.6$$

Antilog —8.6 = .00000004

.006×.00000004 =.000000000240 or 240 MICROMICRO-WATTS

(b) This example differs somewhat from that of the above in that the mantissas are added differently.—A low powered amplifier has an input signal level of —17.3DB. How many milliwatts does this value represent?

Solution by equation (6)

$$—N_{db} = \frac{—17.3}{10} = —2.33$$

```
  —17 .  3
  + 3 . 30
  --------
  —20 . 33
```

(The mantissas were added as 30 plus 3, and NOT .3 plus .30)

Antilog —2.33 = .0398

.006 × .0398 = .0002388 or **.24 MILLIWATTS.**

Voltage Amplifiers: When plans are being drafted contemplating the design of power amplifiers it is essential that the following data be determined: First, the input and output signal levels to be used; second, the size of the power tubes that would adequately deliver an undistorted output; and third, the input signal voltage that must be applied to the amplifier to deliver the desired output. This last requirement is the most important in the design of voltage amplifiers as it is the ratio of the input signal voltage to the output signal voltage that governs the amount of amplification.

The voltage step-up in a transformer-coupled amplifier depends chiefly upon the mu of the tubes and the turns ratio of the inter-stage coupling transformers. The step-up value in any amplifier is calculated by multiplying the step-up factor of each voltage amplifying or step-up device. Thus for example, if an amplifier were designed having an output transformer with a ratio of 3:1 coupled to a tube having a mu of

7, the voltage step-up would be approximately 3 times 7, or 21. It is seldom that the total product will be exactly the figure derived because it is not quite possible to obtain the full mu of the tube.

From the voltage gain in an amplifier it is possible to calculate the input and output signal levels and at the same time be able to determine at what level the input signal must be in order to obtain the desired output. By converting voltage ratios into decibels, power levels can be determined. Hence, to find the gain in DB when the input and output voltages are known, the following expression is used:

$$\text{(gain) } N_{db} = 20 \text{ Log } \frac{E_1}{E_2} \qquad (7)$$

where E_1, is the output voltage; and E_2, the input voltage.

Employing the above equation to a practical problem, note that the logarithm is multiplied by 20 instead of by 10 as in previous examples.

A certain one-stage amplifier consisted of the following parts: 1 input transformer, ratio 2:1; and 1 output tube having a mu of 95. Determine the gain in decibels with an input voltage of 1 volt.

Solution by equation (7)

$$2 \times 95 = 190 \text{ voltage gain}$$

$$\text{therefore, } \frac{E_1}{E_2} = \frac{190}{1} = 190$$

$$\text{Log } 190 = 2.278$$

$20 \times 2.278 =$ **45.56 DECIBEL GAIN**

To reverse the above and convert decibels to voltage ratios, use the following expression:

$$E \text{ (gain)} = \text{antilog } \frac{N_{db}}{20} \qquad (8)$$

where E is the voltage gain (power ratio); N_{db}, the decibels; and 20, the divisor.

To find the gain, simply divide the decibels by 20, then extract the antilog from the quotient, the result gives the voltage ratio.

Input Voltages: In designing power amplifiers, it is paramount to have **exact** knowledge of the magnitude of the input signal voltage necessary to drive the output power tubes to maximum undistorted output. Without this information it would largely be a matter of guesswork in determining whether or not the power stages were being worked overloaded or underloaded.

To determine the input voltage take the **peak voltage** necessary to drive the grid of the output tube to maximum and divide this figure by the total overall gain **preceding the last stage.**

Microphone Levels: Practically all acoustic-electric apparatus energizing amplifiers have output levels rated in decibels. The output signal levels of these devices vary considerably from each other as may be noted from the table above:

	DECIBELS	AVERAGE
Phonograph pickup	0 to —30	—15
Carbon microphones	—30 to —60	—45
Piezo-elec. microphones	—70 to —80	—45
Dynamic microphones	—75 to —95	—85
Condenser microphones	—95 to —100	—97
Velocity microphones	—100 to —110	—105

In general, the lower the output signal level, the higher will be the acoustic fidelity over the entire audio spectrum. On the other hand, the higher the input signal level, the lower will be the overall fidelity.

The output levels of microphones and phonograph pickups have the same power values ascribed to them as those derived from calculating power output levels of amplifiers. Therefore, the same equations employed in connection with power ratios are similarly applied when converting output signal levels to power levels.

Computing Specifications: From the preceding explanations the following data can be computed with a very high degree of accuracy:

(1) Voltage amplification
(2) Overall gain in DB
(3) Output signal level in DB
(4) Input signal level in DB
(5) Input signal level in watts
(6) Input signal voltage

Push-Pull Amplifiers: To double the output of any cascade amplifier it is only necessary to push-pull the last amplifying stage and replace the interstage and output transformers with push-pull types.

To determine the voltage step (voltage ratio) of a push-pull amplifier take the ratio of one **half** of the secondary winding of the push-pull transformer and multiply it by the mu of one of the output tubes in the push-pull stage; the product, **when doubled,** will be the voltage amplification or step-up.

Note: Doubling the output power of any amplifier will not double the output signal level. In general, doubling the power adds about only 3 DB.

Acoustically, that is from the loudspeaker standpoint, it takes approximately one DB to note any appreciable change in the volume of sound. This is because the intensity of sound as heard by the ear varies logarithmically with the acoustic power. For practical purposes it is only necessary to remember, that if two sounds differ in physical intensity by less than one DB they usually sound alike. If they are much more than one decibel apart one sounds slightly louder than the other. This quantitative data is also applicable to amplifiers in that the output signal levels must differ by at least 1.5 DB in order to note any change in volume.

Pre-amplifiers: Pre-amplifiers are employed to raise low input signal levels up to some required input level of another intermediate or succeeding amplifier. For example: If an amplifier was designed to operate at an input level of —30DB, and instead, a considerably lower input level was used, a pre-amplifier would then have to be designed to bring the low input signal up to the rated input signal level of —30DB to obtain the full undistorted output from the power tubes in the main amplifier. The amount of gain necessary to raise a low input signal level up to another level may be determined by the following equation:

$$E(\text{gain}) = \text{antilog}\,\frac{N_{db_1} - N_{db_2}}{20} \quad (9)$$

where E is the voltage step-up or gain; N_{db_1}, the input signal level of the pre-amplifier or the new input signal level; N_{db_2}, the input signal level to the intermediate amplifier; and 20, the divisor.

To apply the equation, take the following example: If a 7-watt amplifier had an input signal level of —32.8DB and a microphone had an output signal level of —60DB which was exciting the amplifier, how much voltage amplification will be necessary to raise the gain up to —32.8DB so that the amplifier will work at full output?

$$\frac{N_{db_1} - N_{db_2}}{20} = \frac{60 - 32.8}{20} = 1.355$$

Antilog 1.355 = **22.6 VOLTAGE GAIN**

The additional gain can be obtained by designing a pre-amplifier having an input transformer with a ratio of 2.5:1 coupled to a tube having a mu of 9.

"Radio" Handbook Data Charts

Radio data charts provide designers of amateur radio equipment with a ready and convenient means of solving problems without having recourse to complicated formula and mathematics.

To properly use the chart and to prevent disfiguring the page, simply place a piece of tracing paper, celluloid, or waxed paper over the scales, then, the index line which intercepts the scales may be drawn with a hard pencil and a straight edge.

The first chart which is a **logarithmic alignment nomogram** will solve many problems encountered in ordinary practice.

Voltage Drop Calculation in Resistors: To find the voltage drop for a certain bias for a self-biased tube, add three ciphers to the value desired, seek this value on scale A; next, search for the value which corresponds to the plate current (cathode current) on the B scale, now, drawing a line between these two points will intersect a point on C, this corresponds to the ohmage. Hence, a resistance required to produce 9 volts bias for a triode which operates at 3 MA plate current is: on the A scale, 9 plus three ciphers equals 9000; on the B scale 3MA; and the ohmage 3000, is found on C.

Wattage or Heat Capacity in Resistors: To find the power liberated in watts by a certain resistor when ohmage and voltage is known, proceed as follows: On C find the voltage, on A, the resistance; draw a line connecting these two points over to the B scale, next, find the voltage (for the second time) on the A scale and draw a line from point B through A, the wattage will be given on C. See the auxiliary Figure for an example.

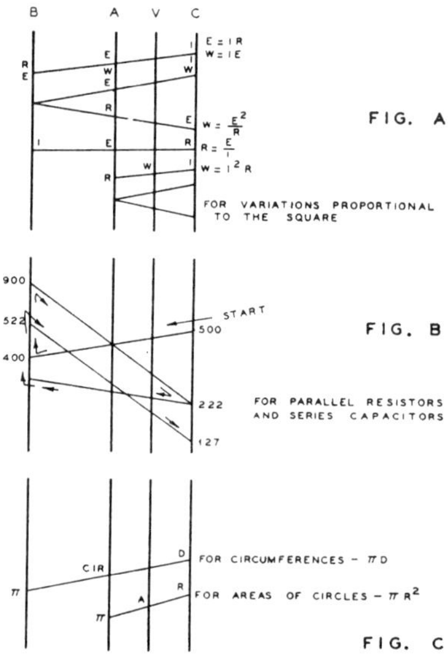

Showing how to use the Calculator Chart on facing page.

If the current instead of the voltage is known in the above procedure, the technique is as follows: On C find the value of current, on A, the resistance, a line drawn connecting these points will intercept the wattage rating on V.

Series Capacity Calculations: To determine the value of any **two** series-capacities, find one of the values on C and the other on B, draw a line to connect these two points; next, **add** the values of the capacities on B, then from this new point, draw a line to intersect A, and the series value will be read on C.

If three series capacities are to be employed, the value of any two of them is found as above, and then this is treated as a single capacity and its value combined with the third can be found by repeating the process which can be carried on indefinitely.

An example illustrating the method is shown in auxiliary Figure (b).

Note: Raise the A scale the distance from 1 to 10 when the reading is beyond the bottom, or by taking a piece of tracing paper and tracing the A scale so as to extend it another length of 1 to 10.

Parallel Resistor Calculations: These are treated exactly as series capacities, and the above explanation will solve all values.

Coil Winding Considerations: It is often necessary to know, in coil winding procedure, the resistance of coils when re-wound to the same volume with a wire larger or smaller than the coil was originally wound

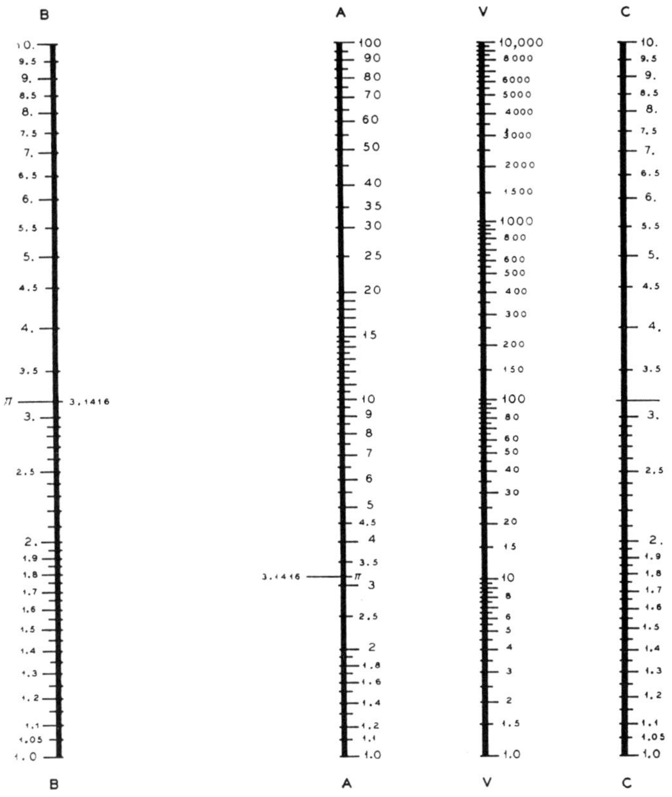

"RADIO" Handbook Logarithmic Alignment Nomogram.

with; hence, by knowing the size of the wires, the circular-mil areas, and the ohmage of the original winding, the ohmic value of the new winding can be found as follows: select the value corresponding to the circular-mil area of the new winding on the C scale, select the value corresponding to the ohmage of the original winding on the V scale, draw a line to A intersecting these points; next, on C find the value corresponding to the circular-mil area of the old winding, a line drawn from A to C will intercept the ohmage of the new winding on V.

These calculations are based on the principle that resistances are inversely proportional to the squares of the wire sizes. This is sensibly true for a change of a few sizes, but the error increases with the range of sizes which should not be over five.

Sound and Light Calculations: The intensity of sound and light, on a surface varies inversely as the square of the distance from the source. Variations proportional to the square is of importance in the sound, light, magnetic and gravitational fields. By employing the considerations given in the preceding paragraph all problems enunciated in this topic can be solved. By using the example shown in auxiliary Figure (a) it will be found that if a surface 12.7 feet distant from a light source receives an intensity of 100 foot-candles when moved 20.2 feet from the light, it will receive 39.5 foot-candles. (After C.P. Nachod, N&US Sig. Co.)

Areas of Circles: The area of any circle can be found by placing the Pi constant (3.1416) on A and the diameter on C, the area will be found on V when a line intersects V drawn through A and C.

Another method for finding the area is to place the constant 0.785 on A, the diameter on C and the area will be found on V when a line is drawn through A, C.

Circumference of Circles: The circumference of any circle can be found by placing the Pi constant (3.1416) on B and the diameter on C, a line connecting B, C will intercept the circumference on A.

Multiplication: The multiplication table is represented on scales A, B, and C. A line drawn through values on scales B, C will intersect the product on A.

Division: To divide the process of multiplication is reversed, the values for the divisor will appear on the C scale, and the dividend on A and a line drawn through these points will intersect the quotient on B.

Square Root: To extract the square root of any number, seek the number on the A scale and the root will appear horizontally on either the B or C scales.

Powers of Numbers: To raise a number to the fourth power (3 × 3 × 3 ×3 = 81), select the number in both the B and C columns, a horizontal line drawn through these points intersects the fourth power on the A scale.

Universal Vacuum Tube Characteristic Chart

The characteristic of **any** vacuum tube such as amplification factor, mutual conductance, plate resistance and the figure of merit can be found quickly by referring to the right-angle logarithmic nomogram in Figure 30.

By connecting the ordinates and abscissas (horizontal and vertical logarithmic scales) by means of a straight edge at any point to form a right angle in the central scale, will indicate at the vertex of the angle the plate resistance (R_p) and the figure of merit. Similarly, selecting any point in the center scales corresponding to R_p and the figure of merit, will give the amplification factor (μ) and the grid-to-plate transconductance (G_m) when a line is drawn at right-angles to the left and bottom scales.

The vacuum tube parameters found from the chart are those which correspond to the actual operating conditions.

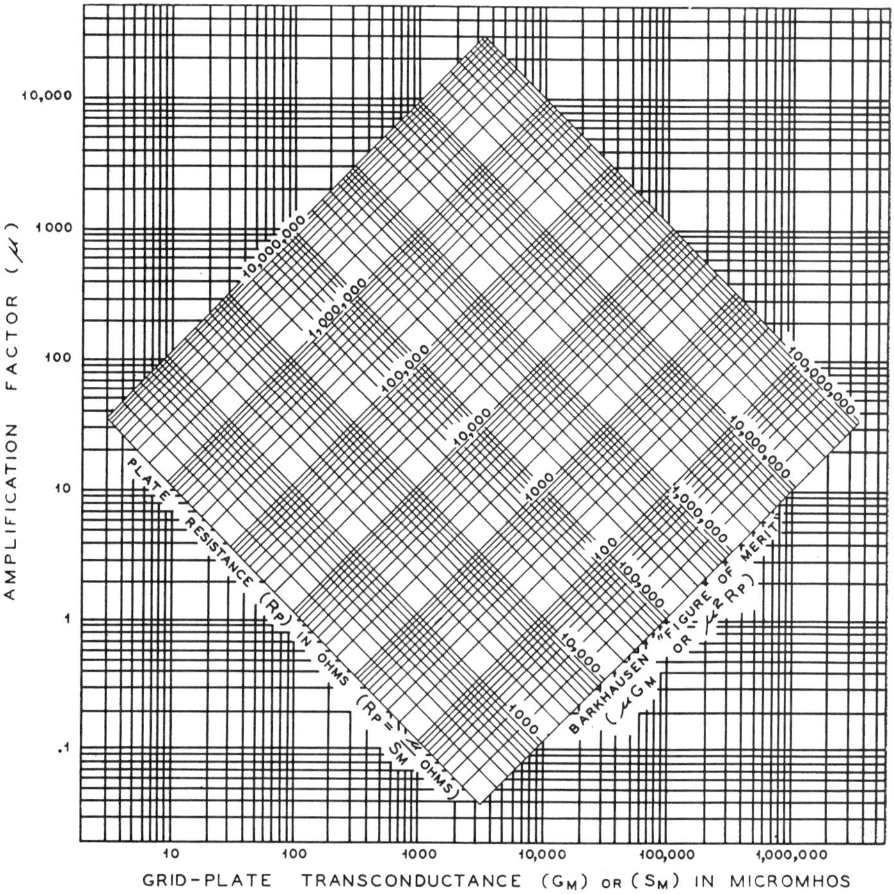

FIG. 30—Universal Vacuum Tube Characteristics Chart.

FEDERAL COMMUNICATIONS COMMISSION

Rules Governing Amateur Radio Stations and Operators

(As in effect June 1, 1935)

The following excerpts from the Commission's rules include all that deal solely with the amateur service and certain others that apply generally.

PART I. GENERAL RULES AND REGULATIONS

1. **Prescribed application forms.**—Each application for an instrument of authorization shall be made in writing on the appropriate form prescribed by the Commission for the purpose. Separate application shall be filed for each instrument of authorization. The required forms except as provided in paragraph 408 for amateur applicants, may be obtained from the Commission or from the office of any inspector. For a list of such offices and related geographical districts, see paragraph 30.

2. **Filing of application.**—Each application for station license shall be submitted as follows:

Class of station	Number of application forms required and method of filing
a. All classes of Alaskan stations, except broadcast and amateur.	3 copies via inspector in charge, radio district No. 14 Seattle, Wash.
b. Aircraft	1 copy direct to Washington, D. C.
c. Broadcast pick-up	Do.
d. Geophysical	Do.
e. Portable (all classes, except amateur)	Do.
f. Ship	Do.
g. All other classes, except amateur	2 copies direct to Washington, D. C.
h. Each application for amateur facilities shall be filed in accordance with the following instructions:	
(1) Applications for amateur station and/or operators' licenses from applicants residing within 125 miles of Washington, D. C., a radio district office of the Commission, or an examining city (see par. 30).	1 copy to the inspector in charge of the radio district in which the applicant resides.
(2) Applications for amateur station and/or operators' licenses from applicants residing more than 125 miles from Washington, D. C., a radio district office of the Commission, or an examining city (see par. 30).	1 copy direct to the Federal Communications Commission, Washington, D. C., in accordance with the instructions specifically set forth on the application form.

129539°—35——1

4. **Amendments.**—Any amendment of a pending application shall be verified in the same manner as was the original application and shall be sent direct to the Commission.

7. **Necessity for construction permit and the filing of applications therefor.**—With the exceptions hereinafter noted no license will be granted by the Commission for the operation of any station unless a permit for its construction has been granted previously by the Commission upon written application therefor. The exceptions to the foregoing are:

a. Amateur stations.
b. Stations upon mobile vessels.
c. Stations upon railroad rolling stock.
d. Stations upon aircraft.
e. Stations the construction of which was completed prior to February 23, 1927.

14. **License where construction permit is not required.**—Each application for new license, where a construction permit is not prerequisite thereto, shall be filed at least 60 days prior to the contemplated operation of the station.

16. **Renewal of license.**—Unless otherwise directed by the Commission, each application for renewal of license shall be filed at least 60 days prior to the expiration date of the license sought to be renewed.

17. **Application called for by Commission.**—Whenever the Commission regards an application for a renewal of license as essential to the proper conduct of a hearing or investigation and specifically directs that the same be filed by a date certain, such application shall be filed within the time thus specified. If the licensee fails to file such application within the prescribed time, or such extension thereof as the Commission may grant upon proper showing, the hearing or investigation shall proceed as if such renewal application had been received.

18. **Extension of license.**—When there is pending before the Commission any application or proceeding that after hearing might lead to or make necessary the modification of, revocation of, or the refusal to renew an existing license the Commission may, at its discretion, grant an extension of

such license. No extension shall be construed as a finding by the Commission that would prejudice or restrict the Commission's liberty of action on any pending application or proceeding.

20. **Penalty for transfer of license without the consent of Commission.**—The transfer of a construction permit or any of the rights granted thereunder without consent of the Commission shall be sufficient ground for refusal of a station license. Likewise, the transfer of a station license or any of the rights granted thereunder without consent of the Commission shall be sufficient ground for revocation of such license or denial of any application for its renewal.

22. **Special authorizations.**—The Commission may grant special authority to the licensee of an existing station authorizing the operation of such station for a limited time in a manner, to an extent or for a service other or beyond that authorized in the license.

23. **Emergencies.**—Where an emergency exists affecting safety of life or property the Commission may, in its discretion, waive any part or all of its regulations governing the filing of applications.

24. **Answering notice of violation.**—Any licensee receiving official notice of a violation of Federal laws, the Commission's rules and regulations, or the terms and conditions of a license shall, within 3 days from such receipt, send a written reply direct to the Federal Communications Commission at Washington, D. C. The answer to each notice shall be complete in itself and shall not be abbreviated by reference to other communications or answers to other notices. If the notice relates to some violation that may be due to the physical or electrical characteristics of the transmitting apparatus, the answer shall state fully what steps, if any, are taken to prevent future violations, and if any new apparatus is to be installed, the date such apparatus was ordered, the name of the manufacturer, and promised date of delivery.

26. If the notice of violation relates to some lack of attention or improper operation of the transmitter, the name and license number of the operator in charge shall be given.

27. **Normal license periods.**—All station licenses will be issued so as to expire at the hour of 3 a. m., eastern standard time.

e. The licenses for amateur stations will be issued for a normal license period of 3 years from the date of expiration of old license or the date of granting a new license or modification of a license.

28. **Designation of call signals.**—Insofar as practicable, call signals of radio stations will be designated in alphabetical order from groups available for assignment, depending upon the class of station to be licensed. Because of the large number of amateur stations, calls will be assigned thereto in regular order and requests for particular calls will not be considered except on formal application the Commission may reassign calls to the last holders of record.

29. **Deletion of call signals.**—Call signals of stations will be deleted in each of the following cases:

a. Where an existing instrument of authorization has expired and no application for renewal or extension thereof has been filed.

b. Where a license has been revoked.

c. Where a license is surrendered or canceled.

d. Other cause, such as death, loss of citizenship, or adjudged insanity of the station licensee. Such occurrences coming to notice should be reported to the Commission, preferably accompanied by the station license for cancelation, if available.

30. **Radio districts.**—The following list of the radio districts gives the address of each field office of the Federal Communications Commission and the territory embraced in each district:

Radio dis- trict	Address of the inspector in charge	Territory within district	
		States	Counties
1	Customhouse, Boston, Mass.	Connecticut	All counties
		Maine	Do.
		Massachusetts	Do.
		New Hampshire	Do.
		Rhode Island	Do.
		Vermont	Do.
2	Federal Building, New York, N. Y.	New York	Albany, Bronx, Columbia, Delaware, Dutchess, Greene, Kings, Nassau, New York, Orange, Putnam, Queens, Rensselaer, Richmond, Rockland, Schenectady, Suffolk, Sullivan, Ulster, and Westchester.
		New Jersey	Bergen, Essex, Hudson, Hunterdon, Mercer, Middlesex, Monmouth, Morris, Passaic, Somerset, Sussex, Union, and Warren.
3	New United States Customhouse, Philadelphia, Pa.	Pennsylvania	Adams, Berks, Bucks, Carbon, Chester Cumberland, Dauphin, Delaware, Lancaster, Lebanon, Lehigh, Monroe, Montgomery, Northampton, Perry, Philadelphia, Schuylkill, and York.
		New Jersey	Atlantic, Burlington, Camden, Cape May, Cumberland, Gloucester, Ocean, and Salem.
		Delaware	Newcastle.

Radio district	Address of the inspector in charge	Territory within district	
		States	Counties
4	Fort McHenry, Baltimore, Md	Maryland	All counties.
		District of Columbia	Do.
		Virginia	Arlington, Clark, Fairfax, Fauquier, Frederick, Loudoun, Page, Prince William, Rappahannock, Shenandoah, and Warren.
		Delaware	Kent and Sussex.
5	New Post Office Building, Norfolk, Va.	Virginia	All except district 4.
		North Carolina	All except district 6.
6	New Post Office Building, Atlanta, Ga.	Alabama	All counties.
		Georgia	Do.
		South Carolina	Do.
		Tennessee	Do.
		North Carolina	Ashe, Avery, Buncombe, Burke, Caldwell, Cherokee, Clay, Cleveland, Graham, Haywood, Henderson, Jackson, McDowell, Macon, Madison, Mitchell, Polk, Rutherford, Swain, Transylvania, Watauga, and Yancey.
7	Post Office Box 150, Miami, Fla	Florida	All counties.
		Puerto Rico	Do.
		Virgin Islands	Do.
8	Customhouse, New Orleans, La	Arkansas	Do.
		Louisiana	Do.
		Mississippi	Do.
		Texas	City of Texarkana only.
9	Prudential Building, Galveston, Tex.	Texas	Arkansas, Brazoria, Brooks, Calhoun, Cameron, Chambers, Fort Bend, Galveston, Goliad, Harris, Hidalgo, Jackson, Jefferson, Jim Wells, Kenedy, Kleberg, Matagorda, Nueces, Refugio, San Patricio, Victoria, Wharton, and Willacy.
10	Federal Building, Dallas, Tex	Texas	All except district 9 and the city of Texarkana.
		Oklahoma	All counties.
		New Mexico	Do.
11	Rives-Strong Building, Los Angeles, Calif.	Arizona	Do.
		Nevada	Clarke.
		California	Imperial, Kern, Kings, Los Angeles, Monterey, Orange, Riverside, San Bernardino, San Diego, San Luis Obispo, Santa Barbara, Tulare, and Ventura.
12	Customhouse, San Francisco, Calif.	California	All except district 11.
		Nevada	All except Clarke.
		Guam	All counties.
		American Samoa	Do.
3	New United States Courthouse, Portland, Oreg.	Oregon	Do.
		Idaho	All except district 14.
14	Federal Office Building, Seattle, Wash.	Alaska	All counties.
		Washington	Do.
		Idaho	Benewah, Bonner, Boundary, Clearwater, Idaho, Kootenai, Latah, Lewis, Nez Perce, and Shoshone.
		Montana	Beaverhead, Broadwater, Cascade, Deerlodge, Flathead, Gallatin, Glacier, Granite, Jefferson, Lake, Lewis and Clark, Lincoln, Madison, Meagher, Mineral, Missoula, Pondera, Powell, Ravalli, Sanders, Silver Bow, Teton, and Toole.
15	Customhouse, Denver, Colo	Colorado	All counties.
		Utah	Do.
		Wyoming	Do.
		Montana	Except district 14.
16	New Main Post Office Building, St. Paul, Minn.	North Dakota	All counties.
		South Dakota	Do.
		Minnesota	Do.
		Michigan	Alger, Baraga, Chippewa, Delta, Dickinson, Gogebic, Houghton, Iron, Keweenaw, Luce, Mackinac, Marquette, Menominee, Ontonagon, and Schoolcraft.
		Wisconsin	All except district 18.
17	Federal Building, Kansas City, Mo.	Nebraska	All counties.
		Kansas	Do.
		Missouri	Do.
		Iowa	All except district 18.

Radio district	Address of the inspector in charge	Territory within district	
		States	Counties
18	Engineering Building, Chicago, Ill.	Indiana	All counties.
		Illinois	Do.
		Iowa	Allamakee, Buchanan, Cedar, Clayton, Clinton, Delaware, Des Moines, Dubuque, Fayette, Henry, Jackson, Johnson, Jones, Lee, Linn, Louisa, Muscatine, Scott, Washington, and Winneshiek.
		Wisconsin	Columbia, Crawford, Dane, Dodge, Grant, Green, Iowa, Jefferson, Kenosha, Lafayette, Milwaukee, Ozaukee, Racine, Richland, Rock, Sauk, Walworth, Washington, and Waukesha.
19	New Federal Building, Detroit, Mich.	Michigan	All except district 16.
		Ohio	All counties.
		Kentucky	Do.
		West Virginia	Do.
20	Federal Building, Buffalo, N. Y.	New York	All except district 2.
		Pennsylvania	All except district 3.
21	Aloha Tower, Honolulu, Territory of Hawaii.	Territory of Hawaii	

a. **Additional examining cities.**—The following is a list of the cities where examinations will be held for radio operators' licenses in addition to Washington, D. C. and the radio district offices of the Commission. Other cities may also be designated from time to time for the purpose of conducting commercial operators' examinations only (see pars. 2, 404, and 408) and class A amateur:

Schenectady, N. Y.	St. Louis, Mo.
Winston-Salem, N. C.	Pittsburgh, Pa.
Nashville, Tenn.	Cleveland, Ohio
San Antonio, Tex.	Cincinnati, Ohio
Oklahoma City, Okla.	Columbus, Ohio
Des Moines, Iowa	

Examinations for commercial and class A amateur privileges will be conducted not more than twice per year in the following cities, which are not to be construed as examining cities under the rules which apply for class B and C amateur privileges:

Albuquerque, N. Mex.	Jacksonville, Fla.
Billings, Mont.	Little Rock, Ark.
Bismarck, N. Dak.	Phoenix, Ariz.
Boise, Idaho	Salt Lake City, Utah
Butte, Mont.	Spokane, Wash.

b. **Amateur call areas.**—The following is a list of the amateur call areas, showing the territory embraced in each area:

Call area	States	Counties
1	Connecticut	All counties.
	Maine	Do.
	Massachusetts	Do.
	New Hampshire	Do.
	Rhode Island	Do.
	Vermont	Do.
2	New Jersey	Bergen, Essex, Hudson, Middlesex, Monmouth, Ocean, Passaic, and Union.
	New York	Albany, Bronx, Columbia, Dutchess, Greene, Kings, Nassau, New York, Orange, Putnam, Queens, Rensselaer, Richmond, Rockland, Schenectady, Suffolk, Ulster, and Westchester.
3	Delaware	All counties.
	District of Columbia	Do.
	Maryland	Do.
	New Jersey	Atlantic, Burlington, Camden, Cape May, Cumberland, Gloucester, Hunterdon, Mercer, Morris, Salem, Somerset, Sussex, and Warren.
	Pennsylvania	Adams, Berks, Bucks, Chester, Cumberland, Dauphin, Delaware, Franklin, Lancaster, Lebanon, Lehigh, Montgomery, Northampton, Philadelphia, and York.
	Virginia	All counties.
4	Alabama	Do.
	Florida	Do.
	Georgia	Do.
	North Carolina	Do.
	Puerto Rico	Do.
	South Carolina	Do.
	Tennessee	Do.
	Virgin Islands	Do.
5	Arkansas	Do.
	Louisiana	Do.
	Mississippi	Do.
	New Mexico	Do.
	Oklahoma	Do.
	Texas	Do.
6	Arizona	Do.
	California	Do.
	Hawaii	Do.
	Nevada	Do.
	Utah	Do.

Call area	States	Counties
7	Alaska	All Counties.
	Idaho	Do.
	Montana	Do.
	Oregon	Do.
	Washington	Do.
	Wyoming	Do.
8	Michigan	Alcona, Allegan, Alpena, Antrim, Arenac, Barry, Bay, Benzie, Berrien, Branch, Calhoun, Cass, Charlevoix, Cheboygan, Clare, Clinton, Crawford, Eaton, Emmet, Genesee, Gladwin, Grand, Traverse, Gratiot, Hillsdale, Huron, Ingham, Ionia, Iosca, Isabella, Jackson, Kalamazoo, Kalkaska, Kent, Lake, Lapeer, Leelanau, Lenawee, Livingston, Macomb, Manistee, Mason, Mecosta, Midland, Missaukee, Monroe, Montcalm, Montmorency, Muskegon, Newaygo, Oakland, Oceana, Ogemau, Osceola, Oscoda, Otsego, Ottawa, Presque Isle, Roscommon, St. Clair, St. Joseph, Saginaw, Sanilac, Shawassee, Tuscola, Van Buren, Washtenaw, Wayne, and Wexford.
	New York	Allegany, Broome, Cattaraugus, Cayuga, Chautauqua, Chemung, Chenango, Clinton, Cortland, Delaware, Erie, Essex, Franklin, Fulton, Genessee, Hamilton, Herkimer, Jefferson, Lewis, Livingston, Madison, Monroe, Montgomery, Niagara, Oneida, Onondaga, Ontario, Orleans, Oswego, Otsego, St. Lawrence, Saratoga, Schoharie, Schuyler, Seneca, Steuben, Sullivan, Tioga, Tompkins, Warren, Washington, Wayne, Wyoming, and Yates.
	Ohio	All counties.
	Pennsylvania	Allegheny, Armstrong, Beaver, Bedford, Blair, Bradford, Butler, Cambria, Cameron, Carbon, Center, Clarion, Clearfield, Clinton, Columbia, Crawford, Elk, Erie, Fayette, Forest, Fulton, Greene, Huntingdon, Indiana, Jefferson, Juniata, Lackawanna, Lawrence, Luzerne, Lycoming, McKean, Mercer, Mifflin, Monroe, Montour, Northumberland, Perry, Pike, Potter, Schuylkill, Snyder, Somerset, Sullivan, Susquehanna, Tioga, Union, Venango, Warren, Washington, Wayne, Westmoreland, and Wyoming.
	West Virginia	All counties.
9	Colorado	Do.
	Illinois	Do.
	Indiana	Do.
	Iowa	Do.
	Kansas	Do.
	Kentucky	Do.
	Michigan	Alger, Baraga, Chippewa, Delta, Dickinson, Gogebic, Houghton, Iron, Keweenaw Luce, Mackinac, Marquette, Menominee, Ontonagon, and Schoolcraft.
	Minnesota	All counties.
	Missouri	Do.
	Nebraska	Do.
	North Dakota	Do.
	South Dakota	Do.
	Wisconsin	Do.

PART IV. SERVICES OTHER THAN BROADCAST

GENERAL REGULATIONS

182. **Scope.**—Whenever in these regulations the words "services other than broadcast" are used without specific restriction, or further qualifications, the words shall be taken to refer to all radio services, frequencies, etc., except broadcast service in the band of frequencies between 550 and 1,500 kilocycles, inclusive.[1]

184. **Waves designated by frequency.**—In these regulations and instruments of authorization issued by the Commission a wave referred to is designated by its frequency in kilocycles per second.

185. **Total spectrum of waves.**—The total spectrum of waves shall be construed as extending in frequency from 10 to 500,000 kilocycles, inclusive. This provision, however, shall not be interpreted as precluding authority of the Commission over the use of waves having frequencies less than 10 kilocycles or more than 500,000 kilocycles if and when such waves, by reason of progress in the art, become available for radio communication either practically or experimentally, nor as precluding the Commission from issuing instruments of authorization with respect to the use of such waves.

186. **Division of total spectrum into major bands.**—The total spectrum of waves as hereinbefore defined is hereby divided into six major bands as follows:

a. Low-frequency: 10 to 100 kilocycles.

b. Medium-frequency: 100 to 550 kilocycles.

c. Broadcast: 550 to 1,500 kilocycles.

d. Medium high-frequency: 1,500 to 6,000 kilocycles.

e. High-frequency: 6,000 to 30,000 kilocycles.

f. Very high-frequency: Above 30,000 kilocycles.

187. **Definitions.**—The following definitions shall apply generally to all services (see also International Telecommunication Convention).

188. **Station.**—The term "station" means all of the radio-transmitting apparatus used at a particular location for one class

[1] *International conventions.*—In addition to the regulations herein contained licensees must observe the provisions of the following insofar as they apply:

a. The Telecommunication Convention and General Radio Regulations annexed thereto (State Department Treaty Series No. 867).

b. North American Radio Agreement of 1929 (State Department Treaty Series No. 777-A).

Copies may be obtained from the Superintendent of Documents, Washington, D. C. Treaty Series No. 867, price 20 cents, and Treaty Series No. 777-A, price 10 cents.

of service and operated under a single instrument of authorization. In the case of every station other than broadcast, the location of the station shall be considered as that of the radiating antenna.

189. **Mobile station.**—The term "mobile station" means a station that is capable of being moved and ordinarily does move.

190. **Fixed station.**—The term "fixed station" means a station, other than an amateur station, not capable of being moved, and communicating by radio with one or more stations similarly established.

191. **Land station.**—The term "land station" means a station not capable of being moved, carrying on a mobile service.

192. **Portable station.**—The term "portable station" means a station so constructed that it may conveniently be moved about from place to place for communication and that is in fact so moved about from time to time, but not used while in motion.

a. **Portable-mobile station.** The term "portable-mobile station" means a station so constructed that it may conveniently be moved from one mobile unit to another for communication, and that is, in fact, so moved about from time to time and ordinarily used while in motion.

193. **Mobile service.**—The term "mobile service" means a radio-communication service carried on between mobile and land stations and by mobile stations, communicating among themselves, special services being excluded.

194. **Fixed service.**—The term "fixed service" means a service carrying on radio-communication of any kind between fixed points, excluding broadcasting services and special services.

195. **Special service.**—The term "special service" means a radio-communication service carried on especially for the needs of a specific service of general interest and not open to public correspondence, such as a service of radiobeacons, radio direction finding, time signals, regular meteorological bulletins, notices to navigators, press messages addressed to all, medical notices (medical consultation by radio), standard frequencies, emissions for scientific purposes, etc.

196. **International service.**—The term "international service" means a radio-communication service between offices or stations under the jurisdiction of different countries, or between stations of the mobile service except when the latter are of the same nationality and are within the limits of the country to which they belong. An internal or national radio-communication service which is likely to cause interference with other services beyond the limits of the country in which it operates shall be considered as an international service from the standpoint of interference.

197. **General communication service.**—The term "general communication service" is used in these regulations only with respect to the service allocation of certain frequencies, and means that such frequencies have not been assigned to any specific service, and are available for future allocation to services which will be designated.

198. **Public service.**—The term "public service" means a service for the use of the public in general.

199. **Public correspondence.**—The term "public correspondence" means any radio communication where the offices and stations, by reason of their being at the disposal of the public, must accept for transmission.

200. **Private service.**—The term "private service" means a radio-communication service which is not open to public correspondence and which may be used only by specified persons for either general or specific purposes.

201. **Private enterprise.**—The term "private enterprise" means any person, company, or corporation which operates one or more stations for radio communication.

202. **Limited service.**—The term "limited service" means a service which can be used only by specified persons or for special purposes.

203. **Radio operating signals.**—The term "radio operating signals" means a letter, figure, or combination of letters and figures, or both, designed to facilitate the conduct of communications; for example, the list of abbreviations to be used in Radio Transmission, appendix 9 to the General Radio Regulations annexed to the International Telecommunication Convention.

204. **Allocation of bands of frequencies to services.**—Allocations of bands of frequencies to services, such as mobile, fixed, broadcast, amateur, etc., are set forth in article 7 of the General Radio Regulations annexed to the International Telecommunication Convention and in the North American Radio Agreement. These allocations will be adhered to in all assignments to stations capable of causing international interference.

205. **Frequency standard.**—The national standard of radiofrequency maintained by the Bureau of Standards, Department of Commerce, shall be the basis for all frequency measurements, and assignments will be made on the basis of this standard.

206. **Frequency checking.**—The licensee of each station, except amateur, shall provide for measurement of the station frequency and establish procedure for checking it regularly. These measurements of station frequency shall be made by means independent of the frequency control of the transmitter and shall be of such an accuracy that the limit of error is within the frequency tolerance allowed the station.

207. **Interference, prevention of.**—Licensees shall use radio transmitters the emissions of which do not cause interference, outside the authorized band, that is detrimental to traffic and programs of other authorized stations.

208. **Emissions.**—Except for amateur stations, each license and construction permit will specify the type or types of emission that the station is authorized to use, in conformity with the definitions given in article 7 of the General Radio Regulations annexed to the International Telecommunication Convention, but special types of emission not specifically defined therein will be designated in instruments of authorization

with reference to the nature of service the station is authorized to render.

209. **Damped waves.**—Except for ship stations under the conditions hereinafter specified, no license will be issued for the operation of any station using, or proposing to use, transmitting apparatus employing damped wave emissions.

210. **Distress messages.**—Radio communications or signals relating to ships or aircraft in distress shall be given absolute priority. Upon notice from any station, Government or commercial, all other transmission shall cease on such frequencies and for such time as may, in any way, interfere with the reception of distress signals or related traffic.

211. No station shall resume operation until the need for distress traffic no longer exists, or it is determined that said station will not interfere with distress traffic as it is then being routed and said station shall again discontinue if the routing of distress traffic is so changed that said station will interfere. The status of distress traffic may be ascertained by communication with Government and commercial stations.

212. The Commission may require at certain stations an effective continuous watch on the distress frequency, 500 kilocycles (410 kilocycles in the Great Lakes area).

213. **Operators.**—One or more licensed operators of the grade specified by these regulations shall be on duty at the place where the transmitting apparatus of each station is located and whenever it is being operated; Provided, however, That for a station licensed for service other than broadcasting, and remote control is used, the Commission may modify the foregoing requirement, upon proper application and showing being made, so that such operator or operators may be on duty at the control station in lieu of the place where the transmitting apparatus is located. Such modification shall be subject to the following conditions:

a. The transmitter shall be capable of operation and shall be operated in accordance with the terms of the station license.

b. The transmitter shall be monitored from the control station with apparatus that will permit placing the transmitter in an inoperative condition in the event there is a deviation from the terms of the license, in which case the radiation of the transmitter shall be suspended immediately until corrective measures are effectively applied to place the transmitter in proper condition for operation in accordance with the terms of the station license.

c. The transmitter shall be so located or housed that it is not accessible to other than duly authorized persons.

214. **Licensed operator required.**—Only an operator holding a radiotelegraph class of operators' license may manipulate the transmitting key of a manually operated coastal telegraph or mobile telegraph station in the international service; and only a licensed amateur operator may manipulate the transmitting key at a manually operated amateur station. The licensees of other stations operated under the constant supervision of duly licensed operators may permit any person or persons, whether licensed or not, to transmit by voice or otherwise, in accordance with the types of emission specified by the respective licenses.[1]

220. **Maintenance tests.**—Licensees of stations other than broadcast stations are authorized to carry on such routine tests as may be required for the proper maintenance of the stations: Provided, however, That these tests shall be so conducted as not to cause interference with the service of other stations.

221. **Licenses, posting of.**—The original of each station license, except amateur, portable, and portable-mobile stations shall be posted by the licensee in a conspicuous place in the room in which the transmitter is located. In the case of amateur, portable, and portable-mobile stations the original license, or a photostat copy thereof, shall be similarly posted or kept in the personal possession of the operator on duty.

a. The original license of each station operator, except amateur and aircraft radio station operators, and operators of portable and portable-mobile stations, shall be posted in a conspicuous place in the room occupied by such operator while on duty. In the case of an amateur or aircraft radio operator, and operators of portable or portable-mobile stations, the original operator's license shall be similarly posted or kept in his personal possession and available for inspection at all times while the operator is on duty.

b. When an operator's license cannot be posted because it has been mailed to an office of the Federal Communications Commission for endorsement or other change, such operator may continue to operate stations in accordance with the class of license held, for a period not to exceed 60 days, but in no case beyond the date of expiration of the license.

222. **Day frequencies.**—In all cases where the word "day" or "daylight" occurs in connection with a specific frequency, such use of the word shall be construed to mean that period of time included between 2 hours after local sunrise and 2 hours before local sunset.

---

[1]With reference to rule 214 the expression "constant supervision of duly licensed operators" shall for the time being and until further notice, be construed to mean:

a. For stations licensed to use frequencies below 30,000 kilocycles an operator of the grade required under rules 403, 420, and 443 shall be on duty at the transmitter location, whenever the transmitter is being operated, or at the remote control point if authorized in accordance with rule 213.

b. For stations licensed to use frequencies above 30,000 kilocycles only, the operator shall be similarly employed as in (*a*) above, provided, however, in the case of two or more stations licensed in the name of the same individual or organization, except amateur, a licensed radio operator of any class except amateur and radiotelephone third class who has the stations within his effective control, may be on duty at any point within the communication range of such stations in lieu of the transmitter location or control point during the actual operation of the transmitting apparatus, and shall supervise the emissions of all such stations so as to insure proper operation in accordance with the station license(s).

AMATEUR SERVICE

361. **Definitions, amateur service.**—The term "amateur service" means a radio service carried on by amateur stations.

362. **Definition, amateur station.**—The term "amateur station" means a station used by an "amateur", that is, a duly authorized person interested in radio technique solely with a personal aim and without pecuniary interest.

363. Deleted. (See pars. 362 and 364.)

364. **Definition, amateur operator.**—The term "amateur radio operator" means a person holding a valid license issued by the Federal Communications Commission who is authorized under the regulations to operate amateur radio stations.

365. **Definition, amateur radio communication.**—The term "amateur radio communication" means radio communication between amateur radio stations solely with a personal aim and without pecuniary interest.

366. **Station licenses.**—An amateur station license may be issued only to a licensed amateur radio operator who has made a satisfactory showing of ownership or control of proper transmitting apparatus: Provided, however, That in the case of a military or naval reserve radio station located in approved public quarters and established for training purposes, but not operated by the United States Government, a station license may be issued to the person in charge of such station who may not possess an amateur operator's license.

a. **Operator's license.**—An amateur operator's license may be granted to a person who does not desire an amateur station license, provided such applicant waives his right to apply for an amateur station license for 90 days subsequent to the date of application for operator's license.

367. **Eligibility for license.**—Amateur radio station licenses shall not be issued to corporations, associations, or other organizations; Provided however, That in the case of a bona fide amateur radio society a station license may be issued to a licensed amateur radio operator as trustee for such society.

368. **Mobile stations.**—Licenses for mobile stations and portable mobile stations will not be granted to amateurs for operation on frequencies below 28,000 kilocycles. However, the licensee of a fixed amateur station may operate portable amateur stations (rule 192) in accordance with the provisions of rules 384, 386, and 387; and also portable and portable-mobile amateur stations (rules 192 and 192a) on authorized amateur frequencies above 28,000 kilocycles in accordance with rules 384 and 386, but without regard to rule 387.

369. Deleted. Paragraph 213 applies.

370. **Points of communication.**—Amateur stations shall be used only for amateur service, except that in emergencies or for testing purposes they may be used also for communication with commercial or Government radio stations. In addition, amateur stations may communicate with any mobile radio station which is licensed by the Commission to communicate with amateur stations, and with stations of expeditions which may also be authorized to communicate with amateur stations.

371. **Amateur stations not to be used for broadcasting.**—Amateur stations shall not be used for broadcasting any form of entertainment, nor for the simultaneous retransmission by automatic means of programs or signals emanating from any class of station other than amateur.

372. **Radiotelephone tests.**—Amateur stations may be used for the transmission of music for test purposes of short duration in connection with the development of experimental radiotelephone equipment.

373. **Amateur stations not for hire.**—Amateur radio stations shall not be used to transmit or receive messages for hire, nor for communication for material compensation, direct or indirect, paid or promised.

374. The following bands of frequencies are allocated exclusively for use by amateur stations:

1,715 to 2,000 kilcoycles
3,500 to 4,000 kilocycles
7,000 to 7,300 kilocycles
14,000 to 14,400 kilocycles
28,000 to 30,000 kilocycles
56,000 to 60,000 kilocycles
400,000 to 401,000 kilocycles

a. The licensee of an amateur station may, subject to change upon further order, operate amateur stations on any frequency above 110,000 kilocycles, without separate licenses therefor, provided:

(1) That such operation in every respect complies with the Commission's rules governing the operation of amateur stations in the amateur service.

(2) That records are maintained of all transmissions in accordance with the provisions of rule 386.

375. **Types of emission.**—All bands of frequencies so assigned may be used for radiotelegraphy, type A-1 emission. Type A-2 emission may be used in the following bands of frequencies only:

28,000 to 30,000 kilocycles
56,000 to 60,000 kilocycles
400,000 to 401,000 kilocycles

376. **Frequency bands for telephony.**—The following bands of frequencies are allocated for use by amateur stations using radiotelephony, type A-3 emission:

1,800 to 2,000 kilocycles
28,000 to 29,000 kilocycles
56,000 to 60,000 kilocycles
400,000 to 401,000 kilocycles

377. **Additional bands for telephony.**—Provided the station shall be operated by a person who holds an amateur operator's license endorsed for class A privileges, an amateur radio station may use radiotelephony, type A-3 emission, in the following additional bands of frequencies:

3,900 to 4,000 kilocycles
14,150 to 14,250 kilocycles

378. **Amateur television, facsimile, and picture transmission.**—The following bands of frequencies are allocated for use by amateur stations for television, facsimile, and picture transmission:

1,715 to 2,000 kilocycles
56,000 to 60,000 kilocycles

379. **Licenses will not specify individual frequencies.**—Transmissions by an amateur station may be on any frequency within an amateur band above assigned.

380. **Aliens.**—An amateur radio station shall not be located upon premises controlled by an alien.

381. **Prevention of interference.**—"Spurious radiations from an amateur transmitter operating on a frequency below 30,000 kilocycles shall be reduced or eliminated in accordance with good engineering practice and shall not be of sufficient intensity to cause interference on receiving sets of modern design which are tuned outside the frequency band of emission normally required for the type of emission employed. In the case of A-3 emission, the transmitter shall not be modulated in excess of its modulation capability to the extent that interfering spurious radiations occur, and in no case shall the emitted carrier be amplitude-modulated in excess of 100 per cent. Means shall be employed to insure that the transmitter is not modulated in excess of its modulation capability. A spurious radiation is any radiation from a transmitter which is outside the frequency band of emission normal for the type of transmission employed, including any component whose frequency is an integral multiple or sub-multiple of the carrier frequency (harmonics and sub-harmonics), spurious modulation products, key clicks and other transient effects, and parasitic oscillations".

382. **Power supply to transmitter.**—Licensees of amateur stations using frequencies below 30,000 kilocycles shall use adequately filtered direct-current power supply for the transmitting equipment to minimize frequency modulation and to prevent the emission of broad signals.

383. **Authorized power.**—Licensees of amateur stations are authorized to use a maximum power input of 1 kilowatt to the plate circuit of the final amplifier stage of an oscillator-amplifier transmitter or to the plate circuit of an oscillator transmitter.

384. **Transmission of call.**—An operator of an amateur station shall transmit its assigned call at least once during each 15 minutes of operation and at the end of each transmission. In addition, an operator of an amateur portable or portable-mobile radiotelegraph station shall transmit immediately after the call of the station the break sign (BT) followed by the number of the amateur call area in which the portable or portable-mobile amateur station is then operating, as for example:

Example 1. Portable or portable-mobile amateur station operating in the third amateur call area calls a fixed amateur station: W1ABC W1ABC W1ABC DE W2DEF BT3 W2DEF BT3 W2DEF BT3 AR

Example 2. Fixed amateur station answers the portable or portable-mobile amateur station: W2DEF W2DEF W2DEF DE W1ABC W1ABC W1ABC K

Example 3. Portable or portable-mobile amateur station calls a portable or portable-mobile amateur station: W3GHI W3GHI W3GHI DE W4JKL BT4 W4JKL BT4 W4JKL BT4 AR.

If telephony is used, the call sign of the station shall be followed by an announcement of the amateur call area in which the portable or portable-mobile station is operating.

a. In the case of an amateur licensee whose station is licensed to a regularly commissioned or enlisted member of the United States Naval Reserve, the commandant of the naval district in which such reservist resides may authorize in his discretion the use of the call-letter prefix N in lieu of the prefix W or K, assigned in the license issued by the Commission, provided that such N prefix shall be used only when operating in the frequency bands 1,715-2,000 kilocycles and 3,500-4,000 kilocycles in accordance with instructions to be issued by the Navy Department.

385. **Quiet hours.**—In the event that the operation of an amateur radio station causes general interference to the reception of broadcast programs with receivers of modern design, that amateur station shall not operate during the hours from 8 o'clock p. m. to 10:30 p. m., local time, and on Sunday from 10:30 a. m. until 1 p. m., local time, upon such frequency or frequencies as cause such interference.

386. **Logs.**—Each licensee of an amateur station shall keep an accurate log of station operation to be made available upon request by authorized Government representatives, as follows:

a. The date and time of each transmission. (The date need only be entered once for each day's operation. The expression "time of each transmission" means the time of making a call and need not be repeated during the sequence of communication which immediately follows; however, an entry shall be made in the log when "signing off" so as to show the period during which communication was carried on.)

b. The name of the person manipulating the transmitting key of a radiotelegraph transmitter or the name of the person operating a transmitter of any other type (type A-3 or A-4 emission) with statement as to type of emission. (The name need only be entered once in the log provided the log contains a statement to the effect that all transmissions were made by the person named except where otherwise stated. The name of any other person who operates the station shall be entered in the proper space for his transmissions.)

c. Call letters of the station called. (This entry need not be repeated for calls made to the same station during any sequence of communication, provided the time of "signing off" is given.)

d. The input power to the oscillator, or to the final amplifier stage where an oscillator-amplifier transmitter is employed. (This need be entered only once, provided the input power is not changed.)

e. The frequency band used. (This information need be entered only once in the log for all transmissions until there is a change in frequency to another amateur band.)

f. The location of a portable or portable-mobile station at the time of each transmission. (This need be entered only once, provided the location of the station is not changed. However, suitable entry shall be

[1]*Waiver of rule 381.*—The following order was adopted by the Telegraph Division on Aug. 1, 1934: "Until further notice, the provisions of rule 381 shall not be construed to apply to amateur operation on frequencies above 56,000 kilocycles."

made in the log upon changing location, showing the type of vehicle or mobile unit in which the station is operated and the approximate geographical location of the station at the time of operation.)

g. The message traffic handled. (If record communications are handled in regular message form, a copy of each message sent and received shall be entered in the log or retained on file for at least 1 year.)

387. **Portable stations.**—Advance notice of all locations in which portable amateur stations will be operated shall be given by the licensee to the inspector in charge of the district in which the station is to be operated. Such notices shall be made by letter or other means prior to any operation contemplated and shall state the station call, name of licensee, the date of proposed operation, and the approximate locations, as by city, town, or county. An amateur station operating under this rule shall not be operated during any period exceeding 30 days without giving further notice to the inspector in charge of the radio inspection district in which the station will be operated. This rule does not apply to the operation of portable or portable-mobile amateur stations on frequencies above 28,000 kc authorized to be used by amateur stations (see rule 368).

PART VI. RADIO OPERATORS

AMATEUR OPERATORS

400. **Only amateur operators may operate amateur stations.**—An amateur station may be operated only by a person holding a valid amateur operator's license, and then only to the extent provided for by the class of privileges for which the operator's license is endorsed.

401. **Validity of operator's license.**—Amateur operators' licenses are valid only for the operation of licensed amateur stations, provided, however, any person holding a valid radio operator's license of any class may operate stations in the experimental service licensed for, and operating on, frequencies above 30,000 kilocycles.

402. **Proof of use.**—Amateur station licenses and/or amateur operator licenses may, upon proper application, be renewed provided: (1) The applicant has used his station to communicate by radio with at least three other amateur stations during the 3-month period prior to the date of submitting the application, or (2) in the case of an applicant possessing only an operator's license, that he has similarly communicated with amateur stations during the same period. Proof of such communication must be included in the application by stating the call letters of the stations with which communication was carried on and the time and date of each communication. Lacking such proof, the applicant will be ineligible for a license for a period of 90 days.

This rule shall not prevent renewal of an amateur station license to an applicant who has recently qualified for license as an amateur operator.

403. **Class of operator and privileges.**—There shall be but one main class of amateur operator's license, to be known as "amateur class", but each such license shall be limited in scope by the signature of the examining officer opposite the particular class or classes of privileges which apply as follows:

Class A.—Unlimited privileges.

Class B.—Unlimited radiotelegraph privileges. Limited in the operation of radiotelephone amateur stations to the following bands of frequencies: 1,800 to 2,000 kilocycles; 28,000 to 28,500 kilocycles; 56,000 to 60,000 kilocycles; 400,000 to 401,000 kilocycles.

Class C.—Same as class B privileges, except that the Commission may require the licensee to appear at an examining point for a supervisory written examination and practical code test during the license term. Failing to appear for examination, the license held will be canceled and the holder thereof will not be issued another license for the class C privileges.

404. **Scope and places of examinations.**—The scope of examinations for amateur operators' licenses shall be based on the class of privileges the applicant desires, as follows:

Class A.—To be eligible for examination for the class A amateur operator's privileges the applicant must have been a licensed amateur operator for at least 1 year and must personally appear at one of the Commission's examining offices, and take the supervisory written examination and code test. (See pars. 2 h (1), 30, and 408.) Examinations will be conducted at Washington, D. C., on Thursday of each week, and at each radio district office of the Commission on the days designated by the inspector in charge of such office. In addition, examinations will be held quarterly in the examining cities listed in paragraph 30 on the dates to be designated by the inspector in charge of the radio district in which the examining city is situated. The examination will include the following:

a. Applicant's ability to send and receive in plain language messages in the Continental Morse Code (5 characters to the word) at a speed of not less than 10 words per minute.

b. Technical knowledge of amateur radio apparatus, both telegraph and telephone.

c. Knowledge of the provisions of the Communications Act of 1934, subsequent acts, treaties, and rules and regulations of the Federal Communications Commission, affecting amateur licensees.

Class B.—The requirements for class B amateur operators' privileges are similar to those for the class A, except that no experience is required and the questions on radiotelephone apparatus are not so comprehensive in scope.

Class C.—The requirements for class C amateur operators' privileges shall be the same as for the class B except the examination will be given by mail. Applicants for Class C privileges must reside more than 125 miles airline from the nearest examining point for Class B privileges, or in a camp of the Civilian Conservation Corps, or be in the regular military or naval service of the United States at a military post or naval station; or be shown by physician's certificate to be unable to appear for examination due to protracted disability.

405. **Recognition of other classes of licenses.**—An applicant for any class of amateur operator's privileges who has held a radiotelephone second-class operator's license or higher, or an equivalent commercial grade license, or who has been accorded unlimited amateur radiotelephone privileges, within 5 years of the date of application may only be required to submit additional proof as to code ability and/or knowledge of the laws, treaties, and regulations affecting amateur licensees.

406. An applicant for the class B or C amateur operator's privileges who has held a radiotelegraph third-class operator's license or higher, or an equivalent commercial grade license, or who has held an amateur extra first-class license within 5 years of the date of application may be accorded a license by passing an examination in laws, treaties, and regulations affecting amateur licensees.

407. **Code ability to be certified by licensed operator.**—An applicant for the class C amateur operator's privileges must have his application signed in the presence of a person authorized to administer oaths by (1) a licensed radiotelegraph operator other than an amateur operator possessing only the class C privileges or former temporary amateur class license, or (2) by a person who can show evidence of employment as a radiotelegraph operator in the Government service of the United States. In either case the radiotelegraph code examiner shall attest to the applicant's ability to send and receive messages in plain language in the continental Morse code (5 characters to the word) at a speed of not less than 10 words per minute. The code certification may be omitted if the applicant can show proof of code ability in accordance with the preceding rule.

408. **Application forms.**—Forms for amateur station and/or operator license shall be obtained by calling or writing to the inspector in charge of the radio inspection district in which the applicant resides. Upon completion of the forms, they shall be sent back to the same office where the final arrangements will be made for the examination: Provided, however, in the case of applicants for the class C amateur operator's privileges, the forms and examination papers when completed shall be mailed direct to the Federal Communications Commission, Washington, D. C.

409. **Grading of examinations.**—The percentage that must be obtained as a passing mark in each examination is 75 out of a possible 100. No credit will be given in the grading of papers for experience or knowledge of the code. If an applicant answers only the questions relating to laws, treaties, and regulations by reason of his right to omit other subjects because of having held a recognized class of license, a percentage of 75 out of a possible 100 must be obtained on the questions answered.

410. **Operator's and station licenses to run concurrently.**—An amateur station license shall be issued so as to run concurrently with the amateur operator's license and both licenses shall run for 3 years from the date of issuance. If either the station license or the operator's license is modified during the license term, both licenses shall be reissued for the full 3-year term: Provided, however, if an operator's license is modified only with respect to the class of operator's privileges, the old license may be endorsed, in which case the expiration date will not change.

411. **Eligibility for reexamination.** An applicant who fails examination for amateur privileges may not take another examination for such privileges within 90 days, except that this Rule shall not apply to successive examinations at a point named in Rule 30-a.

412. **Penalty.**—Any attempt to obtain an operator's license by fraudulent means, or by attempting to impersonate another, or copying or divulging questions used in examinations, or, if found unqualified or unfit, will constitute a violation of the regulations for which the licensee may suffer suspension of license or be refused a license and/or debarment from further examination for a period not exceeding 2 years at the discretion of the licensing authority.

413. **Duplicate licenses.**—Any licensee applying for a duplicate license to replace an original which has been lost, mutilated, or destroyed, shall submit an affidavit to the Commission attesting to the facts regarding the manner in which the original was lost. Duplicates will be issued in exact conformity with the original, and will be marked "duplicate" on the face of the license.

414. **Oath of secrecy.**—Licenses are not valid until the oath of secrecy has been executed and the signature of the licensee affixed thereto.

415. **Examination to be written in longhand.**—All examinations, including the code test, must be written in longhand by the applicant.

General Information

Any licensee receiving notice of violation of radio laws shall reply to said notice in writing to the FCC at Washington.

Requests for special call-letters will not be considered.

The person manipulating the telegraph key of an amateur station must be a duly licensed operator.

Tests can be conducted by amateurs if such tests do not interfere with services of other stations.

The original license shall be posted in the station or kept in the personal possession of the operator on duty, except when it has been mailed to an office of the FCC for endorsement or change before date of its expiration.

Amateur stations must not be used to handle messages for pecuniary interests, direct or indirect, paid or promised.

Amateur transmissions must be free from harmonics. Loosely-coupled circuits must be used, or devices that will result in giving equivalent effects to minimize keying impacts, clicks, harmonics and parasitics.

1KW power input to the stage which feeds the antenna is the maximum permissible power for amateur operation.

Amateur operators must transmit their assigned call letters at the end of each transmission, or at least once during each 15 minutes of operation. If an amateur transmitter causes general interference with reception of broadcast signals in receivers of modern design, that amateur station shall not operate during the hours from 8 p.m. to 10:30 p.m., local time, and on Sundays from 10:30 a.m. until 1 p.m., local time, upon such frequency or frequencies as cause such interference.

Each licensee of an amateur station must keep an accurate LOG of station operation, name of person operating the transmitter, with statement as to the nature of transmission. The call letters of the station, the input power to the stage which feeds the antenna, the frequency band used, the location of the station if portable operation is used, must all be entered in the station LOG. A copy of each message sent and received must be kept on file for at least one year. This information must be available on request by authorized representatives of the Government of the United States. The station may be operated only to the extent provided by the class of privileges for which the operator's license is endorsed.

Distress Signals

The International Distress Signal is ...— — —... (three dots, three dashes, three dots). The distress signal is **NOT** SOS; it is an easily-recognized group of **characters** of three dots, three dashes, three dots. For radiotelephony distress calls the signal is **MAYDAY.** All communications must cease when a distress call is heard. Communication must not be resumed until it has been definitely determined that all is clear again. When you hear a distress call, notify the nearest source from which aid can be secured.

Other Important Laws and Regulations

It is unlawful to send fraudulent signals of distress or communications relating thereto; to maliciously interfere with any other radio communications. Distress calls have precedence over all others. Minimum power must be used to effect reliable communication. The use of profane language is prohibited. The contents or meaning of a message must be kept secret, except to an authorized agency which takes part in the forwarding of the message, or to the addressee or his agent, or upon the demand of a court of competent jurisdiction or authority.

Secrecy provisions do not apply to broadcasts for public use, or to distress calls. In the event of a national emergency the station can be ordered closed.

Third-party messages cannot be transmitted by amateur stations unless special arrangements have been made between the governments of the countries concerned. Amateur stations cannot be used to broadcast entertainment. In the event of an emergency an amateur station is permitted to communicate with stations other than amateur.

"AR" denotes the end of a message. "SK" denotes the end of a communication.

Penalties

The penalty for violating the provisions of the Communications Act of 1934 is $10,000, or imprisonment not to exceed 2 years, or both, for each offense. The operator's license is liable to suspension for 2 years if a conviction is secured. The station license can also be revoked.

For violation of any of the regulations of the Federal Communications Commission a fine not to exceed $500 can be imposed for each day of such offense. If the convicted person is a licensed operator his license can be suspended for a period not to exceed 2 years. The station license can also be revoked. The penalty for not keeping a station log is the same as related above. For malicious interference with distress communications the maximum penalty of $10,000 and 2 years can be imposed. For malicious interference with other than distress communications the license can be suspended for up to 2 years. An amateur who accepts material compensation for any services rendered by his station is subject to a fine of not more than $500 for each day of such offense. His license can also be suspended for as long as 2 years.

READABILITY—SIGNAL STRENGTH REPORTS

The "QSA-R" System

A short-cut method which indicates signal strength from 1 to 5. 1 indicates the hardest signal to copy, 5 the easiest to copy. The "R" portion indicates signal STRENGTH, from R1 to R9, R1 being the weakest, R9 the loudest. Thus QSA-1, R1 would denote the most difficult signal to read, and the WEAKEST signal, as well. QSA-5, R-9 would denote the EASIEST signal to copy, plus the MAXIMUM signal strength.

THE "R-S-T" SYSTEM.

READIBILITY:	SIGNAL STRENGTH:		TONE:	
1—Not readable	1—Very faint	6—Good	1—Very rough	6—Modulated, slight whistle
2—Barely distinguishable	2—Very weak	7—Moderately strong	2—Rough AC note	7—Nearly DC note
3—Readable with difficulty	3—Weak	8—Strong	3—Fairly rough AC	8—Good DC note
4—Easily readable	4—Fair	9—Extremely strong	4—Rather rough AC	9—Perfectly-pure DC note
5—Perfectly readable	5—Fairly good		5—Musical—fairly good	

Q SIGNALS

Abbreviation	Question	Answer
QRA	What is the name of your station?	The name of my station is......
QRB	What is your approximate distance from my station?	The approximate distance between our stations is......
QRC	By what private company (or Government administration) are the accounts for charges of your station liquidated?	The accounts for charges of my station are liquidated by the private company...... (or by the...... government administration).
QRD	Where are you going and where do you come from?	I am going to......and I am coming from......
QRG	Will you indicate to me my exact frequency (or wavelength) in kc. (or m.)?	Your exact frequency in kc. (or wavelength in m.) is......
QRH	Does my frequency (or wavelength) vary?	Your frequency (or wavelength) varies.
QRI	Is my note steady?	Your note varies.
QRJ	Are you receiving me badly? Are my signals weak?	I can not receive you. Your signals are too weak.
QRK	Are you receiving me well? Are my signals good?	I receive you well. Your signals are good.
QRL	Are you busy?	I am busy (or I am busy with......). Please do not interfere.
QRM	Are you being interfered with?	I am being interfered with.
QRN	Are you bothered by static?	I am bothered by static.
QRO	Shall I increase power?	Increase power.
QRP	Shall I decrease power?	Decrease power.
QRQ	Shall I send faster?	Send faster.
QRS	Shall I send slower?	Send slower.
QRT	Shall I stop sending?	Stop sending.
QRU	Have you something for me?	I have nothing for you.
QRV	Are you ready?	I am ready.
QRW	Shall I tell......you are calling him on......kc. (or m.)?	Please tell......I am calling him on......kc. (or m.)
QRX	Shall I wait? When will you call me again?	Wait. (or Wait until I have finished communicating with......) I shall call you at......o'clock (or soon).
QRY	Which is my turn?	Your turn is number......(or after every other call).
QRZ	Who is calling me?	You are called by......
QSA	What is the strength of my signals (1 to 5)?	The strength of your signals is......(1 to 5).
QSB	Does the strength of my signals vary?	The strength of your signals varies.
QSD	Is my keying accurate? Are my signals distinct?	Your keying is inaccurate. Your signals are bad.
QSG	Shall I transmit......telegrams (or one telegram) at once?	Transmit......telegrams (or one telegram) at once.
QSJ	What is the charge per word for......, including your interior telegraph charge?	The charge per word for......is......francs, including my interior telegraph charge.
QSK	Shall I continue the transmission of all my traffic? I can use break-in operation	Continue the transmission of all your traffic. I shall break you if necessary.
QSL	Can you give me acknowledgment of receipt?	I give you acknowledgment of receipt.
QSM	Shall I repeat the last telegram I sent to you?	Repeat the last telegram you sent to me.
QSO	Can you communicate with......directly (or through......)?	I can communicate with......directly (or through......).
QSP	Will you relay to......free of charge?	I will relay to......free of charge.
QSR	Has the distress call from......been attended to?	The distress call received from......has been attended to by......
QSU	Shall I transmit (or reply) on......kc. (or......m.) and/or with waves of type A1, A2, A3, or B?	Transmit (or Reply) on......kc. (or......m.) and/or with waves of type A1, A2, A3 or B.
QSV	Shall I transmit a series of VVV......?	Transmit a series of VVV......
QSW	Do you wish to transmit on......kc (or......m.) and/or with waves of type A1, A2, A3 or B?	I am going to transmit (or I shall transmit) on......kc. (or......m.) and/or with waves of type A1, A2, A3 or B.
QSX	Do you wish to hear......(call signal) on......kc. (or......m.)?	I hear......(call signal) on......kc. (or......m.)
QSY	Shall I change to transmission on......kc. (or......m.) without changing the type of wave? or Shall I change to transmission on another wave?	Change to transmission on......kc. (or......m.) without changing the type of wave, or Change to transmission on another wave.
QSZ	Shall I send word or group twice?	Send each word or group twice.
QTA	Shall I cancel telegram number......as if it had not been sent?	Cancel telegram number......as if it had not been sent.
QTB	Do you agree with my word count?	I do not agree with your word count; I herewith repeat the first letter of each word and the first figure of each number.
QTC	How many telegrams have you sent?	I have......telegrams for you (or for......).
QTE	What is my true bearing relative you? or What is my true bearing relative to......(call signals)? or What is the true bearing of......(call signal) to......?	Your true bearing relative to me is......degrees, or Your true bearing relative to......(call signal) is......degrees at......o'clock, or The true bearing of......(call signal) relative to......(call signal) is......degrees at......o'clock.
QTF	Will you give me the position of my station based on bearings taken by radiocompass stations you control?	The position of your station as based on radiocompass stations that I control is......latitude......longitude.
QTG	Will you transmit your call signal for fifty seconds, ending with a dash of ten seconds, on......kc. (or......m.) so that I can take your radiocompass bearing?	I am going to transmit my call signal for forty seconds, ending with a dash of ten seconds, on......kc. (or......m.) so that you can take my radiocompass bearing.
QTH	What is your position in latitude and longitude (or according to any other indication?	My position is......latitude,......longitude (or according to any other indication).
QTI	What is your true course?	My true course is......degrees.
QTJ	What is your speed?	My speed is......knots (or......kilometers) per hour.
QTM	Send radio signals and submarine sound signals so that I can determine my bearing and my distance.	I am sending radio signals and submarine signals so that you can determine your bearing and your distance.
QTO	Are you leaving the dock (or the port)?	I am going to leave the dock (or the port).
QTP	Are you going to enter the dock (or the port)?	I am going to enter the dock (or the port).
QTQ	Can you communicate with my station by means of the International Signal Code?	I am going to communicate with your station by means of the International Signal Code.
QTR	What is the correct time?	The correct time is......
QTU	What are the working hours of your station?	The working hours of my station are from......to......
QUA	Have you any news of......(call signal of mobile station)?	The news of......(call signal of mobile station) is......
QUB	Can you give me, in this order, information concerning: visibility, height of clouds, and ground wind for......(place of observation)?	Following are the weather details requested:......
QUC	What is the last message received by you from......(call signal of mobile station)?	The last message received by me from......(call signal of mobile station) is......
QUD	Have you received the urgency signal made by......(call signal of mobile station)?	I have received the urgency signal made by......(call signal of mobile station) at......(o'clock).
QUF	Have you received the distress signal made by......(call signal of mobile station)?	I have received the distress signal made by......at......o'clock.
QUG	Are you going to be forced to alight at sea (or on land)?	I am going to be forced to alight at sea (or on land) at......(place).
QUH	Will you give me the barometric pressure at sea level?	The barometric pressure at sea level is......(units).
QUJ	Will you give me the true head to follow, with no wind, for directing me to come to you?	The true head to follow, with no wind, for directing you to come to me, is from......degrees at......o'clock.

The signal series of QA, QB, QC, QD, QF and QG are reserved for special aeronautical codes.

INDEX

Time TURNS AN EXPERIMENT INTO *Reality*

IN any industry, leadership usually rests upon the firm base of Time. Leaders result from years of experiment, developments and unceasing struggle for perfection!

TIME has turned the dream of C-D engineers into a reality. Today—the proud results are C-D capacitors, the world's leading line of oil, paper, mica and electrolytic condensers!

Symbolic are the Type 9 and Type 86 mica transmitting capacitors.

The TYPE 9, moulded in a special bakelite composition and constructed of the finest India mica available—are the result of twenty-six years of manufacturing experience and development.

Type 9

The TYPE 86, enclosed in a porcelain container, supplied with sturdy mounting flanges, utilizes mica, instead of ordinary flint glass as a dielectric. By the use of mica, the loss of power flowing through the capacitor is 1/20th that of the glass dielectric condensers.

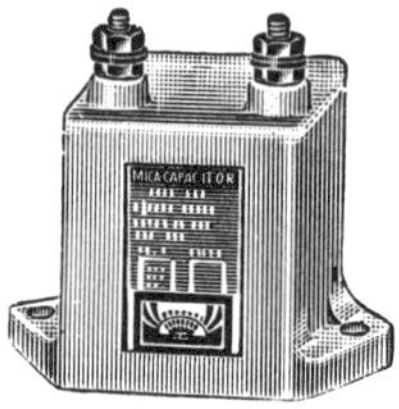

Type 86

These and other quality transmitting condensers are fully illustrated in catalog 127. Write for your copy today. See opposite page of this book on the C-D Dykanol oil filled and impregnated, hermetically sealed, non-inflammable and non-explosive transmitting capacitors.

CORNELL-DUBILIER
CORPORATION
4372 BRONX BOULEVARD
NEW YORK

LEADERS

For 26 Years

For 1936 Cornell-Dubilier offers the amateurs the world's most complete transmitting condenser line—still foremost, after 26 years of leadership.

One of the outstanding developments introduced by the CORNELL - DUBILIER laboratories, in the early part of 1934, **was the oil transmitting condenser.** In line with the C-D creed of *"Ever forward,"* we now point with pride to one of the outstanding developments of 1935 in the condenser industry—the DYKANOL CAPACITOR.

The still higher dielectric strength of non-inflammable Dykanol has permitted the further reduction in size and weight of high voltage bypass, and filter condensers.

Keep in step with the times. Bring your rig up-to-date by installing C-D Dykanol transmitting condensers. **Available in a complete capacity range from .05 mfd. at 600** V.D.C. to 2 mfd. at 25,000 V.D.C.

Have your local C-D authorized jobber show you the new features of these guaranteed condensers. Note their sturdy construction—**high porcelain insulators and** convenient mounting flanges—also consider their hermetic sealing in non-corrosive containers.

These and other transmitting condensers are listed in Catalog 128. Your copy is ready.

CORNELL-DUBILIER
CORPORATION
4372 BRONX BOULEVARD NEW YORK, N.Y.

UTC class B audio and power components are available for every type of transmitter. Thorough research, expert engineering design and exacting laboratory checkups are responsible for the enviable reputation of UTC products.

The output transformer illustrated, designed either for 204A's or 849'S in class B, has a linear response from 30 to 12,000 cycles. The unit is oil immersed and shielded in a heavy gauge high permeability casting. All leads are terminated in high tension ceramic bushings. The transformer is insulated for 20,000 volts and will handle 1000 watts audio power.

INPUT TRANSFORMERS

Type	Description	LIST	NET
PA-238	Push pull parallel 2A3, 45, 50 plates to four 838 or 203 Class B grids.....	17.50	10.50
PA-239	Four 2A3, 45, 50 or two 211, 845 plates to Class B 204 or 849 grids	17.50	10.50
PA-501T	Four 2A3, 45, 50 or two 50T, 211, 845 plates to Class B, HF 300 or 150T grids	17.50	10.50
PA-512	500, 200 or 50 ohm line to two 150T, 204, HF 300 or 849 grids	20.00	12.00

OUTPUT TRANSFORMERS

Type	Description	LIST	NET
PA-601	Primary arranged to match 838 or 203 Class B plates to 6500, 5000 to 3500 ohm RF load..	32.50	19.50
PA-438	Primary arranged to match push pull parallel 838 or 203 Class B plates to 6,000, 4,000 or 3,000 ohm RF load......	50.00	30.00
PA-489	Primary arranged to match 204A, HF 300 or 849 Class B plates to 6500, 5000 or 3400 ohm RF load	125.00	75.00
PA-150T	Primary arranged to match Class B 150 T's or HK 354's in A prime or Class B at various plate voltages. Specify primary plate voltage and secondary RF load when ordering	125.00	75.00

UNITED TRANSFORMER CORP.

72 SPRING STREET NEW YORK, N. Y.

EXPORT DIVISION – 15 LAIGHT STREET, NEW YORK, N. Y.

BRUSH *Spherical* MICROPHONE

● A specially designed, general purpose microphone for commercial interstation and amateur transmission work. Low in price . . . but built to Brush's traditionally high mechanical and electrical standards. Wide frequency response. Non-directional. No diaphragms. No distortion from close speaking. Trouble-free operation. No button current and no input transformer to cause hum. Beautifully finished in dull chromium. Size only $2\frac{1}{8}$ inches in diameter. Weight 5 oz. Output level minus 66 D.B. Locking type plug and socket connector for either suspension or stand mounting furnished at no extra cost. Full details, Data Sheet No. 13. Free. Send for one.

BRUSH *Headphones*

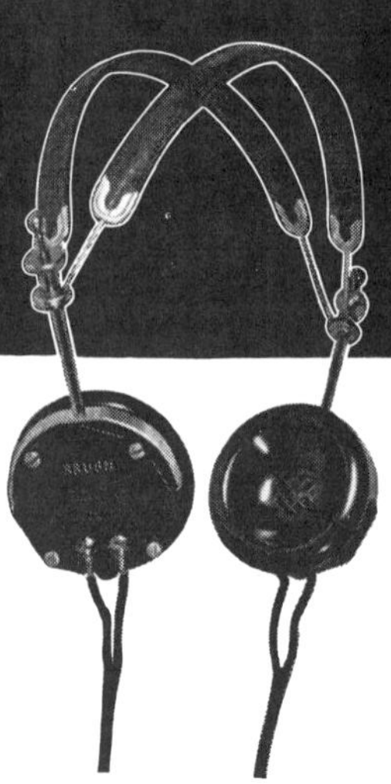

Meet every headphone requirements. They bring in weak signals strong and clear and will handle excessive volume without overloading. Response 60 to 10,000 cycles. No magnets to cause diaphragm chatter. Specially designed cases minimize breakage. Light in weight. Only 6 oz., complete with headband and cords. A quality product at a low price. Details, Data Sheet No. 10. Copies on request. Send for one.

The **BRUSH** DEVELOPMENT COMPANY

PIEZO ELECTRIC

1895 E. 40th St. CLEVELAND, O.

MICROPHONES • MIKE STANDS • TWEETERS • HEAD PHONES • LOUD SPEAKERS

IMPORTANT CHARACTERISTICS...RAYTHEON AMATEUR TUBES

Type		Filament Volts	Filament Amps		Capacities Cgp	Capacities Cgf	Capacities Cpf	Purpose	Rated Voltages Plate	Rated Voltages Grid	Rated Voltages Scr. G	Rated Voltages Supp. G	Rated Milliamps. Plate	Rated Milliamps. Grid	Rated Milliamps. Scr. G	Power Pl. Diss	Power Output	Driving Power	Over All Height	
RK-10	Triode	7.5	1.25	Thor.	7uuf	4uuf	3uuf	Osc-Rf. Amp.	450	−120			60	10–15		12W	15W	3W	5 3/8"	RK-10
								Class B Mod.	425	−50			8*			12W	25W*			
RK-15	Triode	2.5	1.75	Oxide	7.5	2.7	5.0	Rf. Amp.-Doubler	400	−50			50	3–5		10W	12W	2W	6"	RK-15
								Class B Mod.	400	−0			12*			10W	20W*			
RK-16	Triode	2.5	2.0	K	7.5	3.8	6.0	Audio Amp.	250	−28	250		26		6.5	10W	3W		5 1/2"	RK-16
RK-17	Rf. Pentode	2.5	2.0	K	1.0	7.5	16.0	Osc.-Rf. Amp.	300	−50	125		40	3-5	20	10W	7W	2W	6"	RK-17
RK-18	Triode	7.5	3.0	Thor.	4.8	4.6	2.9	Osc.-Rf. Amp.	1000	−150			85	12–15		35W	40W	3W	8 1/2"	RK-18
								Class B Mod.	1000	−50			28*			35W	100W*			
RK-19	Full W. Rect.	7.5	2.5	K				HIGH VACUUM—LOW DROP 1250V. R.M.S. PER ANODE 600 MA Peak CURR.											6 5/8"	RK-19
RK-20	Rf. Pentode	7.5	3.0	Thor.	.012	11.0	10.0	Osc.-Rf. Amp.	1250	−100	300	O	80	7–10	37	40W	64W	1.0W	8 3/4"	RK-20
								Osc.-Rf. Amp.	1250	−100	300	+45	92	7–10	32	40W	80W	1.0W		
								Supp. Mod. Phone	1250	−100	300	−45	43	7–10	36	40W	18W	1.0W		
RK-21	Half W. Rect.	2.5	4.0	K				HIGH VACUUM—LOW DROP 1250V. R.M.S. PER ANODE 600 MA PEAK CURR.											6 5/8"	RK-21
RK-22	Full W. Rect.	2.5	8.0	K															6 5/8"	RK-22
RK-23	Rf. Pentode	2.5	2.0	K	.02	10.0	10.0	Osc.-Rf. Amp.	500	−90	200	O	50	6–8	40	10W	18W	.8W	6"	RK-23
								Osc.-Rf. Amp.	500	−90	200	+45	55	6–8	35	10W	24W	.8W		
								Supp. Mod. Phone	500	−90	200	−45	32	6–8	40	10W	5.5W	.8W		
RK-24	Triode	2.0	0.12	Oxide	5.5	3.5	3.0	5 Meter Osc.	180	−45			20			1.5W	1.2W		4 1/4"	RK-24
RK-25	Rf. Pentode	6.3	0.8	K	.02	10.0	10.0	CHARACTERISTICS SAME AS RK-23 SHOWN ABOVE											6"	RK-25
RK-28	Rf. Pentode	10.0	5.0	Thor.	.02	15.5	5.5	Osc.-Rf. Amp.	2000	−100	400	O	120	10–12	75	100W	160W	1.8W	9 1/2"	RK-28
								Osc.-Rf. Amp.	2000	−100	400	+45	140	10–12	60	100W	200W	1.8W		
								Supp. Mod. Phone	2000	−100	400	−50	80	10–12	85	100W	60W	2.7W		
RK-30	Triode	7.5	3.25	Thor.	2.5	2.75	2.75	Osc.-Rf. Amp.	1250	−175			70	12–15		35W	65W	4.0W	6 7/8"	RK-30
								Class B Mod.	1250	−70			30*			35W	106W*			
RK-31	Triode	7.5	3.0	Thor.				Class B Mod.	1250	O			30*			40W	140W*		8 1/2"	RK-31
RK-32	Triode	7.5	3.25	Thor.	3.0	2.0	1.0	Ultra H. F. Oscill-Amp.	1250	−250			100	18–20		50W	65W	7.5W	6 7/8"	RK-32
RK-34	Dual Triode	6.3	0.8	K	2.7	4.2	2.1	Osc.-Rf. Amp	300	−36			80	16–18		10W	14W	2.5W	5 1/16"	RK-34
								Class B Mod.	300	−15			30*			10W	12W*			
RK-100	Gas Triode	6.3	0.9	K	19.0	23.0	3.0	Rf. Oscillator	110	−25			80	50	150**	10W	3.5W		6 1/4"	RK-100
								Rf. Amplifier	110	−25			250	50	250**	10W	12W	3.0W		
841	Triode	7.5	1.25	Thor.	7.0	4.0	3.0	Rf. Amplifier	450	−32			50	10–12		12W	14W	1.5W	5 3/8"	841
842	Triode	7.5	1.25	Thor.	7.0	4.0	3.0	Aud. Modulator	425	−100			28			12W	3W		5 3/8"	842
866A	Half W. Rect.	2.5	5.0	Oxide				MERCURY VAPOR MAX INVERSE PEAK 10000 V. PEAK CURRENT—600 MA.											6 5/8"	866A
872A	Half W. Rect.	5.0	10.0	Oxide				" " " " " 10000 V. " " 5.0 A.											8 1/2"	872A

* Indicates Value for Two Tubes. ** Indicates Cathanode Current.

RAYTHEON PRODUCTION CORP., 30 East 42nd St., New York, N. Y.

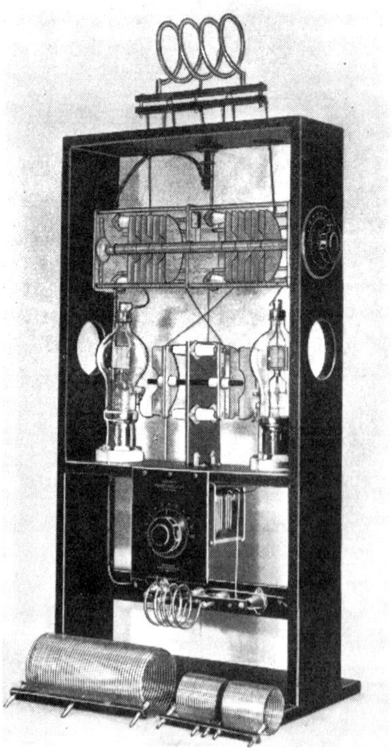

1 KW on 28 mc.

—Phone or C.W.—

You can increase power on any band by adding this amplifier. Conservatively rated at 1 KW using 150Ts or 354s, requiring only 60 watts grid excitation. Designed for 100% plate modulation of the 1 KW input. It is an improved model of the amplifier described in November 1934 RADIO. Coils shown are for 28mc. (in place) and 3.5mc. The 14 and 7mc coils are similar. Power requirements are 5 volts at 20 amps for filaments, and 2500 volts at 400ma. for plates. Excitation is coupled through the link furnished on each grid coil.

The amplifier may be obtained wired and tested, ready to go; or in kit form for home assembly. The price complete with coils for one band, less tubes, is $100.00, f.o.b. Los Angeles. As above, in kit form, $75.00 including wire and incidentals. For those desiring individual parts, an itemized list is available on request.

HIGH POWER CONDENSERS

Commercial Type Transmitting condensers at Amateur prices. Especially designed for high power, these condensers are the only ones built to handle present day high voltage operation—ask the high power gang. Construction is extremely rugged; steel tie rods and shaft with heavy brass spacers, all cadmium plated. Plates have machine rounded edges. Coorsite high frequency insulation insures lowest losses. See new catalog for more details and complete list of models. We also build special condensers to fit your every requirement of space, voltage, and capacity.

On the right is the famous WS1502035, 35mmf 15000 volts per section for 150Ts, 354s, 852s, etc., up to 5000 volts plate. $15.00 net. For lower voltages use the WS75205, 50mmf at 7500 volts for plate voltages up to 2500—$12.00 net. If higher capacity is desired, use the WS75210, 100 mmf 7500 volts per section —$13.33 net. A popular single section condenser is the WS15010 100mmf. at 15000 volts —$13.33 net. For high power neutralizers, the single WSN-852 at $3.33 net and the twin at $6.66 net are the berries.

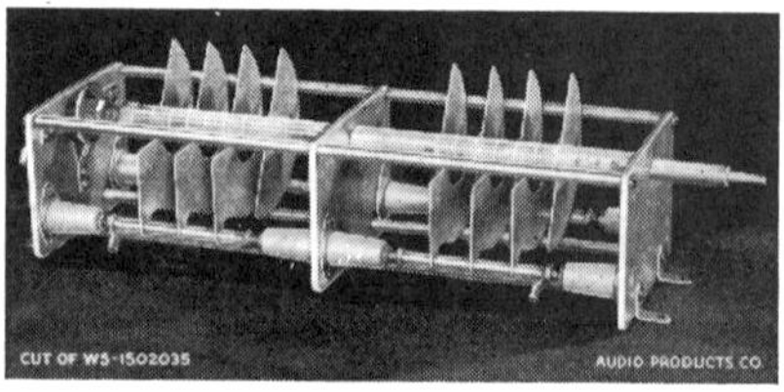

NEW "TUBE & CONDENSER" DATA AVAILABLE

A series of information sheets telling you how to get the most out of your tube. Drop us a card for one of these sheets and include your tube type; it gives data on circuit, bias, excitation, plate voltages, etc., for all types of operation including high efficiency and power output. Order direct from this page. Goods shipped C. O. D., express collect, unless otherwise requested. Include 50% deposit with your order.

AUDIO PRODUCTS CO.

A. J. EDGCOMB

4185-7-9 W. SECOND ST. *Since 1907* LOS ANGELES, CALIF.

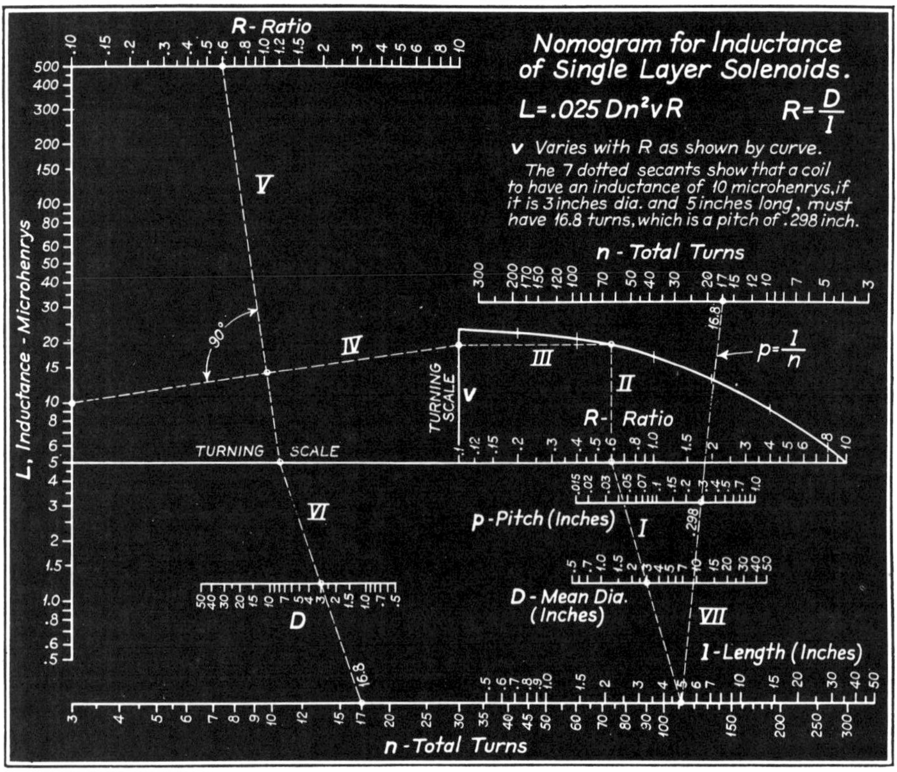

MONOGRAM FOR SINGLE-LAYER SOLENOIDS

COURTESY OF: COMMUNICATION AND BROADCAST ENGINEERING

The inductance of a single-layer solenoid or helix may be computed by the well-known Nagaoka formula $L = .0.025\ D\ n^2\ v\ R$, as shown on the nomogram. In the Nagaoka formula, L is the inductance in microhenrys, D is the mean diameter of the coil in inches, l is the coil length in inches, $D/l = R$ is the ratio of the diameter to the length, and n is the total number of turns. In this formula, also, v is a function of R. A table of assumed values of R and the corresponding values of v is given in the U. S. Bureau of Standards Circular 74, page 252. This circular is now out of print, but it may be consulted at most public libraries.

By drawing six lines or secants, as shown, the inductance L may be computed from the nomogram, if D, l and n are given; or n may be found, if L, D and l are given. The secants are numbered to show their consecutive order for the computation of n, secant VII determining the pitch p, or $p = l/n$. The curve is plotted from the above mentioned tabular relationship between R and v, for which no equation is given. Secants II and III are the coordinates of a point in this curve.

The scales n and l, at the bottom, overlap at their inner ends. However, graduations are above for l and below for n, so there should be no confusion.

The square nomogram at the upper left is best solved by drawing a pair of right-angle lines on transparent paper. These lines are manipulated so that three of the arms pass through the desired points causing the fourth arm to pass through the value to be found. Each line of the cross must pass through **opposite** sides. The left side for L has been extended to give a larger range, and the intersection may be made outside the square.

All scales are logarithmic. The modulus if the scale n is 2m, while the modulus of the p and the two D scales is m/2. For all other scales the modulus is π. (Communication and Broadcast Engineering—Dec., 1935).